Carbon Sequestration in Agricultural Ecosystems

Klaus Lorenz · Rattan Lal

Carbon Sequestration in Agricultural Ecosystems

Springer

Klaus Lorenz
Carbon Management and Sequestration Center, School of Environment and Natural Resources
The Ohio State University
Columbus, OH
USA

Rattan Lal
Carbon Management and Sequestration Center, School of Environment and Natural Resources
The Ohio State University
Columbus, OH
USA

ISBN 978-3-030-06405-1 ISBN 978-3-319-92318-5 (eBook)
https://doi.org/10.1007/978-3-319-92318-5

Softcover re-print of the Hardcover 1st edition 2018

Printed on acid-free paper

This Springer imprint is published by the registered company Springer International Publishing AG part of Springer Nature
The registered company address is: Gewerbestrasse 11, 6330 Cham, Switzerland

Preface

We live on a planet Earth which has been drastically altered and shaped by anthropogenic activities. Paul Crutzen and others argue that anthropogenic changes to Earth's climate, land, oceans, and biosphere have become so drastic and so rapid that a new geological epoch has begun and is defined by the action of humans which is equivalent to that of any geological force, i.e., the Anthropocene. Large areas of terrestrial ecosystems, in particular, have been converted into agricultural ecosystems driven by ever-increasing human demand for food, feed, fiber, and fuel, as well as into urban ecosystems for housing, transportation, and industrial infrastructures. About 12 millennia ago, humans began the agricultural land-use and land-cover change (LULCC). Now, agriculture is the dominant global land use with about 40% of the ice-free land area covered in total by croplands and grasslands. This drastic change has left its footprint on the global carbon (C) cycle and its dynamic, and on the global climate.

In natural terrestrial ecosystems, the C input from plant photosynthesis is balanced by C losses from vegetation and soil resulting in a characteristic equilibrium ecosystem C stock determined by climate, eco-graphic region, and soil type. However, goal of agricultural activities is to produce as much biomass on an area of land as possible, and remove biomass C with harvestable products and other plant material. Some biomass is also lost by voluntary and managed fire. Thus, standing biomass C stocks of agroecosystems are profoundly smaller than those of natural ecosystems. In particular, deforestation and other land-cover changes to make room for agriculture, and land management (i.e., grazing) have massive effects on biomass C stocks. Drastic changes are also occurring in the soil. On the one hand, more photosynthetically fixed C may enter the soil due to increased net primary production (NPP) by cultivating high-yielding plant varieties managed by fertilization, irrigation, and application of plant protection products. On the other hand, tillage continues to be a widely used practice for beneficial effects on topsoil properties, planting and seedbed preparation, mixing crop residues into soil, and for weed control. As a consequence, soil organic carbon (SOC) is lost as carbon dioxide (CO_2) by decomposition and some as methane (CH_4), and also removed through erosion by water and wind. Together with extended periods of time outside the main vegetation period when the soil is non-vegetated with agricultural species, this has resulted in lower equilibrium SOC stocks of agroecosystems compared to those of natural terrestrial ecosystems they replace. More uncertain are the effects of agriculture on the other principal soil C stock, especially the soil inorganic carbon (SIC) stock. Specifically, soils of agroecosystems developed on carbonate-containing rocks may contain appreciable amounts of SIC. Further, certain practices such as application of acidifying fertilizers and liming may alter dynamics of SIC stocks. Also, cycling of SIC and cycling of SOC are interconnected.

The hidden treasure SOC is one of Earths' most precious natural resources. Soil-based agricultural production depends on an adequate SOC stock. Aside playing a central role in soil health, fertility, quality, and productivity, the SOC stock supports critically important soil-based ecosystem services (ESs) including the supporting ES of nutrient cycling; the provisioning ES of food, freshwater, wood, and fiber; and the regulating ESs of flood control and attenuation of water quantity. Increase in SOC stock may enhance soil health, reduce soil erosion, increase soil biota, improve soil storage functions, strengthen filtering and transformation of pollutants, and increase sequestration of CO_2. However, negative effects of relatively lower SOC stocks on agroecosystem productivity compared to natural ecosystems are easily masked by inputs of fertilizers and other amendments. Thus, resource-poor smallholders in developing countries may particularly benefit from SOC-conserving and -enhancing soil and land-use management practices. Now is the time to implement SOC-enhancing agricultural practices.

Without LULCC, the ice-free land area would hypothetically store significantly more C in biomass and soil. About fourteen times more C would be stored in cropland biomass, and about three times more C in grassland and grazing land biomass without land use under current climate conditions. Historically, agriculture has also depleted the SOC stock by about 130 Pg (1 Pg = 1 Gt = 10^{15}g) C. Currently, about 40% of the global SOC stock to 1 m depth (1,500 Pg C) is stored in croplands, temperate grasslands/shrublands, and tropical grasslands/savannahs (609 Pg C). However, even more challenging and daunting task is assessing the effects of agroecosystems on deep (<2 m) SOC and SIC stocks. Nevertheless, agricultural soil and land use have and are contributing to anthropogenic increases in atmospheric concentrations of the important greenhouse gases (GHGs): CO_2, CH_4 and nitrous oxide (N_2O). Thus, agroecosystems have contributed to the observed increase in global average surface temperature and, especially, CO_2 already emitted into the atmosphere from agricultural soils will contribute to warm the climate for millennia to come. Adaptation of agricultural production to climate change is, therefore, necessary because negative effects of climate change on crop production are already observed. Maintaining and increasing the SOC stock of agroecosystems is a low-hanging fruit to increase the resilience of food, feed, fiber, and fuel production toward climate perturbations. Sequestration of C in more healthy soils with higher SOC stocks will contribute to soil and food security to meet the increasing demands of an ever-increasing, more affluent, and more animal products consuming global population.

There is a lack of reference and textbooks for graduate and undergraduate students interested in understanding basic processes of soil C dynamics in agroecosystems and the underlying factors and causes which determine the technical and economic potential of soil C sequestration. This book provides both the information on SIC and SOC dynamics. It illustrates the topic for different types of agroecosystems and lists a set of questions at the end of each chapter to stimulate thinking and promote academic dialogue. Each chapter provides up-to-date references on the current issues and summarizes the current understanding while identifying the knowledge gaps for future research.

As the biomass C stocks of agroecosystems are profoundly smaller than the soil C stocks, the aim of this book is to discuss soil C stocks, their dynamics and management for food security and climate. This book is the first to describe both SIC and SOC sequestration in different agroecosystems, and how various processes can be enhanced to maintain and increase soil C stocks. After an introduction on agricultural land uses and their effects on the global C cycle, processes, factors, and causes influencing SIC and SOC dynamics are presented in detail based on up-to-date scientific knowledge. Then, SIC and SOC sequestration processes in soils of croplands and grasslands, and how agricultural practices affect those are discussed. This is followed by a description of SOC sequestration in wetland soils under agricultural use, and those under agroforestry systems. Then, how the increasing biomass production for bioenergy and biofuel affects the SOC stocks of agroecosystems is presented. This is followed by a critical discussion on the importance of biochar for soil C sequestration in agroecosystems. The book concludes with an outlook at the importance of soils of agroecosystems for climate change policy. Focus of the book is on scientific understanding of SIC and SOC sequestration in agroecosystems. Thus, this book is a valuable source of information intended for use by graduate and undergraduate students, scientists, land managers, business, policy makers, and the interested public.

It is difficult to acknowledge listing names of all those who contributed directly or indirectly toward bringing this book to fruition. Nonetheless, all the cited materials are appropriately referenced in the bibliography at the end of each chapter. The material presented in this volume builds upon the outstanding contributions of soil and plant scientists, biogeochemists, geologists, ecologists, agricultural engineers, and climatologists whose research is cited throughout the book.

Columbus, OH, USA | Klaus Lorenz
June 2018 | Rattan Lal

Contents

Chapter 1
Agricultural Land Use and the Global Carbon Cycle

Contents

Abstract Earth's soils and the global carbon (C) cycle have been profoundly affected by agriculture. Agricultural land-use and land-cover change (LULCC) date back to the early Holocene ~12,000 years ago when hunter-gatherers in the Fertile Crescent region of the eastern Mediterranean began using agricultural practices to manage soils. Since then, agriculture has spread across the globe and is now the dominant global land use with about 40% of the ice-free land area covered by croplands and grasslands. Especially, clearing of natural vegetation to make room for cultivating crops has released up to one-third of the soil organic carbon (SOC) stock from the top meter of soil. However, effects on the soil inorganic carbon (SIC) stock continue to be overlooked. Further, methane (CH_4) is produced in anaerobic environments under agricultural practices, such as the sediments of wetlands, peatlands, and rice (*Oryza sativa* L.) paddies as well as by livestock production. Globally, up to 357 Pg (1 Pg = 1 Gt = 10^{15}g) C pre-1850 and 168 Pg C post-1850 may have been released by agricultural land-use changes. To 2 m depth, about 133 Pg SOC may have been lost since the early Holocene. LULCC emissions from 1850 to 2015 have been estimated at 98.4 and 16.3 Pg C for crop and pasture land uses, respectively. This net C release together with emissions of CH_4 and nitrous oxide (N_2O) has contributed to increasing atmospheric greenhouse gas (GHG) concentrations and accelerating climate change. Climate change interferes

K. Lorenz and R. Lal, *Carbon Sequestration in Agricultural Ecosystems*,
https://doi.org/10.1007/978-3-319-92318-5_1

with agriculture with severe negative effects. Thus, climate change adaptation and mitigation are necessary to sustainably intensify production amidst the increasing challenge of satisfying the demands for food, feed, fiber, and fuel of a growing, more affluent, and more animal products consuming population. Food-related GHG emissions are lower for plant-based diets. Improved soil management in agroecosystems can substantially reduce GHG emissions and sequester some of the atmospheric carbon dioxide (CO_2) as SOC and oxidize some of its CH_4. Additional benefits of agroecosystems with increased SOC stocks are more healthy and resilient soils. The hidden treasure of SOC has finally been recognized by policy-makers. For example, to help address food security and climate change issues, the 4 per Thousand Initiative (4p1000) was proposed in December 2015 at COP21 in Paris to enhance the soil C stock on a large portion of the world's managed soils by an average annual increase of 0.4% in 0–40 cm depth.

Keywords Land-use and land-cover change · Green revolution
Sustainable intensification · Soil organic carbon · Soil inorganic carbon
Historic carbon loss · Climate change · Adaptation and mitigation
Carbon sequestration

1.1 History of Agricultural Land Use

Except for some regions in Europe, the human use of land before the mid-1700s was insignificant when compared with contemporary changes in the Earth's ecosystems (UNCCD 2017). Now, human land use activities are a global geophysical force shaping Earth's land surface (Ramankutty et al. 2008). Some are arguing that anthropogenic changes to Earth's climate, land, oceans, and biosphere have become so drastic and so rapid that a new geological epoch has begun defined by the action of humans, i.e., the Anthropocene (Crutzen and Stoermer 2000; Zalasiewicz et al. 2011). In particular, the global land-use and land-cover change (LULCC) alters ecosystem structure and functioning, and diminishing the ability of ecosystems to continue providing valuable resources such as food, freshwater, and forest resources, and services such as regulation of climate, air quality, water quality, and soil resources (Ramankutty et al. 2008). Among LULCC, agricultural activities are responsible for a vast majority of land-use-related ecosystem alterations. Between 1985 and 2013, the global cropland and pasture area have increased by 16.7% (Borrelli et al. 2017). Agriculture is now the dominant global land use with about 40% of the ice-free land surface used for agricultural practices, and much of cropland has replaced forests, savannas, and grasslands (Ramankutty et al. 2008). This agricultural land transformation started a few thousand years ago when hunter-gatherers engaged in agricultural activities (domestication of animals and plants) began modifying Earth's land surface with consequences to the global carbon (C) cycle (Ellis et al. 2013). However, timing, extent, and impacts of pre-industrial agricultural expansion are uncertain (Stocker et al. 2017). A few hundred years ago, the systematic use of crop rotations and other improvements in farming practice evolved (Walter et al. 2017). The Green Revolution (GR) included systematic breeding and the widespread use of man-made fertilizers and pesticides a few decades ago. Walter et al. (2017) suggest that agriculture is now undergoing a fourth revolution triggered by the exponentially increasing use of information and communication technology (ICT).

1.1.1 Early Developments in Agriculture

Humans have been altering tropical forests for at least 45,000 yrs through techniques ranging from controlled burning of sections of forest to plant and animal management to clear-cutting (Roberts et al. 2017). Profound long-term terrestrial ecosystem changes started with the Neolithic revolution by domestication of crops and livestock thousands of years ago in widely scattered global regions (Larson et al. 2014). About 20,000 years ago, the last ice age was at its peak. Then, the climate warmed suddenly about 14,500 years ago at the onset of the Holocene which began about 11,700 years ago (Balter 2010). The Holocene has been perceived as the most stable climate period of the last 130,000 years (Blümel 2009). Global average surface temperature in the Holocene has fluctuated plus or minus 1 °C (Marcott et al. 2013). At its onset, hunter-gatherers in the Fertile Crescent region of the eastern Mediterranean what is now Israel, Jordan, Syria, and Lebanon began with agricultural activities as the warm and moist environmental conditions were conducive to agriculture (Gupta 2004; Richerson et al. 2001). However, a number of important issues remain unresolved, including why hunter-gatherers adopted agriculture and how crops were domesticated to depend on people (Kluyver et al. 2017).

The earliest archaeological evidence for cultivating plants and herding animals dates back 10,000–12,000 years ago in the Fertile Crescent (Gibbons 2016). However, about 11,000 years before Neolithic farming, earliest, small-scale attempt to cultivate wild cereals may have occurred at Ohalo II, a 23,000-year-old hunter-gatherers' sedentary camp on the shore of the Sea of Galilee, Israel (Snir et al. 2015). By 7,700 year ago, the agricultural activities had spread across most of Europe and had reached all of its arable regions by 5,500 years ago (Ruddiman 2013). The earliest farming population in Central Europe shared an affinity with the modern-day Near East and Anatolia, supporting a major genetic input from this area during the advent of farming in Europe (Haak et al. 2010). Farming spread into and across Europe mainly via migration of people and not the dissemination of ideas. For example, Hofmanová et al. (2016) showed a direct genetic link between Mediterranean and Central European early farmers and those of Greece and Anatolia. This extends the European Neolithic migratory chain all the way back to southwestern Asia. However, important gaps remain in the understanding of the spread of farming into Europe (Pereira et al. 2017). For example, did the farmers wipe out the pre-existing European hunter-gatherers or did they slowly out-compete them over time? Lipson et al. (2017) suggested that these groups coexisted side-by-side for some time after the early farmers spread across Europe. Eventually, the farming populations slowly integrated local hunter-gatherers, showing more assimilation of the hunter-gatherers into the farming populations with time.

Important centers of initial plant domestication and agricultural development included southwestern Asia, a large part of central Africa, India, eastern Asia, and Central and South Americas (Gupta 2004). In the New World, the earliest crops, such as lerén (*Calathea allouia*), arrowroot (*Maranta arundinacea*), and *Cucurbita moschata* squash first appeared 10,200 years ago in northern South America, and maize (*Zea mays* L.) was domesticated in Mexico by 9,000 years ago (Piperno 2011). The identification of the specific nature of ancient agricultural land and livestock management practices is, however, challenging. For example, Amazonian rainforests once thought to be pristine wildernesses are now increasingly known to have been inhabited by large populations even before European contact. In Amazonia, incidental plant domestication started earlier than 8,000 years ago (Levis et al. 2017).

Over 6,000 years, small, temporary clearings in Amazonian forests were made for agroforestry practices including cultivation of palm trees (Arecaceae), maize and squash (*Cucurbita* sp.), and for dwelling space (Watling et al. 2017). In Europe, early Neolithic farmers used livestock manure and water management to enhance crop yields (Bogaard et al. 2013). In contrast to the Neolithic of Europe and Eurasia, a reliance on cattle (*Bos taurus*), sheep (*Ovis aries*) and goats (*Capra aegagrus hircus*) emerged as a stable and widespread way of life in the prehistoric green Sahara of Holocene, North Africa, long before the first evidence for domesticated plants or settled village farming communities (Dunne et al. 2012).

Agricultural innovations during the Neolithic transition took place independently in some regions, while they spread in other regions. For example, a high level of genetic continuity of hunter-gatherers in East Asia over the last 7,700 years indicates a lack of major genetic turnover by exogenous farming populations (Siska et al. 2017). The local hunter-gatherers may have progressively added food-producing practices to their original lifestyle. In contrast, sophisticated social networks linked the Neolithic front in southern Europe to the Mesolithic peoples of northern Europe (Smith et al. 2015). About 8,000 years BP, the hunter-gatherer economy in the British Isles was replaced by arable farming of crops such as einkorn (*Triticum monococcum* L.), emmer (*T. dicoccum* Schrank ex Schübl), and barley (*Hordeum vulgare* L.). Groups of farmers from the Continent independently colonized England and Scotland and introduced farming to Britain (Collard et al. 2010). Similarly, agriculture was practiced in North America and Australia only after it was brought from elsewhere. In China, millet, soybeans (*Glycine max* (L.) Merr.), and pigs (*Sus scrofa domesticus* Erxleben, 1777) were domesticated in the early–middle Holocene. Rice (*Oryza sativa* L.) domestication may have begun in China at least 9,400 years ago (Zuo et al. 2017). Irrigated rice was adapted across China by 5,000 years ago, and by 1,000 years ago, livestock tending was widespread and irrigated rice was grown in every region of southern Asia. Because Europe and China have long accounted for 40% of Earth's human population, humans may have, therefore, had large early effects on Earth's surface as effects of early agriculture became widespread thousands of years before the industrial era (Ruddiman 2013).

1.1.2 Recent Changes in Agricultural Land Use

Several promising agricultural transitions have occurred globally in the recent past (Solbrig and Solbrig 1994). They include the intercontinental 'Columbian exchange' of economic plants beginning 500 years ago and the entrance of tropical crops such as coffea (*Coffea arabica* L.), sugarcane (*Saccharum officinarum* L.), rubber (*Hevea brasiliensis* Müll.Arg.), and cocoa (*Theobroma cacao* L.) into world markets starting 150 years ago (Solbrig and Solbrig 1994). During the twentieth century, significant changes in agriculture took place in developing countries between the 1930s and the late 1960s by the GR (Gaud 1968; Borlaug 2007). The same period witnessed a remarkable process of industrialization and globalization of food and agriculture (FAO 2017). The main strategy of GR was to adapt crop genetic improvements for wheat (*Triticum* L.), rice, and maize achieved in the developed world to the conditions of developing countries (Pingali 2012). Thus, together with high rates of investment in crop research, infrastructure, and market development and appropriate policy support, production of cereal crops tripled in the past 50 years in developing countries with only a 30% increase in land area cultivated. However, sub-Saharan Africa (SSA) continues to be the exception to this trend (Pingali 2012). A central part of the GR strategy was to develop institutional capacity, particularly in plant breeding.

A major issue is how the benefits of biotechnology can be shared with small-scale farmers in the developing world. To enhance cooperation, the Consultative Group on International Agricultural Research (CGIAR) was established in 1971 with the specific goals to reduce poverty, improve food security, and improve natural resources (Editorial Nature Plants 2017). Fifteen independent and non-profit centers are now grouped under CGIAR. Among the centers are the International Maize and Wheat Improvement Centre (CIMMYT) in Mexico, the International Rice Research Institute (IRRI) in the Philippines, and the International Potato Center (CIP) in Peru focussing on potato (*Solanum tuberosum* L.) and sweet potato (*Ipomoea batatas* (L.) Lam.). Further, the International Center for Tropical Agriculture (CIAT) in Colombia is about tropical agriculture and grain legumes. The International Center for Agricultural Research in the Dry Areas (ICARDA) in Lebanon specializes in dry region agriculture with crops like chickpea (*Cicer arietinum* L.) and lentil (*Lens culinaris* Medikus), while the International Crops Research Institute for the Semi-Arid Tropics (ICRISAT) in India works in semiarid crop improvement and focuses on locally important crops such as millet and sorghum. Further, AfricaRice is about rice while Bioversity International in Italy works to safeguard plant biodiversity. Individually and collectively the centers have produced countless impressive stories of education and technology dissemination

(Editorial Nature Plants 2017). However, the significant gains in ecological efficiency achieved in industrial agriculture in some developed and middle-income countries need to be recognized and applied more broadly in the tropics using technologies adapted for smallholder farmers (Sayer and Cassman 2013). Fortunately, there has been renewed interest in agricultural investment (Pingali 2012). Crop research must be integrated globally as current yield trends and agri-food systems are inadequate to match projected demand (Ray et al. 2013). Further, a refinement of research infrastructure and better leverage of global expertise and technologies are needed (Reynolds et al. 2017). This may be achieved by a Global Crop Improvement Network (GCIN) encompassing most staple food crops, providing access to well-controlled 'field laboratories,' while harmonizing research practices and sharing data. Combined with socioeconomic and cropping systems research, this network may revolutionize the ability to understand and model crop responses to environments globally and accelerate adoption of vital technologies (Reynolds et al. 2017).

The agricultural intensification by GR saved some new land from conversion to agriculture, allowed the release of marginal land, and potentially reduced greenhouse gas (GHG) emissions (Pingali 2012). Borlaug (2007) argued that between 1950 and 2000, >1 million km^2 was spared from being brought into agricultural production partly as a result of the technological change made possible by GR. In contrast, it was estimated that the total cropland area in 2004 would have been only 0.179–0.267 million km^2 larger in a world that had not benefited from crop germplasm improvement since 1965 (Stevenson et al. 2013). Of these, 0.120–0.177 million km^2 would have been in developing countries, displacing pastures and resulting in an estimated 0.02 million km^2 of additional deforestation. However, unintended consequences of GR in water use, soil degradation, and chemical runoff have had serious environmental impacts beyond the areas cultivated (Burney et al. 2010). Soil degradation, in particular, is responsible for the slowdown in yield since the mid-1980s (Pingali 2012). Further, while 'land-saving' benefits of GR are lower than previously suggested, continued investments in agricultural productivity in the traditional areas for crop production, away from the forest frontier, should remain an important part of the global efforts in containing agricultural expansion (Stevenson et al. 2013).

Major changes in agricultural land use continue to occur globally. For example, from 1998 to 2013, 20% of the Earth's vegetated land surface showed persistent declining trends in productivity, apparent in 20% of cropland, 19% of grassland, and 27% of rangeland (UNCCD 2017). Specifically, the growing global population increases the need for more food, fuel and shelter, and increasing affluence results in changing food consumption patterns toward commodities that are more land-intensive to supply such as meat and dairy products (Foley et al. 2011). A more affluent population generates also more food waste as an increase in per capita gross domestic product (GDP) is associated with an increase in per capita food waste (Xue et al. 2017). For example, in 2010, food loss and wastes of the average US American's diet accounted for 35% of energy use, 34% of blue water use, 34% of GHG emissions, 31% of land use, and 35% of fertilizer use related to an individual's food-related resource consumption (Birney et al. 2017). Food loss and waste use approximately one-third of all resources in the US food system.

Agricultural land area changes for food production can be linked to diet, population and yields drivers (Alexander et al. 2015). Total area of agricultural land currently used for livestock feed production at global level is 25 million km^2 (Mottet et al. 2017). The production of animal products dominated agricultural LULCC from 1961 to 2011, accounting for 65% of land-use change. Modest improvements in feed use efficiency can reduce further expansion (Mottet et al. 2017). Otherwise, bioenergy accounted for less than a tenth of the increase in demand for agricultural land. The consumption of cereals and oilseeds for the production of biofuels has particularly increased, as has the use of biomass as a substitute for petrochemicals (FAO 2017). Overall, population expansion has been the largest driver for agricultural land-use change, but dietary changes are now a significant and growing driver (Alexander et al. 2015). Dietary changes may also result in higher demand for food produced by organic agriculture (OA) as it is perceived by many as being healthier than conventionally produced food. Although currently practiced on only 1% of the global agricultural land area, OA is, therefore, a fast-growing land-use sector (Willer and Lernoud 2017). However, organic crop yields are ~20% lower than those under conventional management (Ponisio et al. 2015). What the increase in OA means for environmental performance and delivery of ecosystem services (ESs) by agroecosystem is discussed controversially (Reganold and Wachter 2016; Seufert and Ramankutty 2017). OA systems require more land, cause more eutrophication, use less energy, but emit similar GHGs per unit of food as conventional systems (Clark and Tilman 2017). In contrast, intensive and industrialized systems show the lowest GHG emissions per unit of agricultural production (Bennetzen et al. 2016b). Muller et al. (2017) emphasized that OA can only contribute to providing sufficient food for the 2050 population and simultaneously reducing environmental impacts from agriculture, if it is implemented in a well-designed food system in which animal feeding rations, and as a consequence reduced animal numbers and animal product consumption, and food wastage are addressed.

As more than half of the global population is now urban (United Nations 2014), the physical expansion of urban areas and changes in diet of urban dwellers also has implications for agriculture (Seto and Ramankutty 2016). Regional consequence of urbanization is the loss of productive agricultural soils. Historically, cities were often established close to high-quality agricultural soils, leading to competition for land between urban and agricultural uses as urban populations and need for housing grow (Martellozzo et al. 2015). For example, between 2000 and 2012, artificial surface in EU-25 expanded, largely associated with urban sprawl, taking much of the arable land (Robinson et al. 2017). The greatest population increases are occurring in tropical developing nations (Laurance et al. 2014), where future expansion of built-up areas will particularly encroach on productive agricultural soils. Loss of the most productive agricultural soils leads to decreased cropland productivity and may result in expansion of crop production into marginal and fragile lands, as well as other ecosystems. This process may also intensify deforestation in the tropics to make room for agriculture (Seto and Ramankutty 2016). For the year 2000, 2.13 million km^2 were classified as urban land, which was 2.06% of the Earth's ice-free land area (van Vliet et al. 2017). However, this urban land was more than proportionally located on soils that were suitable and available for crop production. By 2030, urban expansion is projected to result in a 1.8–2.4% loss of global croplands (d'Amour et al. 2017). About 80% of this loss will take place in Asia and Africa where much of the cropland encroached is more than twice as productive as national averages. Asia will experience the highest absolute loss, whereas African countries will experience the highest percentage loss of cropland. Globally, the croplands that are likely to be lost to urban expansion were responsible for 3–4% of worldwide crop production in 2000. Further, urban expansion is expected to take place on cropland that is 1.77 times more productive than the global average (d'Amour et al. 2017). In the year 2040, 6.21 million km^2 or 4.72% of all the Earth's surface may be urban land, and the increase in urban land between 2000 and 2040 will also be more than proportionally located on land that is suitable and available for crop production, further limiting food production capacity (van Vliet et al. 2017). Thus, urban growth from 2000 to 2040 may cause the displacement of almost 65 Mton (million ton or Tg) of crop production, which could lead to an expansion of up to 0.35 million km^2 of new cropland (van Vliet et al. 2017).

The location of food production is also changing due to the globalization of food supply and increasing international trade in agricultural commodities (Foley et al. 2011). Corporations and external investors began to recognize the potential for high financial returns and started large-scale acquisitions of agricultural land especially in tropical developing countries (D'Odorico and Rulli 2014). Governments also strengthen their food security by purchasing agricultural land in foreign countries. Globally, about 0.48 million km^2 of cropland has been acquired by foreign investors (landmatrix.org), about twice the area of agricultural land of Spain. The top target countries were Papua New Guinea, Indonesia, South Sudan, the Democratic Republic of Congo, Mozambique, and Brazil. Together, these countries accounted for 45% of the land sold. This land is expected to be used primarily for agriculture and forestry. However, only about 12% of the land acquired for agriculture has been cultivated. In many cases, the land remains unused and simply serves as an option for investors in development (D'Odorico and Rulli 2014).

Globally, agricultural land use has developed in a contradictory and divergent manner (Mazoyer and Roudart 2006). Many farmers have adopted more technology in their pursuit of greater yields and manage large-scale operations (King 2017). As a consequence, small-scale farming is under increasing economic pressure. Thus, agriculture is now occurring in some regions on a reduced number of farms where continually more capital is accumulated concentrating on the most productive crops and livestock systems, and new market shares are explored. This type of agriculture is able to be profitable through an excess of inputs and means (Mazoyer and Roudart 2006). For example, throughout Asia and in parts of Latin America, expanding seed and fertilizer use has been accompanied by investments in irrigation, rural roads, marketing infrastructure, and financial services (FAO 2015).

Otherwise, extensive global regions are cultivated by the majority of the world's farmers, i.e., smallholders in developing countries (Mazoyer and Roudart 2006). About 475 million smallholder farm families are in developing countries (FAO 2016). The majority of people in developing countries live in rural areas, and most of them depend on agriculture for their livelihoods (FAO 2015). Those communities are often excluded from the world economy and no longer renew the fertility of the environment they exploit as resources are lacking. For example, in regions such as SSA, the uptake of agricultural inputs is relatively low because it is often cheaper to expand cropland to have higher production (FAO 2015). Further, SSA and South Asia have seen flat or declining rates of investment per worker in agriculture over the past thirty years which contributes to the challenges for improving agricultural systems in those regions.

1.1.3 Future Agricultural Land Use

Agriculture is now facing the challenge of dealing with the multiple drivers of population growth, dietary shifts, increasing *per capita* consumption, energy insecurity, and accelerated climate change (Pretty et al. 2010). For example, the projected rapid population growth may be the leading cause of food insecurity and widespread undernourishment across Africa (Hall et al. 2017). In contrast, very little to no difference in undernourishment projections were found for future scenarios with and without the effects of climate change. Challenging is globally the increase in the production of high-quality food with reduced inputs in the face of global environmental change (Tester and Langridge 2010). Most planetary boundaries for land system change, climate change and the global nitrogen (N) and phosphorus (P) cycles were already transgressed in 2010 (Conijn et al. 2018). By 2050, transgression of the boundaries for agricultural area requirement, GHG emission, and P flow into the ocean may be prevented by reduction of waste, less consumption of animal products, higher feed conversion efficiency, higher crop and grassland yields, reduction of N and P losses from agricultural land, and reduction of ammonia (NH_3) volatilization. However, global mineral N and P fertilizer inputs, and total N loss to air and water may still exceed their boundaries in 2050. Conijin et al. (2018) highlighted the necessity of eliminating the dependency of food production on finite P reserves. Previously not explicitly considered, the vadose zone is an important store of nitrate (NO_3^-) that should be included in future budgets for effective policymaking (Ascott et al. 2017). Extensive historical agricultural practices and thick vadose zones, in particular, have contributed to high NO_3^- storage density in the vadose zone in North America, China, and Europe. Long travel times in the vadose zone in these areas may delay the impact of changes in agricultural practices on groundwater NO_3^- concentrations. Thus, vadose zone NO_3^- storage should be considered when planning groundwater pollution mitigation measures (Ascott et al. 2017).

Among issues that need to be addressed is also how to manage the global human protein supply from livestock production. For example, elevated temperatures, reduced precipitation, rising atmospheric CO_2 concentrations, and sustained nutrient export all have the potential to reduce cattle performance by reducing the nutritional quality of forage (Craine et al. 2017). In general, elevated CO_2 concentrations are associated with lower N concentrations in roots but not in leaves of crops (Sardans et al. 2017). No clear relationship was observed between elevated CO_2 concentrations and N concentrations in leaves and roots in temperate grasslands. However, when elevated CO_2 concentrations and N fertilization coincided, leaves had lower N concentrations, whereas root had higher N concentrations suggesting that more nutrients will be allocated to roots to improve uptake. N fertilization and drought generally increased foliar and root N concentrations, while the effects on P concentrations were less clear (Sardans et al. 2017). With regard to plant protein, future increases in crop production will need to come from increased productivity as the amount of global arable land is decreasing because of urbanization, salinization, limited irrigation water, and soil erosion. Future changes in irrigated crop yields are particularly uncertain and depend on many factors, including plant water use, irrigation efficiency, growing season, heat stress, and antecedent soil moisture (Winter et al. 2017). Plant breeders need to focus on traits with the greatest potential to increase yield and also improve nutritive value of many crops (Tester and Langridge 2010).

In 2016, the dominant GHGs released into Earth's atmosphere—CO_2, methane (CH_4), and nitrous oxide (N_2O)—continued to increase and reached new record highs (Blunden and Arndt 2017). Earth's surface observed record warmth in the same year. Human activities strongly enhance the natural climate warming which otherwise may peak at the next climate optimum after the precarious Little Ice Age pessimum (Blümel 2009). Specifically, warming will lead to more precipitation in some agricultural regions, while all global terrestrial ecosystem margins will be on the move. Some of the agriculture in dryland areas will profit from the shifting desert margins and others not. The frontiers of arable land will shift toward the Polar Regions and higher into the mountains, similar to the post-glacial Megathermal. Most likely, however, there will be more disadvantages than gains for global agriculture (Blümel 2009). On the positive side, agricultural production and GHGs have been steadily decoupled over recent decades and are projected to be further decoupled by 20–55% under business-as-usual by 2050 (Bennetzen et al. 2016a).

Many plant species have an optimum temperature for productivity of <33 °C (CAST 2017). Thus, any crop productivity gains by more favorable higher atmospheric CO_2 concentrations may be offset by lower productivity at higher temperatures. Further, projected higher frequency of climate extremes is another major global concern for crop production (CAST 2017). However, modeling of crops under rising temperature and climate change, and crop yield projections need to be improved for planning adaptation strategies (Wang et al. 2017).

Currently, negative impacts of climate trends on global crop yields are already observed (Lobell and Field 2007). Climate change may have already reduced the growth in crop yields by 1–2% per decade over the twentieth century (Porter et al. 2014). However, climate effects on crop yields differ among regions. For example, while in 1981–2010 a decrease in yield variability for maize, soybean, rice, and wheat was the main trend worldwide across crops, yields in some regions of the world have become more unstable (Izumi and Ramankutty 2016). Specifically, changes in variability of temperatures exceeding the optimal range for agronomic yield of maize, soybean, rice, and wheat were more important to explaining the change in yield variability than temperature below the optimal range for yield formation and soil water deficit (Izumi and Ramankutty 2016). Further, water limitation is a major driver of observed variations in national crop yield variability (Frieler et al. 2017). Specifically, observed weather variations can explain more than 50% of the variability in wheat yields in Australia, Canada, Spain, Hungary, and Romania. For maize, weather sensitivities exceed 50% in seven countries, including the USA. The explained variance exceeds 50% for rice in Japan and South Korea, and for soybean in Argentina. However, the fraction of the observed variability of irrigated crop yields that is due to weather cannot be quantified because national crop-yield variability is dominated by yields under rain-fed conditions (Frieler et al. 2017). For example, climate variability dominated the change trends of corn yield variability in the US Midwest Corn Belt during 1980–2010, while the ability of climate variability in controlling yield variability was low in southeastern and western arid regions (Leng 2017). However, for US regions where separate estimates of irrigated and rain-fed corn yield exist, irrigation has largely reduced the corn yield variability, highlighting the importance of non-climatic factors in governing the changes in corn yield variability (Leng 2017). Further, solar brightening (decadal-scale increases in incident solar radiation) contributed 27% of the US Corn Belt yield trend from 1984 to 2013 (Tollenaar et al. 2017). This calls into question the implicit use of historical yield trends in predicting yields under future climate change scenarios.

Each degree Celsius increase in global mean temperature would reduce global yields of wheat by 6.0%, rice by 3.2%, maize by 7.4%, and soybean by 3.1% without CO_2 fertilization, effective adaptation, and genetic improvement (Zhao et al. 2017). However, the projected changes are highly heterogeneous across crops and geographical areas. Thus, while regional change in yields of main cereals may occur in the future, crop yields are projected to decrease globally because of climate change (FAO 2017). By 2050, the projected reductions in agricultural yields due to climate change are larger for some crops than those estimated for the past half-century, but smaller than projected increases to 2050 due to rising demand and intrinsic productivity growth (Wiebe et al. 2015). Van Meijl et al. (2017) projected that the impact of climate change on agricultural production in 2050 is negative but relatively small at the aggregated global level. However, decreases in crop yields are not only projected to have large impacts on agriculture toward the end of the twenty-first century but already over the next 10 or 20 years because of climate trends (Lobell and Tebaldi 2014). In conclusion, although the full implications of climate change on agriculture are difficult to predict, it is expected that the impacts will be of different levels and of a different nature in each region, ecological zone, and production system. Further, significant knowledge gaps remain about the impacts of climate change on agriculture (FAO 2017).

By 2050, human population is projected to increase to almost 10 billion, while demand for food is projected to double (Manders et al. 2012; Royal Society of London 2009; World Bank 2008). However, largely because of recent production gains, an increase of about 25–70% above current production levels may be sufficient to meet the demand in 2050 (Hunter et al. 2017). Similarly, in a scenario of modest economic growth, 2050 agricultural demand is expected to grow by ~50% compared to 2013 (FAO 2017). Between 2010 and 2100, total demand for crops and grass could increase by ~35–165% (Bijl et al. 2017). The demand for agricultural products is projected to be more influenced by the population developments and assumptions on dietary preferences than by the GDP developments (Van Meijl et al. 2017). However, a healthier diet can also result in an increase in resource use (Birney et al. 2017). For example, nationally recommended diets in poorer middle-income nations are associated with increases in GHG, eutrophication, and land use from 12.4 to 17.0%, 24.5–31.9%, and 8.8–14.8%, respectively (Behrens et al. 2017). In contrast, in high-income nations impacts decrease by 13.0–24.8%, 9.8–21.3%, and 5.7–17.6% for GHG, eutrophication, and land use, respectively. The increased environmental impacts of nationally recommended diets in low- and middle-income nations are associated with increased intake in animal products. Otherwise, the reduced environmental impact in high-income countries is driven by reductions in calories and a change in composition (Behrens et al. 2017).

Future demand growth can be particularly tempered more effectively by replacing animal products than by reducing food waste. Transitioning toward more plant-based diets may reduce global mortality by 6–10% and food-related GHG emissions by 29–70% compared with a reference scenario in 2050 (Springmann et al. 2016). Alexander et al. (2017a) emphasized that influencing consumer behavior, e.g., to consume less animal products or to reduce per capita consumption closer to nutrient requirements, offers substantial potential to improve food security for the rising global population in a sustainable manner. Further, by 2050, a combination of less resource-intensive diets and improved agricultural productivity, the latter especially in Africa, could make it possible to use land and water resources sustainably while also constraining increases in food prices (Springer and Duchin 2014). This may be accompanied by a decisive shift of production and export of agricultural products away from developed countries toward Africa and Latin America.

Resource constraints over water, soil, biodiversity, and land will increasingly affect agricultural systems (Pretty and Bharucha 2014). Further, environmental impact of agricultural production must be reduced. For example, Phalan et al. (2016) proposed to increase food production per unit area (yield) on existing farmland to minimize farmland area and to spare land for habitat conservation or restoration. Reductions in N fertilizer inputs are also needed to reduce eutrophication of water resources by the projected climate change-induced precipitation changes (Sinha et al. 2017). To address the resource constraints, some have called for sustainable intensification (SI) of agriculture (Royal Society of London 2009;

Lal 2016). SI can be defined as a process or system where agricultural yields are increased without adverse environmental impact and without the conversion of additional non-agricultural land (Pretty and Bharucha 2014). However, the term is not uniformly understood, and by many agricultural experts not perceived as a significant departure from current agricultural practices (Petersen and Snapp 2015). Struik and Kuyper (2017) argued that SI is about societal negotiation, institutional innovation, justice, and adaptive management. Thus, Hunter et al. (2017) called for a recalibrated vision of SI in which up-to-date production goals are coupled with quantitative environmental targets for meeting food demand while maintaining functioning ecosystems. For industrialized agricultural systems with low yield gaps such as those in Germany, Techen and Helming (2017) argued that more or less of the same input factors, such as fertilizers, would be a future management option as part of a moderate intensification.

In addition, SI may lack engagement with established principles that are central to sustainability (Loos et al. 2014). Sustainability of agriculture is particularly under threat from accelerated soil erosion. Estimates of human-induced agricultural erosion amount to 25–40 Pg yr^{-1} for water erosion, 5 Pg yr^{-1} for tillage erosion, and 2–3 Pg yr^{-1} for wind erosion (Van Oost et al. 2007; Govers et al. 2014). However, previous annual soil erosion reference values may be too high. For example, Borrelli et al. (2017) estimated that 35.9 and 17 Pg yr^{-1} of soil and cropland soil eroded in 2012, respectively, based on an unprecedently high resolution (250 $\times$ 250 m) global potential soil erosion model. Importantly, farmers hardly have an incentive to adopt land conservation practices to contain soil erosion as costs for changing practices are high and occur immediately, while the benefits spread out over a longer time horizon (Borrelli et al. 2017). Thus, farmers do generally not consider erosion in their management decisions (Auerswald et al. 2018). Otherwise, Govers et al. (2017) argued that smart intensification is a necessary component of a strategy for effective soil conservation, particularly, in the Global South where significant losses in soil and nutrients by erosion continue to occur and fundamental changes in agriculture are needed. The greatest increases in soil erosion are predicted to occur in sub-Saharan Africa, South America, and Southeast Asia (Borrelli et al. 2017). Smart intensification will not only allow for soil conservation to be made more economical, but will also allow for significant gains to be made in terms of soil organic carbon (SOC) storage, water efficiency and biodiversity, while at the same time lowering the overall soil erosion risk (Govers et al. 2017).

Documenting and evaluating evidence from SI projects is a major challenge (Pretty and Bharucha 2014). For example, Pretty et al. (2006) analyzed 286 SI projects in 57 countries. The geometric mean relative yield increase was 64% across a wide range of systems and crop types (Pretty and Bharucha 2014). With regard to land use, the most likely future SI scenario is that more food will need to be produced from the same amount of (or even less) agricultural land (Godfray et al. 2010). Otherwise, environmental impacts of meeting increasing global crop demand depend particularly on how global agriculture expands (Tilman et al.

2011). Future land-cover projections are, however, inherently uncertain with the highest variability occurring in future cropland areas (Alexander et al. 2017c).

If current trends of greater agricultural intensification in more wealthy nations and greater land clearing in poorer nations continue, $\sim$ 10 million km^2 of land may be cleared globally by 2050 (Tilman et al. 2011). However, rapid production growth in recent years may reduce the necessity to convert such a large area (Hunter et al. 2017). For example, some projections indicate a need for less than 1 million km^2 of additional land for agricultural use in 2050 (Alexandratos and Bruinsma 2012). Tilman et al. (2011) argued that attaining high yields on existing croplands of underyielding nations is of particularly great importance if global crop demand is to be met with minimal environmental impacts. As the current distribution of crops around the world neither attains maximum production nor minimum water use, the agricultural landscape may be reconfigured (Davis et al. 2017). By reshaping the global distribution of crops within current rain-fed and irrigated croplands based on total water consumption, an additional 825 million people could be fed while reducing the consumptive use of rainwater and irrigation water by 14 and 12%, respectively. This optimization does not entail a loss of crop diversity, cropland expansion, or impacts on nutrient and feed availability and does not necessarily invoke massive investments in modern technology (Davis et al. 2017). Otherwise, for Germany, future management changes may include higher precision and lightweight machines triggered by ICT and robotics; diversification of crop rotations, including the integration of lignocellulosic crops; inoculation with biota; and new crop varieties (Techen and Helming 2017).

Pingali (2012) has called for a Green Revolution 2.0 to enhance the spread and sustainable adoption of productivity-enhancing technologies. Similarly, Kesavan and Swaminathan (2008) advocated for a more holistic approach with the goal of transcending GR with an Evergreen Revolution by taking a comprehensive farming systems approach that considers land, cultivar improvement, water, biodiversity, and integrated natural resource management. As breeding efforts to improve crop yield have generally focused on aboveground, shoot-related phenotypes, researchers have started 'the second green revolution' to explore the possibility of yield improvements through optimization of root systems (Lynch 2007). Lal (2016) called for a reduction in the land area under cropland through SI and sparing land for nature conservance. Further, Lal (2017) argued about increase in pulse-based diet so that some of the grazing lands can also be set aside for nature conservancy.

The contributions of biodiversity to the sustainability of agriculture and food systems must also be advanced (Zimmerer and de Haan 2017). About 90% of the world's energy demands are fulfilled by just 15 crops, according to estimates by the FAO (Gruber 2017). This dependency on a handful of crops is problematic. Modern crops are susceptible to changing climatic conditions and are heavily affected by pests, which can claim 30–40% of global production of staple crops. The nutritional content of agricultural crops is also declining (Gruber 2017). Thus, the rigorous integration of plant, environmental, social and health sciences is needed to meet agriculture and food systems challenges (Zimmerer and de Haan 2017).

The importance of soil biota for plant productivity and agricultural sustainability is increasingly recognized (Bender et al. 2016). Agricultural sustainability may be enhanced, in particular, by engineering soil organisms and their potential to enhance sustainable ecosystem functioning. This soil ecological engineering would rely more on internal regulatory processes compared with contemporary agriculture. Combining targeted soil biological approaches with state-of-the-art technological advances in agricultural science such as precision agriculture for tillage and weed control or optimized nutrient and water addition may serve to reduce external resource use to a minimum, while yields could be maintained or increased (Bender et al. 2016). Transformation of the soil microbiome is also among the goals of the development of perennial polyculture ecosystems (Baker 2017). This will include perennial crops as a way to make agriculture sustainable and to support ecological intensification (Glover et al. 2010). Perennial crops of rice, wheat, sorghum (*Sorghum bicolor* (L.) Conrad Moench), pigeon pea (*Cajanus cajan* (L.) Millsp.), barley, and sunflowers (*Helianthus* L.) are currently under development (Kane et al. 2016). Among the advantages of long-lived crops are much longer root systems than annuals, allowing the plants to survive for several years without disturbing the soil (Baker 2017). Further, by planting perennial grains along with cover crops, farmers can reduce the need for tillage and for fertilizers and pesticides. Ultimately, perennial crops need to be grown profitably to result in improved ecosystem services—improve soil quality, reduce runoff and N leaching, and improved water quality and wildlife habitat. Perennial crops typically increase belowground C inputs, leading to C sequestration (Paustian et al. 2016). However, major investments are needed to transform agriculture on a wide scale into perennial polyculture systems (Baker 2017).

The approaches of the GR have reached their biological limits as indicated by slowing or stagnating yield improvements (Long et al. 2015). However, photosynthesis, which has been improved little in crops and falls far short of its biological limit, may be the key remaining route to increase the genetic yield potential of major crops. For example, slow induction of photosynthesis on shade to sun transitions in wheat may cost at least 21% of productivity (Taylor and Long 2017). Suppressed shade avoidance response and deetiolation offer the potential for significant enhancement of crop yields (Ganesan et al. 2017). Improved yields often come with the tolerance to abiotic stresses such as drought, salinity, and heavy metal ions. The greater the photosynthetic efficiency that crop plants can be engineered to possess by photobiotechnology, the stronger the resistance to biotic and abiotic stresses (Ganesan et al. 2017). Ort et al. (2015) suggested that increasing the efficiency and productivity of photosynthesis in crop plants will be essential if the grand challenge of substantial improvements in agriculture is to be met. Some recent examples of photosynthesis improvements are encouraging (Bourzac 2017). By altering the photoprotection mechanism of non-photochemical quenching, Kromdijk et al. (2016) achieved 15% higher yield of tobacco (*Nicotiana tabacum* L.) plants under field conditions. This photoprotection boost is now also being tested for cowpea (*Vigna unguiculata* (L.) Walp.), a major source of protein in sub-Saharan Africa, and for rice. However, efforts to improve the performance of the enzyme RuBisCO, which catalyzes the fixation of CO_2 by photosynthesis, have not been successful so far. Further, the introduction of engineered rice on farmer's fields using the more efficient C_4 photosynthetic compared to the common C_3 photosynthetic pathway may still be decades away (Bourzac 2017).

A paradigm shift may be needed to reposition world agriculture from its current role as the world's single largest driver of global environmental change, to become a key contributor of a global transition to a sustainable world within a safe operating space on Earth (Rockström et al. 2017). Such a paradigm shift will make agriculture a part of the revolution toward adaptation and mitigation of climate change and improving the environment. Foley et al. (2011) argued that tremendous progress could be made by halting agricultural expansion, closing 'yield gaps' on under-performing lands, increasing cropping efficiency, shifting diets, and reducing waste. These strategies could increase food production while greatly reducing the environmental impacts of global agriculture. Food production may grow substantially, while, at the same time, agriculture's environmental footprint may shrink dramatically (Foley et al. 2011). Fischer et al. (2017) calculated higher land demand of livestock-based diets compared to crop-based diets for Germany and the EU28. The land footprint of each Germany citizen appropriates 2,693 m^2 cropland (one half for livestock-based diets, one quarter for crop-based diets, and one quarter for non-food products). Additionally, 1,655 m^2 of grassland per capita is used for the consumption of ruminant livestock products. Germany's current net 'cropland imports' amount to 106,000 km^2. In total, half of Germany's 220,000 km^2 cropland footprint relies on domestic cultivation and half on land resources abroad (Fischer et al. 2017).

Clark and Tilman (2017) suggested that dietary shifts toward low-impact foods and increases in agricultural input use efficiency would offer larger environmental benefits than would switches from conventional agricultural systems to alternatives such as OA. Reducing animal agriculture or consumption of animal-derived foods, in particular, may reduce GHGs and enhance food security. Further, alternatives to conventional animal products such as imitation meat and insects have a high land-use efficiency (Alexander et al. 2017b). As an extreme example, removing animals from US agriculture would reduce agricultural GHG emissions, but it would also create a food supply incapable of supporting the US population's nutritional requirements (White and Hall 2017). In addition to shifting diets, policy efforts are needed to reduce food waste, halt cropland expansion, and ensure that the world's poorest people have secure access to nutritious food (Lal 2016; Hunter et al. 2017). High-input, resource-intensive farming systems, in particular, cannot deliver sustainable food and agricultural production (FAO 2017). Thus, innovative systems of SI are needed that protect and enhance the natural resource base while increasing productivity. Such 'holistic' approaches may include agroecology, agroforestry, climate-smart agriculture, and conservation agriculture, which may also build upon indigenous and traditional knowledge (FAO 2017).

The economic model of farming is also going to change by technological advances in robotics and sensing technologies (King 2017). Both large-scale and small-scale producers may profit. Aim of technological improvements is to make agricultural production more efficient and more sustainable. For example, livestock farmers may increasingly use sensing technologies in the future to manage the health and welfare of their animals. Further advances are made to improve monitoring and maintenance of soil quality and to eliminate pests and disease without resorting to indiscriminate use of agrichemicals. Drones, robotics, and autonomous machines may help in the future (i) to reduce collateral damage to wildlife, lower resistance, and cut costs by precise chemical application or pest removal, (ii) for smarter application of N fertilizer, and (iii) sowing different crop varieties in one field to better match shifting soil properties. Many of these disruptive technologies are not ready yet, but the revolution may be coming (King 2017).

1.2 The Global Carbon Cycle and Agricultural Land Use

Biogenic C fluxes from global agricultural production and consumption profoundly affect the global C cycle (Wolf et al. 2015). The agricultural LULCC directly alters the amount of C stored in vegetation and soils and, thus, the C flux between agricultural land and atmosphere (Houghton 2014). The conversion of forest to croplands and pasture dominated the total net LULCC flux in 1901–2012, while the contribution of transitions between non-forest vegetation and agriculture was comparatively small (Li et al. 2017).

The current biomass C stock on cropland is estimated at 10 Pg (1 Pg = 1 Gt = 10^{15}g) C, and 119–121 Pg C on grassland and grazing land (Erb et al. 2018). Without hypothetical land use and under current climate conditions, potential cropland vegetation would store 139–141 Pg C, and potential grassland and grazing land vegetation would store 374–379 Pg C. Irrespective of the climate zone, the difference in biomass between potential and actual stocks mostly follows the pattern of global agriculture (Erb et al. 2018). Importantly, the current biomass C stock of agricultural land is also profoundly lower than the soil C stock. The SOC stocks to 1 m depth of agricultural biomes are estimated at 128–157 Pg C for cropland, 176–295 Pg C for temperate grassland/shrubland, and 198–264 Pg C for tropical grassland/savanna (Stockmann et al. 2013). While effects of changes in agricultural land uses on terrestrial C stocks and, in particular, on SOC stocks vary (e.g., Don et al. 2011; Poeplau et al. 2011), conversion of forests to croplands or pastures causes large losses in vegetation and SOC stocks. For example, 25–30% of the SOC stock stored in the top meter of soil may be released by cultivation of native soils, whether under forest or prairie vegetation (Houghton 2010). The global C debt due to agriculture for 12,000 years has been estimated at 133 Pg C for the top 2 m of soil, with the rate of loss increasing dramatically in the past 200 years (Sanderman et al. 2017). Tropical deforestation to clear land for agriculture is, in particular, the largest contributor to C emissions form land-use change (Le Quéré et al. 2017). In 2016, global emissions from land-use change were estimated as 1.3 Pg C yr^{-1}, while the 2007–2016 average was also 1.3 Pg C yr^{-1}. Between 1990 and 2015, Latin America had the highest proportion of deforestation driven by agriculture

(78%), and Africa had the lowest (62%; Carter et al. 2018). Latin America had the highest emissions from agriculture-driven deforestation, peaking at 0.27 Pg C yr^{-1} in 2000–2005. Africa saw a continuous increase in emissions between 1990 and 2015 (from 0.04–0.11 Pg C yr^{-1}). However, uncertainties for emissions from agriculture-driven deforestation are ±62.4% (average over 1990–2015; Carter et al. 2018). Globally, a key uncertainty is the difficulty in accessing land-cover change data to estimate land-use change C emissions (Le Quéré et al. 2017).

Perturbations of the global C cycle by agriculture did already occur thousands of years ago, but its importance is discussed controversially. For example, in Paleolithic and Neolithic times, C was emitted by land-use changes mainly associated with clearing woody vegetation, first to improve hunting and gathering opportunities, and later to provide agricultural land (Kaplan et al. 2010). Ruddiman (2003) proposed the early anthropogenic hypothesis indicating that important anthropogenic effects on the environment and on global climate; i.e., the global C cycle began thousands of years ago and grew to substantial size long before the industrial era. However, direct evidence for the effect of early agriculture on the global C cycle and the global climate is not available. Particularly contested is the extent of ancient forest burning in Amazonia (Watling et al. 2017). Some have proposed that pre-Columbian (pre-AD 1492) deforestation was on a large enough scale to have influenced the C cycle and global climate. Others argue that large-scale slash-and-burn agriculture is a largely postcontact phenomenon (Watling et al. 2017). Further uncertain are estimates of global land use prior to 1960, as are the data on the global extent of deforestation and crop cultivation (Ruddiman 2013). Also, population levels are only reasonably well known for Europe and China for the past 2,000 yr. Prior to this, most global estimates rely on geometric models of population growth. Accordingly, some suggest that roughly two-thirds of total forest clearance occurred during the industrial era and only one-third prior to 1850. In contrast, others argue that a large majority (perhaps three-fourths) of total forest clearance occurred prior to the start of the industrial era. Otherwise, Ruddiman (2013) suggested that the available historical and archeological evidence of per capita land use supports the early-anthropogenic view rather than the industrial one. In contrast, Stocker et al. (2017) argued that despite high simulated per capita CO_2 emissions from LULCC in early phases of agricultural development, humans emerge as a driver with dominant global C cycle impacts only in the most recent three millennia.

Pre-industrial land-use impacts on biomass C stocks were considerable with estimated losses of 115–425 PgC (Erb et al. 2018). For the industrial period (1850–2000), estimates for net losses of C from global changes in land use ranged between 108 and 188 Pg C (Houghton 2014). Houghton and Nassikas (2017) estimated that cumulative LULCC emissions in 1850–2015 were 98.4 and 16.3 Pg C for crop and pasture land uses, respectively. Mechanization of agriculture in the 1950s has accelerated SOC losses in croplands, but development of C sequestering practices over recent years (e.g., no-till, irrigation, reduced use of bare fallows, land conservation program, increased crop residue incorporation) may have limited SOC losses (Eglin et al. 2010). Currently, croplands do not probably represent a strong source of CO_2 to the atmosphere in Europe, China, and USA. However, regional, systematic inventories of the SOC balance of arable lands are basically lacking (Eglin et al. 2010).

Previous analyses did often not consider changes in agricultural management. For example, adding the effects of shifting cultivation, wood harvesting, regular grazing and crop harvesting, and more realistic cropland management practices increased LULCC emission estimates by 20–30% for 1901–2014 (Arneth et al. 2017). When grazing and crop harvesting were included, LULCC emissions increased from 77.6 to 123.2 Pg C while emissions increased from 154.5 to 168.1 Pg C when cropland management practices such as flexible sowing and harvesting or tillage were considered. Thus, Arneth et al. (2017) argued that terrestrial ecosystems might have a greater potential to sequester C in the future than commonly assumed.

Aside interfering with the CO_2 exchange between agricultural land and the atmosphere, agricultural LULCC may also directly affect the exchange of CH_4, but estimates are more uncertain than those for CO_2. Further, land-use change C emissions are generally considered more important than direct CH_4 emissions from agricultural systems (Burney et al. 2010). Methane is produced in anaerobic environments, such as the sediments of wetlands, peatlands, and rice paddies (Houghton 2014). The net CH_4 flux is determined by both CH_4 production by methanogens and CH_4 consumption in aerobic soil environments by methanotrophs (Tate 2015). Ecosystem-level emissions of CH_4 increase markedly with seasonal increases in temperature (Yvon-Durocher et al. 2014). Among dominant contributors to CH_4 emissions are global rice fields. Already more than 5,000 years ago, development and expansion of wet-rice agriculture may have contributed significantly to a rise in global CH_4 levels (Fuller et al. 2011). Emissions from rice fields are largely dependent on different water schemes (Zhang et al. 2016). For example, emissions ranged from 18.3 Tg CH_4 y^{-1} under intermittent irrigation to 38.8 Tg CH_4 y^{-1} under continuous flooding for global rice fields in the 2000s. In addition, agriculture burning also releases CH_4 into the atmosphere (Lin et al. 2012). Humans use fire to manage cropland and pasture. Pasture-associated fire accounted for over 40% of annual burned area, while cropland reduces fire occurrence (Rabin et al. 2015), except in regions where crop residues are burnt because of the need for a short turn around time to seed another crop such as wheat after rice in the Indo-Gangetic Plains (Fig. 1.1). Annual C emissions (in various chemical forms) from fire have been estimated at 2.5 Pg C y^{-1} for 2001–2009 (Randerson et al. 2012).

Another agricultural CH_4 source is that produced by enteric fermentation in ruminants, with 1.6 Pg CO_2 eq emitted in 2000 (Herrero et al. 2013). Further, manure management resulted in CH_4 emissions of 0.25 Pg CO_2 eq. Cattle, in particular, accounted for 77% of total non-CO_2 GHG emissions from the livestock sector in the year 2000 (Herrero et al. 2013). However, livestock emission factors are uncertain. For example, bottom-up inventories of livestock CH_4 emissions were recently updated including information for cattle and swine (domestic pig) by region, based on reported recent changes in animal body mass, feed quality and

Fig. 1.1 Burning of rice (*Oryza sativa* L.) residues in SE Punjab, India, prior to the wheat (*Triticum* L.) season (Wikimedia Commons)

quantity, milk productivity, and management of animals and manure (Wolf et al. 2017). For 2011, global livestock emissions were estimated at 119.1 Tg CH_4 which is comparable to recently reported top-down global estimates for recent years. Wolf et al. (2017) suggested that livestock CH_4 emissions may be a major contributor to the observed annual CH_4 emissions increases over the 2000s–2010s.

Agricultural land use affects the global C cycle also by accelerated soil erosion associated, in particular, with land clearing and tillage (Lal 2003). Gross SOC displacement by soil water erosion has been estimated at 2.5 Pg C yr^{-1} (Borrelli et al. 2017). However, estimates of erosion effects are uncertain and range from a source of 1 Pg C y^{-1} to a sink of the same magnitude (van Oost et al. 2007). Soil erosion can induce a net C sink or source depending on management practices, the extent to which SOC oxidation and production characteristics change with erosion, and the fate of eroded SOC (Billings et al. 2010). Nevertheless, current accelerated rates of erosion are causing major modifications to the terrestrial C cycle (Quinton et al. 2010).

In addition to SOC, agricultural land use may also affect soil inorganic carbon (SIC) and, thus, the global C cycle. Specifically, LULCC on limed acidic and carbonate-bearing soils of agroecosystems can fundamentally alter C dynamics, creating an important feedback to atmospheric CO_2 concentrations (Ahmad et al. 2015). Transformation of carbonates in such soils and its implication for the global C cycle are, however, largely unknown. Some suggest that fertilization and straw return into soils of agroecosystems may enhance SIC stocks (Wang et al. 2015).

Thus, formation of secondary carbonates and leaching of bicarbonates may be an important form for C sequestration in arid and semiarid regions (Monger et al. 2015) that cover about 35% of the Earth's land surface, and this process may play a large role in the global C cycle. However, when calcium in the carbonate is derived from preexisting carbonate, an equilibrium reaction occurs that does not sequester C in soil carbonate but does sequester C in groundwater until bicarbonate precipitates as $CaCO_3$ (Monger et al. 2015). Otherwise, when calcium is derived from silicate minerals, a unidirectional reaction occurs that sequesters C in both soil carbonate and groundwater. Monger et al. (2015) argued that C sequestration as SIC occurs only when calcium is released directly from silicates in soil. Adding powder of silicate mineral (e.g., olivine) to soils of agroecosystems may also enhance formation of secondary carbonates and mitigate climate change (Hartmann et al. 2013). Agricultural SIC sequestration processes are discussed in detail in Chap. 2.

1.3 Agricultural Land Use and Climate Change

Anthropogenic GHG emissions since the pre-industrial era have driven large increases in the atmospheric concentrations of CO_2, CH_4, and N_2O (Allen et al. 2014). Together with other anthropogenic forcings, anthropogenic increases in GHG concentrations have caused more than half of the observed increase in global average surface temperature from 1951 to 2010 (Allen et al. 2014). In 2016, global average surface temperature reached 1.1 °C above pre-industrial levels, making it the warmest year on record (WMO 2017a). Globally averaged concentrations for CO_2 reached 403.3 parts per million (ppm) in 2016, up from 400.0 ppm in 2015. This is a record annual increase (WMO 2017b). Following three years of almost no growth, CO_2 emissions from fossil fuels and industry are projected to grow by 2.0% in 2017 (Le Quéré et al. 2017). Thus, the world has not reached peak emissions yet.

Warming and elevated atmospheric CO_2 and tropospheric ozone (O_3) concentrations are already affecting crop production (Porter et al. 2014). Specifically, CO_2 has stimulatory effects on crop yields in most cases while elevated O_3 has damaging effects. Further, large negative sensitivity of crop yields to extreme daytime temperatures around 30 °C has been reported. For the major crops (wheat, rice, and maize) in tropical and temperate regions, climate change without adaptation will negatively impact production for local temperature increases of 2 °C or more above late-twentieth-century levels. Worryingly, negative impacts of climate trends on crop production have been more common than positive ones (Porter et al. 2014). For example, the intercomparison of multiple global gridded crop models (GGCMs) indicated strong negative effects of climate change on global crop yields, especially at higher levels of warming and at low latitudes (Rosenzweig et al. 2014). Models that included explicit N stress projected more severe impacts. However, uncertainties in crop yield projections were related to the representation of CO_2, N, and high-temperature effects (Rosenzweig et al. 2014). Otherwise, socioeconomic feedbacks (i.e., technological change, input substitution, and crop switching) must also be considered. For example, while pure climate effects on crop yield in the USA were projected to range from −1 to +6% in 2040, the indirect effects of climate due to socioeconomic feedbacks may range from −12 to +15% (Calvin and Fisher-Vanden 2017).

In China, negative impacts of climate change on agriculture tend to dominate (Ju et al. 2013). Without adaptation, yields of irrigated wheat, corn, and rice in China are projected to decrease by 2.2–6.7%, 0.4–11.9%, and 4.3–12.4%, respectively, in the 2050s compared with the potential yield in 1961–1990 (Ju et al. 2013). With regard to global crop cover, increases and decreases under future climates are projected to cancel each other out (Porfirio et al. 2017). Crop cover in the Northern Hemisphere is projected to be impacted more by future climate than the in Southern Hemisphere. Further, the crop cover in temperate regions is projected to decrease more than in tropical regions. Regions of opportunities identified were East Africa, Asia, and Latin America and may be a priority target for adaptation and investments. In contrast, regions of concern that would be vulnerable under future climate were South Asia and North-West Africa. Further, non-food-insecure regions with a stronger socioeconomic structure, like Europe, the Argentinean Pampas, and central USA, may also see a decrease in crop cover in future climates. These regions may invest in new low C emission technologies to ameliorate the impacts of climate change on crop cover (Porfirio et al. 2017). The economic impacts of climate change on agriculture differ among regions with European farms slightly more sensitive to warming than American farms (Van Passel et al. 2016). For US agriculture, longer run adaptations may have mitigated less than half and more likely none of the large negative short-term impacts of extreme heat on productivity (Burke and Emerick 2016). This limited recent adaptation implies substantial losses for US agriculture under future climate change in the absence of countervailing investments.

The adaptation of agriculture to climate change is necessary. Appropriate strategies include technological advances, government policy and insurance, farm production practices, and farm financial management (Smit and Skinner 2002). Adaptation should be both socially equitable and environmentally sustainable (Luo et al. 2017). In the past, technological innovation has played an important role in reducing some of the risks of weather variability to agriculture (Smithers and Blay-Palmer 2001). However, technological advances alone have not offset the risks associated with weather variability. For example, availability of days suitable for fieldwork is driven by soil temperature and moisture regimes, both of which may be altered by climate change (Tomasek et al. 2017). Thus, adaptations to climate change will require a more integrated approach. This needs to include both adaptations by the production system and socioeconomic adaptations in market risk-taking and policy (Howden et al. 2007). For example, if effective measures are taken in a timely way, then climate change in the next 30–50 years will not have a significant influence on China's food security (Ju et al. 2013). However, the aggregate climate effects on agricultural growth patterns have not been well quantified (Liang et al. 2017). For example, temperature and precipitation in distinct agricultural regions and seasons in the US explain $\sim$70% of variations in total factor productivity (TFP) growth during 1981–2010. TFP is an important indicator of the efficiency of the agricultural system and represents the ratio of measured output (e.g., crops, livestock, and goods and services) per unit of measured inputs (e.g., land, labor, capital, and resources) (Fuglie et al. 2012). To date, effects of regional climate trends in the USA on TFP have been outweighed by improvements in technology (Liang et al. 2017). However, should these relationships continue, the projected climate change could cause TFP to drop by 2.84–4.34% per year under medium to high emissions scenarios (Liang et al. 2017).

Agricultural land use may directly affect the global climate by altering the GHG exchange between land and atmosphere and also by altering the albedo. The CO_2 emissions and uptakes associated with agricultural soils, biomass, and land-use change are accounted for by IPCC in the land use, land-use change and forestry (LULUCF) sector (Richards et al. 2015). Thus, agricultural net emissions include CH_4 and N_2O emissions, and C sequestration resulting from the production of crops, livestock, and agroforestry on farms (Wollenberg et al. 2016).

Already thousands of years before the pre-industrial era, agricultural activities may have altered the land-atmosphere exchange of GHGs and the climate. For example, Ruddiman (2003) argued that GHG emissions from pre-industrial farming were large enough to warm the global climate. Specifically, atmospheric CO_2 and CH_4 increases over the last few millennia were found to be anomalous. Simultaneously, agricultural activities did spread across the continents and emitted anthropogenic GHGs potentially overprinting natural trends (Ruddiman 2003). Paleoecological and archaeological syntheses provide some ground truth evidence on early anthropogenic releases of GHGs by agricultural activities, but more information is needed to constrain the size of the agricultural emissions (Ruddiman et al. 2016).

More accurate assessments of effects of agriculture on the global climate are possible for the industrial era. For example, in 2000–2010, annual GHG emissions from agricultural production were estimated at 5.0–5.8 $GtCO_2$ eq yr^{-1} or $\sim 11\%$ of total anthropogenic GHG emissions, not including those from land-use change (Smith et al. 2014). However, only agricultural non-CO_2 sources are reported by IPCC as anthropogenic GHG emissions as CO_2 emitted is considered neutral, being associated to annual cycles of C fixation and oxidation through photosynthesis. The enteric fermentation and agricultural soils represent together about 70% of annual total non-CO_2 emissions, followed by paddy rice cultivation (9–11%), biomass burning (6–12%), and manure management (7–8%; Smith et al. 2014). Soils, in particular, contribute a major share (37%, mainly as N_2O and CH_4) of agricultural non-CO_2 emissions (Tubiello et al. 2015), and those emissions are increasing.

The emissions of N_2O to the atmosphere increased since the beginning of the industrial era due to increased use of synthetic N fertilizers, supplemented by biological N fixation by leguminous crops and pasture species (Smith 2017). About 65% of the current emissions derive from agricultural soils (Winiwarter et al. 2018). Fertilizer use in agriculture is estimated to be responsible for emission of 5.3 Tg N_2O y^{-1} or 64% of the global anthropogenic N_2O budget (Syakila and Kroeze 2011). Cropland soils, in particular, are the largest anthropogenic source of N_2O

(Wagner-Riddle et al. 2017). However, estimates of N_2O emissions are uncertain as, for example, freeze–thaw events in seasonally frozen croplands alone may be responsible for emissions of about 1 Tg N_2O–N—emissions that have not been considered in previous estimates of GHG emissions from agriculture (Butterbach-Bahl and Wolf 2017). Thus, Wagner-Riddle et al. (2017) argued that the global agricultural N_2O budget might have been previously underestimated by 17–28%. The climate sensitivity of N_2O emissions is also poorly known. For example, under current trends in climate and anthropogenic N use, Griffis et al. (2017) projected a strong positive feedback of regional N_2O emissions to warmer and wetter conditions. Otherwise, the CO_2 fertilization effect may enhance plant N uptake in several regions, which subsequently dampens N_2O emissions (Kanter et al. 2016). Employing precision fertilizer application technology as well as chemical treatment of fertilizers to suppress conversion processes in soil (nitrification inhibitors) can greatly contribute to reductions in N_2O emissions (Winiwarter et al. 2018). However, deeper reductions in emissions require also adjusting human diets toward moderate animal protein consumption.

The CH_4 emissions from rice cultivation have long been recognized as one of the dominant contributors to anthropogenic GHG emissions. Over the past century, global CH_4 emissions from rice cultivation increased by about 85% (Zhang et al. 2016). The expansion of rice fields was the dominant factor for the increasing trends, followed by elevated CO_2 concentration, and N fertilizer use. Most non-CO_2 agricultural emissions come from just a few countries; i.e., 39% of global emissions came from the four countries China, India, Brazil, and the USA (Richards et al. 2015). Regionally, South America, Eastern Asia (including China), and South Asia make the largest contributions to non-CO_2 agricultural emissions, while North America contributed less (Fig. 1.2). Importantly, agricultural non-CO_2 emissions are significant at national levels, contributing an average of 35% of emissions in developing countries and 12% in developed countries (Richards et al. 2015).

Aside directly altering the GHG flux to and from the atmosphere, agricultural land-use activities can either reduce or accelerate climate change by affecting biophysical processes (e.g., evapotranspiration, albedo; Smith et al. 2014). There is robust evidence that human-induced land-use changes have led to an increased surface albedo (Myhre et al. 2013) and, thus, potentially increase air temperature. Otherwise, burn scars resulting from agricultural practices such as slash-and-burn have a lower albedo than unperturbed vegetation. However, at high latitude, burnt areas are more easily covered by snow, which may result in an overall increase of the surface albedo. In contrast, changes in evapotranspiration and surface roughness may counteract the effect of changes in albedo. Further, land-use changes also affect latent heat flux and influence the hydrological cycle. There is, however, currently low agreement on the net biophysical effect of land-use changes on the global mean temperature (Myhre et al. 2013).

Less well known are the effects of climate change on the hundreds of millions of the world's poorest people directly depending on smallholder farming systems (Cohn et al. 2017). However, smallholders may be highly vulnerable to climate impacts. While smallholder farming systems emit a small amount of CO_2 per capita

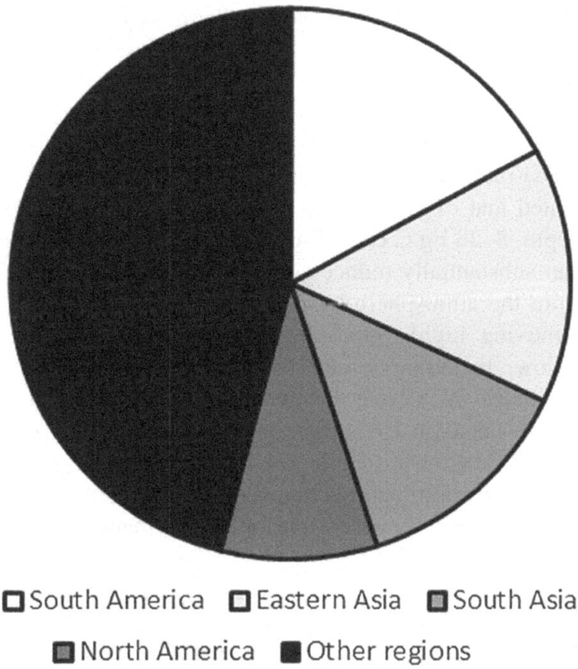

Fig. 1.2 Contributions of global regions to non-CO_2 agricultural emissions (based on data reported in Richards et al. 2015)

and are poor, making GHG regulation unwarranted, they produce GHG intensive food and emit disproportionate quantities of black carbon (BC) through traditional biomass energy. Effectively accounting for smallholders in climate change mitigation and adaption, and sustainable development policies is, thus, critical (Cohn et al. 2017).

1.4 Carbon Sequestration by Agricultural Land Use

The previous sections have highlighted that since a long time, humans have changed the global land use and cover by agricultural activities and, thus, contributed to loss of terrestrial C and increases in atmospheric GHG concentrations. Population growth and a shift toward more meat-intensive diets, in particular, have contributed to an increasing demand for agricultural land (Smith et al. 2010). These factors are expected to continue to be important in the future. Future increases in land area for bioenergy, crops, and livestock are projected, in particular, with forest and other lands otherwise decreasing (Smith et al. 2010). Based on several scenarios, the 2050 projections for change in cropland range from 6% reduction with adoption of SI to an increase of more than 30%, with an average increase of 10–20%. For grazing land, the range of 2050 scenario projections ranges between a 5% contraction to a 25% increase, with most studies showing an increase of 10% or less

(Smith et al. 2010). A major challenge will be how detrimental effects of projected agricultural land-use changes on soil C stocks and atmospheric GHG levels can be reduced.

The future policy decisions in the agriculture sector may have profound effects on land-use with different demands for land to supply multiple ecosystem services (ESs) including C sequestration (Smith et al. 2010). Sanderman et al. (2017) estimated that of the 133 Pg C lost historically from soils of agroecosystems to 2 m depth, 8–28 Pg C can be recaptured. Improved soil management in agroecosystems can substantially reduce GHG emissions and sequester some of the CO_2 removed from the atmosphere by plants as C in SOC (Paustian et al. 2016). For example, removing highly erodible cropland from crop production, eliminating summer fallow, including winter cover crops in crop rotations and widespread use of no-tillage (NT) have the potential to increase SOC storage by 63.7 Tg C yr^{-1} to 30-cm depth in US agricultural soils (Sperow 2016). Some of the SOC must be sequestered for millennia to contribute to climate change mitigation (Mackey et al. 2013). The dominating control of SOC losses is via the input-driven changes, which are more directly accessible to human management than the turnover-driven ones (Nyawira et al. 2017). However, CO_2 removal was not specifically discussed at the 2015 United Nations Climate Change Conference (COP21) in Paris (Williamson 2016). The 196 parties to the United Nations Framework Convention on Climate Change (UNFCCC), however, agreed to balance the human-driven GHG budget sometime between 2050 and 2100. This commitment is intended to limit the increase in global average temperature above pre-industrial levels to well below 2 °C"—and preferably to 1.5 °C. Limiting the global temperature rise to 2 °C would require the removal of ~600 Pg CO_2 over the twenty-first century (Wiltshire and Davies-Barnard 2015). SOC sequestration and biochar provide some of the required 'negative emissions' (Smith 2016). If phasedown of fossil fuel emissions begins soon, improved agricultural practices, including steps to improve soil fertility and increase its SOC content, may provide much of the necessary CO_2 extraction (Hansen et al. 2017). However, a constraint for SOC sequestration is that of finite capacity and sink saturation which occurs after 10–100 years, depending on the SOC sequestration option, soil type, and climate zone (Smith 2016). Further, as sinks derived from SOC sequestration are also reversible, recommended practices need to be maintained, even when the sink is saturated so any yearly costs will persist even after the emission potential has reduced to zero at sink saturation. The same issues associated with sink saturation apply partly to biochar, but equilibrium (if it occurs) would be expected to take much longer as biochar is thought to be more recalcitrant (Smith 2016). However, the potential of biochar to contribute to SOC sequestration in global agricultural soils is uncertain (Lorenz and Lal 2014; Mukherjee and Lal 2014).

The importance of agricultural soils and of soils in general to mitigate GHG emissions has often been overlooked in the past. For example, SOC sequestration is currently not considered in global climate stabilization scenarios due to concerns about the length of time required to build up SOC, the reversibility of sequestered C, competition for soil inputs and difficulties of detecting improvements

(Frank et al. 2017). However, sustainable agricultural practices can reduce emissions and increase C sequestration in healthy, well-managed soils (Figueres et al. 2017). An increase in the SOC stock was promoted at COP21 by the 4 per 1000 Initiative (4p1000) to help address food security and climate change issues (http://4p1000.org/). Increasing SOC sequestration on agricultural land is projected to enhance food security (Frank et al. 2017). The 4p1000 initiative will be discussed in Chap. 9.

In the following chapters, C sequestration in soils of agroecosystems is discussed including distinguishing SIC from SOC stocks. This is followed by chapters on C sequestration in soils of croplands and grasslands and those of agroforestry systems and wetlands. Further presented are issues regarding C sequestration by biomass and bioenergy production and biochar. The final chapter will discuss the climate policy implications of agricultural soil C sequestration.

1.5 Conclusions

Since the dawn of agriculture thousands of years ago, humans have altered the exchange of CO_2, CH_4, and N_2O between soils and the atmosphere. In total, up to 525 Pg C may have been released from natural ecosystems, particular, by clearing land to make room for crop and grasslands. Agriculture is now practiced on 40% of the global ice-free land area and is not only a significant contributor to climate change but is also itself vulnerable to a changing climate. Also in the face of increasing human demands, adaptation and mitigation of agriculture are, therefore, needed. Among practices, increasing the SOC stock has recently received increased attention by addressing climate change together with enhancing food security, for example, by the 4p1000 initiative proposed at COP21 in Paris. Whereas whatever is known must be translated into action, more research is also needed to identify site-specific agricultural practices to enhance SOC stocks and its relation to environmental effects and, particularly, on how to incentivize the large number of smallholders in the developing countries.

1.6 Review Questions

1. Describe the history of agriculture spreading across the globe.
2. What are the consequences of agricultural land-use and land-cover change for global soils and climate—in the past, now, and in the future?
3. What are possible agricultural climate change adaptation and mitigation options?
4. Contrast and compare the effects of dietary choices and food waste on agricultural GHG emissions.

5. How can sustainable intensification be made climate-resilient?
6. What are benefits and challenges of increasing SOC stocks for agroecosystem services?
7. Contrast and compare possible practices to implement the 4p1000 initiative in major global growing regions
8. Are changes in subsoil SOC and SIC stocks relevant to the resilience of agricultural systems and how can those stocks be managed and studied?

References

Ahmad W, Singh B, Dalal RC, Dijkstra FA (2015) Carbon dynamics from carbonate dissolution in Australian agricultural soils. Soil Res 53:144–153
Alexander P, Brown C, Arneth A et al (2017a) Losses, inefficiencies and waste in the global food system. Agric Syst 153:190–200
Alexander P, Brown C, Arneth A et al (2017b) Could consumption of insects, cultured meat or imitation meat reduce global agricultural land use? Glob Food Sec 15:22–32
Alexander P, Prestele R, Verburg PH et al (2017c) Assessing uncertainties in land cover projections. Glob Change Biol 23:767–781
Alexander P, Rounsevell MDA, Dislich C et al (2015) Drivers for global agricultural land use change: the nexus of diet, population, yield and bioenergy. Glob Environ Change 35:138–147
Alexandratos N, Bruinsma J (2012) World agriculture towards 2030/2050: the 2012 revision. ESA Working Paper No. 12–03. FAO, Rome
Allen MR, Barros VR, Broome J, et al (2014) In: Aldunce P, Downing T, Joussaume S, et al (eds) Climate change 2014: synthesis report. Approved Summary for Policymakers, Fifth Assessment Report of the Intergovernmental Panel on Climate Change. Cambridge University Press, Cambridge
Arneth A, Sitch S, Pongratz J et al (2017) Historical carbon dioxide emissions caused by land-use changes are possibly larger than assumed. Nat Geosci 10:79–86
Ascott MJ, Gooddy DC, Wang L et al (2017) Global patterns of nitrate storage in the vadose zone. Nat Commun 8:1416. https://doi.org/10.1038/s41467-017-01321-w
Auerswald K, Fischer FK, Kistler M et al (2018) Behavior of farmers in regard to erosion by water as reflected by their farming practices. Sci Tot Environ 613–614:1–9
Baker B (2017) Can modern agriculture be sustainable. BioSci 67:325–331
Balter M (2010) The tangled roots of agriculture. Science 327:404–406
Behrens P, Kiefte-de Jong JC, Bosker T et al (2017) Evaluating the environmental impacts of dietary recommendations. Proc Natl Acad Sci USA 114:13412–13417
Bender SF, Wagg C, van der Heijden MGA (2016) An underground revolution: biodiversity and soil ecological engineering for agricultural sustainability. Trends Ecol Evol 31:440–452
Bennetzen EH, Smith P, Porter JR (2016a) Decoupling of greenhouse gas emissions from global agricultural production: 1970–2050. Glob Change Biol 22:763–781. https://doi.org/10.1111/gcb.13120
Bennetzen EH, Smith P, Porter JR (2016b) Agricultural production and greenhouse gas emissions from world regions—the major trends over 40 years. Glob Environ Change 37:43–55
Bijl DL, Bogaart PW, Dekker SC et al (2017) A physically-based model of long-term food demand. Glob Environ Change 45:47–62

Billings SA, Buddemeier RW, deB Richter D, Van Oost K, Bohling G (2010) A simple method for estimating the influence of eroding soil profiles on atmospheric CO_2. Global Biogeochem Cyc 24, GB2001, https://doi.org/10.1029/2009gb003560

Birney CI, Franklin KF, Davidson FT, Webber ME (2017) An assessment of individual foodprints attributed to diets and food waste in the United States. Environ Res Lett 12:105008

Blümel WD (2009) Natural climatic variations in the Holocene: past impacts on cultural history, human welfare and crisis. In: Brauch HG, et al (eds) Facing global environmental change. Environmental Human, Energy, Food, Health and Water Security Concepts. Springer Verlag, Berlin, Germany, pp 103–116

Blunden J, Arndt DS (eds) (2017) State of the climate in 2016. Bull Amer Meteor Soc 98:Si–S277. https://doi.org/10.1175/2017bamsstateoftheclimate.1

Bogaard A, Fraser R, Heaton THE et al (2013) Crop manuring and intensive land management by Europe's first farmers. Proc Natl Acad Sci USA 110:12589–12594

Borlaug N (2007) Feeding a hungry world. Science 318:359

Borrelli P, Robinson DA, Fleischer LR et al (2017) An assessment of the global impact of 21st century land use change on soil erosion. Nat Commun 8:2013. https://doi.org/10.1038/s41467-017-02142-7

Bourzac K (2017) Solar upgrade. Nature 544:S11–S13

Burke M, Emerick K (2016) Adaptation to climate change: evidence from US agriculture. Am Econ J Econ Pol 8:106–140

Burney JA, Davis SJ, Lobell DB (2010) Greenhouse gas mitigation by agricultural intensification. Proc Natl Acad Sci USA 107:12052–12057

Butterbach-Bahl K, Wolf B (2017) Warming from freezing soils. Nat Geosci 10:248–249

Calvin K, Fisher-Vanden K (2017) Quantifying the indirect impacts of climate on agriculture: an inter-method comparison. Environ Res Lett 12:115004

Carter S, Herold M, Avitabile V et al (2018) Agriculture-driven deforestation in the tropics from 1990–2015: emissions, trends and uncertainties. Environ Res Lett 13:014002

Clark M, Tilman D (2017) Comparative analysis of environmental impacts of agricultural production systems, agricultural input efficiency, and food choice. Environ Res Lett 12:064016

Cohn AS, Newton P, Gil JDB et al (2017) Smallholder agriculture and climate change. Annu Rev Environ Resour 42:347–375

Conijn JG, Bindraban PS, Schröder JJ, Jongschaap REE (2018) Can our global food system meet food demand within planetary boundaries? Agric Ecosyst Environ 251:244–256

Council for Agricultural Science and Technology (CAST) (2017) Plant breeding and genetics—A paper in the series on the need for agricultural innovation to sustainably feed the world by 2050. Issue Paper 57. CAST, Ames, Iowa

Collard M, Edinborough K, Shennan S, Thomas MG (2010) Radiocarbon evidence indicates that migrants introduced farming to Britain. J Archaeolog Sci 37:866–870

Craine JM, Elmore A, Angerer JP (2017) Long-term declines in dietary nutritional quality for North American cattle. Environ Res Lett 12:044019

Crutzen PJ, Stoermer EF (2000) The "Anthropocene" IGBP Newsl 41:12–14

d'Amour CB, Reitsma F, Baiocchi G et al (2017) Future urban land expansion and implications for global croplands. Proc Natl Acad Sci USA 114:8939–8944

Davis KF, Rulli MC, Seveso A, D'Odorico P (2017) Increased food production and reduced water use through optimized crop distribution. Nat Geosci 10:919–924. https://doi.org/10.1038/s41561-017-0004-5

D'Odorico P, Rulli MC (2014) The land and its people. Nat Geosci 7:324–325

Don A, Schumacher J, Freibauer A (2011) Impact of tropical land-use change on soil organic carbon stocks—a meta-analysis. Glob Change Biol 17:1658–1670

Dunne J, Evershed RP, Salque M et al (2012) First dairying in green Saharan Africa in the fifth millennium BC. Nature 486:390–394

Editorial Nature Plants (2017) Saving the world. Nat Plants 3:17069. https://doi.org/10.1038/nplants.2017.69

Eglin T, Ciais P, Piao SL et al (2010) Historical and future perspectives of global soil carbon response to climate and land-use changes. Tellus 62B:700–718
Ellis EC, Fuller DQ, Kaplan JO, Lutters WG (2013) Dating the Anthropocene: towards an empirical global history of human transformation of the terrestrial biosphere. Elem Sci Anth 1:18. http://doi.org/10.12952/journal.elementa.000018
Erb KH, Kastner T, Plutzar C et al (2018) Unexpectedly large impact of forest management and grazing on global vegetation biomass. Nature 553:73–76
FAO Food and Agriculture Organization of the United Nations (2015) Statistical Pocketbook 2015. FAO, Rome
FAO Food and Agriculture Organization of the United Nations (2016) The state of food and agriculture 2016. FAO, Rome
FAO Food and Agriculture Organization of the United Nations (2017) The future of food and agriculture—trends and challenges. FAO, Rome
Figueres C, Schellnhuber HJ, Whiteman G et al (2017) Three years to safeguard our climate. Nature 546:593–595
Fischer G, Tramberend S, Bruckner M, Lieber M (2017) Quantifying the land footprint of Germany and the EU using a hybrid accounting model. Umweltbundesamt [Ferderal Environment Agency] TEXTE 78/2017. https://www.umweltbundesamt.de/publikationen/quantifying-the-land-footprint-of-germany-the-eu
Foley JA, Ramankutty N, Brauman KA et al (2011) Solutions for a cultivated planet. Nature 478:337–342
Frank S, Havlík P, Soussana JF et al (2017) Reducing greenhouse gas emissions in agriculture without compromising food security? Environ Res Lett 12:105004
Frieler K, Schauberger B, Arneth A et al (2017) Understanding the weather signal in national crop-yield variability. Earth's Future 5:605–616. https://doi.org/10.1002/2016EF000525
Fuglie KO, Ball VE, Wang SL, eds (2012) Productivity growth in agriculture: an international perspective. CABI, Cambridge, MA
Fuller DQ, van Etten J, Manning K et al (2011) The contribution of rice agriculture and livestock to prehistoric methane levels: An archeological assessment. Holocene 21:743–759
Ganesan M, Lee HY, Kim JI, Song PS (2017) Development of transgenic crops based on photobiotechnology. Plant Cell Environ 40:2469–2486. https://doi.org/10.1111/pce.12887
Gaud WS (1968) The green revolution: accomplishments and apprehensions. AgBioWorld. http://www.agbioworld.org/biotech-info/topics/borlaug/borlaug-green.html
Gibbons A (2016) First farmers' motley roots. Science 353:207–208
Glover JD, Reganold JP, Bell LW et al (2010) Increased food and ecosystem security via perennial grains. Science 328:1638–1639. https://doi.org/10.1126/science.1188761
Godfray HCJ, Beddington JR, Crute IR et al (2010) Food security: the challenge of feeding 9 billion people. Science 327:812–818
Govers G, Van Oost K, Wang Z (2014) Scratching the critical zone: the global footprint of agricultural soil erosion. Proc Earth Plan Sci 10:313–318
Govers G, Merckx R, van Wesemael B, Van Oost K (2017) Soil conservation in the 21st century: why we need smart agricultural intensification. SOIL 3:45–59
Griffis TJ, Chen Z, Baker JM et al (2017) Nitrous oxide emissions are enhanced in a warmer and wetter world. Proc Natl Acad Sci USA 114:12081–12085
Gruber K (2017) The living library. Nature 544:S8–S10
Gupta AK (2004) Origin of agriculture and domestication of plants and animals linked to early Holocene climate amelioration. Curr Sci 87:54–59
Haak W, Balanovsky O, Sanchez JJ, Koshel S, Zaporozhchenko V, Adler CJ, Der Sarkissian CSI, Brandt G, Schwarz C, Nicklisch N, Dresely V, Fritsch B, Balanovska E, Villems R, Meller H, Alt KW, Cooper A, the Genographic Consortium (2010) Ancient DNA from European early neolithic farmers reveals their near eastern affinities. PLoS Biol 8(11):e1000536. https://doi.org/10.1371/journal.pbio.1000536

Hall C, Dawson TP, Macdiarmid JI, Matthews RB, Smith P (2017) The impact of population growth and climate change on food security in Africa: looking ahead to 2050. Int J Agric Sustain 15:124–135

Hansen J, Sato M, Kharecha P et al (2017) Young people's burden: requirement of negative CO_2 emissions. Earth Syst Dynam 8:577–616

Hartmann J, West AJ, Renforth P et al (2013) Enhanced chemical weathering as a geoengineering strategy to reduce atmospheric carbon dioxide, supply nutrients, and mitigate ocean acidification. Rev Geophys 51:113–149. https://doi.org/10.1002/rog.20004

Herrero M, Havlík P, Valin H et al (2013) Biomass use, production, feed efficiencies, and greenhouse gas emissions from global livestock systems. Proc Natl Acad Sci USA 110:20888–20893

Hofmanová Z, Kreutzer S, Hellenthal G et al (2016) Early farmers from across Europe directly descended from Neolithic Aegeans. Proc Natl Acad Sci USA 113:6886–6891

Houghton RA (2010) How well do we know the flux of CO_2 from land-use change? Tellus 62B:337–351

Houghton RA (2014) The contemporary carbon cycle. In: Holland H, Turekian K (eds) Treatise on geochemistry, 2nd edn. Elsevier, Amsterdam, Netherlands, pp 399–435

Houghton RA, Nassikas AA (2017) Global and regional fluxes of carbon from land use and land cover change 1850–2015. Glob Biogeochem Cycles 31:456–472

Howden SM, Soussana J-F, Tubiello FN et al (2007) Adapting agriculture to climate change. Proc Natl Acad Sci 104:19691–19696

Hunter MC, Smith RG, Schipanski ME, Atwood LW, Mortensen DA (2017) Agriculture in 2050: recalibrating targets for sustainable intensification. Bioscience 67:386–391. https://doi.org/10.1093/biosci/bix010

Izumi T, Ramankutty N (2016) Changes in yield variability of major crops for 1981–2010 explained by climate change. Environ Res Lett 11:034003

Ju H, van der Velde M, Lin E, Xiong W, Li Y (2013) The impacts of climate change on agricultural production systems in China. Clim Change 120:313–324

Kane D, Rogé P, Snapp S (2016) A systemic review of perennial staple crops literature using topic modeling and bibliometric analysis. PLoS ONE 11:e0155788. https://doi.org/10.1371/journal.pone.0155788

Kanter DR, Zhang X, Mauzerall DL, Malyshev S, Shevliakova E (2016) The importance of climate change and nitrogen use efficiency for future nitrous oxide emissions from agriculture. Environ Res Lett 11:094003. https://doi.org/10.1088/1748-9326/11/9/094003

Kaplan JO, Krumhardt KM, Ellis RC, Ruddiman WF, Lemmen C, Klein Goldewijk K (2010) Holocene carbon emissions as a result of anthropogenic land cover change. The Holocene 21:775–791

Kesavan PC, Swaminathan MS (2008) Strategies and models for agricultural sustainability in developing Asian countries. Philos Trans R Soc B 363:877–891

King A (2017) The future of agriculture. Nature 544:S21–S23

Kluyver TA, Jones G, Pujol B et al (2017) Unconscious selection drove seed enlargement in vegetable crops. Evolution Letters 1–2:64–72. https://doi.org/10.1002/evl3.6

Kromdijk J, Głowacka K, Leonelli L et al (2016) Improving photosynthesis and crop productivity by accelerating recovery from photoprotection. Science 354:857–861

Lal R (2003) Soil erosion and the global carbon budget. Environ Int 29:437–450

Lal R (2016) Feeding 11 billion on 0.5 billion hectare of area under cereal crops. Food Energ Sec 5:239–251. https://doi.org/10.1002/fes3.99

Lal R (2017) Improving soil health and human protein nutrition by pulses-based cropping systems. Adv Agron 145:167–204

Larson G, Piperno DR, Allaby RG et al (2014) Current perspectives and the future of domestication studies. Proc Natl Acad Sci USA 111:6139–6146

Laurance WF, Sayer J, Cassman KG (2014) Agricultural expansion and its impacts on tropical nature. Trends Ecol Evol 29:107–116

Leng G (2017) Recent changes in county-level corn yield variability in the United States from observations and crop models. Sci Tot Environ 607–608:683–690. https://doi.org/10.1016/j.scitotenv.2017.07.017
Le Quéré C, Andrew RM, Friedlingstein P et al (2017) Global carbon budget 2017. Earth Syst Sci Data Discuss. https://doi.org/10.5194/essd-2017-123
Levis C, Costa FRC, Bongers F et al (2017) Persistent effects of pre-Columbian plant domestication on Amazonian forest composition. Science 355:925–931
Li W, Ciais P, Peng S et al (2017) Land-use and land-cover change carbon emissions between 1901 and 2012 constrained by biomass observations. Biogeosciences 14:5053–5067
Liang XZ, Wu Y, Chambers RG et al (2017) Determining climate effects on US total agricultural productivity. Proc Natl Acad Sci USA 114:E2285–E2292
Lin HW, Jin Y, Giglio L, Foley JA, Randerson JT (2012) Evaluating greenhouse gas emissions inventories for agricultural burning using satellite observations of active fires. Ecol Appl 22:1345–1364
Lipson M, Szécsényi-Nagy A, Mallick S et al (2017) Parallel palaeogenomic transects reveal complex genetic history of early European farmers. Nature 551:368–372. https://doi.org/10.1038/nature24476
Lobell DB, Field CB (2007) Global scale climate–crop yield relationships and the impacts of recent warming. Environ Res Lett 2:014002
Lobell DB, Tebaldi C (2014) Getting caught with our plants down: the risks of a global crop yield slowdown from climate trends in the next two decades. Environ Res Lett 9:074003
Long SP, Marshall-Colon A, Zhu XG (2015) Meeting the global food demand of the future by engineering crop photosynthesis and yield potential. Cell 161:56–66
Loos J, Abson DJ, Chappell MJ et al (2014) Putting meaning back into "sustainable intensification". Front Ecol Environ 12:356–361
Lorenz K, Lal R (2014) Biochar application to soil for climate change mitigation by soil organic carbon sequestration. J Plant Nutr Soil Sci 177:651–670
Luo XS, Muleta D, Hu Z et al (2017) Inclusive development and agricultural adaptation to climate change. Curr Op Environ Sustain 24:78–83
Lynch JP (2007) Turner review no.14. Roots of the second green revolution. Aust J Bot 55:493–512. https://doi.org/10.1071/BT06118
Mackey B, Prentice IC, Steffen W et al (2013) Untangling the confusion around land carbon science and climate change mitigation policy. Nat Clim Change 3:552–557. https://doi.org/10.1038/NCLIMATE1804
Manders T, Chateau J, Magné B, et al (2012) Socio-economic developments. In: OECD (ed) OECD environmental outlook to 2050: the consequences of inaction. OECD Publishing, Paris
Marcott SA, Shakun JD, Clark PU, Mix AC (2013) A reconstruction of regional and global temperature for the past 11,300 years. Science 339:1198–1201
Martellozzo F, Ramankutty N, Hall RJ et al (2015) Urbanization and the loss of prime farmland: a case study in the Calgary-Edmonton corridor of Alberta. Reg Environ Change 15:881–893
Mazoyer M, Roudart L (2006) A history of world agriculture. Monthly Review Press, New York
Monger HC, Kraimer RA, Khresat S, Cole DR, Wang X, Wang J (2015) Sequestration of inorganic carbon in soil and groundwater. Geology 43:375–378
Mottet A, de Haan C, Falcucci A et al (2017) Livestock: on our plates or eating at our table? A new analysis of the feed/food debate. Glob Food Secur 14:1–8
Mukherjee A, Lal R (2014) The biochar dilemma. Soil Res 52:217–230
Muller A, Schader C, El-Hage Scialabba N et al (2017) Strategies for feeding the world more sustainably with organic agriculture. Nat Commun 8:1290. https://doi.org/10.1038/s41467-017-01410-w
Myhre G, Shindell D, Bréon FM, et al. (2013) Anthropogenic and natural radiative forcing. In: Stocker TF, Qin D, Plattner GK, et al (eds) Climate change 2013: the physical science basis. Contribution of Working Group I to the Fifth Assessment Report of the Intergovernmental

Panel on Climate Change. Cambridge University Press, Cambridge, United Kingdom and New York, NY, USA
Nyawira SS, Nabel JMES, Brovkin V, Pongratz J (2017) Input-driven versus turnover-driven controls of simulated changes in soil carbon due to land-use change. Environ Res Lett 12:084015
Ort DR, Merchant SS, Alric J et al (2015) Redesigning photosynthesis to sustainably meet global food and bioenergy demand. Proc Natl Acad Sci USA 112:8529–8536
Paustian K, Lehmann J, Ogle S et al (2016) Climate-smart soils. Nature 532:49–57
Pereira JB, Costa MD, Vieira D et al (2017) Reconciling evidence from ancient and contemporary genomes: a major source for the European Neolithic within Mediterranean Europe. Proc R Soc B 284:20161976
Petersen B, Snapp S (2015) What is sustainable intensification? Views from experts. Land Use Pol 46:1–10
Phalan B, Green RE, Dicks LV et al (2016) How can higher-yield farming help to spare nature? Science 351:450–451. https://doi.org/10.1126/science.aad0055
Pingali PL (2012) Green Revolution: Impacts, limits, and the path ahead. Proc Natl Acad Sci USA 109:12302–12308
Piperno DR (2011) The origins of plant cultivation and domestication in the New World Tropics: Patterns, process, and new developments. Curr Anthropol 52:S453–S470
Poeplau C, Don A, Vesterdal L, Leifeld J, Van Wesemael B, Schumacher J, Gensior A (2011) Temporal dynamics of soil organic carbon after land-use change in the temperate zone—carbon response functions as a model approach. Glob Change Biol 17:2415–2427
Ponisio LC, M'Gonigle LK, Mace KC et al (2015) Diversification practices reduce organic to conventional yield gap. Proc R Soc B 282:20141396. https://doi.org/10.1098/rspb.2014.1396
Porfirio LL, Newth D, Harman IN, Finnigan JJ, Cai Yiyong (2017) Patterns of crop cover under future climates. Ambio 46:265–276
Porter JR, Xie L, Challinor AJ, et al. (2014) Food security and food production systems. In: IPCC. 2014. Climate Change 2014: Impacts, adaptation, and vulnerability. Part A: Global and sectoral aspects. Contribution of Working Group II to the Fifth Assessment Report of the Intergovernmental Panel on Climate Change. Cambridge University Press, Cambridge, UK and New York, USA, pp 485–533
Pretty J, Bharucha ZP (2014) Sustainable intensification in agricultural systems. Ann Bot 114:1571–1596
Pretty J, Noble AD, Bossio D, Dixon J et al (2006) Resource-conserving agriculture increases yields in developing countries. Environ Sci Technol 3:24–43
Pretty J, Sutherland WJ, Ashby J et al (2010) The top 100 questions of importance to the future of global agriculture. Int J Agric Sustain 8:219–236
Quinton JN, Govers G, Van Oost K, Bardgett RD (2010) The impact of agricultural soil erosion on biogeochemical cycling. Nature Geosci 3:311–314
Rabin SS, Magi BI, Shevliakova E, Pacala SW (2015) Quantifying regional, time-varying effects of cropland and pasture on vegetation fire. Biogeosciences 12:6591–6604
Ramankutty N, Evan AT, Monfreda C, Foley JA (2008) Farming the planet: 1. Geographic distribution of global agricultural lands in the year 2000. Global Biogeochem Cyc 22:GB1003. https://doi.org/10.1029/2007gb002952
Randerson JT, Chen Y, van der Werf GR, Rogers BM, Morton DC (2012) Global burned area and biomass burning emissions from small fires. J Geophys Res 117:G04012
Ray DK, Mueller ND, West PC, Foley JA (2013) Crop yield trends are insufficient to double global food production by 2050. PLoSONE 8:e66428
Reganold JP, Wachter JM (2016) Organic agriculture in the twenty-first century. Nat Plants 2. http://dx.doi.org/10.1038/NPlants.2015.221
Reynolds MP, Braun HJ, Cavalieri AJ et al (2017) Improving global integration of crop research. Science 357:359–360

Richards MB, Wollenberg E, Buglion-Gluck S (2015) Agriculture's contributions to national emissions. CGIAR Research Program on Climate Change, Agriculture and Food Security (CCAFS), Copenhagen

Richerson PJ, Boyd R, Bettinger RL (2001) Was agriculture impossible during the Pleistocene but mandatory during the Holocene? A climate change hypothesis. Am Antiq 66:387–411

Roberts P, Hunt C, Arroyo-Kalin M, Evans D, Boivin N (2017) The deep human prehistory of global tropical forests and its relevance for modern conservation. Nature Plants 3:17093. https://doi.org/10.1038/nplants.2017.93

Robinson DA, Panagos P, Borrelli P et al (2017) Soil natural capital in Europe; a framework for state and change assessment. Sci Rep. https://doi.org/10.1038/s41598-017-06819-3

Rockström J, Williams J, Daily G et al (2017) Sustainable intensification of agriculture for human prosperity and global sustainability. Ambio 46:4–17

Rosenzweig C, Elliott J, Deryn D et al (2014) Assessing agricultural risks of climate change in the 21st century in a global gridded crop model intercomparison. Proc Natl Acad Sci USA 111:3268–3273

Royal Society of London (2009) Reaping the benefits: Science and the sustainable intensification of global agriculture. Royal Society, London

Ruddiman WF (2003) The anthropogenic greenhouse area began thousands of years ago. Clim Change 61:261–293

Ruddiman WF (2013) The anthropocene. Annu Rev Earth Planet Sci 41:45–68

Ruddiman WF, Fuller DQ, Kutzbach JE et al (2016) Late Holocene climate: natural or anthropogenic? Rev Geophys 54:93–118

Sanderman J, Hengl T, Fiske GJ (2017) Soil carbon debt of 12,000 years of human land use. Proc Natl Acad Sci USA 114:9575–9580

Sardans J, Grau O, Chen HYH et al (2017) Changes in nutrient concentrations of leaves and roots in response to global change factors. Glob Change Biol 23:3849–3856. https://doi.org/10.1111/gcb.13721

Sayer J, Cassman KG (2013) Agricultural innovation to protect the environment. Proc Natl Acad Sci USA 110:8345–8348

Seto KC, Ramankutty N (2016) Hidden linkages between urbanization and food systems. Science 352:943–945

Seufert V, Ramankutty N (2017) Many shades of gray—the context-dependent performance of organic agriculture. Sci Adv 3:e1602638

Sinha E, Michalak AM, Balaji V (2017) Eutrophication will increase during the 21st century as a result of precipitation changes. Science 357:405–408

Siska V, Jones ER, Jeon S et al (2017) Genome-wide data from two early Neolithic East Asian individuals dating to 7700 years ago. Sci Adv 3:e1601877

Smit B, Skinner MW (2002) Adaptation options in agriculture to climate change: a typology. Mitig Adapt Strateg Glob Change 7:85–114

Smith KA (2017) Changing views of nitrous oxide emissions from agricultural soil: key controlling processes and assessment at different spatial scales. Europ J Soil Sci 68:137–155

Smith O, Momber G, Bates R et al (2015) Sedimentary DNA from a submerged site reveals wheat in the British Isles 8000 years ago. Science 347:998–1001

Smith P (2016) Soil carbon sequestration and biochar as negative emission technologies. Glob Change Biol 22:1315–1324

Smith P, Bustamante M, Ahammad H, et al. (2014) Agriculture, forestry and other land use (AFOLU). In: Edenhofer O, Pichs-Madruga R, Sokona Y et al (eds) Climate change 2014: mitigation of climate change. Contribution of working group III to the fifth assessment report of the intergovernmental panel on climate change. Cambridge University Press, Cambridge, pp 811–922

Smith P, Gregory PJ, van Vuuren D et al (2010) Competition for land. Phil Trans R Soc B 365:2941–2957

Smithers J, Blay-Palmer A (2001) Technology innovation as a strategy for climate adaptation in agriculture. Appl Geogr 21:175–197

Snir A, Nadel D, Groman-Yaroslavski I et al (2015) The origin of cultivation and proto-weeds, long before Neolithic farming. PLoS ONE 10(7):e0131422. https://doi.org/10.1371/journal.pone.0131422

Solbrig OT, Solbrig DJ (1994) So shall you reap: farming and crops in human affairs. Island Press, Washington, DC

Sperow M (2016) Estimating carbon sequestration potential on U.S. agricultural topsoils. Soil Till Res 155:390–400

Springer NP, Duchin F (2014) Feeding nine billion people sustainably: conserving land and water through shifting diets and changes in technologies. Environ Sci Technol 48:4444–4451

Springmann M, Godfray HCJ, Rayner M, Scarborough P (2016) Analysis and valuation of the health and climate change cobenefits of dietary change. Proc Natl Acad Sci USA 113:4146–4151

Stevenson JR, Villoria N, Byerlee D, Kelley T, Maredia M (2013) Green Revolution research saved an estimated 18 to 27 million hectares from being brought into agricultural production. Proc Natl Acad Sci USA 110:8363–8368

Stocker BD, Yu Z, Massa C, Joos F (2017) Holocene peatland and ice-core data constraints on the timing and magnitude of CO_2 emissions from past land use. Proc Natl Acad Sci USA 114:1492–1497

Stockmann U, Adams MA, Crawford JW et al (2013) The knowns, known unknowns and unknowns of sequestration of soil organic carbon. Agric Ecosyst Environ 164:80–99

Struik PC, Kuyper TW (2017) Sustainable intensification in agriculture: the richer shade of green. A review. Agron Sustain Dev 37:39. https://doi.org/10.1007/s13593-017-0445-7

Syakila A, Kroeze C (2011) The global nitrous oxide budget revisited. Greenh Gas Meas Manag 1:17–26

Tate KR (2015) Soil methane oxidation and land-use change—from process to mitigation. Soil Biol Biochem 80:260–272

Taylor SH, Long SP (2017) Slow induction of photosynthesis on shade to sun transitions in wheat may cost at least 21% of productivity. Phil Trans R Soc B 372:20160543. https://doi.org/10.1098/rstb.2016.0543

Techen AK, Helming K (2017) Pressures on soil functions from soil management in Germany. A foresight review. Agron Sustain Dev 37:64. https://doi.org/10.1007/s13593-017-0473-3

Tester M, Langridge P (2010) Breeding technologies to increase crop production in a changing world. Science 327:818–822

Tilman D, Balzer C, Hill J, Befort BL (2011) Global food demand and the sustainable intensification of agriculture. Proc Natl Acad Sci USA 108:20260–20264

Tollenaar M, Fridgen J, Tyagi P, Stackhouse PW, Kumudini S (2017) The contribution of solar brightening to the US maize yield trend. Nat Clim Change 7:275–278. https://doi.org/10.1038/NCLIMATE3234

Tomasek BJ, Williams MM, Davis AS (2017) Changes in field workability and drought risk from projected climate change drive spatially variable risks in Illinois cropping systems. PLoS ONE 12:e0172301

Tubiello FN, Salvatore M, Ferrara AF et al (2015) The contribution of agriculture, forestry and other land use activities to global warming, 1990–2012. Glob Change Biol 21:2655–2660

United Nations Convention to Combat Desertification (UNCCD) (2017) The Global Land Outlook, 1st edn. Bonn, Germany

United Nations (2014) 2014 Revision of World urbanization prospects. New York

Van Meijl H, Havlik P, Lotze-Campen P, et al. (2017) Challenges of global agriculture in a climate change context by 2050 (AgCLIM50). JRC Science for Policy Report, EUR 28649 EN, https://doi.org/10.2760/772445

Van Oost K, Quine TA, Govers G et al (2007) The impact of agricultural soil erosion on the global carbon cycle. Science 318:626–629

Van Passel S, Massetti E, Mendelsohn R (2016) A ricardian analysis of the impact of climate change on European agriculture. DOI, Environ Resource Econ. https://doi.org/10.1007/s10640-016-0001-y

van Vliet J, Eitelberg DA, Verburg PH (2017) A global analysis of land take in cropland areas and production displacement from urbanization. Glob Environ Change 43:107–115

Wagner-Riddle C, Congreves KA, Abalos D et al (2017) Globally important nitrous oxide emissions from croplands induced by freeze–thaw cycles. Nat Geosci 10:279–286

Walter A, Finger R, Huber R, Buchmann N (2017) Smart farming is key to developing sustainable agriculture. Proc Natl Acad Sci USA 114:6148–6150

Wang E, Martre P, Zhao Z et al (2017) The uncertainty of crop yield projections is reduced by improved temperature response functions. Nat Plants 3:17102. https://doi.org/10.1038/nplants.2017.102

Wang X, Wang J, Xu M et al (2015) Carbon accumulation in arid croplands of northwest China: pedogenic carbonate exceeding organic carbon. Sci Rep 5:11439

Watling J, Iriarte J, Mayle FE et al (2017) Impact of pre-Columbian "geoglyph" builders on Amazonian forests. Proc Natl Acad Sci USA 114:1868–1873

White RR, Hall MB (2017) Nutritional and greenhouse gas impacts of removing animals from US agriculture. Proc Natl Acad Sci USA E10301–E10308. doi/https://doi.org/10.1073/pnas.1707322114

Wiebe K, Lotze-Campen H, Sands R et al (2015) Climate change impacts on agriculture in 2050 under a range of plausible socioeconomic and emissions scenarios. Environ Res Lett 10:085010

Willer H, Lernoud J (eds) (2017) The world of organic agriculture. statistics and emerging trends 2017. Research Institute of Organic Agriculture (FiBL), Frick, and IFOAM—Organics International Bonn

Williamson P (2016) Scrutinize CO_2 removal methods. Nature 530:153–155

Wiltshire A, Davies-Barnard T (2015) Planetary limits to BECCS negative emissions. Avoid2

Winiwarter W, Höglund-Isaksson L, Klimont Z, Schöpp W, Amann M (2018) Technical opportunities to reduce global anthropogenic emissions of nitrous oxide. Environ Res Lett 13:014011

Winter JM, Lopez JR, Ruane AC et al (2017) Representing water scarcity in future agricultural assessments. Anthropocene 18:15–26

WMO World Meteorological Organization (2017a) WMO statement on the state of the global climate in 2016. WMO-No. 1189. Geneva, Switzerland

WMO World Meteorological Organization (2017b) Greenhouse gas concentrations surge to new record. https://public.wmo.int/en/media/press-release/greenhouse-gas-concentrations-surge-new-record

Wolf J, Asrar GR, West TO (2017) Revised methane emissions factors and spatially distributed annual carbon fluxes for global livestock. Carbon Balance Manage 12:16. https://doi.org/10.1186/s13021-017-0084-y

Wolf J, West TO, Le Page Y et al (2015) Biogenic carbon fluxes from global agricultural pro-duction and consumption. Global Biogeochem Cycles 29:1617–1639. https://doi.org/10.1002/2015GB005119

Wollenberg E, Richards M, Smith P et al (2016) Reducing emissions from agriculture to meet the 2 °C target. Global Change Biol 22:3859–3864

World Bank (2008) World development report 2008. Agriculture for Development (World Bank), Washington, DC

Xue L, Liu G, Parfitt J et al (2017) Missing food, missing data? A critical review of global food losses and food waste data. Environ Sci Technol 51:6618–6633. https://doi.org/10.1021/acs.est.7b00401

Yvon-Durocher G, Allen AP, Bastviken D, Conrad R, Gudasz C et al (2014) Methane fluxes show consistent temperature dependence across microbial to ecosystem scales. Nature 507:488–495

Zalasiewicz J, Williams M, Haywood A, Ellis M (2011) The Anthropocene: a new epoch of geological time? Phil Trans R Soc A 369:835–841

Zhang B, Tian H, Ren W et al (2016) Methane emissions from global rice fields: Magnitude, spatiotemporal patterns, and environmental controls. Global Biogeochem Cyc 30:1246–1263

Zhao C, Liu B, Piao S et al (2017) Temperature increase reduces global yields of major crops in four independent estimates. Proc Natl Acad Sci USA 114:9326–9331

Zuo X, Lu H, Jiang L et al (2017) Dating rice remains through phytolith carbon-14 study reveals domestication at the beginning of the Holocene. Proc Natl Acad Sci USA 114:6486–6491

Zimmerer KS, de Haan S (2017) Agrobiodiversity and a sustainable food future. Nat Plants 3:17047. https://doi.org/10.1038/nplants.2017.47

Chapter 2
Soil Carbon Stock

Contents

Abstract Total soil carbon (C) stock comprises of the soil organic C (SOC) and the soil inorganic C (SIC) components. The global SOC stock of ice-free land contains about 1,325–1,500 Pg (1 Pg = 10^{15} g) C in the top 1 m, 2,300 Pg C in the top 3 m, and 3,000 Pg C in the soil profiles. Up to 716 Pg SOC may be stored to 1 m depth in cropland, temperate grassland/shrubland, and tropical grassland/savannah. However, estimates of global terrestrial inventories have large uncertainties because of the fewer studies and lack of credible estimates of the SOC stocks in permafrost, peatlands, and subsoil horizons. The SIC stock, primarily occurring in soils of the arid regions, is estimated at 700–1,700 Pg C in the top 1-m of soil. The SIC stocks, probably more in soils of the temperate regions and in deeper layers, are not widely studied. The SIC consists of lithogenic inorganic C (LIC) or primary carbonates derived from soil parent material, and pedogenic inorganic C (PIC) or secondary carbonates formed through pedogenic processes. Climate, geology, and land management practices are principal controls of the magnitude of the soil C stock as they are determinants of the soil and vegetation type. The SIC stock can be sink, source, or neither relative to the atmospheric

K. Lorenz and R. Lal, *Carbon Sequestration in Agricultural Ecosystems*,
https://doi.org/10.1007/978-3-319-92318-5_2

carbon dioxide (CO_2). For example, an increase in SIC stock occurs after weathering of soluble Ca/Mg-bearing silicates followed by the precipitation of pedogenic carbonates. Ecosystems with annual net Ca-carbonate (calcite) dissolution are local geochemical sinks of atmospheric CO_2 as bicarbonates move into the groundwater. With the mean residence time (MRT) of PIC at 85,000 years, it is much less dynamic than the SOC stock with MRT or mean turnover time of about 35 years. However, SIC and SOC interact with each other but underlying mechanisms are less well known. Carbon is sequestered in the SOC stock via the C inputs from photosynthetic fixation of atmospheric CO_2 by vegetation, deposition, and the accumulation of stabilized SOC fractions, and the input of black carbon (BC) with charred biomass. The main C input into the soil is net primary production (NPP) as a major fraction of the CO_2 fixed during plant photosynthesis by gross primary production (GPP) is respired autotrophically and returned back to the atmosphere. NPP enters soil by rhizodeposition and decomposition of plant litter, and a large fraction is converted back to CO_2 by soil respiration and some lost as methane (CH_4). Aside microbial decomposition, C losses from soils of agroecosystems are also associated with fire, erosion, leaching, and harvest. Thus, a small amount of fixed C remains in the soil and accumulates in the SOC stock through a combination of short- and long-term stabilization processes. Important among stabilization processes include physical protection of organic matter (OM) against decomposers and their enzymes, stabilization by organo-mineral complexes and organo-metal interactions, and some as biochemically recalcitrant BC. Soil aggregation and formation of organo-mineral complexes may be the most important stabilization process in topsoils of agroecosystems. Site-specific factors including climate, physicochemical characteristics, soil and vegetation management determine the balance between C input and losses. However, it is unclear whether and how SOC saturation may occur in soil profiles of agroecosystems. This chapter begins with a discussion about SIC dynamics and its sequestration in soils of agroecosystems. Then, the effects of agricultural practices on SIC are compared, and indirect effects of carbonates on soil C sequestration are discussed. The second subsection of this chapter discusses the SOC stock and its dynamics. First, photosynthesis and SOC input processes are discussed followed by a comparison of processes contributing to SOC loss from agricultural soils. The dynamics of SOC in the mineral soil, and stabilization and decomposition/destabilization processes are presented in detail. The importance of BC is discussed in the chapter about biochar. The concluding section discusses the importance of SOC to soil quality, ecosystem services, and food security. Research needs and some pertinent questions are summarized at the end of the chapter.

Keywords Global soil carbon stock · Lithogenic inorganic carbon
Pedogenic inorganic carbon · Soil organic carbon · Mean turnover time
Photosynthesis · Gross primary production · Net primary production
Soil respiration · Stabilization processes · Soil organic matter models

2.1 Soil Inorganic Carbon

The soil inorganic C (SIC) stock mainly consists of carbonates and bicarbonates. High stability of carbonates, as reported in studies of ancient soils, is indicated by the mean residence time (MRT) of up to 2.6 Ga (Watanabe et al. 2004). Carbonates are particularly common in soils of subhumid to arid regions (Ming 2006). While most available studies of soil C are focused on soil organic carbon (SOC), soils of arid and semiarid regions may contain SIC stock up to ten times larger than the SOC stock (Eswaran et al. 2000). The accumulation of SIC may also be an important process for C sequestration in some soils of non-arid regions (Goddard et al. 2009). For example, estimates of the SIC stocks to 30 cm depth in France amount to about one-third of the SOC stocks (Marchant et al. 2015).

Globally, the SIC stock amounts to an estimated 700–1,700 Pg C (Pg = 10^5 g = Gt) in the top 1-m of soil (Fig. 2.1; Batjes 1996; Eswaran et al. 2000). However, the total SIC stock may be much higher because large amounts of SIC occur below 1 m depth in subsoil of the arid and semiarid regions (Díaz-Hernández 2010). Further, SIC stored in gravel coatings is generally excluded from chemical analyses of soil carbonates, and thus, the global SIC stock may be underestimated. For example, 13% of the total SIC in a watershed in Idaho, USA, was stored as carbonate coatings within the gravel fraction (Stanbery et al. 2017). Other forms of inorganic C soils include carbon dioxide (CO_2) in soil pore spaces and dissolved inorganic carbon (DIC) in the form of bicarbonate (HCO_3^-), dissolved CO_2 ($CO_{2\,(aq)}$), and carbonate (CO_3^{2-}) in the soil solution. However, the stock of these other SIC forms is relatively small (Schlesinger 2006). The DIC in soil water is derived either from respiration, dissolution of carbonate minerals, or from atmospheric CO_2 (Kindler et al. 2011). Cavities in karst systems which develop in regions with carbonate rocks can be considered temporal depots for CO_2 originating mainly from weathering and respiration processes (Serrano-Ortiz et al. 2010). The maximum possible gas-phase storage in karst systems has been estimated at 2 Pg C, but this

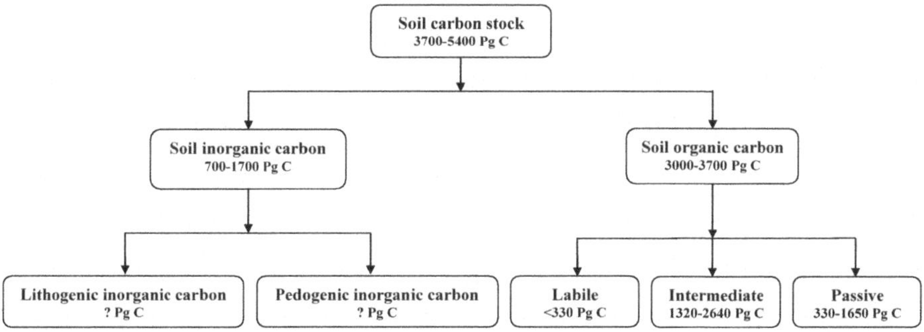

Fig. 2.1 Soil carbon stocks (based on estimates for SOC stocks to soil profile depth or 3 m depth + permafrost C + peatland C + SIC to 1 m depth; Eswaran et al. 2000; Jobbágy and Jackson 2000; Yu et al. 2010; Hugelius et al. 2014; Köchy et al. 2015; no data available on global LIC and PIC stocks separately; labile, intermediate, and passive SOC stocks to 3 m depth based on a conceptual SOM model in Trumbore 1997)

storage may have little influence on the decadal CO_2 cycling. However, the reported anomalous CO_2 uptake using eddy covariance techniques in arid and semiarid ecosystems are questionable (Schlesinger 2017). Also, soil carbonates may be sources of CO_2 emissions in alkaline desert soils (Soper et al. 2016).

Soil carbonates store about one-third of the total soil C stock in the top 1 m and may play an important role in the global C cycle (Ming 2006). The SIC may be sink, source, or neither relative to atmospheric CO_2 depending upon the mechanism of its formation (Lal 2001). The SIC stocks in agricultural soils may change by as much as 1 Mg ha^{-1} yr^{-1} (Sanderman 2012). However, the role and importance of SIC in terrestrial C sequestration are not well understood (Lal and Kimble 2000). The rate of soil C sequestration as SIC ranges from about 5 to 15 kg C ha^{-1} yr^{-1} but from 100 to 1,000 kg C ha^{-1} yr^{-1} for SOC depending on land use, soil properties, landscape position, climate, and cropping farming systems (Lal and Follett 2009a). Further, the global MRT of pedogenic inorganic C (PIC) (i.e., the amount of time C resides in the PIC pool before circulating to other pools) is about 85,000 years with a range of 30,000–90,000 years (Lal and Kimble 2000; Schlesinger 2006). In contrast, the global MRT of SOC is 35 years. Thus, the formation of pedogenic carbonates in the soil can be an effective C sequestration mechanism over a geological timescale (Lal and Kimble 2000). The sequestration of SIC occurs, in particular, when Ca^{2+} is released directly from silicates (Monger et al. 2015). However, the annual flux of atmospheric C into pedogenic carbonates is estimated to be only 0.023 Pg C yr^{-1} (Schlesinger 1997). Otherwise, the weathering processes in carbonate-bearing soils and those in arid ecosystems with a negative water balance and inputs of Ca^{2+} that form pedogenic carbonates or caliche may be responsible for 3–8% of the total annual atmospheric CO_2 sink (Gombert 2002; Liu and Zhao 2000). The weathering of carbonate rocks accounts for about 40% of the global atmospheric/soil CO_2 uptake which is estimated at 0.237 and 0.288 Pg C yr^{-1} (Hartmann et al. 2009). The remaining 60% of the CO_2 uptake is attributed to silicate rock weathering (Amiotte Suchet et al. 2003). Furthermore, the global bicarbonate flux due to natural weathering is estimated at 0.26 Pg C yr^{-1} (Schlesinger 2006).

In summary, while CO_2 is sequestered from the atmosphere during the formation of pedogenic carbonates with Ca^{2+} released directly from silicates, PIC may not become a major sink or source of atmospheric CO_2 over the next several centuries. For example, exhumed petrocalcic horizons which develop from pedogenic carbonates do not emit CO_2 at rates significantly different from adjacent soils (Serna-Pérez et al. 2006). Thus, C stored in petrocalcic horizons can be considered a recalcitrant reservoir within the time frame of terrestrial C sequestration policies aimed at an additional transfer of C from the atmosphere to land and, thus, a genuine contribution to climate change mitigation (Powlson et al. 2011). As up to 35% of the CO_2 emitted remains in the atmosphere up to 20 millennia (Mackey et al. 2013), some of the soil C has to be sequestered for millennia for climate change mitigation. Recalcitrance implies resistance to microbially mediated degradation within a soil due to inherent molecular characteristics (Torn et al. 2009).

2.1.1 The Dynamics of Soil Inorganic Carbon

The most common rock-forming carbonates are calcite ($CaCO_3$) and dolomite [$CaMg(CO_3)_2$]. The most common carbonate in soils is calcite but dolomite, aragonite (crystal form of $CaCO_3$), and siderite ($FeCO_3$) also occur (Ming 2006). Soil carbonates are formed as a result of lithogenic or pedogenic processes or some combination of these two (Schlesinger 2006). Two principal components of SIC include the lithogenic inorganic C (LIC) containing C derived from soil parent material and the PIC formed by pedogenic processes. Because LIC is inherited from carbonate parent material, it has no impact on the net C budget (Goddard et al. 2009).

Formation of PIC may be the result of weathering/dissolution of carbonate parent materials (i.e., LIC) through reaction with carbonic acid followed by pedogenic carbonate precipitation of weathering products at deeper depths in the soil profile (Wu et al. 2009). Pedogenic carbonates are formed through reaction with Ca^{2+} or Mg^{2+} under xeric or aridic conditions, and with Na^{+} under the most alkaline conditions (Macías and Arbestain 2010). Formation of PIC is CO_2 neutral when one molecule of atmospheric CO_2 is consumed during dissolution (Eq. 2.1 to the right) and one is released during precipitation (Eq. 2.1 to the left).

$$CaCO_3 + H_2O + CO_2 \rightleftharpoons Ca^{2+} + 2HCO_3^- \tag{2.1}$$

Warm conditions and high concentrations of CO_2 accelerate weathering of carbonate minerals (Macías and Arbestain 2010). The reaction rates of dissolution and/or precipitation of carbonates are rapid. For example, only weeks or months are needed for fresh, undersaturated waters to reach equilibrium by calcite dissolution (Serrano-Ortiz et al. 2010). The inorganic C in the PIC originates mainly from CO_2 generated through the respiration of plant roots and associated microorganisms and is, thus, 'new' C compared to the 'old' C in carbonate parent material (Cerling 1984). However, the role of carbonate weathering processes in CO_2 cycling is debatable (Eshel et al. 2007). Specifically, precipitation of solid carbonates is sometimes considered as a CO_2 sequestration process, and sometimes as a CO_2 source while dissolution of solid carbonate is considered a mechanism of CO_2 uptake (Serrano-Ortiz et al. 2010). In summary, precipitation of calcite implies net C accumulation in soil on long-term basis, but CO_2 release from the substrate to the atmosphere over a short-time horizon. On the other hand, dissolution implies CO_2 uptake from soil air (reducing respiratory CO_2 release to the atmosphere) and at the same time net C release by rocks into the soil solution in the form of bicarbonate (Serrano-Ortiz et al. 2010). These anomalies need to be resolved through additional research.

PIC may also accumulate by weathering of soluble Ca/Mg-bearing silicates followed by precipitation of pedogenic carbonates (Eqs. 2.2 and 2.3).

$$CaSiO_3 + 3H_2O + 2CO_2 \rightarrow Ca^{2+} + 2HCO_3^- + H_4SiO_4 \quad (2.2)$$

$$Ca^{2+} + 2HCO_3^- \rightarrow CaCO_3 \downarrow + H_2O + CO_2 \quad (2.3)$$

The formation of PIC from silicate parent materials consumes two molecules of atmospheric CO_2 for every molecule of CO_2 released and, thus, sequesters atmospheric CO_2 in the soil (Schlesinger 2006; Monger et al. 2015). However, the rates of chemical weathering of most silicates in surface environments are slow with mean lifetimes from 10^2 to 10^7 years (Serrano-Ortiz et al. 2010). Thus, Ca^{2+} and Mg^{2+} released during dissolution of silicates can form a long-term sink for CO_2 (Jansen et al. 2010). Calcite and Mg-calcite are the two carbonates that commonly precipitate (Ming 2006). With time, PIC in the subsoil may progress into carbonate-plugged (calcic) horizons and further into petrocalcic horizons (Birkeland 1999). In leaching environments, however, DIC in the form of bicarbonates derived from both silicates and carbonates may also remain in solution in groundwater and rivers representing a net sink for atmospheric CO_2 (Nordt et al. 2000). Groundwater contains about 1,400 Pg C as bicarbonate (Monger et al. 2015). Basins that do not drain to external bodies of water represent an important global C sink by accumulating 0.152 Pg C annually as DIC (Li et al. 2017). Ultimately, some bicarbonate may reach the oceans and be approximately balanced by precipitation of carbonate minerals on geological timescales (Berner 2003).

A number of environmental and geologic factors affect the dissolution and precipitation of carbonates. Important among these include temperature, concentrations of Ca^{2+} and Mg^{2+}, partial pressure of dissolved CO_2, alkalinity and pH of the soil solution (Ming 2006). For example, the amount of CO_2 dissolved in the soil solution ($CO_{2(aq)}$) is directly proportional to the partial pressure of CO_2 in equilibrium with the solution (Serrano-Ortiz et al. 2010). The CO_2 in soil air reacts rapidly in the soil solution to form DIC (Eq. 2.4).

$$CO_{2(aq)} + H_2O \leftrightarrow H_2CO_3 \rightleftharpoons H^+ + HCO_3^- \leftrightarrow 2H^+ + CO_3^{2-} \quad (2.4)$$

These process occurs in all soils while, in addition, the dissolution or precipitation of calcite takes place in carbonate-bearing soils (Eq. 2.1). Thus, increasing CO_2 pressure and decreasing pH favor dissolution of carbonates. In contrast, degassing or removal of CO_2 from the solution or increase in pH favors precipitation of carbonates (Ming 2006). The process of weathering is amplified, in particular, when plant activity and that of associated organisms (e.g., fungal mycorrhiza, lichens, and a range of free living soil organisms) increase the concentration of CO_2 in the soil pore space (Lenton and Britton 2006; Schlesinger 2006). Further, dissociated protons from organic acids in the microenvironment of the rhizosphere (i.e., the volume of soil around living plant roots that is influenced by root activity; Hinsinger et al. 2005) may increase dissolution of calcite (Sposito 1989). Otherwise, precipitation of carbonates and formation of bicarbonates depend on supply of CO_2 to the soil (Kuzyakov et al. 2006). Possible sources of CO_2 are

(i) CO_2 dissolved in rainfall, (ii) CO_2 evolved in soil by microbial breakdown of soil organic matter (SOM) and dissolved in the soil solution, (iii) CO_2, CO_3^{2-}, or HCO_3^- evolved by plant respiration or nutrient transport processes and dissolved in the soil solution, and (iv) dissolution of soil carbonate minerals by acid precipitation or soil solution containing dissolved CO_2 (Cihacek and Ulmer 2002). Thus, the CO_2-enriched water in the pore space may also precipitate Ca^{2+} and Mg^{2+} released during decomposition of plant residues aside reacting with soil-derived Ca^{2+} and Mg^{2+} (Ming 2006). An increase in concentration of SOC may also amplify carbonate weathering and precipitation by enhancing microbial activity which in turn increases the release of CO_2, Ca^{2+}, and Mg^{2+} (Bronick and Lal 2005). Precipitation of PIC occurs, in particular, during seasonal periods of drought as enhanced evapotranspiration removes water (Eq. 2.1 to the left; Schlesinger 2006). Similarly, seasonal reduction of CO_2 in the soil pore space accentuates deposition of $CaCO_3$ from the aqueous solution and the release of CO_2 to the atmosphere (Serrano-Ortiz et al. 2010). However, sequestration of SIC is important in predominant soil types in deserts and arid and semiarid regions but also those in non-arid regions (Goddard et al. 2009). Specifically, soils and environments that receive high amounts of rainfall (humid and perhumid climates) may accumulate PIC at subsoil depths.

The formation of new carbonates requires outside sources of Ca^{2+} and/or Mg^{2+} such as deposition of wind-blown carbonates at the soil surface and subsequent translocation within the soil by solution and precipitation (Ming 2006; Goddard et al. 2009). Thus, buildup of SIC occurs when cations are supplied from outside the ecosystem via dust, fertilizers, lime, or other soil amendments (Nordt et al. 2000). For example, carbonate minerals are commonly observed in atmospheric aerosols (Van Pelt and Zobeck 2009). In particular, deserts in North Africa, Asia, and the eastern Mediterranean are important source areas for carbonate-rich Aeolian materials. Other common sources and processes for carbonate occurrence in soils are (i) precipitation by interaction of Ca^{2+} and Mg^{2+} in rainwater entering the soil and combining with HCO_3^- from CO_2-enriched water, (ii) precipitation from CO_2-enriched natural surface or irrigation water containing Ca^{2+} and/or Mg^{2+}, and (iii) precipitation in soils from groundwater that has moved through carbonate-containing soils or sediments. Aside environmental variables and biological CO_2 production, soil microorganisms may also directly induce precipitation of carbonates by the biomineralization processes (Boquet et al. 1973; Braissant et al. 2002). For example, soil bacteria transform Ca-oxalate originating from tropical iroko trees (*Milicia excelsa* (Welw.) C. C. Berg) and those originating from fungi into calcite (Cailleau et al. 2004). The SIC of biotic origin is called 'biogenic carbonates' vis-à-vis geogenic carbonates formed by the chemical processes (Monger and Gallegos 2000).

The ecosystems with annual net calcite dissolution are local geochemical sinks for atmospheric CO_2 (Serrano-Ortiz et al. 2010). Specifically, soil CO_2 dissolves the carbonate rock (Eqs. 2.1 and 2.4 to the right) which decreases CO_2 emissions relative to the atmosphere. In addition, calcite dissolution implies loss of C from soil as calcite C enters the soil solution and is leached down as bicarbonate. Therefore, a net loss of C from the soil C stock occurs despite the local uptake of CO_2. Nonetheless, the local atmospheric CO_2 geochemical sink is compensated elsewhere by the reverse reaction (calcite precipitation) in aquatic ecosystems over long periods. Thus, over geological timescales and globally, the dissolution of carbonate outcrops on the continents is balanced by precipitation of carbonate mineral in the aquatic ecosystems, and this process is CO_2-neutral relative to the atmosphere (Berner 2003). However, formation of caliche in arid ecosystems may contribute to soil C sequestration over long timescales although it implies local CO_2 release over short timescales (Serrano-Ortiz et al. 2010). Two molecules of bicarbonate, previously formed during the wet growing season by dissolution of two molecules of CO_2 (Eq. 2.4 to the right), react with one molecule of Ca^{2+} in the dry season to form one molecule of calcite and release one molecule of CO_2 (Eq. 2.3). Caliche formation is negligible at annual timescales in terms of the ecosystem C balance as carbonate accumulation rates are less than 1 g C m^{-2} yr^{-1} (Marion et al. 2008). Weathering processes in ecosystems with carbonate soils along with similar processes in arid ecosystems with a negative water balance and inputs of Ca^{2+} that form pedogenic carbonates (caliche) may contribute 3–8% of the total annual net atmospheric CO_2 sink but are balanced globally and over long term as described above (Liu and Zhao 2000; Gombert 2002; Berner 2003).

2.1.2 Carbonates and Carbon Sequestration in Agricultural Soils

The contribution of agricultural SIC to C sequestration is unclear. Some consider pedogenic carbonate precipitation as a sequestration process while others argue that carbonate dissolution should be considered sequestration (Sanderman 2012; Monger et al. 2015). Soil carbonates are persistent in agricultural soils but carbonate precipitation may not directly affect net C sequestration (Eswaran et al. 2000). Otherwise, there may be an antagonistic interaction between SIC and SOC (Dong et al. 2017). Further, agricultural practices may alter processes which affect SIC stocks and fluxes (Suarez 2000). Among those are enhanced mineral weathering as a result of organic acids present in agricultural soils, calcite precipitation/dissolution, dolomite dissolution, and changes in dissolved CO_2 related to changes in soil pH. For example, desert grasslands on calciferous soils can permanently release CO_2 into the atmosphere when respiratory efflux exceeds the assimilatory uptake (Emmerich 2003; Gilmanov et al. 2010). The leaching of DIC, in particular, depends on the type of agricultural land use (Kindler et al. 2011). For example, leaching of biogenic DIC has been estimated at about 24 and 15 g C m^{-2} yr^{-1} for European grasslands and croplands, respectively. Further, changing the rates of weathering of carbonate minerals and subsequent reprecipitation may play an important role in the balancing of atmospheric CO_2 (Hamilton et al. 2007). Indirectly, soil carbonates may also affect soil C sequestration by altering net primary production (NPP) and soil C inputs. Specifically, carbonates strongly influence soil chemical and physical properties such as soil pH, adsorption–desorption processes, and soil cementation (i.e., binding together of soil particles; Ming 2006).

2.1.2.1 The Dynamics of Soil Carbonates

The process of silicate weathering (Eq. 2.2) and the SIC balance is relatively insensitive to decreases in pH as agricultural soils are less likely to be highly acidic (Suarez 2000). In the pH range of 4–8, for example, there is only slight pH dependence for the dissolution of silicates. Weathering of silicates becomes increasingly more sensitive to changes in pH with decrease in solution pH, but the anthropogenic increase in atmospheric CO_2 concentration results only in a relatively small decrease in pH (Suarez 2000).

In contrast to silicate weathering, carbonic acid weathering is either a net source or a sink for CO_2 (Hamilton et al. 2007). The carbonic acid weathering is the major natural process of limestone weathering. Specifically, carbonic acid reacts with solid carbonates and exists in equilibrium with dissolved CO_2 from microbial and root respiration (Eqs. 2.4 and 2.5).

$$CaMg(CO_3)_2 + 2H_2CO_3 \rightleftharpoons Ca^{2+} + Mg^{2+} + 4HCO_3^- \qquad (2.5)$$

The processes of dissolution or precipitation of calcite play an important role in buffering carbonate soils against acidification or basification (Serrano-Ortiz et al. 2010). For every molecule of limestone-derived C that dissolves, two molecules of HCO_3^- alkalinity are produced. Thus, carbonic acid weathering is a sink for soil CO_2. Agricultural activities that promote high biological activity such as manuring, retention of crop residues, and cover cropping enhance this process as respiratory CO_2 accumulates. In contrast, the dissolution of carbonate minerals in the presence of strong acids (such as HNO_3) is a CO_2 source (Eq. 2.6).

$$CaMg(CO_3)_2 + 4HNO_3 \rightarrow Ca^{2+} + Mg^{2+} + 4NO_3^- + 2CO_2 + 2H_2O \qquad (2.6)$$

HNO_3 may be produced in agricultural soils, for example, by nitrification of ammonia-based N fertilizers which is the major source of soil acidity in agricultural soils (Eq. 2.7 to the right; Fisher et al. 2003).

$$NH_4^+ + 2O_2 \rightleftharpoons NO_3^- + 2H^+ + H_2O \qquad (2.7)$$

Strong soil acids are responsible for between 12 and 38% of carbonate dissolution on a catchment scale (West and McBride 2005; Hamilton et al. 2007). However, carbonate dissolution by strong acids is only important at $pH < 5$ which is less likely to occur in agricultural soils because of the liming. However, H^+-ions produced during acidification processes in agricultural soils are buffered by conversion into non-dissolved forms. Specifically, calcite is converted into Ca-bicarbonate and HCO_3^- is leached (Eq. 2.8 to the right).

$$CaCO_3 + H^+ \rightleftharpoons HCO_3^- + Ca^{2+} \tag{2.8}$$

The buffer capacity of carbonate soils is, however, sufficient to neutralize the entire H^+ input. Thus, the soil pH remains around 7 or 8 as long as reactive carbonates are present in agricultural soils (Serrano-Ortiz et al. 2010). The leached bicarbonate eventually reprecipitates in subsoil layers (Eq. 2.3). In leaching environments (e.g., under humid and subhumid climates), however, the dissolved inorganic C in the form of carbonate alkalinity is subsequently transported through groundwater flow paths and eventually into streams and rivers and ultimately the oceans with little reprecipitation (Hamilton et al. 2007).

2.1.2.2 Effects of Agricultural Practices on Soil Inorganic Carbon

The formation of pedogenic carbonates can be enhanced through increase in biogenic processes by soil application of organic amendments, materials rich in cations, lime, and amendments or conditioners that conserve water in the soil column (Lal and Follett 2009b). However, the effects of various soil cultivation systems (e.g., fertilizer formulation and rate) as well as management practices (e.g., till, no-till, liming, irrigation frequency, and other management) on $CaCO_3$ dissolution and CO_2 efflux are not completely understood (Zamanian et al. 2016). Upon dissolution, CO_2 in soil air forms bicarbonate and carbonic acid (Eq. 2.4) which can be leached into groundwater or precipitated in the soil as Ca or Mg-carbonate, or Na-carbonate under most alkaline conditions (Eq. 2.3). Thus, carbonates precipitate particularly when soil moisture content or partial pressure of CO_2 is decreased or the levels of Ca^{2+}, Mg^{2+}, Na^+, or bicarbonate are increased by agricultural land use (Denef et al. 2008). External sources of cations include eolian dust, Ca^{2+} dissolved in rainwater, sea salts and sprays, use of fertilizers, application of biosolids and soil amendments (Nordt et al. 2000; Lal 2004a).

Fertilization, crop cultivation, and harvest

Soil acidification in agricultural production system is caused by the transformation of C, N, and S, which releases H^+ into soil solution (Kunhikrishnan et al. 2016). In particular, addition of N and S fertilizers, the cultivation of N-fixing crops, and crop harvest may acidify soil (Fisher et al. 2003). According to the equilibrium Eq. 2.4, acidification may release CO_2 and promote carbonate dissolution. For example, the application of N fertilizers that produce NH_4^+ acidifies soil as one molecule NH_4^+ is oxidized by microbial activity generating one molecule NO_3^- (Eq. 2.7 to the right). Furthermore, for one molecule NO_3^- taken up by plants one molecule H^+ is released, yet only one of the two molecules H^+ released by nitrification is neutralized. Thus, the soil is acidified by NH_4^+-producing fertilizers. In total, one molecule of CO_2 for every molecule of NH_4^+ is released when ammonia-based N fertilizers are applied in neutral to alkaline soils (Suarez 2000). The loss of alkalinity relative to the Ca and Mg contents relates to weathering of carbonates by protons from N fertilizers nitrification, which is a net source of CO_2 to the atmosphere. Globally, this may not be negligible because it may amount to 6–15% of CO_2 uptake by natural silicate weathering and partly counterbalances natural CO_2 sinks (Perrin et al. 2008). Long-term N fertilization may also affect the SIC concentration in particle-size fractions. For example, fertilization with diammonium phosphate and urea, and with chicken manure over 22 years resulted in lower SIC concentrations in sand-, silt-, and clay-size fractions to 20 cm depth (Dong et al. 2017). This decrease occurred also in 40–60 cm depth by fertilization with diammonium phosphate and urea. However, in 80–100 cm depth, only the SIC concentration in the clay fraction decreased by fertilization (Dong et al. 2017).

The activity of crop roots may lead to bicarbonate formation due to root respiration in the upper parts of the soil profile at neutral pH (Denef et al. 2008). Deeper in the profile and under alkaline conditions, $CaCO_3$ may precipitate due to root excretion of Ca^{2+}, HCO_3^- and acid exudates, and localized soil drying near the roots (Kuzyakov et al. 2006).

Crop harvest contributes also to the acidification of agricultural soils. Specifically, plant tissues produce salts of weak organic acids as more cations than anions are accumulated during preferential plant NH_4^+ uptake. These salts are returned to the soil with plant litter and neutralize the excess H^+. However, with crop harvest, the neutralizing salts are removed, and thus, the excess of H^+ remains in the soil causing acidification. Further sources of soil acidity are the decomposition of N-containing OM of N-fixing crops and those of S-containing OM, and organic acids produced during decomposition. Sources of acidity also include simple organic acids (i.e., citric acid and oxalic acid) which are released from plant roots and also added as intermediate products of OM decomposition. However, these acids are decomposed quickly by microorganisms and may, thus, not contribute to acidification in most soils.

The addition of above- and belowground biomass is critical to the formation of PIC in agricultural soils as this increases the CO_2 partial pressure indirectly by promoting CO_2 production upon decomposition, and directly increases Ca^{2+} levels (Lal 2004a). Biomass addition may also enhance termite activity which may result in biogenic carbonate formation (Monger and Gallegos 2000). In general, techniques that enhance soil fertility may lead to the formation of PIC following increased root and shoot biomass production (Lal 2004a).

Liming

As discussed previously, cultivation of agricultural soils causes acidification. Thus, liming is an important agricultural practice for improving nutrient availability in acid soils by increasing base saturation and soil pH, and by reducing Al^{3+} availability (Paradelo et al. 2015). Most common amendments applied to agricultural soils are Ca-carbonate (lime), Ca/Mg-carbonate (dolomitic lime), Ca-oxide, or Ca-hydroxide. Soil liming causes the release of alkalinity by reaction with carbonic acid (Eq. 2.5 to the right), and the release of CO_2 at $pH < 5$ by reaction with strong acids (Eq. 2.9).

$$CaCO_3 + 2H^+ \rightleftharpoons Ca^{2+} + H_2O + CO_2 \quad (2.9)$$

Dissolution of liming materials can be either a net source or a sink for CO_2 (Kunhikrishnan et al. 2016). Lime-derived CO_2 reacts with microbial respiration derived H_2CO_3 in soils to produce carbonate material, which is a sink of CO_2 in soil. Agricultural lime ($CaCO_3$) is a net sink for CO_2 in calcareous soils with high pH, but it is a net source of CO_2 in acidic soils. Further, adding lime to soils can increase CH_4 oxidation and reduce greenhouse gas (GHG) emission (Kunhikrishnan et al. 2016). Thus, depending on whether the reaction of lime in soil is with strong acids or carbonic acid, agricultural lime can be either a source or a sink for CO_2 (Fig. 2.2; Hamilton et al. 2007). West and McBride (2005) estimated the net CO_2 emissions of 0.059 Mg C per Mg limestone and 0.064 Mg C per Mg dolomite applied.

Liming and soil organic carbon

The impacts of liming on SOC storage are variable, and strongly relate to soil type, land use, climate, and multiple management factors (Holland et al. 2017). Liming can stimulate SOM decomposition mainly by affecting soil pH (Ahmad et al. 2015). Specifically, liming may temporally decrease the stability of macroaggregates through decomposition of POC. This may also result in increased leaching of DOC from limed soils (Ahmad et al. 2015). Further, in some soils, liming may reduce SOC mineralization due to increased microbial C-use efficiency (Grover et al. 2017). Other effects of liming on SOC mineralization are (i) increased SOC solubility due to increase in pH and subsequent SOC stabilization via Ca^{2+} bridging, (ii) increased crop productivity resulting in higher soil C inputs, (iii) improved soil

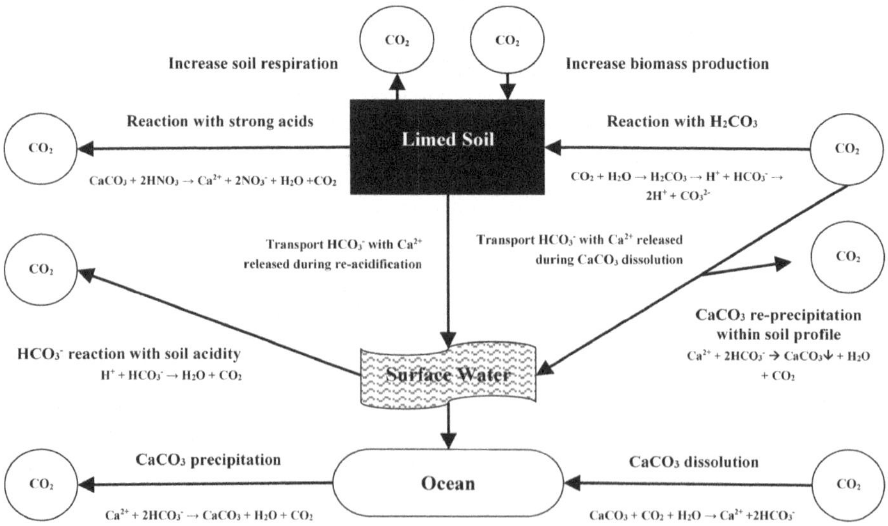

Fig. 2.2 Carbon dioxide flux after lime application to soil with equations (modified from Page et al. 2009)

structure enhancing physical protection of SOC, and (iv) altered size, activity and/or composition of the soil microbial community (Grover et al. 2017).

Mineral fertilizer

The addition of Ca^{2+}, Mg^{2+}, or Na^{+} with fertilizers may increase the formation of PIC (Lal 2004a). As discussed previously, sequestration of inorganic C in soils may occur when these cations are supplied from outside the agricultural ecosystem via fertilizers or other soil amendments (Nordt et al. 2000).

Enhanced weathering

Applying crushed silicate rock as a soil amendment is a method for sequestering CO_2 as bicarbonate and carbonate minerals in soils while increasing nutrient availability to plants (Kantola et al. 2017). The choice of source rocks and material comminution technique dominate the CO_2 efficiency of enhanced weathering (Moosdorf et al. 2014). The CO_2 emissions from transport amount to only 0.5–3% of potentially sequestered CO_2, and emissions of material mining and application are negligible. About 0.5–1.0 Mg CO_2 can be sequestered per Mg of rock. However, to control or reduce atmospheric CO_2 concentrations substantially with enhanced weathering would require very large amounts of rock (Moosdorf et al. 2014). Conventional and perennial bioenergy agroecosystems, in particular, may benefit from enhanced weathering (Kantola et al. 2017). Further, warm tropical climates and productive crops will substantially enhance weathering rates, with potential co-benefits including decreased soil acidification and increased phosphorus (P) supply promoting higher crop yields (Edwards et al. 2017). However, potential pitfalls include the impacts of mining operations on deforestation, and the erosion of silicates into rivers and coral reefs that increases inorganic turbidity, sedimentation, and pH, with unknown impacts for biodiversity. Thus, uncertainties remain in the long-term effects and global implications of large-scale enhanced weathering efforts (Kantola et al. 2017). In particular, major issues of cost, social acceptability, and potential unanticipated consequences will limit utilization of enhanced weathering (Taylor et al. 2016).

In addition to SIC, applying silicate rock to agroecosystems may also affect SOC. Specifically, the cationic species (e.g., Ca^{2+}, Mg^{2+}) released following dissolution of silicate rock can accumulate on the surface of spent mineral particles and attract SOC through cation-bridging reactions (Singh et al. 2018). The cations increase the cation exchange capacity (CEC) of the soil, and an increase of CEC and base saturation generally improves SOC sequestration. However, increased SOC formation from long-term application of pulverized silicate rocks to agricultural soils has yet to be fully understood or quantified (Singh et al. 2018). Directly, the colloidal-sized rock mineral particles may also sorb DOC and prevent them from loss or microbial decomposition (Yu et al. 2017). Eventually, new non-crystalline or poorly crystalline mineral phases are formed. These new mineral phases may also participate in SOC sequestration through ligand exchange and electrostatic attraction (Yu et al. 2017).

Irrigation

About 25% of the global harvested area was irrigated in 2000 (Portmann et al. 2010). Specifically, 1 million km^2 was irrigated for rice (*Oryza sativa* L.), 0.7 million km^2 for wheat (*Triticum* spp.), and 0.3 million km^2 for maize (*Zea mays* L.) production. Irrigation may affect the SIC stock by drastic increase in biomass addition to the soil (Lal and Kimble 2000). Irrigation may enhance carbonate formation by stimulating plant and root growth and soil biotic activity in cultivated dryland ecosystems (Denef et al. 2008). Also, Ca^{2+} and Mg^{2+} applied with the irrigation water may promote the formation of new soil carbonates. For example, groundwater in arid regions is usually rich in dissolved Ca^{2+} and carbonates which typically favors $CaCO_3$ precipitation (Eq. 2.1 to the left). However, the effects of irrigation on changes in SIC stocks of principal soils of different ecoregions are not known (Lal and Kimble 2000). Irrigation can either dissolve carbonates and leach them into the groundwater or precipitate $CaCO_3$ depending on the amount and quality of irrigation water and the application method (Denef et al. 2008). For example, significant carbonate depletions may occur in the top 2 m in irrigated agricultural soils in an arid region compared to non-irrigated soils (Eshel et al. 2007). However, the irrigation water quality also affects SIC dynamics. Specifically, irrigation with treated effluent water inhibits carbonate dissolution compared to soils irrigated with freshwater. Also, irrigation with Ca-rich irrigation water and agricultural land use has been shown to decrease SIC and calcite concentrations in 0–50 cm depth in arid soils compared to desert soils (Sartori et al. 2007). Otherwise, cultivation and irrigation resulted in an increase in SIC stocks in 0–75 cm depth in two semiarid agroecosystems compared to non-cultivated native grassland systems (Denef et al. 2008).

In addition to $CaCO_3$ precipitation (Eq. 2.1 to the left), groundwater will release CO_2 as it reequilibrates with the atmosphere if used for irrigation. Thus, a significant and previously unrecognized CO_2 contribution arises from groundwater depletion (net removal from storage; Wood and Hyndman 2017). The total amount of groundwater depletion is not well constrained globally but is estimated to be between 145 $km^3\ yr^{-1}$ (Konikow 2011) and 189 $km^3\ yr^{-1}$ (Wada 2016). The global CO_2 flux from this source is likely between 8 and 17 Tg yr^{-1} (Wood and Hyndman 2017). However, additional data synthesis and analysis are needed to better estimate global groundwater depletion rates and bicarbonate concentrations.

2.1.2.3 Indirect Effects of Carbonates on Soil Carbon Sequestration

Soil reaction

Agricultural soils which contain $CaCO_3$ have a pH in the range of 7.1–8.3 as $CaCO_3$ acts as a pH buffer (Ming 2006). Soil carbonates contribute to buffering acidity from wet deposition, fertilizer addition, and irrigation. Plant growth varies with soil pH as the relationship between biomass production and soil pH follows an optimum curve in accord with the availability of exchangeable Al^{3+} and Mn^{3+}

which can be harmful to plants. In general, pH values around 6 are beneficial to yield. Thus, soil carbonates may also enhance NPP and accentuate SOC sequestration by contributing to optimal soil pH for plant growth. However, the optimum pH for soil respiration and the aboveground biomass production depend also on the composition of plant and soil microbial communities (Page et al. 2009).

Lime is sometimes applied to raise the soil pH, and to increase the fertility of crop- and grasslands (West and McBride 2005). Specifically, OM is readily transformed into inorganic nutrients at optimal soil pH values of 6.6–8.0 for bacteria-mediated, and 5.5–6.5 for fungal-mediated decomposition (Whalen and Sampedro 2009). Liming may enhance biomass production and CO_2 input into soils, in particular, where plant growth is limited by acidity as nutrient toxicities (Al, Mn) and deficiencies (Ca, Mg) may be reduced by liming (Fig. 2.2; Page et al. 2009). For most arable crops, there is a positive yield response to liming (Holland et al. 2017). Further, repeated lime applications tend to improve grassland biomass production. A net SOC sequestration occurs when this plant biomass is returned to the soil at a higher rate than can be consumed by the decomposer community and an additional transfer of C from the atmosphere to the land occurs (Powlson et al. 2011). Further, increased availability of labile material to the microbial biomass after lime application may result in increased CO_2 emissions but these may soon decrease once the labile material or N is lacking (Page et al. 2009). However, the effects of liming on plant biomass production are variable and depend on the severity of acidity effects on plant growth and the plant species (Page et al. 2009). Similarly, liming only increases rates of soil respiration and, thus, CO_2 emissions when soil acidity is limiting soil biotic and root respiration (Fig. 2.2). In summary, the net effect of lime application on agricultural SOC stocks is less well known (Paradelo et al. 2015).

Water-holding characteristics

The precipitation of carbonates on soil particles and pores in arid and semiarid regions may form calcic horizons which restrict downward soil water movement (Duniway et al. 2007). Furthermore, carbonates tend to be leached or removed from upper soil horizons with increase in leaching or rainfall. The leached carbonates may precipitate in deeper soil horizons and form carbonate cemented layers (e.g., petrocalcic horizons or caliches; Díaz-Hernández 2010). These layers are characterized by high bulk densities and carbonate contents and may dramatically alter water-holding characteristics. For example, Duniway et al. (2007) reported that petrocalcic horizons had high water-holding capacities and released much of the retained water at plant-available soil water potentials. The petrocalcic horizons may also force soil water to move laterally which reduces water loss to deeper horizons and, thus, sustains water supply for plant growth (Ming 2006). Thus, petrocalcic horizons are potentially important as a plant water source, especially during drought periods. Plant productivity in arid and semiarid regions, in particular, may benefit by water supply from petrocalcic horizons. Otherwise, root growth may not be restricted by calcic or petrocalcic horizons as roots and fungal hyphae are able to penetrate them.

Soil organic matter stabilization

Carbonates may interfere with soil structure formation and stabilization, cycling of soil aggregates, and, thus, SOM stabilization (Virto et al. 2018). However, the mechanisms of interaction of soil aggregation and SOM stabilization in calcareous soils and their relationship to SOC preservation are not completely understood. Among involved processes may be (i) mineral–mineral and organic–mineral interactions promoted by the presence of Ca, (ii) reprecipitation of dissolved carbonates, or (iii) alterations in the biotic degradation of SOM. Carbonates may act as agents of aggregate stabilization independently from SOM. They may also act as rate modifiers of SOM cycling in the short term. This may be more related to the existence of a more cemented fabric than to the preferential accumulation of clay minerals. Thus, physical protection due to carbonate precipitation is a possible explanation for the observed greater stability of SOM in calcareous soils (Virto et al. 2018).

2.1.2.4 Sequestration of Inorganic Carbon in Agricultural Soils

The formation of PIC can be a significant process leading to sequestration of atmospheric CO_2 (Lal and Kimble 2000), especially when Ca^{2+} is released directly from silicates (Monger et al. 2015). For example, Eswaran et al. (2000) estimated that less than 10% of the 947 Pg SIC stored globally to 100 cm depth is the result of sequestration of atmospheric CO_2. However, as discussed previously, the exact size of the PIC stock and the rate of its formation are not well known (Monger et al. 2015). Among the reasons is that routine soil carbonate analyses do not differentiate between PIC and LIC (Eswaran et al. 2000). Another reason is the paucity of carbonate data for deep soil layers (Batjes 1996). The transport of bicarbonate as DIC to groundwater and fluvial systems in agricultural landscapes as well as to the oceans can be regarded as sequestration of CO_2 from the atmosphere (Nordt et al. 2000). For example, the riverine DIC flux is estimated at 0.4 Pg C yr^{-1} (Richey 2004). A significant portion of this global DIC export may be attributable to agricultural activities but the impacts on CO_2 sequestration are difficult to quantify (Barnes and Raymond 2009). Net storage of atmospheric CO_2 may occur in arid, non-leaching environments where bicarbonates are mainly formed through biotic activity (Eq. 2.4 to the right), even though CO_2 is released in the precipitation step (Eq. 2.3; Denef et al. 2008). Some examples of more detailed studies on carbonates in agricultural soils are given below.

Between 161 and 176 Mg PIC ha^{-1} may be stored to 120 cm depth in boreal grasslands, and 91–117 Mg C ha^{-1} in the form of SOC (Landi et al. 2003). Furthermore, 0.99–1.37 g C m^{-2} yr^{-1} accumulates as PIC in these ecosystems which correspond to 0.67% of annual NPP. In contrast, only 0.29% of annual NPP accumulates as SOC which highlights the importance of PIC for C sequestration in boreal grassland soils (Landi et al. 2003). When soils of the Northern Great Plains are till, the SIC stocks to 100 cm depth have been shown to increase by 10 Mg C

ha^{-1} (Cihacek and Ulmer 2002). Possible reasons are increased infiltration and storage of precipitation, and freezing of the soil profile. Otherwise, erosion caused by long-term tillage management can also increase the proportion of SIC in soil profiles as surface SOC may be removed and the calcic subsoil become exposed at the soil surface by truncation (Papiernik et al. 2007). However, it is important to study soil profiles to assess management effects on SIC stocks. For example, in cropping treatments in the Russian Chernozem, SIC stocks between 25 and 125 cm depths increased compared to fallow treatment (Mikhailova and Post 2006). Thus, in this region soil profiles should be sampled to at least 200 cm depth to capture the SIC stock changes caused by cultivation. The PIC in Russian Chernozems is leached from the top 100-cm, redeposited deeper in the soil profile and, to some degree, also discharged to surface water (Lapenis et al. 2008). However, the most likely explanation for the loss in PIC may not be cultivation but an increase in downward flux of soil water due to changes in the steppe climate. Thus, Chernozems are net sinks for atmospheric CO_2 due to leaching of PIC. On the other hand, Mollisols in grassland ecosystems may also be a significant sink for SIC through PIC formation (Mikhailova et al. 2009).

Irrigation of agricultural soils may also alter SIC dynamics. For example, in arid, non-leaching environments, carbonate precipitation following irrigation with Ca- or Mg-rich water may result in the net release of CO_2 (Eq. 2.1 to the left; Denef et al. 2008). However, SIC stocks to 75 cm depth in two semiarid agroecosystems were not significant different among cultivated, and cultivated and irrigated soils (27.6 vs. 20.5 Mg C ha^{-1}, and 57.0 vs. 54.6 Mg C ha^{-1}, respectively; Denef et al. 2008). Otherwise, irrigation of a cultivated soil in an arid region with freshwater for more than 75 years sequestered 71.5 Mg SIC ha^{-1} to 4 m depth, and between 9 and 24 Mg SIC ha^{-1} by irrigation with treated effluent (Eshel et al. 2007). Compared to the non-irrigated soil profile, 54 Mg SIC ha^{-1} were lost to 4 m depth from the soil irrigated with freshwater. Smaller losses of 9–16 Mg SIC ha^{-1} were observed in the soil profile irrigated with treated effluent.

The dissolution of liming material may also sequester CO_2 equal to roughly 25–50% of its C content in contrast to the prevailing assumption that all lime C becomes CO_2 (Hamilton et al. 2007). Specifically, analyses of infiltration waters, tile drainage waters, and streams draining agricultural catchments indicate a net CO_2 uptake via the generation of bicarbonate and its transport to groundwater and fluvial systems (Fig. 2.2; Oh and Raymond 2006). Even when the reaction of lime with strong acids releases CO_2, the Ca^{2+} from agricultural lime may still sequester CO_2 (Oh and Raymond 2006). However, agricultural lime may become a net CO_2 source as nitrate concentrations may also increase in infiltrating waters (Hamilton et al. 2007). When reprecipitation of $CaCO_3$ in oceanic waters is considered, liming a whole catchment is also a net CO_2 source (West and McBride 2005). Further, the amount of CO_2 released during the production, transport, and spreading of lime can almost be equal to the on-site net atmospheric C sequestration by liming agricultural catchments. Thus, agricultural liming appears not to be attractive as a C sequestration strategy (Oh and Raymond 2006). However, it is unclear whether the CO_2 flux once the products of agricultural lime reaches the ocean must be included for GHG accounting of agricultural liming (Page et al. 2009). If leached carbonates remain in the lower soil profile or reside in rivers and groundwater for hundreds of years, this C may be considered sequestered (West and McBride 2005). Definite conclusions regarding the effects of agricultural liming on SIC sequestration are, however, not possible because only a few studies have examined all processes involved (Fig. 2.2; Page et al. 2009).

2.2 Soil Organic Carbon

The SOC stock rapidly exchanges C with the atmosphere and is, thus, important as potential sink for mitigating the ever-increasing anthropogenic CO_2 emissions. Global SOC stocks to 1 and 3 m depth have been estimated at 1,500 and 2,300 Pg C, respectively, and permafrost and peatlands may contain an additional 1,400 Pg C (Fig. 2.1; Jobbágy and Jackson 2000; Yu et al. 2010; Hugelius et al. 2014). Improved estimates based on the Harmonized World Soil Database indicate that soils may contain 1,325–1,408 Pg SOC in the upper 1 m, and 3,000 Pg SOC in soil profiles (Köchy et al. 2015; Batjes 2016). The global SOC stock in the upper 2 m of soil was estimated at 2,273 Pg C based on a global soil database, and at 2,800 Pg C to a depth of 3 m under various assumptions (Jackson et al. 2017). However, global soil databases have no data for soil deeper than 2 m, and deeper (>3 m) deposits may contain appreciable stocks of SOC. Further, the extent to which shallow bedrock restricts the depth of SOC stocks is unclear (Jackson et al. 2017).

Agricultural biomes may contain about 609 Pg SOC to 1 m depth (Stockmann et al. 2013). Organic C is found in soil in the form of plant, microbial, and faunal residues at various stages of decomposition, collectively called SOM (Kleber and Johnson 2010), and often known as humus—Earth's most important natural resource (Paul 2016). The SOM is understood today as the non-living product of the decomposition of plant and animal substances (Manlay et al. 2007). Three periods in the scientific perception of SOM can be distinguished: 'the humic period' (before 1840), the 'mineralist period' (1840–1940), and the 'ecological period' from 1940 up to the present time (Manlay et al. 2007).

Amount and composition of SOM in natural ecosystems are determined by the long-term balance between plant-derived soil inputs of C fixed from the atmosphere, and C losses by microbial decomposition and those associated with fire, erosion, and leaching (Torn et al. 2009). In addition, photosynthetically fixed C is removed from agroecosystems by harvest. Physicochemical and biological influences from the surrounding environment reduce the rate of decomposition, thereby allowing SOM to persist (Schmidt et al. 2011). The gross primary production (GPP) is the total amount of C fixed by photosynthesis in an area of land, and the NPP what is left after accounting for C losses by plant or autotrophic respiration (R_a). The bulk of NPP is allocated to the production of foliage, shoots, and roots (Ciais et al. 2010b). The net biome production (NBP) describes the total rate of organic C accumulation in (or loss from) an agroecosystem integrated over time and space. Ultimatively, the fraction of NBP entering the soil (NBP_{soil}) determines the amount of organic C sequestered in SOM (Schulze et al. 2010).

The plant tissue composition and soil mineralogy affect decomposition and SOM stabilization at local scales (Torn et al. 2009). In ecosystems dominated by silicon (Si)-accumulating plants such as grasslands and agroecosystems under rice cultivation, Si may be a significant factor in SOC cycling by controlling lignin biosynthesis (Klotzbücher et al. 2018). At the regional scale, environmental factors (e.g., climate and soil parent material) interact with SOM dynamics as those affect plant productivity and soil development (Barré et al. 2017). The dynamic accumulation and distribution of SOM are the result of stabilization and destabilization processes that in turn are influenced by biotic, abiotic, and anthropogenic factors (Jackson et al. 2017). Biotic factors, i.e., the amount, chemical composition, and relative allocation of plant inputs, and the soil food web are critical regulators of SOM. Important abiotic factors include climate, mineralogy, landscape position (e.g., slope, aspect), and the abundance of fires. Increasingly, anthropogenic factors control and alter SOM distributions. Anthropogenic factors include fire, N deposition, climate change, land use, and tillage and other management practices that alter soil biota and aggregates (Jackson et al. 2017).

Sources of SOC and its dynamic are discussed in the following sections with reference to agroecosystems. First, factors limiting C fixation by plant photosynthesis and NPP are discussed in detail as both are major sources of soil C inputs. Then, stabilization and chemical composition of SOM, and environmental factors explaining differences among soils are elaborated. The section concludes with a discussion how agricultural practices can contribute to SOC sequestration by enhancing both its size and stability (i.e., for millennia; Mackey et al. 2013), resulting in an additional transfer of C from the atmosphere to land (Powlson et al. 2011). The term SOC is used preferably as no universal factor exists to convert SOC content into SOM content (Pribyl 2010). The conversion factors range from 1.9 based on empirical studies to 2.0 based on theoretical considerations (i.e., assuming 50% of SOM is SOC). Traditionally, the incorrect factor of 1.724 is used based on the assumption that 58% of SOM is SOC. Thus, in the following section the term SOM is only used when the molecular structure is discussed in more detail.

2.2.1 The Dynamics of Soil Organic Carbon

Various mechanisms have been proposed to explain SOC dynamics over space and time (Dungait et al. 2012; Lehmann and Kleber 2015; Schmidt et al. 2011). Among the main groups of factors affecting SOC dynamics are (1) climatic variables such as precipitation and temperature, (2) soil conditions including various physicochemical properties, and (3) biotic properties consisting mainly of the quantity and quality of C inputs into soil (Luo et al. 2017). These factors interact together to regulate SOC dynamics.

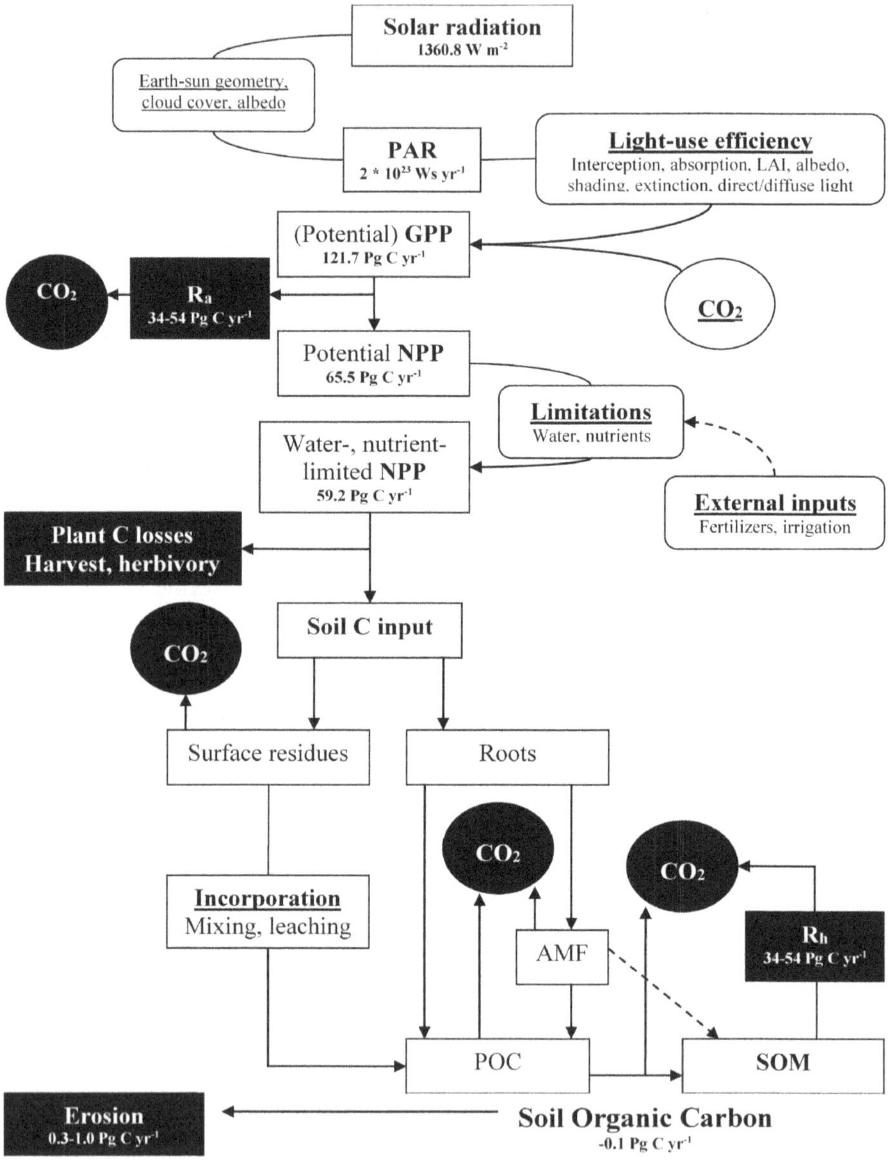

2.2.1.1 Plant Photosynthesis as the Major Source of Soil Organic Carbon

Land-based plants' photosynthesis underpins food production for virtually all land species, provides roughly half of atmospheric oxygen (Walker 1980), and removes roughly 25% of anthropogenic CO_2 from the atmosphere (Le Quéré et al. 2017).

◂**Fig. 2.3** Carbon flow in terrestrial ecosystems. Limitations to carbon assimilation, net primary production (NPP), and soil organic carbon (SOC) formation are underlined. Carbon losses are in black boxes. Negative carbon fluxes describe sinks. (PAR = Photosynthetically active radiation, LAI = Leaf area index, GPP = Gross primary production, R_a = Autotrophic respiration, AMF = Arbuscular mycorrhizal fungi, R_h = Heterotrophic respiration, POC = Particulate organic carbon, SOM = Soil organic matter) (modified from Sanderman et al. 2010 with data from Beer et al. 2010; Billings et al. 2010; Chappell et al. 2016; Denman et al. 2007; Gifford 2003; Haberl et al. 2007; Houghton 2007; Ito and Oikawa 2004; Kopp and Lean 2011; Raich et al. 2002; Wang et al. 2015) (Assuming half of Rs of 68–108 Pg C y^{-1} is root-derived; Hursh et al. 2017; Hashimoto 2012)

Most importantly, it is the major source of SOC in terrestrial ecosystems (Fig. 2.3). The photosynthetic uptake of atmospheric CO_2 or GPP is not exactly known as it can only be estimated indirectly from net CO_2 flux using appropriate separation algorithms as direct measurements are not possible (Reichstein et al. 2005). Photosynthesis takes place in plant chloroplasts and is driven by the availability of solar radiation, CO_2, temperature, and the availability of N which is required to produce photosynthetic enzymes. Thus, GPP is not constant but changes diurnally and seasonally as environmental factors are also dynamic. For example, Anav et al. (2015) reported a large range in the mean global GPP estimates (112–169 Pg C yr^{-1}) for the period 1990–2009 based on simulations by ten models. Also, global GPP simulated by eight biome models ranged from 98 to 141 Pg C yr^{-1} (1981–2000 mean; Ito et al. 2017). Chen et al. (2017) observed large uncertainties in spatiotemporal variability of GPP at the regional and global levels across eight models, which are probably due to the differences in the formulation of underlying photosynthetic processes. Generally, simulations indicated incremental trends of annual GPP, seasonal-cycle amplitude, radiation-use efficiency (RUE), and water-use efficiency (WUE), mainly caused by the CO_2 fertilization effect. For example, Cheng et al. (2017) suggested that the rising atmospheric CO_2 concentration has caused a shift in terrestrial water economics of C uptake from 1982 to 2011 simulated by using a combination of ground-based and remotely sensed land and atmospheric observations. The terrestrial C uptake (i.e., GPP) increase of 0.83 Pg C yr^{-1} was not accompanied by a proportional increase in water use (i.e., evapotranspiration) but was largely (about 90%) driven by increased C uptake per unit of water use, i.e., WUE. The increased WUE was positively related to rising CO_2 concentration and increased canopy leaf area index (LAI), and negatively influenced by increased vapor pressure deficits (Cheng et al. 2017). Otherwise, based on long-term atmospheric carbonyl sulfide (COS) records, Campbell et al. (2017) estimated that GPP may have increased by 31% during the twentieth century. However, the improved WUE due to enhanced CO_2 concentration is unlikely to ease regional droughts in croplands because of increases in evapotranspiration, which are associated with the enhanced GPP due to increased anthropogenic N application (He et al. 2017).

For 1998–2005, estimates of GPP were 31.3, 8.5, and 14.8 Pg C yr^{-1} for tropical savannahs and grasslands, temperate grasslands and shrublands, and croplands, respectively, based on eddy covariance flux data and diagnostic models (Beer et al.

2010). For 2001–2005, the GPP of grasslands and croplands was estimated at 6.1 and 15.4 Pg C yr^{-1}, respectively, based on light-use efficiency (LUE) using a random forest regression model (Wei et al. 2017). Liang et al. (2017b) reported a higher average grassland GPP of 11 Pg C yr^{-1} for 1982–2011 by application of a model tree ensemble trained on local eddy covariance data and using as predictors gridded climate and management intensity field (grazing and cutting). Guanter et al. (2014) reported a higher global cropland GPP of 17 Pg C yr^{-1} using spaceborne measurements of solar-induced chlorophyll fluorescence (SIF). SIF may be a powerful proxy for GPP (Sun et al. 2017).

In the year 2000, maize contributed most (1.55 Pg C yr^{-1}) to the global cropland GPP with Asia contributing in total 38.9% of global cropland GPP (Chen et al. 2014). A novel step in this analysis was to assign a maximum LUE estimate to each of the 26 crop types studied instead of taking a uniform value as done previously. Based on a Moderate Resolution Imaging Spectroradiometer (MODIS) data model, Chang et al. (2016) estimated a mean grassland GPP of 537 g C m^{-2} yr^{-1} while a mean value of 796 g C m^{-2} yr^{-1} was simulated for global grasslands using a process-based model. Minor contributions to agroecosystem GPP originate also from photosynthetic C assimilation by other eukarya (i.e., algae), bacteria (e.g., cyanobacteria, purple and green bacteria), and archaea (Thauer 2007). This may occur, for example, in microbiotic crusts on soil, rocks, and plants (Elbert et al. 2009). The basic equation of oxygenic or O_2-releasing photosynthesis in plants, algae, and cyanobacteria is (Eq. 2.10):

$$6\ CO_2 + 12\ H_2O + \text{Light energy} \rightarrow C_6H_{12}O_6 + 6\ H_2O + 6\ O_2 \qquad (2.10)$$

The main products of photosynthesis are sucrose and starch and not glucose has previously suggested (Amthor 2010; Heldt 2005).

Light energy is needed for oxygenic photosynthesis. Total solar irradiance is Earth's dominant energy input with a most probable value of 1360.8 W m^{-2} representative of the solar minimum (Kopp and Lean 2011). About 48.7% of the irradiance at Earth's surface in the 400–700 nm waveband designated photosynthetically active radiation (PAR), and 2×10^{23} J is captured annually by photosynthesis (Amthor 2010; Zhu et al. 2010; Ito and Oikawa 2004). The PAR is the energy source for synthesis of the energy-rich molecules adenosine triphosphate (ATP) and nicotinamide adenine dinucleotide (NADPH). ATP and NADPH power the synthesis of sugars that plants depend on to grow and multiply (Chapin et al. 2002). Plants use the chemical energy stored in ATP and NADPH to fix CO_2 into sucrose, starch, and other organic compounds through light-independent reactions. The overall equation for C fixation in green plants can be written as Eq. (2.11):

$$\begin{aligned}&3\ CO_2 + 9\ ATP + 6\ NADPH + 6\ H^+ \\ &\quad \rightarrow C_3H_6O_3\text{-phosphate} + 9\ ADP + 8\ P_i + 6\ NADP^+ + 3\ H_2O\end{aligned} \qquad (2.11)$$

The formation of $C_3H_6O_3$-phosphate by condensation of CO_2 with ribulose-1,5-bisphosphate (RuBP) (i.e., the carboxylation reaction) is catalyzed by the enzyme ribulose-1,5-bisphosphate carboxylase/oxygenase (RuBisCO). This reaction occurs in the so-called C_3 photosynthetic pathway because the primary carboxylation product is a sugar containing three C atoms (Zhu et al. 2010). Nutrient limitation may be the driver of global maximum carboxylation rate distributions (Walker et al. 2017). Specifically, decreasing leaf N is associated with increasing SOC based on increasing costs of N uptake as the high concentrations of RuBisCO in leaves make up a large portion of whole-plant N demand.

The C_3 photosynthetic pathway occurs in 95% of all plant species and in nearly all plants of cold climates including many economically important agricultural species such as wheat, rice, and soybean [*Glycine max* (L.) Merr.] (Schulze et al. 2005). However, RuBisCO also catalyzes the initial step of photorespiration or oxygenation of RuBP which results in the loss of 30% of synthesized sugars at 25 °C and current atmospheric CO_2 concentrations. Thus, photorespiration needs to be decreased to increase crop yields. However, this process is also a valuable advantage for crops growing in a competitive environment (Busch et al. 2017). For example, the high rates of photorespiration enable C_3 plants to efficiently take up transiently available nitrogen (N), and assimilate it into organic compounds at high rates. In conclusion, photorespiration can improve photosynthetic performance despite reducing the efficiency of RuBisCO carboxylation (Busch et al. 2017).

The synthesized sugars are transported mainly in the form of sucrose from the net sugar-exporting (source) sites (i.e., mature leaves) to the net sugar-importing (sink) sites (i.e., branches, stems, seeds, and roots for storage, meristematic growth, or cell wall synthesis; Jansson et al. 2010). As the main photosynthetic activity occurs inside of plant leaves, CO_2 has to enter the leaf through small actively regulated pores called stomata (Heldt 2005). However, the uptake of CO_2 is accompanied by a significant escape of water vapor through transpiration. Plant leaves loose water vapor through stomata and the cuticle (Hanson et al. 2016). For example, about 97% of the water absorbed by roots is transpired in C_3 plants (Raven and Edwards 2001).

Thus, to reduce water loss by transpiration, plants have also evolved the C_4 and the crassulacean acid metabolism (CAM) photosyntheses. Phosphoenolpyruvate (PEP) carboxylase is the primary carboxylase in C_4 plants, and the primary carboxylation product in light is a compound containing four C atoms (Zhu et al. 2010). RuBisCO is a secondary carboxylase in C_4 plants that functions in a high-CO_2 environment by suppressing oxygenation and photorespiration. Major advantages of the C_4 compared to C_3 photosynthetic pathway are that less water is lost and less N is required to maintain the photosynthetic rate (Chapin et al. 2002). Many warm-climate plants and about 5% of all plant species are C_4 plants. For example, half of the world's grass species dominating grassland and savanna ecosystems employ C_4 photosynthesis (Osborne and Freckleton 2009). Economically important crops such as corn (*Zea mays* L.), sugarcane (*Saccharum* spp.), millet and *Sorghum* spp., and biofuel crops such as switchgrass (*Panicum virgatum* L.) and *Miscanthus* spp. also use the C_4 photosynthesis. Globally,

C_4 plants account for 30% of terrestrial C fixation despite representing only 5% of Earth's biomass (Bond et al. 2005; Osborne and Beerling 2006). This high efficiency is also indicated by a detailed stepwise biophysical and biochemical analysis of energy transduction from interception of radiation to carbohydrate formation (Fig. 2.4; Zhu et al. 2010). The maximum conversion efficiencies of solar radiation into biomass are 4.6% (C_3) and 6.0% (C_4) at 30 °C, or 9.4 and 12.3% of PAR, respectively. Thus, increasing the proportion of C_4 plants is among the options to enhance climate change mitigation by biosequestration of atmospheric CO_2 in plants.

About 3% of all plant species is CAM plants (Chapin et al. 2002). The CAM photosynthetic pathway enables terrestrial ecosystems to gain C under extremely dry conditions. Specifically, high WUE (mol CO_2 assimilated per mol H_2O respired) is achieved through certain plant features including nocturnal stomatal opening, rapid rectifier-like root hydraulic conductivity responses to fluctuating soil water potential, and the capacity to buffer against periods of drought (Owen et al. 2016). Some CAM plants are epiphytic species but typical CAM plants are desert succulents (Keeley and Rundel 2003). The commercially most important CAM plant is pineapple [*Ananas comosus* (L.) Merr] (Borland et al. 2009). In addition, some CAM plants such as *Agave tequilana* F.A.C. Weber and *Opuntia ficus-indica* (L.) Mill. have also received increased attention for producing bioenergy feedstock on degraded land (Owen et al. 2016).

Photosynthesis at the leaf level is generally limited by CO_2 concentration, availability of N and P, light, temperature, air humidity, and water potential (Farquhar 1989). The sink strength has also a dominant influence on photosynthesis at the plant level (Jansson et al. 2010). Sink strength is governed by sucrose metabolism which channels C into plant storage or structural components. At the stand level, photosynthesis is further limited by phenology, leaf area density, and canopy structure (Farquhar 1989). Photosynthesis at the scale of ecosystems (i.e., GPP) varies diurnally and seasonally in response to changes in light, temperature, and N supply (Chapin et al. 2002). Differences in annual GPP among ecosystems depend primarily on the quantity of leaf area and the duration of its photosynthetic activity. Both leaf area and photosynthetic season are determined by the availability of soil resources (water and nutrients), climate, and also time since disturbance (Chapin et al. 2002).

Determinants of GPP of agroecosystems can also result from the interaction of its collective genome (i.e., plant traits + microbial capabilities) with (i) environmental characteristics (e.g., soil, altitude, latitude, hydrology), (ii) climate variables (e.g., solar radiation, CO_2, temperature, precipitation amount and timing, ozone, length of seasons, atmospheric deposition), (iii) developmental history, and (iv) nutrient availability (US DOE 2008). For example, the spatial variation of GPP is associated with precipitation in 69% of the area of temperate grasslands and shrublands, in 55% of the area of tropical savannahs and grasslands, and in 51% of the area of croplands, respectively (Beer et al. 2010). Otherwise, irrigation in some agricultural regions may result in the independence of GPP from precipitation (Walker et al. 2017). Thus, irrigation, precipitation, and associated water

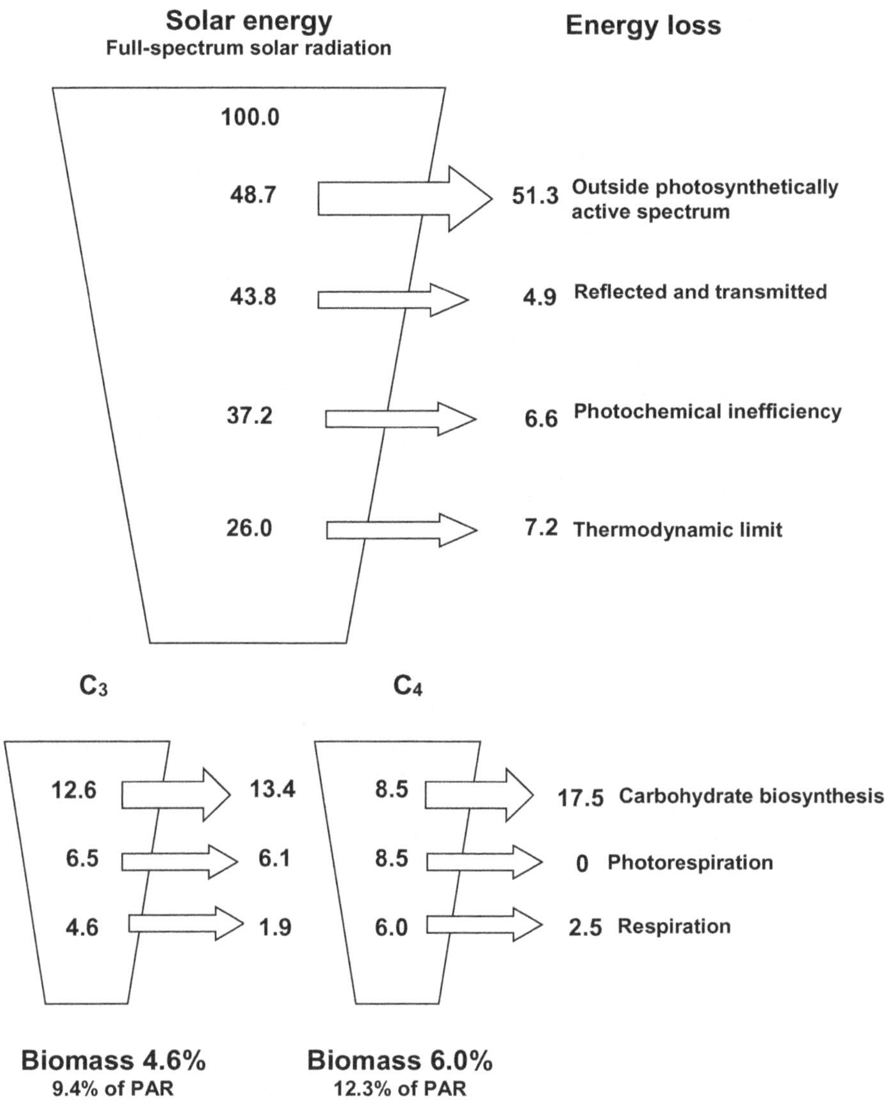

Fig. 2.4 Maximum energy losses from 100% total full-spectrum solar radiation for photosynthetic energy transduction stages from sunlight incident on a leaf to plant biomass for both C_3 and C_4 (NADP-malic enzyme type) photosynthesis. (PAR = Photosynthetically active radiation; modified from Zhu et al. 2010)

availability have a large effect on GPP in many agroecosystems. In addition, GPP is negatively correlated with incoming shortwave radiation in savannas, the Mediterranean, and Central Asian grasslands which may indicate an additional indirect effect of radiation or temperature on GPP by the water balance. However, it may also indicate a direct effect of temperature on vegetation due to heat stress as well as increased levels of diffuse radiation associated with lower levels of radiation (Beer et al. 2010). Further, measuring and modeling the fully coupled plant–microbe–soil ecosystem may be critical to understanding and predicting agroecosystem GPP (US DOE 2008).

As indicated previously, the synthesis of organic compounds from assimilated C or GPP in plants and the maintenance of living plant cells are associated with respiratory C losses by R_a (Fig. 2.3). The fraction of assimilated C remaining is the NPP (Eq. 2.12):

$$\text{NPP} = \text{GPP} - R_a \tag{2.12}$$

The NPP was originally defined as the amount of photosynthetically fixed C available to the first trophic level in an ecosystem (Lindeman 1942). Otherwise, NPP can be defined as the total amount of OM produced annually in an ecosystem. Agroecosystems are moderately productive with respect to NPP per unit area in comparison to other ecosystems (Whalen and Sampedro 2009). The bulk of agroecosystem NPP is allocated to the production of biomass in foliage, shoots, and roots (Ciais et al. 2010b). About 10 Pg C are stored in cropland biomass, and 119–121 Pg C in grassland and grazing land vegetation (Erb et al. 2018). However, not all of NPP remains in the agroecosystem as some plant biomass is lost by harvest and herbivory. Further losses occur during weed and seed production, by the emission of volatile organic compounds (VOCs), by exudation from roots (i.e., the loss of soluble organic compounds that diffuse or are secreted by roots into the soil), and by C transfer to root symbionts. For example, emissions from biogenic VOCs amount to 0.76 Pg C yr^{-1} globally (Sindelarova et al. 2014). However, these losses are rarely measured. Thus, the sum of all components is the agroecosystem NPP but the sum of $\text{NPP}_{\text{foliage}}$, $\text{NPP}_{\text{shoots}}$ and $\text{NPP}_{\text{roots}}$ can serve as its approximation (Ciais et al. 2010b). The conversion of GPP into NPP can be characterized by the carbon use efficiency (CUE; Eq. 2.13):

$$\text{CUE} = \text{NPP}/\text{GPP} \tag{2.13}$$

The CUE for annual crops grown without stress or environmental fluctuations in controlled environments ranges from 0.55 to 0.63 (Gifford 2003). Further, CUE varies with plant age as respiration is high compared to photosynthesis soon after seed germination. Then, CUE increases for a couple of weeks after germination as leaves become established and seed reserves are consumed. Gradually, CUE may decline with age or plant size for long-lived plants (Gifford 2003). On a larger scale, the CUE for croplands in geographic Europe (i.e., Europe bordered in the east by the Ural Mountains, the Caspian sea, the Caucasus, and the Black Sea) was 0.49 and that for grasslands was 0.56 (calculated from Schulze et al. 2010).

A small fraction of NPP is directly transferred belowground to the roots via the phloem (Kutsch et al. 2009a). When arbuscular mycorrhizal (AM) fungi are present, some C may flow to them through roots and, thus, enter the soil (Fig. 2.3;

Sanderman et al. 2010). This mycorrhizal (literally meaning 'fungus-root') symbiosis between roots and fungi is widespread with about 90% of land plants engaged in this beneficial association including also many plants of agroecosystems (Hodge et al. 2009; van der Heijden et al. 2015). Some C also enters the soil as organic compounds excreted by roots. However, the major NPP fraction is converted to biomass, and some of it enters the soil with surface residues or root litter in the form of particulate organic C (POC). The decomposition of POC and other soil C inputs is associated with heterotrophic respiration R_h, and Net Ecosystem Exchange (NEE) is the difference between GPP and respiratory losses by R_a and R_h (Fig. 2.3; Eq. 2.14).

$$\mathrm{NEE} = \mathrm{GPP} - R_a - R_h \tag{2.14}$$

The long-term C gain at the agricultural landscape level is further reduced by biomass harvest and fire, and by C export into aquatic systems. The remaining NBP describes the net storage of C in agroecosystems and, thus, determines whether agroecosystems contribute to reducing atmospheric CO_2 levels by C sequestration or act as C source by increasing CO_2 levels (Churkina et al. 2010).

The storage of OM as SOC is the key component of NBP (NBP_{soil}; Schulze et al. 2010). For example, in geographic Europe, NBP_{soil} was −57 g C m^{-2} yr^{-1} for grasslands and +10 g C m^{-2} yr^{-1} for croplands between 2000 and 2005. Thus, while grasslands accumulated C, croplands were a C source. An ongoing C loss from croplands in Europe was also observed by a cropland flux network (Kutsch et al. 2010). However, generalizations are problematic as global annual values for NBP have varied considerably during the last decades (Jansson et al. 2010). In summary, only a small fraction of GPP initially fixed during photosynthesis enters the soil and is potentially accumulating in the SOC stock (e.g., NBP is 3% of GPP for geographic Europe; Schulze et al. 2010). However, SOC stock represents the major accumulation of C, and the most important long-term C storage pool of agroecosystems as plant biomass is harvested periodically.

2.2.1.2 Organic Carbon Input

Plant Production

The plant biomass production or NPP determines the upper limit for the C input into soils of agroecosystems (Fig. 2.3; Sanderman et al. 2010). As shown previously, the capture of solar radiation by plants and the conversion of GPP to chemical energy used for production of sugars are constrained by solar radiation, water and nutrients, and C losses resulting from R_a (Eq. 2.12). Among the most important determinants of biomass production are the efficiencies of interception of solar radiation by plants and of solar energy conversion into biomass (Jansson et al. 2010).

Carbon Assimilation and Net Primary Production

The solar radiation reaching Earth's surface is the primary driver of plant photosynthesis (Mercado et al. 2009). However, only the blue-to-red visible wavelengths of incident solar radiation (400–700 nm) or PAR reaching the plant canopy are energy source for photosynthesis (Sanderman et al. 2010). Healthy crops typically absorb 90% of PAR but emit most of the near infrared radiation (>700 nm), which comprises about half of the energy of sunlight (Zhu et al. 2010). Aside the incident radiation, canopy structure and the optical properties of plants and soil surface determine the amount of radiation absorbed. The fixation of one CO_2 molecule necessitates a quantum requirement of ten (or more) quanta of light absorbed (Whalen and Sampedro 2009). The net photosynthetic efficiency ranges between 3 and 6% of total solar radiation which is lower compared with a theoretical maximum efficiency of 11%. The relationship between GPP and absorbed photosynthetically active radiation (APAR) can be described by the average photosynthetic LUE (Eq. 2.15; Coops et al. 2010).

$$\text{GPP} = \text{APAR} * \text{photosynthetic LUE} \tag{2.15}$$

The LUE represents the actual use of absorbed radiation energy by a plant to produce biomass (Coops et al. 2010). LUE of plants is driven by abiotic and biotic stress factors that reduce the photosynthetic capacity such as photoinhibition. Among long-term stressors for LUE may be soil water and temperature. Crops maintain a constant LUE over the entire range of naturally occurring irradiance (Ruimy et al. 1996). However, values for LUE vary among agroecosystems and depend on cloudiness, phenology, and biome specific parameters (Turner et al. 2003). Further, soil nutrients also affect LUE adjusted for the effects of temperature and water stress (Huston and Wolverton 2009). Specifically, grasslands exhibit low, C_3 croplands intermediate and C_4 croplands high LUE values (Bradford et al. 2005). For example, Ruimy et al. (1994) reported values of 1.26, 2.71, and 3.51 g MJ^{-1} for temperate grasslands, C_3 crops, and C_4 crops, respectively. However, these are only approximations as establishing accurate values is difficult (Metting et al. 2001; Coops et al. 2010). Nevertheless, in cool, moist conditions under normal light intensity, the C_3 pathway is more efficient than both the C_4 pathway and CAM (Sanderman et al. 2010). However, under high light intensity and temperature, C_4 plants photosynthesize faster and have higher WUE than C_3 plants. Further, under arid conditions, CAM plants have up to sixfold and up to threefold higher WUE than C_3 and C_4 plants, respectively (Borland et al. 2009).

Among the major ecophysiological parameters for describing light-response functions of photosynthesis is quantum yield α (i.e., apparent quantum efficiency). The quantum yield describes the rate of increase in assimilation or net C uptake A with irradiance at levels below those at which CO_2 has an effect (Gilmanov et al. 2010). The mean value of α for non-forest sites is 33.3 mmol mol^{-1} with 1 mol equivalent to the molecular mass of CO_2 fixed and equivalent to 4.6 μmol photons of absorbed PAR. Intensively managed grasslands exhibit quantum yields of up to

75 mmol mol^{-1} which is higher than the mean value for forests (37 mmol mol^{-1}), but differences between forest and non-forest ecosystems are not significant (Gilmanov et al. 2010).

Other parameters for describing photosynthesis are maximum photosynthesis A_{max}, the light saturation point, and the LAI (Chapin et al. 2002; Gilmanov et al. 2010). The photosynthetic capacity is represented by A_{max}. For non-forest ecosystems, the mean A_{max} value is 1.2 mg CO_2 m^{-2} s^{-1} with values ranging between 0.2 and 3.4 mg CO_2 m^{-2} s^{-1} for mixed prairie during drought and for irrigated continuous corn, respectively (Gilmanov et al. 2010). The intensity at which light is no longer the factor limiting the overall rate of photosynthesis is the light saturation point. This point is reached by C_3 plants at about 25% of maximum full sunlight (Jansson et al. 2010). At moderate-to-high light intensities, the carboxylation reaction catalyzed by RuBisCO is the rate-limiting step in C_3 photosynthesis. In contrast, C_4 plants have considerably higher light saturation points and are more efficient than C_3 plants in converting light energy to biomass. However, the overall productivity can be higher for C_3 plants as C_4 photosynthesis at lower temperatures comes with an extra cost (Jansson et al. 2010).

As indicated previously, RuBisCO also contributes to the loss of sugars by binding of O_2 instead of CO_2 by catalyzing the oxygenation reaction. This photorespiration is energetically costly and, thus, reduces the photosynthetic rate of C_3 crops by 10–30% but beneficial features may have decreased the evolutionary pressure to eliminate this process (Busch et al. 2017). The LAI is a measure of canopy foliage content defined as the amount of leaf area in a canopy per unit ground area (Asner et al. 2003). The LAI can be used to obtain the fraction of PAR absorbed by vegetation (Sanderman et al. 2010). Values for LAI range between about 1.7, 2.1, and 3.6 for grasslands, shrublands, and croplands, respectively, but with high variations for grasslands and shrublands.

Non-photosynthetic Assimilation of Carbon Dioxide

The most abundant CO_2-fixation pathway is catalyzed by RuBisCO, and numerous prokaryotes harbor this enzyme aside photosynthetic eukaryotes (Atomi 2002). For example, a rich functional diversity of RuBisCO-coding genes was detected in agricultural soils indicating that a wide range of soil bacteria is capable of autotrophic CO_2 fixation (Selesi et al. 2005). Thus, non-phototropic CO_2 fixation may represent an important microbial activity in soils but all processes involved have not yet been characterized (Miltner et al. 2005).

Plant Biomass and Net Primary Production

A large but variable fraction (i.e., 25–75%) of all C fixed in assimilates daily by higher plants is expended by respiration (Lambers et al. 2005). Only the smaller fraction of assimilated C or GPP remaining after accounting for C losses by R_a is stored in new plant biomass (NPP; Eq. 2.12). This conversion of GPP into NPP (i.e., CUE) is varying among agroecosystems as it depends on the climatic characteristics such as temperature and precipitation, and geographic factors such as latitude and altitude (Zhang et al. 2009). The major components of NPP are generally new plant biomass (40–70% NPP), root secretions (20–40% NPP), losses to herbivory and mortality (1–40%), and volatile emissions (0–5% NPP; Chapin et al. 2002). However, these values are uncertain as all NPP components have never been measured in a single study. For example, most field measurements of NPP document only the new plant biomass produced and, thus, probably underestimate the true NPP by at least 30% (Chapin and Eviner 2014). The NPP is generally constrained by the supply of belowground resources whereas other factors such as CO_2 concentration directly enhance the NPP in the most productive ecosystems (Chapin et al. 2002). The presence of earthworms in agroecosystems, for example, leads to a 25% increase in crop yield and a 23% increase in aboveground biomass (van Groenigen et al. 2014). Supposedly, earthworms stimulate plant growth predominantly through releasing N locked away in residue and SOM.

A greater genetic diversity among plant species of agroecosystems enables increased, more stable, and more 'sustainable' NPP as was observed, for example, for grassland ecosystems (Hooper et al. 2005; Balvanera et al. 2006). This greater genetic diversity yields, in particular, greater variation in phenotypic traits among species which are potentially important for NPP (Loreau et al. 2001). Phenotypic traits include: (i) photosynthesis (C assimilation, capacity), (ii) growth (biomass accumulation, phenology), (iii) allocation (C partitioning between biomass and respiration; biomass partitioning among leaf, stem, root and seed), (iv) resource (water, N, P, and other nutrients) acquisition [uptake, WUE, nutrient-use efficiency (NUE)], (v) root architecture, chemical composition, and longevity, (vi) life history strategy (annual vs. perennial) and longevity, (vii) stress tolerance, (viii) microbial patterns, (ix) responses to environmental factors, and (x) leaf phenology (US DOE 2008). In general, more diverse communities are characterized by greater growth and resource acquisition and allocation compared to less diverse communities (Cardinale et al. 2006). Diversity of phenotypic traits within species populations impacts community NPP (Whitham et al. 2006). Diversity-enabled and survival mechanisms functioning at the species level enhance NPP. Particularly for communities in agroecosystems, i.e., communities with few species or those dominated by a small number of species that are the biggest contributors to NPP, trait variation within populations may be important (Fridley et al. 2007). Some global estimates for cropland and grassland NPP are presented in the following sections, and region- and crop-specific estimates are discussed in Chaps. 3 and 4.

Net Primary Production of Croplands

Direct measurements of cropland NPP are not possible as, for example, not all of the biomass produced remains in croplands as some is removed by harvest and herbivory by insects and mammals (Ciais et al. 2010b). In addition, other components of cropland NPP are rarely measured. This includes weed production, seed production, emission of VOCs to the atmosphere, exudation from roots, and C transfer to root symbionts. However, cropland NPP can be estimated from yield statistics by using process-oriented or remote sensing models.

Wolf et al. (2015) included the harvested portion of the plant (e.g., grain, fruit, or other plant part) versus the residue and belowground biomass, and crop biomass associated with harvest losses in estimates of crop NPP. Global crop NPP in 2011 was estimated at 5.25 Pg C yr^{-1}. Based on biomass harvest data, global cropland NPP was estimated to be 3.9 Pg C yr^{-1}, and on average 610 g C m^{-2} yr^{-1} (Chapin and Eviner 2014). For 1991, cropland NPP was estimated at 344 g C m^{-2} yr^{-1} derived from agricultural production data through application of crop-specific coefficients such as dry matter content, harvest index (HI), root production, and C and N content (Goudriaan et al. 2001). In another study, mean global cropland NPP was estimated at 397 g C m^{-2} yr^{-1} for the year 2000, based on HI and harvest statistics (Haberl et al. 2007). Monfreda et al. (2008) applied crop-specific factors (i.e., HI, root:shoot ratio) to agricultural production data (i.e., metric tons of economic yield per hectare, dry proportion of the economic yield, C content) and adapted them to gridded crop data to produce a global map of cropland NPP. The crop functional types of global crop- and grasslands included annual/perennial, herbaceous/shrub/tree, C_3/C_4, and leguminous/non-leguminous. The mean global crop NPP was 420 g C m^{-2} yr^{-1} for the year 2000. The average yields approximated the distribution of global cropland NPP (Monfreda et al. 2008).

Net Primary Production of Grasslands

Similar to croplands, direct measurements of grassland NPP are not possible given permanent biomass removal by grazing and mowing, uncertainties in rhizodeposition, and production of VOCs lost to the atmosphere (Ciais et al. 2010a). Based on biomass harvest data, tropical savanna/grassland and temperate grassland NPP were estimated at 14.0 and 5.3 Pg C yr^{-1}, with mean values of 1,080 and 750 g C m^{-2} yr^{-1}, respectively (Chapin and Eviner 2014). Further, the measured aboveground NPP of climatically determined grasslands (i.e., those not resulting from human action) ranged from 50 to 800 g C m^{-2} yr^{-1} (Sala 2001). Assuming a C content of dry matter of 0.45, the NPP of tropical grasslands and savannas ranged from about 280 to 4,470 g C m^{-2} yr^{-1} (House and Hall 2001). The GPP Data Initiative reported global mean NPP values of 353 g C m^{-2} yr^{-1} for grasslands, and of 916 g C m^{-2} yr^{-1} for savanna (Cramer et al. 2001). Further, grazing land NPP was 433 g C m^{-2} yr^{-1} in the year 2000 (Haberl et al. 2007). A lower global NPP for croplands (304 g C m^{-2} yr^{-1}) compared to tropical savannas and grasslands (540 g C m^{-2} yr^{-1}), and those of temperate grasslands and shrublands (393 g C m^{-2} yr^{-1}) may be explained by the prevalence of low-yield agriculture in developing countries and the low belowground productivity of crops (Roy et al. 2001).

Limitations to Agroecosystem Net Primary Production

Many factors determine the NPP of plants (Schulze et al. 2005). These include the C balance over a growing season (mean rate of net photosynthesis × mean leaf area × length of the growing season), distribution of assimilates to different organs, respiration in existing and growing biomass, and storage. Further, light, climate (i.e., temperature, precipitation, air humidity), atmospheric CO_2 concentration, macro- and micronutrients, plant water status, plant age, and disturbances affect NPP directly though their effect on photosynthesis, respiration, and plant growth (Ciais et al. 2001; Huston and Wolverton 2009). For example, photosynthetic rates increase asymptotically with increase in precipitation and have a temperature optimum (Ciais et al. 2001). Stimulating effects of increases in CO_2 concentrations on NPP are mostly indirect, and reflect the effects of CO_2 on water and nutrient availability (Chapin et al. 2002). Over the long-term, belowground resources (i.e., water and nutrients) limit NPP of most terrestrial ecosystems (Chapin and Eviner 2014). The nature of environmental limitation varies among years, being most moisture-limited in dry years, and most nutrient-limited in wet years. Thus, moisture limitation of NPP occurs most frequently in dry sites, and nutrient limitation in mesic sites (Chapin and Eviner 2014).

The harvestable yield of crops is limited by (i) the light interception efficiency of PAR determined by the speed of canopy development and closure, leaf absorbance, canopy longevity, size, and architecture, (ii) the conversion efficiency, i.e., the combined gross photosynthesis of all leaves within the canopy, minus all plant respiratory losses, and (iii) the partitioning efficiency or HI which is the amount of total biomass energy partitioned into the harvested portion of the crop (Zhu et al. 2010). The potential harvestable yield for non-oil grains is 18, and 35–40 MJ g^{-1} for oil-rich seeds. The increases in potential yield over the past 50 years have largely resulted from increases in both HI (i.e., dwarfing of the stem, increase in potential number of seeds set) and light interception efficiency (i.e., development of larger-leaved cultivars, more rapid soil coverage after germination). Dwarfing improved also the standing power of the crop to adverse weather conditions such as rain, wind, and/or hail, and, thus, indirectly improved realized light interception efficiency. However, overcoming future yield limitations for the major crops, corn, rice, wheat, and soybean, may only be possible by extending the growing season or increasing the photosynthetic conversion efficiency (Zhu et al. 2010). For example, *Miscanthus* x *giganteus* produces 60% more biomass in the US Midwest than its relative corn by developing a closed canopy four weeks before corn and maintaining it four weeks after corn has senesced while conversion efficiencies among both plant species are comparable. The realized light interception and partitioning efficiencies determining grain crop yield potential are close to their theoretical maximum. Thus, improving the conversion efficiency can potentially increase the crop yield potential. Near-term opportunities to achieve this include improving the display of leaves in crop canopies to avoid light saturation of individual leaves and optimizing a photorespiratory bypass. Long-term opportunities include engineering carboxylases that are better adapted to current and projected CO_2 concentrations, and the molecular optimization of resource investment among the components of the photosynthetic apparatus, to maximize C gain without increasing crop inputs. Such changes can more than double the yield potential of major crops (Zhu et al. 2010).

The long-term trend of cropland NPP is driven more by technological changes (>90%) than by those in climate and atmospheric CO_2 concentration (<10%; Ciais et al. 2010b). However, the interannual variability of cropland NPP is determined by climatic conditions. In contrast, the grassland NPP depends on radiation, temperature, precipitation, and length of the growing season (Ciais et al. 2010a; Gilmanov et al. 2010). Water, in particular, appears to be the factor that most strongly controls grassland NPP (Chapin et al. 2002). The indirect effects of climate on terrestrial NPP arise from the influence of temperature and precipitation on chemical and biological processes that form soil by dissolving and transforming soil minerals in rocks (Huston and Wolverton 2009). These processes reduce soil fertility as soils are aging and, thus, decrease the resources available to support NPP. For example, most old and highly weathered soils are highly acidic and have low availabilities of P and the basic cations Ca^{2+}, K^+, Mg^{2+}, and Na^+. Further, NPP of tropical systems is limited by P and that of temperate systems by N (Huston and Wolverton 2009). In contrast to natural terrestrial ecosystems, mineral and organic fertilizers, and pH-moderating compounds (such as agricultural lime) are applied to sustain NPP in agroecosystems (Marschner 1995). Aside increasing genetic yield potential (i.e., defined as the yield a crop can attain under optimal management practices in the absence of biotic and abiotic stresses), crop yields have also been increased through better fertilization and improved disease protection (Zhu et al. 2010). In summary, maximizing agroecosystem NPP can be realized by large-scale irrigation, increase in the availability of soluble nutrient sources (fertilizers), agricultural mechanization, improved varieties and application of agrochemicals for pest control (Whalen and Sampedro 2009).

Fate of Recently Fixed Carbon

The fate of C recently fixed by photosynthesis is a major determinant of the SOC stock (Sanderman et al. 2010). Specifically, a substantial portion of fixed C is removed from agroecosytems by harvest. Herbivory by insects and mammals, and emissions of VOCs also contribute to biomass loss from agroecosystems (Ciais et al. 2010b). Otherwise, animals and their excreta provide significant C inputs to the soil with animal manure, for example, containing about 50% organic C (Nieder and Benbi 2008; Whalen and Sampedro 2009). Annual SOC sequestration was increased by 699.6 and 401.4 kg C ha^{-1} yr^{-1} in upland and paddy fields, respectively, when synthetic N fertilizer was substituted by livestock manure applied at equivalent N rates (Xia et al. 2017). Thus, animal excreta are important sources for SOC in livestock production systems.

A fraction of fixed C in non-erosional agroecosystems remaining after accounting for the losses by harvest, herbivory, and VOCs is deposited as above- and belowground residues. Another fraction is transferred into the soil as root and mycorrhizal exudates including turnover of fine roots (i.e., <2 mm in diameter) and fungal hyphae, and as other dead microbial cells. About 18 to >30% of total photosynthetic production is invested in root turnover, and 2–5% of it is released as root exudates in the soil (Adl 2016). Further, a small fraction of aboveground NPP may also be removed from agroecosystems by natural and human-induced fires (Andreae 2004). For example, about 1,200 Tg of residues burn annually in croplands. However, aboveground residues such as stubble not burned but retained at the soil surface make only a small direct contribution to SOC sequestration compared to stubble incorporated into the soil (Sanderman et al. 2010). Especially in the tropics, the ignition of fires is a common human activity to clear land for shifting cultivation by converting forests into agroecosystems (Crutzen and Andreae 1990). Dry vegetation is also removed by human-induced fires to promote agricultural productivity. However, burning of agricultural residues in developed countries has been greatly reduced in the industrial era due to the large-scale expansion of intensively managed agroecosystems combined with active fire management (Marlon et al. 2008). Thus, some agricultural residues are still being burned but comprehensive contemporary global data on biomass burning are lacking (Crutzen and Andreae 1990; Pechony and Shindell 2010).

2.2.1.3 Dynamics of Organic Carbon in the Mineral Soil

Surface residues are incorporated into the mineral soil by physical mixing and solubilization, transport, and subsequent adsorption (Lorenz and Lal 2005). Relative importance of these processes *vs*. root litter and rhizodeposition for profile SOC distribution and dynamics depend on climate, soil, and vegetation types (Rumpel and Kögel-Knabner 2011). However, there is a paucity of studies addressing the contribution of belowground C inputs to SOC sequestration (Denef and Six 2006). Most studies on SOC dynamics in subsoil and, in particular, on dissolved organic carbon (DOC), have been performed in temperate forest soils. By convention, DOC is defined as the organic C that passes through a 0.45-μm-diameter filter (Zsolnay 2003). The organic C contained in the soil aqueous phase represents a pool of highly mobile, predominately solvated molecular fragments of relatively small molecule size (Kleber et al. 2015). The DOC constitutes only a small proportion of SOC but the most mobile and actively cycling SOC fraction (Bolan et al. 2011).

There are only a few studies on DOC in temperate and tropical agroecosystems (Chantigny 2003). For example, the process of DOC movement and retention was responsible for 9% of the total mineral SOC stock to 1 m depth in a temperate grassland soil (Sanderman and Amundson 2008). Also, DOC concentrations tend to be lower in grasslands than in forest with similar rainfall, probably owing to the source strength of the forest litter layer (O-horizon) for organic compounds (Kleber et al. 2015).

Over and above the effects of plant management and translocation of DOC, rooting and belowground biomass also impact SOC sequestration in the subsoil, but the reasons for the size discrepancy between root and SOC stocks are unknown. For example, on average, half of the world's SOC is found below 20 cm depth but only 30% of the world's roots (Jobbágy and Jackson 2000; Rumpel and Kögel-Knabner 2011). An increase in root C:N ratios with increase in soil depth may contribute to the unexpectedly large proportion of SOC at deeper depths (Dietzel et al. 2017). Further, in extensively managed ecosystems, there exist a myriad of soil biota that affect subsoil SOC inputs (Wilkinson et al. 2009). For example, earthworms, termites, ants, arthropods, and roots efficiently bury SOC while forming voids in the form of burrows, nests, chambers, galleries, and root channels. Direct subsoil inputs include litter sequestration into nests, termitaria, borrows, and bioturbator disposal of dead tissues. The earthworms and litter arthropods are effective in fragmenting and mixing surface residues into the soil. Indirect inputs of SOC into subsoils may occur by infilling of biogenic voids with litter, redistribution of SOC, and subsurface mixing and burial (Rumpel and Kögel-Knabner 2011).

Plant Roots

Growing plant roots are major sinks of recently fixed C with new roots representing on average 30–40% of NPP (Chapin et al. 2002). In grasslands, roots constitute >50% of NPP. Root biomass is about 87% of shoot biomass for hay crops (Whalen and Sampedro 2009). However, root biomass data are highly uncertain as not all fluxes and the entire root biomass can be directly measured. Yet, direct measurements of complete root profiles are rarely reported in the primary literature (Schenk and Jackson 2005). Deep roots enhance bedrock weathering, thus regulate the long-term C cycle but deepest roots are likely undersampled (Fan et al. 2017). Roots of annual crops start from seeds each season reaching only shallow depths. Mean rooting depths for annual herb, perennial grass, annual grass, and perennial herb were 0.90, 1.04, 1.06, and 1.20 m, respectively. Roots are highly adaptive to local soil environments with soil hydrology as a globally prevalent force (Fan et al. 2017). Estimates for root production are also uncertain. For example, mean root production ranged from 61 to 1,107 g OM m^{-2} yr^{-1} based on a comparison of 11 methods at a single shortgrass steppe site (Milchunas 2009). In particular, inventory data on root biomass are uncertain due to spatial and temporal heterogeneity, uneven sampling, and methodological differences among studies (Robinson 2007). Thus, root biomass in agroecosystems and, in particular, those of Mediterranean shrubland, tropical savanna and grassland, and temperate grassland may be 75–88, 70–87, and 50–100% larger than previously estimated, respectively (calculated based on Robinson 2007). However, there are only few studies of root biomass in tropical savannas and grasslands (Mokany et al. 2006).

The belowground residues and root turnover are direct C inputs into the subsoil. Thus, the vertical root distribution combined with shoot/root allocations affects the distribution of SOC with depth (Jobbágy and Jackson 2000). However, roots are distributed more shallowly than SOC, primarily as SOC turnover increases with depth. Further reasons for the depth distribution of SOC are that root turnover also increases with depth. In addition, SOC is leached from upper to lower layers and is mixed vertically by bioturbation, and this affects the identification of root effects on SOC depth distribution. However, Jobbágy and Jackson (2000) did not consider agricultural soils because of potential effects of tillage on the vertical SOC distribution. Nevertheless, root C is preferentially stabilized in SOC relative to surface residue C (Rasse et al. 2005). Root inputs at agroecosystems are five times more likely than an equivalent mass of aboveground litter to be stabilized as SOC (Jackson et al. 2017). Thus, it has been hypothesized that roots are a more important sources of SOC in subsoils than shoots. However, the root C flux into soil is poorly understood because accurate measurements of root exudation and root cell sloughing are difficult to perform (Rumpel and Kögel-Knabner 2011).

Belowground C inputs are often estimated from root-to-shoot ratios to overcome difficulties in measurements of root-derived C (Sanderman et al. 2010). However, much current knowledge about root–shoot biomass relations is based on unreliable data. For example, higher estimates for median root-to-shoot ratios were calculated than previously by omitting unreliable data (Mokany et al. 2006). Specifically, median root-to-shoot ratios were 0.322, 0.420, 0.642, 1.837, 1.887, 4.224, and 4.504 for tropical/subtropical/temperate dry woodland, tropical/subtropical moist woodland, savanna, shrubland, tropical/subtropical grassland, temperate grassland, and cool temperate grassland, respectively. Thus, longer-lived species such as perennial grasses invest more resources into maintaining a healthy root system and this contributes to higher root-to-shoot ratios compared to annual species (DuPont et al. 2014). To avoid erroneous results, vegetation-specific root-to-shoot ratios must be used to estimate root biomass (Mokany et al. 2006). Further, the often used static measurement of live root biomass may result in inaccurate estimates because a significant fraction of total root biomass may be short-lived fine roots. For example, turnover times of tree fine roots are one year or less based on minirhizotron experiments and between 1.3 and 32 years based on temporal changes in C isotopic ratios (Pritchard and Strand 2008). However, minirhizotron measurement generally underestimates the turnover whereas estimated fine root turnover times based on isotopic studies are always too high (Guo et al. 2008). In contrast to woody roots, non-woody roots of many plants in agroecosystems which lack secondary growth and cell wall thickening are short-lived (Hodge et al. 2009). For example, the turnover rate (i.e., annual belowground production/maximum belowground standing crop) of whole-tree root systems is 0.10 compared with 0.34 for shrubland species, 0.53 for grasses, and 0.56 for fine roots of trees (Gill and Jackson 2000).

Not all of the net fixed C allocated belowground is used for root growth. About 50% of the C allocated belowground (19% of net fixed C) is retained in root biomass, and roughly 27% of C allocated to roots (11% of the net fixed C) is rhizodeposition (Jones et al. 2009). While there is no common definition for 'rhizodeposition,' Rasmussen (2011) suggested that it should describe processes related to the release of organic compounds by living roots. In practice, it is inherently difficult to separate root exudation from turnover of the finest root parts. Thus, an operational definition of rhizodeposition may include C or N from turnover of root hairs, fine roots, and minor root fragments, but exclude C or N from turnover of decaying roots and larger parts of the root system (Rasmussen 2011). However, most isotopic labeling studies used to quantify the amount of photosynthate partitioned belowground have focused on young plants at a vegetative stage but partitioning is strongly affected by plant age. Further, almost half of the published data on rhizodeposition are for wheat and ryegrass, and 76% of the studies are related to only five crop/grassland species. Thus, the knowledge of C rhizodeposition, especially in mixed plant communities, is scanty (Jones et al. 2009). Also, while soil C inputs in annual crops can be modeled based on root-to-shoot ratios, the above- and belowground biomass in perennial species such as grasses may not be very representative for soil C inputs (Gill et al. 2002).

Rhizodeposition

The C flow at the soil–root interface is bidirectional as C is being lost from the roots and simultaneously taken up by them from the soil (Jones et al. 2009). Rhizodeposition describes the release of organic C compounds by living roots which results in dramatic changes in the biological, chemical, and physical soil properties. For example, rhizodeposition releases highly bioavailable, low-molecular weight organic substrates (Dennis et al. 2010) which fuel the soil microflora leading to an abundant microbial population in the rhizosphere which is responsible for a major fraction of soil biological activity (Bais et al. 2006; Kuzyakov and Blagodatskaya 2015; Nguyen 2003). The enormously complex microbial community in the rhizosphere, the microbiome, consists of symbionts, commensals, and pathogens (Poole 2017). Rhizosphere microorganisms interact with roots, the soil, and one another to form a root holobiont, an assemblage of different species that form an ecological unit. This root holobiont determines much of the agricultural productivity (Poole 2017).

The rhizodeposition processes include (i) loss of root cap and border cell, (ii) death and lysis of root cells, (iii) flow of C to root-associated symbionts living in the soils (e.g., mycorrhiza), (iv) gaseous losses, (v) leakage of solutes from living cells (root exudates), and (vi) insoluble polymer secretion from living cells (mucilage). Most of the C lost during root growth is in the form of complex polymers. Otherwise, root exudates contain a diverse array of C-containing primary and secondary metabolites aside ions, free O_2, water, enzymes, and mucilage (Bais et al. 2006). However, the relative importance of each rhizodeposition process is unknown as it is extremely difficult to discriminate between them both in space and in time. Also, the C flow in the rhizosphere is extremely complex, being highly plant and environment dependent and varying both spatially and temporally along the root. Thus, amount and type of rhizodeposits are highly context specific (Jones et al. 2009).

The SOC sequestration is directly and indirectly influenced by rhizodeposition. Specifically, root exudation may represent a significant C loss for the plant. The magnitude of photosynthates secreted varies with soil type, age, and physiological state of the plant and nutrient availability (Bais et al. 2006). The root exudates may also affect plant–plant, plant–microbe, and plant–nematode interactions. The plant–microbe interactions in the rhizosphere may particularly contribute to C sequestration in plant and soil as micronutrient acquisition is enhanced by the stimulation of microbial communities through exudates. Rhizodeposition may range between less than 10% of the net C assimilation by a plant to 44% of a nutrient-stressed plant's total C (Bais et al. 2006). According to a review by Pausch and Kuzyakov (2018), net rhizodeposition may account for 3 and 5% of recently assimilated C for crops and grasses, respectively. More than half of the rhizodeposits is utilized by microorganisms within a few days. However, many of the published data on C flow in both soils and roots are severely biased (Rees et al. 2005; Kuzyakov 2006). Further, the partitioning of rhizosphere respiration from mycorrhizal respiration is almost impossible. Whether root exudates directly contribute to the SOC stock is also under discussion (Sanderman et al. 2010).

The Priming Effect

The decomposition rate of SOC can increase or decrease substantially after addition of easily decomposable C sources. This process, known as 'the priming effect,' has already been described in the year 1926 but is still poorly understood (van der Wal and de Boer 2017). Various explanations for priming mechanisms have been proposed (Blagodatskaya and Kuzyakov 2008). For example, it was shown recently that low-molecular weight organic substances 'trigger' an acceleration of microbial metabolism and this may explain priming under certain circumstances (Mason-Jones and Kuzyakov 2017).

The rhizosphere priming effect describes the fact that the presence of living roots may suppress SOM decomposition by up to 79% or stimulate it by more than 500% compared to soil incubations without plants (Huo et al. 2017). On average, rhizosphere priming enhances SOC mineralization rate by 59% in soils of agroecosystems. Grasses had higher stimulation levels than crops, and rhizosphere priming may persist beyond months and years in perennial-dominated ecosystems. However, variables poorly explain the total variation in rhizosphere priming pointing to its complex nature (Huo et al. 2017). Among major uncertainties is the response of the subsoil SOC stock to the presence of living roots and aboveground vegetation (Kuzyakov 2002, 2010). Joergensen and Wichern (2018) assume that subsoils must have been predominantly colonized by microorganisms during periods near to the surface. Laboratory studies by Fontaine et al. (2007) indicated that adding an additional energy source to the subsoil similar to those in rhizodeposits prime microbial decomposition of subsoil SOC. However, the stimulation of decomposition of stable subsoil SOC by addition of labile material under field conditions may be small (Sanaullah et al. 2011). Further, a subsoil priming effect is not always observed but an increase in the subsoil SOC stock from inputs of root exudates is also likely (Salomé et al. 2010). Also, more complex compounds derived from root turnover may contribute indirectly to SOC stocks by enhancing aggregation (Rees et al. 2005). Otherwise, subsoil SOC storage may be threatened in the future by the priming effect. For example, labile substrates may increasingly enter deeper soil depths due to climate change-induced increases in above- and belowground biomass production and root exudation, and due to increases in soil N from elevated atmospheric N deposition (Heitkötter et al. 2017).

Mycorrhizal Fungi

Mycorrhizal fungi live in symbiosis with the roots of plants. Most important and globally widespread are ericoid (ERM), ecto-(ECM), and arbuscular mycorrhizal (AM) associations (Smith and Read 2008). The ECM are associations between fungi and woody plant roots while 80–90% of all terrestrial plants are AM associations, i.e., symbioses between plant roots and soil fungi (order Glomeromycota) (Hodge et al. 2009). The symbiosis facilitates the uptake of mineral nutrients such as phosphate and N by the plant from the soil. In return, the fungus receives C from the plant in the form of sugars (hexoses) or alternative C sources. For example, fatty acids synthesized by the host plant may also be transferred to the fungus to sustain mycorrhizal colonization (Jiang et al. 2017). Thus, lipids may also be a major source of organic C delivered to AM fungi in addition to sugars, and this transfer may be necessary for the production of fungal lipids (Luginbuehl et al. 2017).

The AM fungi are ubiquitous in agroecosystems, and most abundant in temperate grasslands and savannas (Treseder and Cross 2006). Most grass species are mycorrhizal plants (Whalen and Sampedro 2009). Synergistic interactions between AM fungi and bacterial species may also play important roles in the establishment and functioning of mycorrhizal associations (Garbaye 1994). Other functional benefits provided by AM fungi are pathogen protection and facilitation of water uptake by the host plant aside the facilitation of nutrient uptake (Verbruggen and Kiers 2010). AM fungi receive about 3–20% of photosynthate from their host plant, and the hyphae outside of the plant root play a major role in C translocation into the soil (Treseder and Cross 2006; Wilson et al. 2004). This process may enhance SOC sequestration as C is removed from the high respiratory activity in the rhizosphere. For example, the large network of hyphae outside of roots may represent up to 15% of the SOC stock (Leake et al. 2004).

The AM fungi produce the glycoprotein glomalin within their hyphal walls (Wright and Upadhyaya 1996). Glomalin is deposited in the soil as the hyphae senesce. It may comprise 0.7–2.4% of the SOC stock in agroecosystems and enhance soil aggregation (Purin and Rillig 2007; Treseder and Turner 2007). Glomalin may represent a reasonably large C influx to SOC in the order of tens to hundreds g C m^{-2} yr^{-1} and relatively slow turnover rates in soils of agroecosystems (Rillig et al. 2003; Treseder and Turner 2007). However, the importance of glomalin for SOC sequestration is not well understood. For example, glomalin cannot be directly isolated and quantified in soil, and all methods for measuring soil glomalin have limitations (Rosier et al. 2006; Schindler et al. 2007). Thus, glomalin may be present in measurable but modest quantities in most soils of agroecosystems but the direct impact of mycorrhiza hyphal turnover on the SOC stock may be small. Indirect effects of mycorrhizal fungi on SOC stocks may be more important. For example, the growth of hyphae improves soil structure and aggregate stability (Sanderman et al. 2010). This is supported by the close correlation between AM fungal hyphal abundance, soil structure, and SOC stocks in agroecosystems (Wilson et al. 2004). In addition to contributing to SOC gain, mycorrhizal fungi also contribute to SOC loss through decomposition (Talbot et al. 2008). Specifically, ERM and ECM fungi possess more extensive enzymatic capabilities than AM fungi and are, thus, efficient decomposers (Smith and Read 2008). Otherwise, AM fungi facilitate shifts in the structure and function of host plant communities by symbiotic interactions (Cameron 2010). In summary, mycorrhizal fungi may moderate SOC dynamics in agroecosystems by enhancing SOC sequestration and by the formation and maintenance of soil aggregates but a better understanding is needed (Talbot et al. 2008; Wilson et al. 2004).

2.2.1.4 Decomposition and Stabilization of Soil Organic Carbon

The net balance between soil C inputs and C losses determines the capacity of a soil to sequester C. The activities of microorganisms and larger fauna contribute to decomposition of OM and its mineralization to CO_2 which is the dominant fate for organic C in most soils (Whalen and Sampedro 2009). Soil microorganisms encompass all organisms smaller than 150–200 μm (Swift et al. 1979), i.e., mainly fungi and bacteria but also archaea, algae, protozoa, rotifers, tardigrades, and small nematodes (Joergensen and Wichern 2018). In addition to microbe-mediated GHG production, soils may also produce CO_2, CH_4, and N_2O through non-microbial pathways (Wang et al. 2017a). The processes discussed include (i) photodegradation, (ii) thermal degradation, (iii) reactive oxidative species (ROS) oxidation, (iv) extracellular oxidative metabolism (EXOMET), and (v) inorganic chemical reactions. The non-microbial GHGs may play key roles in contributing to budgets of GHGs in arid regions, whereas their global importance may be generally enhanced with accelerated global environmental changes (Wang et al. 2017a).

Photodegradation is the decomposition of OM compounds directly by solar irradiance which increases CO_2 fluxes through either microbial facilitation or, more importantly, photochemical mineralization (Rutledge et al. 2010). Exposure to radiation can facilitate microbial decomposition of litter by altering litter chemistry and consequently litter degradability (i.e., photopriming; Lin et al. 2018). Short-term photopriming was recently identified as a novel mechanism behind photodegradation. Whether abiotic processes are primarily responsible for degrading lignin during photodegradation is under discussion (Lin et al. 2018). Photodegradation may be an important contributor to CO_2 loss from agroecosystems where SOC, litter, and/or standing dead material are exposed to solar irradiance (Rutledge et al. 2010). Thus, photodegradation may occur in (i) drylands, (ii) barren peat areas in tundra, (iii) bare burnt areas, (iv) agroecosystems that are sparsely vegetated like shrublands, savannas, grasslands, agroecosystems after cultivation, harvest or during prolonged drought, and (v) those with naturally large amount of exposed standing dead material like peat bogs. Photodegradation of surface OM by ultraviolet (UV; 280–400 nm) radiation and PAR may contribute to decomposition, especially in arid and semiarid ecosystems (i.e., drylands; Barnes et al. 2015). The annual C loss from a temperate peatland in New Zealand by photodegradation was 34–66 g C m^{-2} yr^{-1} but may be lower for other agroecosystems (Rutledge et al. 2010). Photodegradation of SOM may be important in agricultural soils after residues have been removed or where erosion exposes SOM to solar radiation (Feng et al. 2011). In conclusion, photodegradation may be quantitatively important in determining rates of mass loss, nutrient release, and the C balance in a broad range of terrestrial ecosystems (Austin et al. 2016). However, while photodegradation may be important locally, it is probably not very important for the global C budget (Foereid et al. 2011).

Soil erosion may also affect SOC sequestration in agroecosystems. Soil erosion involves (i) detachment and breakdown of soil aggregates, (ii) transport of soil particles by runoff water or wind, (iii) redistribution of eroded material over the landscape, and (iv) deposition of eroded material in depressional sites and protected areas (Lal 2004b). Human activities have accelerated soil erosion rates to up to 100 times above natural levels (Sanderman and Berhe 2017). Typically less than 10% of NPP is removed from a site by erosion (Wang et al. 2017b). Thus, net C losses from agroecosystems may occur globally by soil erosion by inducing a lateral SOC-C flux estimated at 0.5 Pg C yr^{-1} (Quinton et al. 2010). Chappell et al. (2016) estimated global terrestrial SOC erosion to be 0.3–1.0 Pg C yr^{-1}. Otherwise, SOC displacement by soil water erosion has been estimated at 2.5 Pg C yr^{-1} (Borrelli et al. 2017). The net impact of erosion on C cycling has been estimated to act as either a C source or a sink on the order of 1 Pg C y^{-1} (Doetterl et al. 2016). Historically, agricultural soils have lost significant amounts of SOC by erosion (Lal 2003). However, the eroded SOC can also be a net sink of C relative to the atmosphere (Van Oost et al. 2007). Specifically, erosion can induce changes in soil-atmosphere C exchange via altered patterns of oxidation of SOC remaining at the eroding site, and of the SOC originally present at the depositional site, as well as via SOC production in both locations (Billings et al. 2010).

Three key mechanisms need to be addressed to characterize the contribution of erosion to C cycling. The first mechanism is dynamic replacement as the depleted SOC stock at eroding sites can be replaced by newly assimilated C. A continued C input and decrease in SOC available to decomposition can lead to a net gain of SOC at eroding sites (Van Hemelryck et al. 2009). Secondly, the burial of topsoil SOC and a reduced decomposition, i.e., the rate of decomposition of SOC in depositional settings, can be reduced due to a combination of physical and chemical processes, such as increased soil wetness, limited aeration, compaction and physical protection of the deposited soil material within newly formed aggregates, leading to a preservation of buried C. The C burial efficiency has been estimated to amount to 15–19% (Wang et al. 2017b). Third, transport and increased decomposition, i.e., the disruptive energy of forces applied to the soil by water erosion (i.e., raindrop impact, the shearing force of flowing water, and collision with other aggregates), may cause the breakdown of aggregates. This process of disaggregation exposes previously protected SOC to microbial decomposition and, combined with a relatively greater proportion of labile SOC within larger soil aggregates, may lead to rapid mineralization of this easily decomposable C following water erosion (Van Hemelryck et al. 2009). Thus, erosion can induce a net C sink or source depending on management practices, the extent to which SOC oxidation and production characteristics change with erosion, and the fate of eroded SOC (Billings et al. 2010). Over short timescales, the acceleration of erosion due to human disturbance leads to a net sequestration of atmospheric C as SOC (Wang et al. 2015). The translocation and burial of soil reduce also decomposition of SOC and may lead to long-term SOC storage (Quinton et al. 2010).

To sum up, whether lateral C movement across terrestrial land surfaces represents a net atmospheric C source or sink depends on the perspective. Focusing only on the site of erosion, Wang et al. (2017b) estimated that erosion has cumulatively removed 783 Pg C from cropland soils over the last 8,000 years with 92% of it mobilized before the year 1850. In comparison, the total net flux from cropland attributable to land-use and land-cover change (LULCC) was calculated at 98.4 Pg C for the period 1850–2015 (Houghton and Nassikas 2017). Wang et al. (2017b) estimated that the erosion-induced C sink of 78 Pg C has offset 37% of the cumulative C emissions due to anthropogenic land-cover change over the last 8,000 years. This analysis considered lateral transport of soil along the entire fluvial network and used historic land-use reconstructions. Further, the rates of C burial may have increased by a factor of 4.6 since the year 1850. However, this is a conservative assessment of the erosion-induced C sink as only water erosion on croplands was considered (Sanderman and Berhe 2017). Natural erosion processes continually redistribute SOC also in forests, and anthropogenic activities such as overgrazing can greatly increase rates of erosion in grasslands, and in forested ecosystems after logging. However, the importance of wind erosion has been underappreciated. Thus, the total global erosion-induced C sink is likely to be higher (Sanderman and Berhe 2017).

In addition to CO_2, some of the soil C input into agroecosystems may also be lost by emission of CH_4 (Denman et al. 2007). Among the major biogenic CH_4 sources are the anaerobic decomposition of OM in wetland soils (100–231 Tg CH_4 yr^{-1}), flooded soils under rice cultivation (31–112 Tg CH_4 yr^{-1}), and crop residues under very wet field conditions (Greenhouse Gas Working Group 2010). The dominant soil process generating CH_4 is microbial methanogenesis which is carried out by a group of anaerobic Archaea (Singh et al. 2010). These microorganisms convert substrates produced by fermentation of organic macromolecules. A large fraction of the CH_4 produced in drier soil is often consumed by methanotrophic bacteria before escaping into the atmosphere (Reay 2003). However, the components of the global CH_4 cycle are not comprehensively understood (Bousquet et al. 2006).

Gaseous losses of C from soils of agroecosystems may also occur by the emission of VOCs, compounds which may also interact with C cycling in soils (Gray et al. 2010; Insam and Seewald 2010). Plants, microorganisms, and animals are the main producers of VOCs (Laothawornkitkul et al. 2009; Leff and Fierer 2008; Loreto et al. 2008). The decomposing litter including the microbial metabolism, in particular, has been identified as major VOC source in laboratory incubations (Gray et al. 2010). However, the relative contributions of abiotic and biotic sources to soil VOC emissions in the field are not completely known (Leff and Fierer 2008).

Hydrological fluxes may also affect agroecosystem C. In particular, C losses through leaching are important for the C balance of agroecosystems (Kindler et al. 2011). For example, leaching of DOC from European grasslands can amount to 5 g C m^{-2} yr^{-1}. However, the DOC concentrations are much smaller than DIC concentrations, particularly in subsoils (Kindler et al. 2011). The total leaching losses of biogenic C (DOC plus biogenic DIC) from grasslands can be 22% of NEE plus C inputs with fertilization minus C exports by harvest. Further, the average leaching loss from grasslands is about 50% of NBP. The C leaching increases the net C loss from European cropland soils by 25%. In contrast, leaching losses of CH_4 are not relevant for the C balance (Kindler et al. 2011).

Potentially leachable C compounds are mainly derived from plant litter and SOM but also from the canopy in case of trees, root exudates, and the microbial biomass (Kalbitz et al. 2000). Throughfall and stemflow contribute also to DOC in agroforests. In cultivated and pastoral soils, plant residues are major sources for DOC (Bolan et al. 2011). In addition, application of manure and sewage sludge may also increase the amount of DOC. The DOC is largely produced by the activity of the soil biota as litter is physically and chemically altered during decomposition, substances are leached from litter, and soluble compounds are formed. Abiotic processes (i.e., desorption and dissolution from the potentially leachable C) control the DOC concentration in the soil solution. The DOC compounds can be grouped into labile DOC compounds such as simple carbohydrates, low-molecular weight organic acids and proteins, and amino sugars (Marschner and Kalbitz 2003). Further, recalcitrant DOC compounds consist of polysaccharides, other plant compounds and/or microbially derived degradation products. Only the mobile DOC fraction in macro- and mesopores is subjected to convective transport by seepage. In contrast, DOC in micropores is immobile and interacts with the mobile fraction by diffusion. Various processes contribute to decomposition or removal of DOC from the soil solution (Kalbitz et al. 2000). Thus, DOC concentrations strongly decrease with increase in soil depth as the distance to the DOC source increases. Ultimatively, small amounts of leached C are advectively transported into aquatic ecosystems (Sanderman and Amundson 2008). Losses of DOC from subsoils are controlled by sorption to poorly crystalline Fe and Al (hydr)oxides with a high specific surface area (Kindler et al. 2011). In general, recently deposited crop residues and application of organic amendments such as biosolids and manures are the most important sources of DOC in soils of agroecosystems (Bolan et al. 2011). Minor sources are root decay, exudates, and microbial metabolites. The DOC concentrations in grassland soils are higher than those in arable lands mostly due to different vegetation types, but very few studies have been published on boreal and tropical agroecosystems (Chantigny 2003).

Soil Respiration

Belowground C can be lost from agroecosystems by soil respiration (Rs). This is a biological process, primarily consisting of heterotrophic respiration (i.e., OM mineralization) and autotrophic respiration (i.e., root respiration) (Hashimoto et al. 2015). Specifically, about half of the annual Rs flux may be root-derived (Hanson et al. 2000). The Rs is dominated by relatively young C that turns over in a year or less (Gaudinski et al. 2000). Otherwise, the flux of CO_2 from the soil surface to the atmosphere is a physical process (Raich et al. 2014). The soil CO_2 efflux integrates, in particular, CO_2 produced by archaea, bacteria, fungi, single-celled eukaryotes, algae, plant roots, mycorrhizas, invertebrates, fossorial vertebrates, and by geochemical processes (Raich et al. 2014). Among CO_2 sources, the microbial respiration may be effectively locally specialized to moisture as historical rainfall controls both the moisture dependence and sensitivity of respiration (Hawkes et al. 2017). Abiotic sources of soil CO_2 (e.g., soil pedogenic or liquid water carbonates, CH_4 oxidation, volcanic activity) and soil sinks for CO_2 (e.g., carbonate dissolution, C fixation by chemoautotrophic microbes) should be studied to improve understanding of the soil CO_2 efflux (Phillips et al. 2017). Nevertheless, soil respiration is the main contributor to total ecosystem respiration and estimated to range globally between 68 and 108 Pg C y^{-1} although with a large uncertainty (Hursh et al. 2017; Hashimoto 2012). Among agroecosystems, estimates for Rs are 27.7, 11.0, 7.9, and 6.2 Pg C y^{-1} for tropical savannas, temperate grasslands, croplands, and deserts and arid shrublands, respectively (Raich et al. 2014). Among biomes, Rs of croplands, grasslands, savannah, shrublands, and wetlands have been estimated at 13.3., 4.4, 11.2, 17.7, and 3.0 Pg C y^{-1} from 1960 to 2012 (Zhao et al. 2017). On an area basis, annual soil respiration rates of croplands, grasslands, savannah, shrublands, and wetlands were 747, 564, 1.231, 776, and 429 g C m^{-2} y^{-1}.

The Rs is typically estimated by measuring the rate at which CO_2 leaves the soil and enters the overlying atmosphere, i.e., the soil CO_2 efflux. However, the CO_2 efflux from soil is not equal to soil respiration (Angert et al. 2015). For example, a considerable fraction of the CO_2 respired in soil can be dissolved in the soil water, transported in the hydrological system, or take part in reactions of the carbonate system. Further, some of the CO_2 respired by roots can be dissolved in the xylem water and carried upward in the transpiration stream (Aubrey and Teskey 2009). Thus, at least on short timescales (weeks to seasons) and for calcareous soils, estimates of the biological CO_2 efflux need to take into account the large temporal decoupling between soil–gas exchange fluxes and biological soil respiration (Angert et al. 2015). However, this may be less important on annual and longer timescales when the effects of CO_2 storage and release are probably canceled out. The processes in acidic and neutral soils are, however, not well understood (Angert et al. 2015).

The factors determining Rs are under discussion as multiple processes operate at diurnal to annual timescales (Bahn et al. 2010). Soil respiration includes a component derived from C recently assimilated by plants (i.e., respiration of roots, mycorrhizas, and rhizosphere microorganisms) and a component derived from the C respired during the decomposition of dead plant litter, microbial debris, and destabilized SOM. The soil microbial community structure influences SOC cycling and, thus, Rs (Schimel and Schaeffer 2012). Kemmitt et al. (2008) hypothesized that the rate-limiting step in SOC mineralization is independent of the size, community structure, or specific activity of the soil microbial biomass ('The Regulatory Gate Hypothesis'). Accordingly, mineralization of SOC may be a two-stage process: Firstly, non-bioavailable forms are converted abiologically to bioavailable forms (termed the 'Regulatory Gate'), which, only then, undergo a second process, biological mineralization (Brookes et al. 2017). This may have significant implications for theories of SOC dynamics, effects of global warming and soil nutrient cycling. However, the mechanisms driving the 'Regulatory Gate' are largely unknown (Brookes et al. 2017).

The plant activity appears to be a key determinant of Rs (Bahn et al. 2010). Plants produce litter as a major substrate for heterotrophic activity in soil, and directly influence soil respiratory processes via root respiration and by providing photoassimilates that fuel the metabolic processes of mycorrhizae, endophytes, and microbial populations in the rhizosphere (Bahn et al. 2010). At plot scales and within certain biomes, temperature, moisture, OM supply, and other site characteristics are known to affect Rs (Hursh et al. 2017). Globally, however, soil temperature is the dominant driver of Rs whereas within certain biomes soil moisture and SOC may also be dominant. For example, the CO_2 release into the atmosphere increases from C-rich soils (>2% SOC) after intense droughts but C-poor soils may show a net decline in C losses (Canarini et al. 2017). This self-reinforcing mechanism in C-rich soils was explained by (i) high substrate availability that magnify bursts of CO_2 release after drought events and (ii) a shift in microbial community with increased loss of C per unit of biomass. Peyton Smith et al. (2017) highlighted that microbial access to SOC is not solely limited by physical protection, but also by drought or wetting-induced shifts in hydrological connectivity. Otherwise, in croplands and grasslands, NPP emerges as an important predictor of within-biome Rs although NPP remains secondary predictor relative to soil temperature (Hursh et al. 2017).

Estimated annual Rs increased with increases in MAT and MAP in cropland, grassland, shrubland, and wetland biomes (Zhao et al. 2017). Additionally, Rs decreased with increases in MAT in the savannah biome. In grasslands, annual Rs was significantly dependent on temperature, precipitation, and SOC storage while temperature sensitivity of annual Rs was not pronounced in croplands (Chen et al. 2010). The annual precipitation may be the most important factor controlling annual cropland and grassland Rs. Further, topsoil (0–20 cm) SOC storage in croplands and grasslands may also be another important determinant of annual Rs (Chen et al. 2010). Worringly, the climate controls on Rs may translate into an increasing trend in global Rs due to increasing air temperature which further increases the large terrestrial C efflux (Hashimoto et al. 2015).

Decomposition

Decomposition is second only to photosynthesis in driving the quantity of C cycled through terrestrial ecosystems (McGuire and Treseder 2010). Decomposition is a key ecological process for maintaining supply of most plant-essential nutrients (Janzen 2015). For example, nutrient recycling via decomposition may account for >90% of plant-available N and P, and for >70% of plant-available K and Ca in natural ecosystems (Chapin et al. 2002). Subject to decomposition in agroecosystems are mainly the dieback of plants and unharvested remnants such as leaf, stem, and root tissues (Whalen and Sampedro 2009). Decomposition can be defined as the physical and chemical breakdown of dead OM (i.e., dead animal, microbial and plant material) that emit C into the atmosphere and nutrients in forms that can be used for microbial, plant and animal production (Chapin et al. 2002). Specifically, litter decomposition is the result of three interlinked processes: leaching, fragmentation, and catabolism (Cotrufo et al. 2009). Decomposition is driven primarily by bacteria and fungi accounting for up to 90% of the total decomposer biomass. Residues may be decomposed rapidly by a bacterial food chain or slowly by a fungal food chain (Beare et al. 1992). About 85–90% of OM decomposition is mediated microbially, and fungi mainly decompose fresh plant material whereas gram-positive bacteria consume both fresh and older OM (Amelung et al. 2008). Further, about 10–15% of decomposition is done by soil animals (Wolters 2000).

Among decomposition processes, leaching is most important during tissue senescence and when plant litter is deposited at the soil surface. Labile compounds such as low-molecular weight soluble sugars, polyphenols, and amino acids are predominantly leached from plant litter (Berg and McClaugherty 2008). Some of the leached and solubilized material may be absorbed by soil organisms. However, a major portion of leached OM may be transported deeper into the mineral soil depending on adequate water supply, and adsorbed to SOM and soil minerals, or transported into adjacent aquatic ecosystems. Soil fauna also partially solubilize fresh plant residue and facilitate the establishment of microbial decomposers, for example, by mixing residues into the soil for more intimate contact with decomposer microorganisms (Wolters 2000). However, soil fauna has only a limited ability to enzymatically decompose organic residues compared to soil microorganisms (Whalen and Sampedro 2009).

The main initial decomposers are fungi which are mostly concentrated closer to the soil surface (Chapin et al. 2002; Fierer et al. 2003). Fungal enzymes can break down virtually all classes of plant compounds, and hyphal fungi can adjust mycelium growth across considerable distances toward OM substrates (Ritz 1995). Fungi have also higher C assimilation efficiencies (i.e., store more C than they metabolize) than bacteria (Singh et al. 2010). The SOC stock in agroecosystems favoring a fungal-dominated community is enhanced (Six et al. 2006). In contrast, bacteria primarily decompose more labile substrates (Moorhead and Sinsabaugh 2006). Further, bacterial biomass extends also to deeper depths. In contrast to many fungi, bacteria rely on solute transport and soil perturbation for their translocation toward organic residues (Jiang et al. 2005). Thus, the composition of the initial microbial community decomposing a residue may gradually shift through time concurrently with changes in residue composition (Hättenschwiler et al. 2005).

In addition to the primary resources plant OM, the biomass of microorganisms and microfauna (Ø < 0.2 mm) are themselves secondary resources for decomposition. The C polymers in fungal cell walls (chitin, melanin) are more resistant to decomposition than those in bacterial cell membranes and walls (phospholipids, peptidoglycan) (Singh et al. 2010). For example, fungal melanin may influence SOC cycling processes as indicated by the positive correlation between estimated melanin content of soil and SOC content (Siletti et al. 2017). Thus, respiration rates in soils dominated by fungi are typically low which increases the potential for SOC sequestration (Six et al. 2006). Ultimately, however, most natural compounds are fully mineralized to inorganic forms under favorable environmental conditions (Kleber 2010a, b; Marschner et al. 2008). For example, about one-half to two-thirds of plant residues entering the soil may be decomposed within one year (Whalen and Sampedro 2009).

Less well known is the long-term fate of leaf and root residues in soils of agroecosystems as decomposition is often studied only for 1–2 years (Silver and Miya 2001). In contrast, some litter decomposition experiments in forests did last for 10 years or longer (Harmon et al. 2009; Moore et al. 2011). Studies monitoring litter decomposition for several years are important as some litter may decompose very slowly depending on environmental conditions and biochemical characteristics. However, the phenomenon of very slow mass loss in late-stage decomposition may only occur in cold northern latitudes or under other suboptimal conditions as it does not appear to occur under warmer conditions (Prescott 2010). In general, grass litter decomposes faster than tree leaf litter which itself decomposes faster than needle litter (Zhang et al. 2008). Among root litter, grass roots decomposes faster than broadleaf roots which decompose faster than coniferous roots (Silver and Miya 2001). Differences in root composition and cell wall structural characteristics among crop and forage species strongly influence the rate and degree of root tissue degradation and mineralization (White et al. 2018). For example, large, species-dependent variation was observed in the extent of root tissue decomposition during 270 d incubations, ranging from 82.5% of initial mass for alfalfa (*Medicago sativa* L.) to 21.5% for switchgrass. Differences in the extent of decomposition and the nature of the cell wall components degraded among warm season grasses, cool season grasses, and legumes suggested that no broad conclusions regarding SOC cycling and potential contribution to SOC sequestration can be made on this basis (White et al. 2018).

With regard to wood, broadleaf woody deciduous litter decomposes faster than litter from woody evergreen species (Cornwell et al. 2008). The herbaceous litter does not decompose faster than woody litter, i.e., graminoid litter (grasses and grass-like monocots) decomposes much slower than forb (eudicot herbs) litter. Also, gymnosperm wood consistently decomposes slower than angiosperm wood (Weedon et al. 2009).

Most studies of aboveground litter decomposition use the litter bag method to follow the progressive loss of mass from fresh litter through time (Bocock and Gilbert 1957). The method is still widely accepted for quantifying and comparing litter decomposition rates (Torn et al. 2009). However, a compromise on mesh size of the bags is needed based on the experimental question. Thus, the method has its limitations as, depending on the mesh size, soil faunal activities may be excluded (Prescott 2005). The process of litter fragmentation may be prevented, in particular, by studying decomposition in mesh bags (Cotrufo et al. 2009). Another limitation is that decomposition is often studied using single-species leaf litterbags. For example, mixed-species litterbag decomposition studies indicate that ecosystem leaf litter decomposition is not predictable from the decay rates of component leaf species (Gartner and Cardon 2004). In summary, available litter decomposition data should be used with caution and referred to the specific decomposition component process (i.e., leaching, fragmentation, catabolism) addressed (Cotrufo et al. 2009). Nevertheless, the use of litterbags is particularly suitable for testing the effects of plant protection products on OM decomposition in agroecosystems (Knacker et al. 2003).

Organic Matter Protection and Stabilization

Inputs of OM to soils can result in an increase in the SOC stock when they are stabilized and not completely mineralized to CO_2 (Kleber and Johnson 2010). For example, charring of OM may create biochemically recalcitrant BC. However, whether selective preservation of relatively unaltered plant-derived compounds (e.g., lipids; Jansen and Wiesenberg 2017) because of biochemical recalcitrance is an important long-term stabilization mechanism beyond the decadal time frame is discussed controversially (Krull et al. 2003; Lorenz et al. 2007). For example, the highly resilient nature (slow turnover) of long-chain leaf-derived plant waxes (*n*-alkanoic acids and *n*-alkanes) in soils suggests that this SOC pool is relatively insensitive to environmental change (van der Voort et al. 2017). Otherwise bioavailable, reduced organic compounds such as lipids and waxes may be selectively protected in anaerobic microsites (Keiluweit et al. 2017).

Old SOC cannot be viewed as being composed of complex or recalcitrant compounds as the ^{14}C age of SOM fractions is not necessarily related to its molecular structure or thermodynamic stability. From a thermodynamic perspective, SOM should readily turnover, and old and stable SOM is not necessarily biochemically recalcitrant (Kleber et al. 2011). Similarly, microbial-derived OM may not be recalcitrant or composed of complex compounds. Measurements and calculations of the microbial biomass C turnover for largely dormant soil microbial communities are biased by several methodological and conceptual problems (Joergensen and Wichern 2018). Nevertheless, MRTs of microbial biomarkers in soils may not exceed several hundred years (Amelung et al. 2008). However, microbial-derived materials may play a role for SOM stabilization (Kleber et al. 2011). Also, fast incorporation of OM into earthworm cast aggregates and subsequent physical protection from decomposition may play a significant role in SOC stabilization (Vidal et al. 2016).

Mean soil C turnover times of 35.5 years, and of 34.4 and 17.7 years for grasslands and croplands have been calculated by Yan et al. (2017). The SOM turnover appears to be a function of microbial ecology and of the resource availability within a given physical soil environment (Kleber et al. 2011). Liang et al. (2017a) proposed that microorganism-mediated ex vivo modification and in vivo turnover may jointly explain SOC dynamics driven by microbial catabolism and/or anabolism. Thus, microorganisms are an active player in SOC storage. The microbial production of a set of organic compounds is coupled to their further stabilization in soil, which Liang et al. (2017a) define as the entombing effect. This integration captures the cumulative long-term legacy of microbial assimilation on SOM formation, with mechanisms (i.e., physical protection or a lack of activation energy due to chemical composition) that ultimately enable the entombment of microbial-derived C in soils (Liang et al. 2017a).

Processes which slow down mineralization are major centennial-scale stabilization mechanisms (Sanderman et al. 2010). Two important groups of processes contributing to long-term stabilization of OM are (i) processes which lead to physical protection, rendering OM spatially inaccessible to decomposers or their water-soluble degradative enzymes, and (ii) organo-mineral complexes and organo-metal interactions, i.e., interactions of OM with minerals, metal ions, and other organic substances (von Lützow et al. 2006). Physical protection may slow down decomposition for decades to centuries whereas organo-mineral complexes or organo-metal interactions may be responsible for most of the highly stable (i.e., for centuries to millennia) non-charred SOM (Kögel-Knabner et al. 2008). The associations of OM with mineral surfaces, in particular, may determine the long-term retention of SOM (Kleber et al. 2015). While much of the micron- and molecular-scale knowledge about organic–mineral interactions remains largely qualitative, direct measurements of molecular binding provide mechanistic insights into organo-mineral interactions (Newcomb et al. 2017). Force spectroscopy has been used to directly measure the binding between organic ligands with known chemical functionalities and soil minerals in aqueous environments. Evidence has been found that chemistry of both the organic ligand and mineral contribute to values of binding free energy, and that changes in pH and ionic strength produce significant differences in binding energies (Newcomb et al. 2017). Nevertheless, both biochemical recalcitrance and physical protection may allow OM to remain in the soil longer, allowing time for organo-mineral complexes to form (Six et al. 2000). Thus, stabilization of SOM is a combination of both short- and long-term processes. However, any disruption of the stabilization process may result in decomposition of SOM even if it is thousands of years old (Ewing et al. 2006).

Several processes may be reasons why OM is physically inaccessible to decomposers and extracellular enzymes. First, decomposers and OM are sparsely and heterogeneously distributed in soils but the mobility of most decomposer organisms is limited (Ekschmitt et al. 2008). For example, <1% of the soil volume is occupied by microorganisms, and many soil microorganisms tend to aggregate, forming microbial hot spots (Kuzyakov and Blagodatskaya 2015; Young et al. 2008). The spatial separation may control SOC dynamics, in particular in the subsoil (Salomé et al. 2010). Thus, a significant proportion of SOC of any chemical composition located in biologically non-preferred soil spaces is not subject to decomposition (Ekschmitt et al. 2008). Secondly, pores < 0.2 μm in diameter are too small for bacteria (typical dimensions 0.5–1 μm), fungi (3–10 μm), microfauna (diameter < 0.2 mm), mesofauna (0.2–2 mm), and macrofauna (>2 mm) to enter (Wolters 2000; Young et al. 2008). At smaller pore sizes <50 nm, enzymes may also be inhibited from entering and accessing OM (Zimmerman et al. 2004). Third, the hydrophobicity of OM, especially of partially-oxidized OM, may greatly reduce accessibility as microbial decomposition is governed by distribution of and accessibility to water (Bachmann et al. 2008). Fourth, soil aggregation at multiple spatial scales may cause inaccessibility of OM by creating strong diffusional limitations to enzymes, and also to oxygen which is another important controlling factor of decomposition (von Lützow et al. 2006). Keiluweit et al. (2016) suggested that oxygen limitations may be under recognized regulators of SOC turnover in upland soils.

Soil Aggregation

Soil aggregation may be the most important process for spatial separation between decomposers and/or their extracellular enzymes and OM in topsoils of agroecosystems (Sanderman et al. 2010). Thus, SOC losses increase when soil aggregates are disrupted (Rovira and Greacen 1957). Aggregates are formed when mineral particles combine with organic and inorganic substances (Bronick and Lal

2005). The OM chemistry potentially affects the dynamics of soil aggregation (Sarker et al. 2018). Microaggregates (20–250 μm in diameter) are formed from small microaggregates (<20 μm) primarily within macroaggregates (>250 μm; Oades 1984). The smallest microaggregate fraction <53 μm is usually referred to as silt + clay fraction. Some studies follow different designations of size classes. Size classes occur in an aggregate hierarchy. Specifically, macroaggregates are less stable with a life span of years, break up most easily by slaking, have the highest concentration of C and the youngest C on average. In contrast, the smallest microaggregates are the most stable with a life span of decades, have the lowest concentration of C, and contain the oldest C. The OM in microaggregates is relatively stable as their organo-mineral complexes are hydrophobic, and clogging of micropores creates physical barriers for bacteria and enzymes which slow down decomposition (Bachmann et al. 2008). As a measureable indicator, microaggregates-within-macroaggregates were proposed as diagnostic fraction for the C sequestration potential in agroecosystems (Kong et al. 2005). Within aggregates, mineral grains differing in size and mineral composition are arranged in well-defined mineral assemblages (Steffens et al. 2017). The combination of specific mineral assemblages with different quantities and qualities of OM as arranged in distinct microarchitectures has been defined as microdomains. These are characterized by different pore systems and, thus, contribute differently to SOC sequestration (Steffens et al. 2017).

Fungal extracellular polysaccharides and hyphae are primarily responsible for the formation of macroaggregates in agroecosystems (Six et al. 2006). Fine roots contribute also to the physical enmeshment of soil particles in aggregates (Miller and Jastrow 1990). However, the hierarchy of aggregate formation and structure is not observed in Oxisols because the key binding factors are oxides instead of OM. Creating and maintaining the stability of aggregates critically depends on soil fauna, especially earthworms (Edwards and Bohlen 1996). Stabilization of OM within macroaggregates may be restricted to surface horizons in agroecosystems but stabilization within silt-size aggregates may also be important in subsurface horizons (Moni et al. 2010). For example, by occlusion in aggregates in the form of occluded particulate organic matter, root-derived C was protected and stabilized in grasslands at 30, 60, and 90 cm soil depths (Sanaullah et al. 2011). Protection of OM by aggregation against microbial decomposition is primarily effective in soils with a large pool of labile OM (Goebel et al. 2009).

Mineral–organic Association

Other processes contributing to OM stabilization involve sorptive reactions with mineral surfaces (Fe-, Al-, Mn-oxides, phyllosilicates), metal ions, and other organic substances (von Lützow et al. 2006). The formation of mineral–organic associations is facilitated by (i) the presence of water enabling life processes, providing the solvent for reactions and the transport medium for microbes and OM, (ii) the presence of plants producing biomass, dead plant biomass serving as substrate for microbial life, roots physically disintegrating rocks and releasing protons and organic compounds, contributing to soil weathering and to the production of organic molecules, which may directly react with mineral surfaces or become part of the soil solution, (iii) the presence of microbiota with the ability to depolymerize larger plant biopolymers into smaller, more soluble units and to oxidize and, thus, functionalize OM, (iv) the presence of reactive mineral phases with large specific surface area, and (v) a low pH in the soil solution promoting the formation of reactive mineral weathering products and favoring the formation of strong inner-sphere bonds between OM and mineral surfaces with variable charge (e.g., Fe oxides; the edges of clay minerals, allophane, imogolite) (Kleber et al. 2015). Thus, the binding of OM on minerals differs in stability depending on the suite of soil minerals and solution chemistry (Sanderman et al. 2010). For example, the quantity and characteristics of clay minerals, and the presence of polyvalent cations such as Ca^{2+} or Mg^{2+} facilitate sorptive OM stabilization (Whalen and Sampedro 2009). Mineral–organic associations may occur throughout soil profiles in grasslands and be equally important in top- and subsoil horizons (Sanaullah et al. 2011). Multiple layers of OM may range outward from the mineral surface with decreasing strength of association (Kleber et al. 2007). Thus, the outer layers may be most actively cycling mineral-stabilized OM (Torn et al. 2009). The destabilization of SOM at the mineral surface may occur by mineral dissolution or soil evolution rather than by direct degradation. For example, the evolution of metastable non-crystalline minerals into crystalline minerals with lower surface area and charge density during soil development reduces the ability to stabilize OM (Torn et al. 1997).

Stable Soil Organic Matter and Humic Substances

Historically, SOM has been thought to consist of parts with a molecular structure that decomposes easily and other dark, amorphous parts that decompose slowly (Kleber and Johnson 2010). The molecular properties of these dark so-called humus compounds have been thought to render them refractory (Alexander 1965). During the humification process, humus compounds supposedly evolve into polymeric macromolecular humic substances, i.e., the part of OM in soils most resistant to microbial attack because of its complexity. However, it is questionable whether operationally defined humic substances artificially prepared by an alkaline extraction procedure can serve as models for SOM fractions (Baldock and Nelson 2000). Materials extracted from soil with alkali do not occur as such in natural soils, and there is no molecular-level evidence in soils for the existence of distinct humic molecules (Kelleher and Simpson 2006; Lehmann et al. 2008). Thus, SOM can be conceptually defined as a mixture of organic compounds in various stages of decomposition (Kleber and Johnson 2010). Further, the humification process does not create stable organic macromolecules out of labile plant litter precursor materials. Also, SOM is not a stable (recalcitrant, refractory) product of secondary syntheses as previously suggested (Stevenson 1994), and not the true end product of decomposition which is CO_2 (Kleber and Johnson 2010). All plant residues decay rather rapidly in aerobic soils with adequate moisture and temperature, and even lignin is more or less transformed with only parts of lignin accumulating and potentially being stabilized in soils (Hernes et al. 2013; Hofmann et al. 2009; Thevenot et al. 2010; Vancampenhout et al. 2009). A part of the fresh plant C is reworked by the biotic community, and the remnants of soil biota attach to mineral surfaces and can be protected against decomposition for centennial to millennial timescales (Kleber et al. 2015). Seemingly labile OM persists or is stable in soil because of sorptive protection, aggregation, occlusion, and facultative non-utilization (Ekschmitt et al. 2008; Mikutta et al. 2007; Salomé et al. 2010). Thus, whether recalcitrant OM exists in soil (aside BC) due to inherent molecular characteristics is a matter of debate (Kleber 2010a, b; von Lützow and Kögel-Knabner 2010). However, important for SOC sequestration is the fact that simple relatively fresh OM inputs can directly contribute to the stable SOC stock as a long slow aging process to produce stable SOC is apparently not required (Sanderman et al. 2010).

Controls on Organic Matter Decomposition and Stabilization

Organic substrates for decomposition in soils of agroecosystems are plant, microbial and animal residues, rhizodeposits, and animal manure (Whalen and Sampedro 2009). Decomposition is influenced by the physical–chemical environment, decomposer organisms, and substrate quality (Swift et al. 1979). Globally, rates of plant litter decomposition are mainly controlled by variations in litter quality (i.e., N, C:N, lignin, lignin:N, and P, K, Ca, Mg), microclimate, soil properties, and microbial community composition (Zhang et al. 2008). Early rates of litter mass loss, in particular, can be predicted from measures of climate and litter chemistry (Prescott 2010). Litter chemistry has the most direct influence on early decay rates with the acid-unhydrolyzable residue (AUR):N ratio as the most consistent predictor. The AUR was formerly known as 'lignin.' Further, leaf dry matter content and specific leaf area are useful predictors of mass loss rates. For example, litter decomposability in European grasslands is negatively correlated with community-weighted mean leaf dry matter content (Fortunel et al. 2009). Apparently, thresholds exist at which specific factors have an overriding influence on decomposition (Prescott 2010).

In contrast to leaf litter decomposition, root chemistry is the primary controller of root decomposition whereas climate and environmental factors play only secondary roles (Silver and Miya 2001). Most of the plant litter compounds decompose rapidly, whereas the remaining 'recalcitrant' compounds decompose slowly. Plant tissues are, however, not decay-resistant as lignin, cutin, and suberin have residence times of years to decades (Prescott 2010). Nevertheless, a portion of the plant residues may be incorporated into soil components that are stable and persist for many years. The C dynamics in aboveground litter vs. those of SOM in mineral soil bear little resemblance as both litter quality and other physical–chemical mechanisms control SOM protection and degradability (Giardina and Ryan 2000). Thus, plant litter decomposition does not produce SOM with a slow turnover time as the link between litter quality and SOM is not well established (Torn et al. 2009).

Under the same environmental conditions, litter 'quality' controls the rate of its decomposition. Litter of high 'quality' decomposes faster than that of lower 'quality.' However, there is neither a common definition nor a quantitative index of 'quality' (Cotrufo et al. 2009). As high energy and nutrient supply for microbial use is required for fast decomposition, a high-quality litter has high nutrient (i.e., N and P) concentrations, a high proportion of easily degradable C compounds (e.g., sugars), and low concentrations of substances inhibiting microbial activity (Swift et al. 1979). Indices of litter quality under discussion are N, C:N, AUR:N, holo-cellulose:lignocellulose, leaf width, and specific leaf area (Cotrufo et al. 2009; Prescott 2010). In later decomposition stages, factors controlling microbial degradation of lignin such as concentrations of N and Mn become key quality parameters (Berg and McClaugherty 2008). However, a general and quantitative index of litter decomposability is lacking (Cotrufo et al. 2009).

Litter decomposition is a biological process (Prescott 2010). Thus, primary controls of decomposition are microbial activity and ultimately enzymatic activity, except for the abiotic process of photodegradation, and initial litter communition and mixing by soil fauna (Sanderman et al. 2010). Decomposition occurs as a result of the secretion of extracelluar enzymes by microorganisms (Prescott 2010). Thus, decomposition depends on the successful diffusion of enzymes to appropriate OM and successful diffusion of decay products back into a microorganism. Adequate temperature, a water film, aeration, suitable pH, enzymes, diffusion conditions, and an accessible and susceptible substrate in the same place at the same time are requirements (Prescott 2010). The size, diversity, and activity of the microbial community as well as interactions with large soil biota also affect the decomposition rate (Whalen and Sampedro 2009). Soil fauna comprises 10–20% of total soil biomass and controls the abundance and diversity of the microbial community by predation (Cotrufo et al. 2009). For example, abiotic factors and microorganisms may account for over 50% of shortgrass litter decomposition under xeric conditions with microarthropods having an additional effect on decomposition rate (Vossbrinck et al. 1979). However, the effects of microarthropods on litter decomposition are poorly understood (Kampichler and Bruckner 2009). In general, fauna has negative or neutral effects on litter decomposition (Prescott 2010). Otherwise, depending on rainfall and chemical composition, leaching may also have important effects on grass litter decomposition (Koukoura 1998).

The optimal soil pH for bacteria-mediated decomposition is 6.5–8.0 whereas optimal soil pH ranges between 5.5 and 6.5 for fungi-mediated decomposition. However, due to a variety of interacting factors, decomposition is more rapid in neutral than in acidic soils (Chapin et al. 2002). Also, soil bacteria are more tolerant to anaerobic conditions than fungi and are, thus, probably responsible for the majority of decomposition in anaerobic soils. Soil texture determines the habitat available for microorganisms and the amount of predation by larger soil organisms. The activities of microbial decomposer may also be limited by insufficient quantities of essential nutrients such as N, P, K, and others required to sustain microbial growth (Whalen and Sampedro 2009).

Soil fauna mediates decomposition and typically increase rates of mass loss from litter and SOC loss (Ayres et al. 2009). Faunal species in soil include collembolans, mites, enchytraeids, isopods, and earthworms. Soil fauna enhances decomposition by (i) partial digestion of OM and returning it to the soil, (ii) fragmentation of plant litter thereby increasing the surface area available for microbial colonization,

(iii) bringing microbes and OM into direct contact, and (iv) grazing on soil microbes (Bardgett 2005). The relative importance of each of the first three processes for decomposition is less well known, but the effects of grazing on decomposition are better characterized (Ayres et al. 2009). Dead OM is partially degraded in the gut of saprotrophic soil fauna, and some OM is returned to the soil as fecal pellets. The OM returned with collembolan, isopod, and termite feces, and with earthworm casts is typically easier to decompose than the original OM which leads to accelerated rates of decomposition. However, decomposition rates of earthworm casts decline rapidly and over longer timescales, casts may enhance belowground SOC sequestration (Martin 1991). The most important process through which soil macrofauna enhances decomposition is fragmentation resulting in increased surface area available for microbial colonization (Swift et al. 1979). Whether other faunal groups also contribute to this process is less well known. Soil fauna brings OM and soil microbes in close contact (i) during gut passage, and (ii) by dragging litter from the surface into the soil. After gut passage, OM is either readily available or less available for decomposition (Fox et al. 2006; Osler and Sommerkorn 2007). Earthworms, in particular, may increase the amount of OM in soil aggregates and, thus, stabilize SOC (Bossuyt et al. 2005). Earthworms play a key role in decomposition by their partial digestion and fragmentation of OM, by dragging litter into the soil and bringing microbial decomposers into direct contact with OM (Wolters 2000). Also, grazing by microarthropods and nematodes may alter decomposition rates by influencing activity and growth of soil bacteria and fungi, and through selective feeding (Ayres et al. 2009).

Soil properties are secondary controls, but climate has an overriding control on OM decomposition and stabilization (Sanderman et al. 2010). Specifically, temperature and rainfall (soil moisture) exert important direct controls on soil microbial activity. Faster litter decomposition rates are generally measured under warmer and wetter conditions when soil temperature and moisture conditions are ideal for microbial activity. However, which of the climatic indices is the best predictor of decomposition rate is under discussion (Cotrufo et al. 2009). In soils with high moisture content, SOC accumulation is the greatest as decomposition is more restricted than is NPP (Chapin et al. 2002). Further, SOC respiration rate which is indicative of decomposition roughly doubles for a 10 °C warming when microbial activity is not limited by substrate availability or soil moisture (Davidson and Janssens 2006). The carbon-quality-temperature (CQT) theory of the temperature sensitivity of OM decomposition links the temperature sensitivity of old SOM to an increase in molecular complexity. However, old SOM fragments are not large, complex, polymeric humic macromolecules as previously thought. Thus, old SOM may not be particularly sensitive to temperature increases as predicted by the CQT theory (Kleber et al. 2011). Similarly, the limiting step for the decomposition of the more stable SOM pool may be the rate at which decomposable substrate becomes available for decomposition independently of temperature increase (Moinet et al. 2018). However, the temperature sensitivity of SOM decomposition remains controversial and continues to be debated.

Nevertheless, temperature may indirectly affect decomposition by altering soil moisture, and quantity and quality of OM inputs (Chapin et al. 2002). Further, climate (i.e., temperature and soil moisture) exerts the strongest controls on SOC losses from soil over large geographic areas (Sanderman et al. 2010). Huang et al. (2018) reported that the correlation between SOC and soil temperature at the regional scale was negative between 52°N and 40°S parallels, and positive beyond this region. It was assumed that the effect was most likely due to the temperature-dependent SOC formation (photosynthesis) and decomposition (microbial activities and substrate decomposability) processes. The large SOC stocks distributed in the low-temperature areas might increase under increasing temperature while the small SOC stocks found in the high-temperature areas might decrease (Huang et al. 2018).

Any control or processes that constrain microbial and enzymatic activity such as aggregation and association with mineral surfaces affect the decomposition rate (Sanderman et al. 2010). Soil-aggregate formation and SOC stabilization are affected, in particular, by soil type, climate, landscape position, ecology, and anthropogenic factors (Christensen 1996). Soil texture, for example, plays a large role in the number and kind of primary organo-mineral complexes formed. Specifically, soils high in clay content exhibit strong aggregate formation and stability. Thus, clay content is positively correlated with SOC content, and the importance of clay for SOC stabilization increases with increase in soil depth (Jobbágy and Jackson 2000). Texture exerts strong controls on SOC losses by reducing apparent decomposition rates through a range of stabilization mechanisms (von Lützow et al. 2006). Clay minerals, in particular, reduce the decomposition rate of SOM as the higher the mineral surface area the higher may be the amount of SOM that can be adsorbed and be protected from microbial enzymes (Chapin et al. 2002). However, clay content is not the most important determinant of SOC stocks in contrast to common assumptions (Barré et al. 2017; Kutsch et al. 2009a). The role of clay mineralogy in the retention and stabilization of SOC is not completely understood (Singh et al. 2018). Rather than texture per se the reactivity of the mineral surface determines the protection of OM (Kleber et al. 2007). For example, SOC in tropical soils such as Oxisols is poorly correlated with soil aggregates as oxides are the key binding factors (Six et al. 2002). Thus, SOM stocks and turnover are often closely related to the mineral properties, in particular, the poorly crystalline phases (Torn et al. 2009). Further, subsoils have a larger potential for long-term stabilization due to abundance of reactive surfaces, smaller and less diverse biotic communities, and limitations on oxygen diffusion compared to topsoils (Rumpel and Kögel-Knabner 2011). In contrast, environmental conditions in topsoils are often favorable for decomposition because of large and diverse biotic community, adequate moisture, available nutrients, and favorable aeration (Sanderman et al. 2010). Otherwise, the topsoil environment is less favorable for SOM stabilization due to scarcity of reactive mineral surfaces.

Temperature and soil moisture affect OM breakdown in aggregates through their effect on microbial activity. For example, aggregate turnover is slower in cold or dry climates but faster in moist and/or warm climates. However, very wet climates may reduce OM breakdown under anaerobic soil conditions. In clayey soils, wetting and drying cycles can increase aggregation (Horn and Smucker 2005). In temperate climates, freeze–thaw cycles may also promote aggregate formation and SOC stabilization (Chen et al. 1997). Plant species can also affect aggregation in agroecosystems. For example, N-fixing species and mycorrhizal associations may boost microbial populations and result in higher levels of aggregate stability. Further, molecular-level properties of SOM under different plant species may have an impact on resistance to degradation in microaggregates (Bachmann et al. 2008). Higher levels of SOC and aggregation are observed under plants with great root density (Nair et al. 2010).

In summary, decomposition of a substrate and destabilization of SOC stocks are accelerated when (i) conditions are suitable for microbial activity, enzyme production, and diffusion, (ii) OM is vulnerable to enzymatic degradation, (iii) microbial transformations that lead to the creation of recalcitrant products are constrained, (iv) chemical reactions that lead to formation of stabilized SOC stocks are constrained, (v) physical protection of OM in soil aggregates is weak, and (vi) chemical protection of SOM is also weak (Prescott 2010).

2.2.2 *The Net Balance of Soil Organic Carbon*

The soil C inputs and losses under similar soil and vegetation management are approximately balanced at steady state depending on site-specific factors (Sanderman et al. 2010). Previous quantitative descriptions of SOM dynamics were based on the assumption that SOM consists of a single homogenous pool decomposing at varying relative rates (analytical SOM models; Shibu et al. 2006). Models have also been classified by Manzoni and Porporato (Manzoni and Porporato 2009). Simulation models consider SOM as comprising of heterogeneous components and decomposition of the components occurs at different relative rates (Shibu et al. 2006). The components are represented by arbitrary SOM pools. For example, the ‘active pool’ decomposes on timescale of hours to months to years, and may consist of root exudates, microbial cell contents, and some fresh litter compounds (Torn et al. 2009). The ‘intermediate’ or ‘slow’ SOM pool has turnover times in the range of decades to centuries, and may consist of structural components of plants more resistant to decay or OM stabilized by association with soil minerals or aggregate structures. The ‘passive’ or ‘millennial cycling’ SOM pool persists in soils for thousands of years and consists of highly stabilized OM, typically associated with soil minerals or stable aggregates (Torn et al. 2009). However, the molecular model for stable SOM is under discussion (Lehmann and Kleber 2015). Independently of any stabilization process, BC may contribute to the ‘passive’ or ‘millennial cycling’ pool due to its biochemical ‘recalcitrance’.

The simulation SOM models can be further divided into comprehensive and summary models (Shibu et al. 2006). In comprehensive SOM models, essential elements are thoroughly understood based on the scientific knowledge. In contrast, essential aspects are formulated in less detail in summary SOM models, and these models are more suitable for applications and predictive purposes. Input data for different SOM models are summarized in Table 2.1 (Falloon and Smith 2009).

Effects of microbial control on soil biogeochemistry must also be represented in models (Wutzler et al. 2017). The depolymerization and mineralization of SOM, in particular, relies on a microbial decomposer community with a strict homeostatic

Table 2.1 Input parameters of SOM models (modified from Falloon and Smith 2009)

Model	Inputs			
	Timestep	Meteorology	Soil and plant	Management
APSIM	Day	Precipitation, air temperature, irradiation	Soil layers, soil water characteristics, soil C content/ dynamics, soil N content/dynamics, soil bulk density, wilting point, plant growth characteristics, plant species composition	Rotation, tillage practice, inorganic fertilizer applications, irrigation
CENTURY	Month	Precipitation, air temperature	Soil water characteristics, clay content, OM content, pH, soil C content/dynamics, soil N content/ dynamics	Rotation, tillage practice, inorganic fertilizer applications, organic manure applications, residue management, irrigation, atmospheric N inputs
DSSAT	Hour, day, month, year	Precipitation, air temperature, irradiation	Soil description, soil layers, depth of impermeable layer, soil water characteristics, clay content, plant species composition, OM content, pH, soil C content/dynamics, soil N content/ dynamics	Rotation, tillage practice, inorganic fertilizer applications, organic manure applications, residue management, irrigation
ICBM	Day, year	Combination weather and climate	Many desirable, none essential	C inputs to soil

(continued)

Table 2.1 (continued)

Model	Inputs			
	Timestep	Meteorology	Soil and plant	Management
Q-Soil	Year	Optional	Soil C content/ dynamics, soil N content/dynamics	Rotation, inorganic fertilizer applications, organic manure applications, residue management, atmospheric N inputs
RothC	Month	Precipitation, air temperature, evaporation over water	Clay content, soil C content/dynamics, soil inert C content (can be estimated)	Organic manure applications, residue management, irrigation
Sandial	Week	Precipitation, air temperature, evaporation over grass	Depth of impermeable layer, clay content, soil water characteristics, yield	Rotation, inorganic fertilizer applications, organic manure applications, residue management, irrigation, atmospheric N inputs

regulation of their stoichiometry, i.e., their elemental ratio of C/N. Thus, it is important for ecosystem-scale models to account for adaptation of C- and N-use efficiencies in order to represent C-N couplings (Wutzler et al. 2017). There are some processes for which the traditional models are unsuited and microbially explicit representations are needed, e.g., soil priming, mortality, leaching, and stabilization of DOC (Smith et al. 2018). Current C and N dynamics models are predominantly dynamic deterministic compartment models (Manzoni and Porporato 2009), in which SOM transformation is assumed to follow first-order kinetics. SOM is conceptually divided into several homogeneous pools such as pool of OM, microbial biomass, or crop residue, each with its unique decomposition rate. Models have been using increasingly detailed formulations of decomposer biomass and its relationships with organic substrates and inorganic nutrients, but further efforts are necessary (Manzoni and Porporato 2009). For example, some models integrate microbial diversity with discrete compartments representing conceptual functional groups or identified taxonomic groups interacting in a food web (Louis et al. 2016). Half of those models have not been tested against an empirical dataset while the other half mainly consider fixed parameters. This is due to the difficulty to link taxonomic and functional diversity (Louis et al. 2016). Another major gap is the lack of mechanistic representations of small-scale processes. Further, the tendency of more recent models toward more sophisticated (and generally more mathematically complex) approaches is not always paralleled by improved model performance or ability to interpret observed patterns (Manzoni and Porporato 2009). Also, the predictive capability of SOM models developed in recent decades at spatial scales that interact with climate remains poor (Smith et al. 2018).

The Rothamsted carbon model (RothC) and the CENTURY model are two of the most widely used summary SOM models (Coleman and Jenkinson 1995; Parton et al. 1987). RothC is purely concerned with soil processes and not linked to a plant production model (Falloon and Smith 2009). In contrast, CENTURY is part of a larger ecosystem model that simulates crop, grass, and tree growth, and the effects of management practices on both plant production and SOM. Limitations for using the ecosystem scale to formulate SOM models are requirement for site-level data to drive and evaluate the models, and uncertain representation of the mechanistic relationship important at smaller scales (Campbell and Paustian 2015). DAYCENT and Ecosys are other SOM models at the ecosystem scale. On the other hand, microsite models depend on specific soil fractionation methods and are difficult to link to dynamics at large scales. Examples of microsite models are EnzModel, NICA, and INDISIM. Limitations of global SOM models are requirement for global-level data to drive and evaluate the models, and that the model complexity depends on the computational capacity. Examples of global models are the Community Land Model (CLM), IBIS, and TEM (Campbell and Paustian 2015).

In RothC, the quality of residue/litter entering the soil is defined using the ratio of decomposable plant material (DPM) to resistant plant material (RPM) (Falloon and Smith 2009). CENTURY uses the lignin:N ratio to define litter quality. In RothC, the SOC stock is split into four active compartments that decompose by a first-order process, and each have their own characteristic rate constants. In the order of decreasing maximum decomposition rates, the active compartments are DPM, RPM, microbial biomass (BIO), and humified organic matter (HUM). RothC assumes that a small amount of SOC is resistant to decomposition (inert organic matter [IOM]). In contrast, the CENTURY SOM submodel includes three SOM pools, two surface and two subsurface litter pools, and a surface microbial pool. All pools decompose following the first-order kinetics and have characteristic decomposition rate constants. The pools can be ordered in decreasing maximum decomposition rates as the soil metabolic litter pool, the surface metabolic litter pool, the active SOM pool, the surface microbial pool, the soil structural litter pool, the surface structural litter pool, the slow SOM pool, and the passive SOM pool. CENTURY also simulates leaching of OM, whereas RothC is not capable of modeling DOC loss from soils (Falloon and Smith 2009).

Both RothC and CENTURY are applicable to arable, grassland, and forest ecosystems (Falloon and Smith 2009). Both models simulate the effects of farmyard manure (FYM) application on SOM, but CENTURY can also simulate the impact of tillage, harvesting, organic amendments, irrigation, grazing, erosion, fire, and tree removal. RothC and CENTURY have rarely been applied to assess short-term aspects of the C cycle and soil C cycling on a small scale. The majority of applications have focused on evaluating RothC and CENTURY against datasets of changes in SOC over decadal to century timescales. Both models can simulate the effects of management on SOC in the long term. Large-scale applications of both RothC and CENTURY have also been performed (Falloon and Smith 2009).

Both RothC and CENTURY models were originally developed to predict turnover of C in topsoils. However, RothC has been modified into a multilayer model to describe the turnover of C in the top meter of soil (Jenkinson and Coleman 2008). The soil profile cannot be treated as a homogenous unit to predict the acceleration of SOC decomposition by global warming. However, both RothC and CENTURY fail to account for pH effects on SOM turnover (Falloon and Smith 2009). Further, most SOM models are unable to simulate SOC changes in permanently waterlogged, very dry, highly organic, and recent volcanic soils. Current SOM models may be further limited in their applicability to tropical ecosystems. Importantly, there are relatively few long-term experiments related to land-use change for validating SOM models rather than those for land-management changes (Falloon and Smith 2009).

The confidence in SOM model predictions is particulary limited as SOC stock measurements in long-term experiments are rarely replicated (Falloon and Smith 2003). Challenges in measuring SOC stocks also include obtaining representative undisturbed soil cores for different layer depths, accurately measure bulk density for conversion to SOC stock, and addressing the high spatial variation in SOC stocks (Falloon and Smith 2009). In addition, accurately measuring litter lignin concentration which is among the input variables for CENTURY is also challenging (Preston et al. 1997). Most importantly, the SOM pools in models are theoretical without measurable counterparts, and it is challenging to link measured fractions with modeled soil C pools. Herbst et al. (2018) reported that a particle-size fractionation can be used to estimate the RothC pools. Otherwise, whether a totally inert IOM pool exists which receives no C inputs and is resistant to decomposition is unclear as was discussed previously (Hammes et al. 2008; Von Lützow et al. 2006). Further, uncertainties are associated with the use of radiocarbon (^{14}C) age as input data in CENTURY and output data produced by RothC (Falloon and Smith 2009). Specifically, old (^{14}C age) and stable SOM is not necessarily chemically recalcitrant (Kleber et al. 2011). Further, the simplifying assumption that SOC stocks are near equilibrium in many models has been challenged by the observation that some old forest soils continue to accumulate SOC (Wutzler and Reichstein 2007). Similarly, agricultural soils may never reach a theoretical equilibrium SOC level because of changing conditions and partial resets by disturbances such as erosion (Polyakov and Lal 2004). For example, the assumption of SOC turnover equilibrium might not be valid for sites, which are under agricultural use for at least 50 years (Herbst et al. 2018).

Other than by modeling, the capacity of a soil to sequester SOC may be estimated as the difference between the existing SOC stock and the potential SOC saturation level. For example, C saturation deficits of pasture soils of New Zealand have been estimated at 12 and 15 mg C g^{-1} for 0–15 and 15–30 cm depths, respectively (Beare et al. 2014). In a follow-up study, the C saturation deficit to 15 cm depth was estimated as the difference between the stabilization capacity of individual soils and their current C concentration (McNally et al. 2017). A quantile (90th) regression model, based on the specific surface area and extractable aluminum (pyrophosphate) content of soils, provided the best prediction of the upper limit of fine fraction C (i.e., the stabilization capacity). For long-term pastures, the mean saturation deficit of allophanic soils (20.3 mg C g^{-1}) was greater than that of non-allophanic soils (16.3 mg C g^{-1}). Further, the saturation deficit of cropped soils across New Zealand was 1.14–1.89 times that of pasture soils in 0–15 cm depth (McNally et al. 2017).

Similarly, Fujisaki et al. (2018) calculated the SOC storage potential of tropical soils as the SOC saturation deficit, i.e., the difference between the actual SOC content and the maximum stabilized SOC content. They reported a large potential for SOC accumulation in cropland soils. However, changing agricultural practices in croplands did not allow soils to reach the SOC saturation line. Long-term studies on agroforestry, use of cover crops, or conservation agriculture may define new levels for the cropland saturation line. The positive relationship between total SOC content and the relative mass of fine silt + clay may be used as an indicator for the SOC storage potential in tropical soils (Fujisaki et al. 2018).

The SOC stock in a soil may eventually saturate at a maximum C saturation level depending on inherent physicochemical characteristics (Stewart et al. 2007). In soils close to their saturation value, any additional increases in C inputs may then remain as unprotected POM and rapidly cycled back to the atmosphere (Stewart et al. 2008). However, while data from some long-term agricultural field experiments indicate that SOC saturation does occur others did not observe saturation behavior, even after many years of additions of OM by manure application (Stewart et al. 2007; Blair et al. 2006a, b). Theoretically, continuous accumulation of SOC is possible (Wutzler and Reichstein 2007). Otherwise, the chemically and biochemically protected pools may be influenced by C saturation behavior even though the whole soil may not be saturated (Stewart et al. 2009). Once the chemically protected SOC pool is filled, added C may accumulate in the physically protected and in the non-protected fractions. Mineral-associated SOC stocks, in particular, eventually get saturated. The reactive mineral surface area is a finite resource in topsoils, and C saturation may also occur for SOC associated with the mineral phase (Séquaris et al. 2010). Thus, arbitrarily defined soil fractions may have different SOC saturation dynamics (Stewart et al. 2009).

2.3 Importance of Soil Organic Carbon for Soil Quality, Ecosystem Services, and Food Security

Soil is a critically important component of Earth's biosphere, functioning in the production of food, fiber, and fuel but also important to the maintenance of local, regional, and global environmental quality (Glanz 1995). About 40% of the world's agricultural land has been degraded by human activity as the result of soil erosion, atmospheric pollution, extensive soil cultivation, overgrazing, land clearing, salinization, and desertification (Oldeman 1994). Land degradation particularly reduces soil quality in agroecosystems. Soil quality implies its ability to meet multi-functional requirements, to provide ecosystem services and the potential of soil to affect other environmental media (Bone et al. 2010). Soil quality is defined as 'the capacity of a specific kind of soil to function, within natural or managed ecosystem boundaries, to sustain plant and animal productivity, maintain or enhance water and air quality, and support human health and habitation' by the Soil Science Society of America Ad Hoc Committee on soil quality (S-581). In the past, the emphasis of soil management has been on controlling erosion and minimizing the effects of soil loss on productivity but not on the multi-functionality of soil (Karlen et al. 2003).

2.3.1 Soil Organic Carbon, Soil Quality, and Soil Management

The SOC is a key indicator of soil quality as it affects essential biological, chemical, and physical soil functions such as nutrient cycling, pesticide and water retention, and soil structure maintenance (Karlen et al. 1997; Mueller et al. 2010). Further, the SOC stock and some associated physical and biological properties are significant to soil quality as they positively influence soil performance (Gregorich et al. 1994). Thus, SOC is most useful biologically when it decays (Janzen 2006). Soil quality depends to a great extent on the amount and quality of the SOC stock in the root zone (Lal 2004c), although quantitative evidence for the relation between root zone SOC stock and crop yield is scanty (Loveland and Webb 2003). Increases in the SOC stock improve soil quality by enhancing soil structure, reducing erosion, increasing water and nutrient reserves (Lal 2007). For example, compost as a proxy for SOC stock has a lower bulk density and a higher water holding capacity than mineral soil components which have lower SOC contents (Wander and Drinkwater 2000). Further, changes in SOM stocks affect aggregate stability, slaking, soil porosity, aeration, and erosion potential. Growth of both roots and shoots benefits from higher SOM stocks. The biologically active and slowly cycling SOC fractions are significant to farmers interested in soil quality (Wander and Drinkwater 2000). For example, active SOC fractions influence nutrient cycling, biological activity, and biologically mediated soil physical properties. Otherwise, passive SOC fractions contribute to soil physical condition and habitat quality. In particular, POM is a promising index of SOM status and because of its contribution to soil quality (Gregorich et al. 1994). The POM may be positively correlated with biologically active SOC fractions and aggregation (Wander and Drinkwater 2000). Thus, the SOC stock is among the proposed minimum dataset of indicators for soil quality at the field, farm, and watershed scale (Karlen et al. 1997). Further, trends in SOC stocks are among the minimum dataset of indicators for soil quality at regional or national scales. Indeed, SOC is a global key indicator of the quality of soils of agroecosystems (Mueller et al. 2010).

The numerous direct and indirect positive impacts of SOC on soil quality are altered by changes in SOC stocks (Lal 2007). Conversion of natural to agroecosystems causes the depletion of the SOC stock of up to 60% in temperate and up to 75% in tropical regions (Lal 2004c). Agriculture caused an estimated loss of 133 Pg SOC to 2 m depth over 12,000 year with losses of 26.1, 21.0, and 20.2 Pg SOC for grasslands, savannas, and croplands, respectively (Sanderman et al. 2017). About 25–30% of the SOC stored in the top meter of soil may be released by cultivation of native soils, whether under forest or prairie vegetation (Houghton 2010). For example, SOC stocks to 1 m depth decrease by 42% when native forest and by 59% when pasture is converted to cropland, respectively (Guo and Gifford 2002). In temperate regions, conversion from forest to cropland and from grassland to cropland may cause the loss of 31% SOC stock to 28.5 cm depth and of 36% SOC stock to 27.1 cm depth, respectively (Poeplau et al. 2011). In tropical regions, SOC losses of 25% to 36 cm depth and of 30% to 48 cm depth occur by conversion of primary forest to cropland or perennial crops, respectively (Don et al. 2011). Losses of 21% SOC stock to 39 cm depth occur when secondary tropical forest is converted to cropland, but no changes in SOC stock to 51 cm depth are observed when converted to perennial crops. When tropical grassland is converted to cropland, 10.4% of SOC stock is lost to 38 cm depth (Don et al. 2011). In the long-term, conversion of native vegetation to cropping results in a 43% decrease in SOC stocks to a depth of 17 cm (Kopittke et al. 2017). In contrast, conversion of native vegetation to pasture results in a long-term decrease in the SOC stock of only 11% to 27 cm depth. Further, SOC stocks increase by +9.2% (no-till) to 52 cm depth, +8.3% (organic agriculture) to 23 cm depth, and +26% (organic amendment) to 24 cm depth compared to conventional cropping systems (Kopittke et al. 2017).

The cropping practices, irrigation, use of fertilizers and different types of tillage affect the SOC stock. For example, conventional tillage management resulted in SOC losses to 15 cm depth in the US Corn Belt compared to conservation practices (Senthilkumar et al. 2009). Otherwise, little to no SOC is lost from pastures as these are generally not cultivated (Hopkins et al. 2009). However, intensively grazed pastures on flat land in New Zealand can lose large amounts of SOC to 1 m depth (Schipper et al. 2007). For example, flat dairy pastures lost SOC to 30 cm depth compared to flat pastures grazed by sheep and beef or grazed tussock grassland (Schipper et al. 2010). The SOC losses under dairying are probably related to high stocking rates, fertilization, and C export with products. Thus, both conversions of native soils to agriculture and cultivation of agricultural soils may reduce soil quality by decreasing SOC stocks. In degraded soils, SOC stocks can be partially restored through adoption of recommended management practices (RMPs) such as conservation tillage, residue mulching and use of cover crops, practices which all contribute to SOC accumulation and sequestration by an additional transfer of C from the atmosphere to the soil (Lal 2007; Powlson et al. 2011). Increasing SOC improves soil quality by improving soil structure and tilth, reducing oil erosion, increasing plant-available water capacity, storing plant nutrients, providing energy for soil biota, purifying water, denaturing pollutants, increasing soil biodiversity, improving crop/biomass yields, and moderating climate (Lal 2007). With an increase in soil C inputs, some of the more labile SOC fractions and soil microbial biomass initially increase (Herrick and Wander 1998). Eventually, changes in these biologically active SOC fractions result in an increase in total SOC stocks, and, thus, enhance soil quality (Monreal et al. 1998).

2.3.2 Soil Organic Carbon and Ecosystem Services

Human well-being depends on ecosystem services (ESs; MEA 2005), including (i) provisioning services such as food, water, timber, and fiber; (ii) regulating services that affect climate, floods, disease, wastes, and water quality; (iii) cultural services that provide recreational, aesthetic, and spiritual benefits; and (iv) supporting services such as soil formation, photosynthesis, and nutrient cycling. Agroecosystems are both providers and consumers of ESs (Power 2010). Specifically, agroecosystems provide humans with food, forage, bioenergy, and pharmaceuticals, and are essential to human well-being. Further, agroecosystems provide a variety of ESs including regulation of soil and water quality, C sequestration, support for biodiversity and cultural services. However, agroecosystems rely on services provided by natural, unmanaged ecosystems such as pollination, biological pest control, maintenance of soil structure and fertility, nutrient cycling, and hydrological services. The flow of certain ESs is affected by both conversion of natural ecosystems to agriculture and by the intensification of agriculture (Matson et al. 1997). Depending on agricultural management, trade-offs for ESs may occur such as the loss of wildlife habitat, nutrient runoff, sedimentation of waterways, GHG emissions, and pesticide poisoning of humans and non-target species. Thus, appropriate agricultural management practices are critical to realizing the benefits of agroecosystem services (Power 2010).

The ability of agroecosystems to provide ESs depends on soil quality (Lal 2010a). Specifically, the cycling of SOC can be managed to amplify ESs, improve water quality, increase biodiversity, and increase NPP (Lal 2010b). ESs are improved with increase in SOC stocks (Franzluebbers 2010). However, maximizing SOC 'stocks' is less critical than maintaining SOC 'flows' to sustain the manifold functions performed by agroecosystems (Janzen 2015). Most degraded and depleted soils of agroecosystems contain lower SOC stocks than those under natural ecosystems (Lal 2010c). By enhancing SOC sequestration, soil biological, chemical, and physical quality can be improved which in turn improves ESs (Lal 2009). Thus, restoring the SOC stock is essential to enhancing numerous agroecosystem services. Agricultural systems can be specifically managed to enhance SOC sequestration (Power 2010). For example, a net gain of the SOC stock can be achieved by conversion of plow tillage (PT) to no-till (NT) and other conservation tillage practices, along with crop residue mulch or cover crops. These conversions provide important ESs such as erosion control, water conservation, reduction in non-point source pollution, stabilization of crop yields against drought, and decrease in the C footprint of the agricultural system (Lal 2009). The climate services provided by ecosystems may be assessed by the greenhouse gas value (GHGV; Anderson-Teixeira and DeLucia 2011). The GHGV considers contributions from both flux and storage of C, uses proper accounting for timing of C emissions over a multiple-year of time frame, and allows for the quantification of probable C exchange arising from disturbance. Thus, GHGV may be an appropriate method to quantify the full GHG effects of land-use or land-cover change to assess the climate services provided by agroecosystems (Anderson-Teixeira and DeLucia 2011). To increase the supply of ESs from agroecosystems, payment for ecosystem services (PES) has been suggested as mechanism to provide farmers with incentives (Antle and Stoorvogel 2009). In particular in developing countries, PES may contribute to poverty alleviation, food security, and sustainability. Payments for agricultural SOC sequestration would potentially reduce poverty and enhance the sustainability of agricultural systems (Antle and Stoorvogel 2009).

2.3.3 Soil Organic Carbon and Food Security

Food security refers to socioeconomic conditions ‘when all people at all times have access to sufficient, safe, nutritious food to maintain a healthy and active life’ (FAO 1996). Many people are food insecure, i.e., more than 1 billion people lack sufficient dietary energy availability, and at least 2 billion people suffer micronutrient deficiencies (Barrett 2010). Thus, food security provides a useful goal toward which the world should strive (Pinstrup-Andersen 2009). Food security can be improved, particularly, by increasing the SOC concentration in the root zone as this increase is associated with gains in agronomic productivity (Lal 2010a; Loveland and Webb 2003). An SOC concentration of about 1.1% in the root zone may be essential for the creation of optimal agronomic conditions (Aune and Lal 1997). Positive impacts of SOC concentration on increase in agronomic yield of wheat, barley, maize, rice, cowpeas (*Vigna unguiculata* [L.] Walp.), and other food crops have been reported for a wide range of ecoregions (Lal 2010a). Thus, there is a potential correlation between SOC concentration in the root zone and grain/straw yields of food grains or other food staples (root crops). However, field experiments specifically conducted to test this relationship are scanty (Loveland and Webb 2003). The contribution of SOC concentration in the root zone to agronomic productivity is particularly high in soils (i) of coarse rather than heavy-texture, (ii) with lower than higher antecedent SOC concentrations, (iii) receiving lower rather than higher rates of chemical fertilizers, (iv) managed under rainfed rather than irrigated conditions, and (v) of poor rather than good quality (Lal 2010a).

In general, the level of SOC within a particular soil can have profound influences on the capacity of the soil to produce food, feed, fiber, and biofuel (Franzluebbers 2010). The expected rise in human population and the need for even more food to be produced on already stressed landscapes requires the widespread adoption of conservation agricultural systems. The reduction in tillage operations may result in higher surface SOC stocks which enhance agricultural productivity also due to non-nutrient attributes of SOM (Franzluebbers 2010). A major improvement in food security is possible by expansion of conservation agriculture which is currently practiced on only about 7% of the global cropland area (Kassam et al. 2009).

2.4 Mitigation of Climate Change

Climate change mitigation means implementing policies which reduce GHG emissions and enhance sinks (IPCC Glossary Working Group III). Thus, agroecosystems mitigate climate change when their use and management reduce the net anthropogenic GHG emissions. Mitigation can be achieved by enhancing sinks through maximizing the land-surface uptake of GHGs (Schulze et al. 2009). This can be expressed by the net greenhouse gas balance (NGB) after including non-CO_2 GHGs such as CH_4, N_2O, and CO in the net ecosystem carbon balance (NECB). Thus, a negative NGB of an agroecosystem implies that climate change is mitigated by a net uptake of GHGs.

Biogeochemical simulation models may be used for describing and quantifying the contribution of agricultural (crop and grassland) systems to C sequestration and GHG source/sink status (Brilli et al. 2017). However, predictions from different models show large variability. Based on a comparison of a range of biogeochemical models, Brilli et al. (2017) made recommendations how to improve model performance. Among them were the explicit representation of soil microbial biomass to drive SOM turnover, the effect of N shortage on SOM decomposition, the improvements related to the production and consumption of GHGs, and an adequate simulations of GHG transport in soil.

Humans have altered the natural exchange of GHGs between the land surface and the atmosphere. For example, C emissions by anthropogenic land use over the preindustrial Holocene have a profound impact on the global C cycle (Kaplan et al. 2011). Even before 1000 BC (3 ka), up to 102 Pg C, and by AD 1850, about 360 Pg C may have been emitted into the atmosphere by clearing woody vegetation to provide cropland (Ruddiman 2003, 2005). Over the last centuries and decades, changes in land use (i.e., conversion of native lands to croplands or pasture) have also caused a net release of C to the atmosphere (Houghton 2010). For example, from 1850 to 2000, land use and land-use change released an estimated 108–188 Pg C to the atmosphere (summarized in Houghton 2010). However, major unknowns are the rates of growth of global cropland and pasture area over the last centuries and decades. The dynamics of global agricultural land area is mainly driven by changes in pasture area, but this is not very well documented. When changes in agricultural management such as flexible sowing and harvesting, or tillage are included, net C releases in 1901–2014 are 20–30% higher than previously assumed (Arneth et al. 2017). However, current assessments may significantly overestimate both past and future anthropogenic C emissions from land by not properly addressing the erosion-induced SOC sink (Wang et al. 2017b). Also, past emissions of non-CO_2 GHGs from agroecosystems are not well known. In 2000, agricultural emissions outside of land-use change were (i) N_2O released from soils related to the application of N fertilizer (38% of total anthropogenic GHG emissions), (ii) CH_4 from livestock enteric fermentation, (iii) CH_4 and N_2O from manure management (38%), (iv) CH_4 from cultivation of rice (11%), and (v) CH_4 and N_2O from burning of savannah, forest, and agricultural residues (13%) (US EPA 2006). In total, agriculture accounted for 52 and 84% of global anthropogenic CH_4 and N_2O emissions, respectively (Smith et al. 2008).

Many agricultural practices can mitigate GHG emissions. The greatest potential exists by improving cropland and grazing land management, and restoring degraded lands and cultivated organic soils (Smith et al. 2008). In comparison, there is a lower mitigation potential through water and rice management, set-aside, land-use change and agroforestry, livestock management and manure management. With respect to cropland management, set-aside and land-use change have positive mitigation effects on CO_2, CH_4, and N_2O emissions with a high relative agreement in the literature and a high relative amount of data supporting the mitigation potential (Smith et al. 2008). Positive mitigation effects on CO_2 emissions with a high agreement were reported for agronomy, nutrient management, and agroforestry. However, there is only a medium potential through agronomy and nutrient management, and even lower through agroforestry. Cropland nutrient management has also a positive mitigation effect on N_2O emissions with a high agreement but medium evidence. Tillage/residue management has positive mitigation effects on CO_2 emissions, but agreement or consensus in the literature and evidence are both only medium (Smith et al. 2008).

Grazing land management/pasture improvement has a lower potential for mitigating GHG emissions than cropland management practices (Smith et al. 2008). For example, nutrient management, the measure with the highest potential, has a positive mitigation effect only on CO_2 emissions but agreement and evidence are both only medium. Increased productivity (e.g., fertilization), fire management and species introduction (including legumes) also mitigate CO_2 emissions from grazing land/pasture, but agreement in the literature is medium for increased productivity, and is low for both fire management and species introduction. The evidence is medium for species introduction but low for both increased productivity and fire management. The management of organic soils (e.g., avoided drainage of wetlands) positively mitigates CO_2 emissions at medium agreement and medium evidence. However, management of organic soils has a negative mitigation effect on CH_4 emissions. The restoration of degraded lands (e.g., erosion control, organic amendments, nutrient amendments) has a positive effect on CO_2 emissions and agreement is high and evidence medium (Smith et al. 2008). With respect to livestock management, improved feeding practices have the highest potential followed by specific agents and dietary additives, and longer-term structural and management changes and animal breeding. However, livestock management is only suitable for mitigating CH_4 emissions. Improved storage and handling, and anaerobic digestion of manure and biosolids have also positive mitigation effects on

CH_4 emissions with a high agreement but variable degree of evidence. Further, more efficient use of manure and biosolids as nutrient sources has positive mitigation effects on both CO_2 and N_2O emissions, and agreement is high and evidence medium. Finally, bioenergy (e.g., energy crops, solid, liquid, biogas, residues) has a positive mitigation effect on CO_2 emissions, and agreement is high and evidence medium (Smith et al. 2008).

Land-use change emissions from agricultural systems are generally more important than direct emissions of CH_4 and N_2O (Burney et al. 2010). However, the net effect of higher yields has avoided global emissions of up to 161 Pg C since 1961. Specifically, climatic impacts of historical agricultural intensification have been more significant compared to those of agricultural systems with lower inputs that instead expanded cropland to meet global demand for food. Thus, for mitigating agriculture's future contributions to climate change, continuing improvement of crop yields is paramount (Burney et al. 2010).

Enhancing GHG removals is another opportunity for mitigating GHGs in agriculture aside reducing GHG emissions (Smith et al. 2008). The potential of agricultural soils to remove CH_4 from the atmosphere by oxidation is, however, small compared with other GHG fluxes such as CO_2 (Smith and Conen 2004). Thus, strategies for mitigating and adaptation to the current climate forcing can be generated by accelerating the transfer of C from the atmosphere into both SIC and SOC for enhancement of soil C sinks (Macías and Arbestain 2010). This may be achieved by: (i) favouring growth of biomass which is the major source for SOC, (ii) promoting and facilitating carbonation processes to increase the SIC stock, (iii) reducing erosional C loss and favouring pedogenesis for buildup of the soil profile C stock, (iv) developing OM-rich horizons, (v) recovering degraded or contaminated soils to restore the soil C sink, and/or (vi) managing waste by use of systems that minimize emissions of GHGs. Some of the C lost in the past from agricultural soils by changes in land use and cultivation can be recovered through improved management, thereby withdrawing atmospheric CO_2 (Smith et al. 2008). Most agricultural soils contain 30–75% less C than their potential capacity as determined by soil, climate, terrain, drainage, land use, and soil and crop management practices (Lal and Follett 2009b). The world's agricultural and degraded soils may have a C sink capacity of 50–66% of the historic C loss (Lal 2004c). The term 'soil C sequestration' implies that the total C stock in the soil profile is increasing through managerial interventions aimed at transferring atmospheric CO_2 to the soil C stock by moderating either organic and/or inorganic transformations (Lal and Follett 2009b). Thus, practices that retains or returns more of the C captured by growing plants cause an increase in the soil C stock (Sanderman and Baldock 2010). Carbon sequestration in soil implies an additional transfer of C from the atmosphere to the soil and, thus, a genuine contribution to climate change mitigation (Powlson et al. 2011). However, as SOM contains also nutrients, SOC sequestration must also address limitations imposed by nutrient availability (van Groenigen et al. 2017; Janzen 2015).

The rate of C sequestration in agroecosystems with adoption of RMPs depends on soil texture and structure, rainfall, temperature, farming system, and soil management (Lal 2004c). The SOC stock in agroecosystems can be enhanced by increasing use efficiency of input, decreasing losses by erosion and leaching, and improving soil structure (Lal and Follett 2009b). Further, the SIC stock can be enhanced by application of biosolids, liming/application of cations, and conserving water in the root zone. However, the potential of SIC sequestration by pedogenic carbonate formation is less well known. Soils of range lands, grasslands and irrigated lands may sequester both SIC and SOC. The rate of soil C sequestration ranges from about 100 to 1,000 kg ha^{-1} yr^{-1} for SOC and 5–15 kg ha^{-1} yr^{-1} for SIC (Lal and Follett 2009a). However, some agricultural field trials indicate that the relative increase in soil C stock with adoption of RMPs is not an actual increase but rather due to a reduction or cessation of soil C losses (Sanderman and Baldock 2010). The global C sink capacity of agricultural pools is about 78 Pg C and can be filled at the potential maximum rate of about 1 Pg C yr^{-1}. However, the attainable and actual cumulative global rate of soil C sequestration may be lower because of managerial, economic, and policy constraints (Lal and Follett 2009a). The duration of soil C sequestration is 25–50 year (Lal 2004c). Intensifying agricultural activity for mitigating GHG emissions carries also substantial environmental costs (i.e., C costs) which must be weighed against the benefit of a net uptake of GHGs, and, in particular, the benefit of increasing the soil C stock (Matson et al. 1997; Schlesinger 2000). Also, with increasing saturation of the soil C stock, the sink activity diminishes and other strategies are required to mitigate GHG emissions by agricultural land use and management.

2.5 Adaptation to Climate Change

Adaptation measures are also required in agroecosystems to mitigate GHG emissions as the climate is changing and becoming more variable. Adaptation to climate change is defined as initiatives and measures to reduce the vulnerability of natural and human systems against actual or expected climate change effects (IPCC Glossary Working Group III). Climate change interacts with soil C dynamics by affecting (i) soil C input processes (e.g., photosynthesis, NPP, NBP), (ii) soil C stabilization processes (e.g., aggregation, interaction with the soil mineral phase, depth distribution), and (iii) soil C losses (e.g., erosion, decomposition, volatilization). Increase in atmospheric CO_2 directly alters biomass production whereas changes in climate regimes (i.e., temperature, precipitation) have direct effects on soil C cycling processes.

Adaptation of C stocks in agricultural soils to climate change is possible at (i) the field scale (e.g., changing machinery, timing of operation, use of different crops, introducing irrigation), (ii) the farm scale (e.g., socioeconomic changes affecting farm sizes and diversification of land uses), and (iii) the regional/national scale (e.g., agricultural policy responses; Rounsevell et al. 1999). The best strategies for adapting agricultural soils and, in particular, SIC and SOC sequestration to climate change are RMPs (Rounsevell et al. 1999; Lal 2004c). Common RMPs for SOC sequestration are mulch farming, conservation tillage, agroforestry and diverse cropping systems, cover crops, and integrated nutrient management, including the use of manure, compost, biosolids, improved grazing, and forest management (Lal 2004c).

Irrigation can be used to buffer against soil moisture deficits caused by climate change, and to sustain agricultural productivity. Where irrigation is constrained, the selection of drought-resistant crops or the introduction of RMPs that conserve soil water are adaptive measures. Further, adapting crop production to increase in temperature may also be required. For example, North American wheat production has been adapted in the past as wheat is now cultivated in environments once considered too arid, too variable, and too harsh to cultivate (Olmstead and Rhode 2011). This adaptation required new biological techniques (the Green Revolution), but further improvements in plant traits are required to adapt to climate change. For example, the development of new corn varieties that are high-temperature tolerant and have high thermal requirements may be beneficial to adapt corn production in the North China Plain (Tao and Zhang 2010).

Rising atmospheric CO_2 concentrations may directly alter crop yield (Ainsworth and McGrath 2010). Free-air CO_2 enrichment (FACE) experiments indicate that soybean and rice grain yield may increase by 13% at 550 ppm CO_2. In contrast, grain yield of sorghum and corn are not expected to increase at elevated CO_2 when water supply is adequate. Increased plant productivity at elevated CO_2 may result in an increase in SOC stocks at abundant N availability (de Graaff et al. 2006). However, when fertility levels are too low, progressive N limitation may limit production and soil C input, and when fertility is too high SOC may decrease by priming microbial decomposers. The type of plant mycorrhizal association and N-fixing capability also determines the response to elevated CO_2 (Terrer et al. 2018). Specifically, the additional growth in ectomycorrhizal plants at elevated CO_2 is partly offset by decreases in SOC via priming. The importance of the cost of N-acquisition needs to be considered to better understand the interactions between the C and N cycles under rising CO_2 (Terrer et al. 2018).

In the long term, responses of row crop agroecosystems to management practices may be more important than responses to elevated CO_2 (Moran and Jastrow 2010). Thus, how soil C sinks can be adapted to increasing atmospheric CO_2 concentrations is uncertain. Most FACE studies have been done in temperate regions, and interactions of elevated CO_2 with increased temperatures, drought stress, and rising ozone concentrations are also not accounted for (Ainsworth and McGrath 2010). For example, current tropospheric ozone levels already cause significant losses in crop yields. Thus, the effects of interacting climate change factors on crop production in global agroecosystems are uncertain. Whether crop production can be adapted to increasing CO_2 and ozone levels is also less well known.

Adaptation to the increased variability in temperature and rainfall means increasing the resilience of agricultural production systems (Smith and Olesen 2010). For example, soil water holding capacity can be increased by adding crop residues and manure. Resilience can also be achieved by diversifying crop rotations. Modifying the microclimate by adding shade and shelter such as in agroforestry systems may reduce effects of extreme high temperatures on crop productivity. Adaptation measures that reduce soil degradation, leaching of N and P, and conserve soil moisture can improve SOC storage (Smith and Olesen 2010).

Increases in temperature are among the major effects of climate change. Because of the slow response of the climate system, the equilibrium climate consistent with the current levels of GHGs may not be reached for many centuries (Matthews and Weaver 2010). Thus, global temperature may gradually increase and equilibrate with the current atmospheric imbalance some time in the future (Solomon et al. 2009). However, warming trends in air temperature are associated with those in soil temperature, especially in regions without snow cover (Qian et al. 2011). Thus, soil C fluxes have increased with the observed warming trend in recent decades, but it is unclear whether a net soil C loss has occurred or whether the efflux only represents an increase in the rate of C cycling (Bond-Lamberty and Thomson 2010; Smith and Fang 2010).

The vertical distribution in soil temperatures must also be considered to understand the turnover of SOC. For example, Koven et al. (2017) reported that the climatological temperature control on SOC turnover in the top meter of global soils is more sensitive in cold climates than in warm climates. The depletion of microbially accessible SOC pools, reductions in microbial biomass, a shift in microbial CUE, and changes in microbial community composition may all determine the response of SOC to warming (Melillo et al. 2017). OM decomposition increases with increase in soil temperature when decomposer activity is not limited by substrate availability or soil moisture (Davidson and Janssens 2006). As initially the soil surface warms some time before reaching deeper soil depths (Qian et al. 2011), a deeper SOC depth distribution in agricultural soils may be a potential strategy to adapt the SOC stock to surface temperature increases. This can be achieved, for example, by soil conservation practices to minimize disruption of vertical flow paths for DIC and DOC, and ensuring the continuity of vertical animal burrows and root channels as conduits for C to the subsoil (Lorenz and Lal 2005).

Planting varieties and species with deep and extended root systems and high root turnover may also result in an increase in the subsoil SOC stock (Rasse et al. 2005). A deeper depth distribution may also promote an increase in the proportion of protected and stabilized SOC in soil profiles as relatively more reactive mineral surfaces are available in deeper soil layers for adsorption, and decomposer activity is lower than closer to the soil surface. The SOC in deeper soil layers can accumulate over thousand years before it saturates (if soil C saturation exists; Mathieu et al. 2015). Deeper root growth and enhance exudation in low carbon, high mineral environments may lead to increases in the stock of mineral–organic association (MOA) C, either by coating organic free mineral surfaces or by enhancing primary weathering reactions which lead to fresh mineral production (Lynch and Wojciechowski 2015). However, root exudation in the presence of existing MOAs may have the opposite impact through the dissolution of existing MOAs and the release of the associated C (Porras et al. 2018). Otherwise, the intrinsic 're-calcitrance' of subsoil C may not be important as labile C can also be stabilized for decades to centuries by many mechanisms (Kleber et al. 2011).

However, the temperature sensitivity of subsoil C stabilization processes (i.e., interaction between substrates, microbial actors and abiotic driving variables) is unknown. Ultimately, subsoil layers may also warm following a continued increase in surface temperature and SOC lost by enhanced decomposition (Qian et al. 2011). Thus, in agroecosystems were decomposition is more enhanced by temperature increases than are both NPP and soil C input, the profile SOC stock may potentially decrease under continued warming.

Aside from a gradual increase in temperature, climate extremes have also been observed and are projected to increase in the future (Easterling et al. 2000). For example, droughts (i.e., below-normal precipitation over a period of months to years) have recently intensified over Africa, southern Europe, East and South Asia, and eastern Australia as natural climate variability has been exacerbated by climate change (Dai 2011). Again, a deeper depth distribution of SOC is an adaptive measure to reduce the increased risk of net C loss by wind and water erosion intensified by climate extremes. Further, enhancing aggregation in surface soil layers, keeping a residue cover on the soil surface and evergreen agriculture contribute to reducing the risk of erosional C losses by climate change (Garrity et al. 2010). Growing varieties and species that could better cope with climate extremes (e.g., heat-resistant or drought-resistant plants) is another adaptive measure. Increases in wildfire activity have been reported and/or are predicted for many regions of the world due to climate change (Fauria et al. 2011). Ecologically sustainable fire management practices are required to adapt agroecosystems to this increase and reduce soil C losses (Driscoll et al. 2010). However, major knowledge gaps are (i) a mechanistic understanding of the responses of a range of plants to fire regimes, (ii) how species are influenced by the spatial arrangement and temporal sequences of fires, and (iii) how factors such as herbivory, predation, fragmentation, invasive species, and weather interact with fire to alter species' responses to fire directly, or via changes to the fire regime. Thus, it is currently not possible to make accurate predictions of climate change effects on wildfires (Fauria et al. 2011).

2.6 Research Needs

Soils of agroecosystems play an important, increasingly recognized, and understood role in the global C cycle. However, while most studies have focused on SOC and on topsoil horizons, soil profile studies are also needed to fully address the role of agricultural SIC and SOC stocks in terrestrial C cycling, and the potential effects of climate change (Díaz-Hernández 2010; Jobbágy and Jackson 2000). Aside soils of arid and semiarid regions, sequestration of SIC occurs also in non-arid regions but additional studies are needed to comprehensively address the SIC sequestration potential of agricultural soils. Such studies are also beneficial to assessing the effects of agricultural practices such as soil and vegetation management on the SIC stock. Similarly, the effects of agricultural practices on the SOC stock are not fully known. For example, most studies in the past assessed aboveground C inputs such as litter and harvest residues, but it is now realized that belowground (i.e., roots, rhizodeposition) inputs are more important for the SOC balance of agroecosystems (Rasse et al. 2005; Denef and Six 2006; Johnson et al. 2006; Jones et al. 2009; Rumpel and Kögel-Knabner 2011). Aside inputs of root-derived C, dissolved C (i.e., DIC, DOC) and particulate C are pathways into the subsoil which must also be studied in agroecosystems. Leaching of dissolved C from soils may a critical component of the NECB, but studies for global agroecosystems are urgently needed to improve the understanding about the role of agroecosystem soils in the global C cycle (Kindler et al. 2011). A critical question that needs also to be addressed is the importance of erosion processes for SIC and SOC sequestration (Lal 2003).

The role of soil biota has been underappreciated although SOC sequestration is a biological process (Torn et al. 2009). Specifically, the loss of soil biodiversity strongly impacts SOC cycling processes (de Graaff et al. 2015). For example, loss of diversity of soil organisms reduces soil C respiration and plant tissue decomposition. The loss of microbial diversity, in particular, strongly reduces soil C respiration. The composition of the soil microbial community influences the SOC turnover in mineral soils (Don et al. 2017). Further, plant tissue decomposition is negatively affected by loss of soil faunal diversity. However, very few studies have assessed the relationship between soil biodiversity and SOC stocks (de Graaff et al. 2015). Similarly, relationships between soil food web structure and C cycling in soils are less well understood (Morriën et al. 2017). The soil microorganisms may play a central role in the land-atmosphere C exchange (Bardgett et al. 2008). Microbially derived materials may play a significant role in SOM stabilization (Kleber et al. 2011). How microbial community dynamics influence mineralization or stabilization of SOM needs additional research (Kutsch et al. 2009b). Long-term litter decomposition but more important root decomposition studies in agricultural soils are needed to improve our understanding about the subsoil SOC input processes. While aggregation is a major SOC stabilization mechanism in topsoil horizons, the biological, chemical, and physical stabilization processes in soil profiles are not fully understood (Rumpel and Kögel-Knabner 2011). Reactive mineral soil surface areas are probably needed for long-term SOC stabilization, but these are less well studied in soils of agroecosystems. To manage agricultural soils for C sequestration for mitigating and adapting to climate change, it is critically important to clarify whether saturation processes occur for both the SIC and the SOC stocks, and how both respond to the projected climate change. In this respect, more molecular-level studies on the temperature sensitivity of SOM decomposition are needed (Kleber et al. 2011). Further, nutrient constraints on SOC sequestration must be addressed.

Food security can be improved by gains in agronomic productivity through increases in SOC concentration in the root zone (Lal 2010a). However, long-term field experiments to test this relationship for the major food crops and food staples on the major soil types are needed. Such experiments must also be accompanied by an assessment of the effects of SOC gains in soil profiles on soil quality and ecosystem services. Current SOM models need to be improved by (i) linking the theoretical to measurable SOM pools, (ii) including subsoil SOC cycling processes,

(iii) incorporating the improved knowledge about SOC stabilization processes, (iv) modeling C turnover for major global soil types, land uses, and climates, and (v) establishing long-term field experiments for model validation. Long-term multifactorial field experiments manipulating CO_2 levels, temperature, soil moisture, ozone concentrations, and other environmental factors are needed for major agroecosystems. Such experiments will facilitate the identification of sustainable production systems and crop varieties and species adapted to climate change.

2.7 Conclusions

The global soil C stock contains about 3,000 Pg C in the form of SOC, 1,000 Pg C in permafrost and peat, and about 1,700 Pg C as SIC. These estimates can be improved by obtaining additional soil profile data covering the terrestrial land-surface area. While primary carbonates (LIC) in agricultural soils are derived from the soil parent material, secondary carbonates (PIC) are the result of weathering/dissolution of carbonate or Ca/Mg-bearing silicate parent materials followed by pedogenic carbonate precipitation of weathering products at deeper depths in the soil profile. Further, sequestration of inorganic C in agricultural soils occurs when Ca^{2+} and Mg^{2+} are supplied via fertilizers or other soil amendments. The formation of PIC can be enhanced through increase in biogenic processes by soil application of biosolids, materials rich in cations, lime, and amendments or conditioners that conserve water in the soil column. The PIC stock is a 'recalcitrant' reservoir for C, but the potential for sequestration of atmospheric CO_2 by PIC formation in soils of agroecosystems is not fully known. In contrast, the SOC stock is more rapidly exchanging C with the atmosphere. Major C inputs to SOC in agroecosystems are non-harvested biomass including roots, rhizodeposition, and addition of biosolids such as manure. Major losses occur by soil respiration during decomposition, through erosion, harvest, leaching, and fire. The SOC balance of agricultural soils depends on site-specific factors including climate, physicochemical characteristics, and soil and vegetation management. Formations of aggregates in topsoil horizons, and of organo-mineral complexes and organo-metal interactions in deeper soil horizons affect SOC sequestration in soils of agroecosystems. However, additional research is needed on how the SOC stock of agroecosystems eventually becomes saturated? The predictability of SOM models needs to be improved by validation using data from more long-term experiments in global agroecosystems, linking theoretical to measurable SOM fractions, simulating SOM dynamics in soil profiles, and by adjusting to the improved knowledge about SOM stabilization processes.

2.8 Review Questions

1. Describe the SIC cycle in agroecosystems.
2. How can sequestration of atmospheric CO_2 in the SIC stock of agricultural soils be managed?
3. What are the principal differences in agroecosystem SIC and SOC cycles?
4. Contrast and compare soil C stabilization processes.
5. What are benefits and risk of increasing SOC levels in agroecosystems?
6. How can soil C cycling be managed to enhance climate change mitigation?
7. Describe experiments to identify land use and soil management practices to adapt soil C to climate change.
8. What ecosystem services are specifically linked to SIC and SOC, and why?
9. Explain the relation between the SOC stock and soil health and soil quality.
10. What is the threshold level of the SOC stock in agroecosystems? How can the SOC stock be enhanced to a level above the threshold?

References

Adl S (2016) Rhizosphere, food security, and climate change: a critical role for plant-soil research. Rhizosphere 1:1–3

Ahmad W, Singh B, Dalal RC, Dijkstra FA (2015) Carbon dynamics from carbonate dissolution in Australian agricultural soils. Soil Res 53:144–153

Ainsworth EA, McGrath JM (2010) Direct effects of rising atmospheric carbon dioxide and ozone on crop yields. In: Lobell D, Burke M (eds) Climate change and food security. Springer, Dordrecht, The Netherlands, pp 109–130

Alexander M (1965) Biodegradation: problems of molecular recalcitrance and microbial fallibility. Adv Appl Microbiol 7:35–80

Amelung W, Brodowski S, Sandhage-Hofmann A, Bol R (2008) Combining biomarker with stable isotope analyses for assessing the transformation and turnover of soil organic matter. Adv Agron 100:155–250

Amiotte Suchet PA, Probst JL, Ludwig W (2003) Worldwide distribution of continental rock lithology: implications for the atmospheric/soil CO_2 uptake by continental weathering and alkalinity river transport to the oceans. Glob Biogeochem Cyc 17(2):1038. https://doi.org/10.1029/2002GB001891

Amthor JS (2010) From sunlight to phytomass: on the potential efficiency of converting solar radiation to phyto-energy. New Phytol 188:939–959

Anav A, Friedlingstein P, Beer C et al (2015) Spatiotemporal patterns of terrestrial gross primary production: a review. Rev Geophys 53:785–818. https://doi.org/10.1002/2015RG000483

Anderson-Teixeira KJ, DeLucia EH (2011) The greenhouse gas value of ecosystems. Glob Change Biol 17:425–438

Andreae MO (2004) Assessment of global emissions from vegetation fires. International Forest Fire News 31:112–121

Angert A, Yakir D, Rodeghiero M, Preisler Y, Davidson EA, Weiner T (2015) Using O_2 to study the relationships between soil CO_2 efflux and soil respiration. Biogeosciences 12:2089–2099

Antle JM, Stoorvogel JJ (2009) Payments for ecosystem services, poverty and sustainability: the case of agricultural soil carbon sequestration. In: Lipper L, Sakuyama T, Stringer R,

Zilberman D (eds) Payment for environmental services in agricultural landscapes. Springer, Dordrecht, The Netherlands, pp 133–161
Arneth A, Sitch S, Pongratz J et al (2017) Historical carbon dioxide emissions caused by land-use changes are possibly larger than assumed. Nat Geosci 10:79–86
Asner GP, Scurlock JMO, Hicke JA (2003) Global synthesis of leaf area index observations: implications for ecological and remote sensing studies. Global Ecol Biogeogr 12:191–205
Atomi H (2002) Microbial enzymes involved in carbon dioxide fixation. J Biosci Bioeng 94:497–505
Aubrey DP, Teskey RO (2009) Root-derived CO_2 efflux via xylem stream rivals soil CO_2 efflux. New Phytol 184:35–40
Aune J, Lal R (1997) Agricultural productivity in the tropics and critical limits of properties of oxisols, ultisols and alfisols. Trop Agric (Trinidad) 74:96–103
Austin AT, Méndez MS, Ballaré CL (2016) Photodegradation alleviates the lignin bottleneck for carbon turnover in terrestrial ecosystems. Proc Natl Acad Sci USA 113:4392–4397
Ayres E, Wall DH, Bardgett RD (2009) Trophic interactions and their implications for soil carbon fluxes. In: Kutsch WL, Bahn M, Heinemeyer A (eds) Soil carbon dynamics: an integrated methodology. Cambridge University Press, Cambridge, UK, pp 187–206
Bachmann J, Guggenberger G, Baumgartl Th, Ellerbrock R, Urbanek E, Goebel M-O, Kaiser K, Horn R, Fischer WR (2008) Physical carbon-sequestration mechanisms under special consideration of soil wettability. J Plant Nutr Soil Sci 171:14–26
Bahn M, Janssens IA, Reichstein M, Smith P, Trumbore SE (2010) Soil respiration across scales: towards an integration of patterns and processes. New Phytol 186:292–296
Bais HP, Weir TL, Perry LG, Gilroy S, Vivanco JM (2006) The role of root exudates in rhizosphere interactions with plants and other organisms. Annu Rev Plant Biol 57:233–266
Baldock JA, Nelson PN (2000) Soil organic matter. In: Sumner ME (ed) Handbook of soil science. CRC Press, Boca Raton, FL, pp B25–B84
Balvanera P, Pfisterer AB, Buchmann N, He J-S, Nakashizuka T, Raffaelli D, Schmid B (2006) Quantifying the evidence for biodiversity effects on ecosystem functioning and services. Ecol Lett 9:1146–1156
Bardgett RD (2005) The biology of soil: a community and ecosystem approach. Oxford University Press, Oxford, UK
Bardgett RD, Freeman C, Ostle NJ (2008) Microbial contributions to climate change through carbon cycle feedbacks. ISME J 2:805–814
Barnes RT, Raymond PA (2009) The contribution of agricultural and urban activities to inorganic carbon fluxes within temperate watersheds. Chem Geol 266:318–327
Barnes PW, Throop HL, Archer SR, Breshears DD, McCulley RL, Tobler MA (2015) Sunlight and soil–litter mixing: drivers of litter decomposition in drylands. In: Lüttge U, Beyschlag W (eds) Progress in botany 76. Springer Switzerland, pp 273–302
Barré P, Durand H, Chenu C et al (2017) Geological control of soil organic carbon and nitrogen stocks at the landscape scale. Geoderma 285:50–56
Barrett CB (2010) Measuring food insecurity. Science 327:825–828
Batjes NH (1996) Total carbon and nitrogen in the soils of the world. Eur J Soil Sci 47:151–163
Batjes NH (2016) Harmonized soil property values for broad-scale modelling (WISE30sec) with estimates of global soil carbon stocks. Geoderma 269:61–68
Beare MH, Parmelee RW, Hendrix PF, Cheng W, Coleman DC, Crossley DA (1992) Microbial and faunal interactions and effects on litter nitrogen and decomposition in agroecosystems. Ecol Monogr 62:569–591
Beare MH, McNeill SJ, Curtin D et al (2014) Estimating the organic carbon stabilisation capacity and saturation deficit of soils: a New Zealand case study. Biogeochemistry 120:71–87
Beer C, Reichstein M, Tomelleri E, Ciais P, Jung M, Carvalhais N, Rödenbeck C, Altaf Arain M, Baldocchi D, Bonan GB, Bondeau A, Cescatti A, Lasslop G, Lindroth A, Lomas M, Luyssaert S, Margolis H, Oleson KW, Roupsard O, Veenendal E, Viovy N, Williams C, Ian Woodward F, Papale D (2010) Terrestrial gross carbon dioxide uptake: global distribution and covariation with climate. Science 329:834–838

Berg B, McClaugherty C (2008) Plant litter: decomposition, humus formation, carbon sequestration. Springer, Berlin

Berner RA (2003) The long-term carbon cycle, fossil fuels and atmospheric composition. Nature 426:323–326

Billings SA, Buddemeier RW, deB Richter D, Van Oost K, Bohling G (2010) A simple method for estimating the influence of eroding soil profiles on atmospheric CO_2. Global Biogeochem Cyc 24, GB2001. https://doi.org/10.1029/2009gb003560

Birkeland PW (1999) Soils and geomorphology. Oxford Univ Press, New York

Blagodatskaya E, Kuzyakov Y (2008) Mechanisms of real and apparent priming effects and their dependence on soil microbial biomass and community structure: critical review. Biol Fert Soils 45:115–131

Blair N, Faulkner RD, Till AR, Körschens M, Schulz E (2006a) Long-term management impacts on soil C, N and physical fertility Part II: Bad Lauchstädt static and extreme FYM experiments. Soil Till Res 91:39–47

Blair N, Faulkner RD, Till AR, Poulton PR (2006b) Long-term management impacts on soil C, N and physical fertility Part I: broadbalk experiment. Soil Till Res 91:30–38

Bocock KL, Gilbert O (1957) The disappearance of leaf litter under different woodland conditions. Plant Soil 9:179–185

Bolan NS, Adriano DC, Kunhikrishnan A, James T, McDowell R, Senesi N (2011) Dissolved organic matter: biogeochemistry, dynamics, and environmental significance in soils. Adv Agron 110:1–75

Bond-Lamberty B, Thomson A (2010) Temperature-associated increases in the global soil respiration record. Nature 464:579–583

Bond WJ, Woodward FI, Midgley GF (2005) The global distribution of ecosystems in a world without fire. New Phytol 165:525–538

Bone J, Head M, Barraclough D, Archer M, Scheib C, Flight D, Voulvoulis N (2010) Soil quality assessment under emerging regulatory requirements. Environ Int 36:609–622

Boquet E, Boronat A, Ramos-Cormenzana A (1973) Production of calcite (calcium carbonate) crystals by soil bacteria is a general phenomenon. Nature 246:527–529

Borland AM, Griffiths H, Hartwell J, Smith JAC (2009) Exploiting the potential of plants with crassulacean acid metabolism for bioenergy production on marginal lands. J Exp Bot 60:2879–2896

Borrelli P, Robinson DA, Fleischer LR et al (2017) An assessment of the global impact of 21st century land use change on soil erosion. Nat Commun 8:2013. https://doi.org/10.1038/s41467-017-02142-7

Bossuyt H, Six J, Hendrix PF (2005) Protection of soil carbon by microaggregates within earthworm casts. Soil Biol Biochem 37:251–258

Bousquet P, Ciais P, Miller JB, Dlugokencky EJ, Hauglustaine DA, Prigent C, Van Der Werf GR, Peylin P, Brunke E-G, Carouge C, Langenfelds RL, Lathière J, Papa F, Ramonet M, Schmidt M, Steele LP, Tyler SC, White J (2006) Contribution of anthropogenic and natural sources to atmospheric methane variability. Nature 443:439–443

Bradford JB, Hicke JA, Lauenroth WK (2005) The relative importance of light-use efficiency modifications from environmental conditions and cultivation for estimation of large-scale net primary productivity. Remote Sense Environ 96:246–255

Braissant O, Verrecchia EP, Aragno M (2002) Is the contribution of bacteria to terrestrial carbon budget greatly underestimated? Naturwissenschaften 89:366–370

Brilli L, Luca Bechini L, Bindi M et al (2017) Review and analysis of strengths and weaknesses of agro-ecosystem models for simulating C and N fluxes. Sci Tot Environ 598:445–470

Brookes PC, Chen YF, Chen L et al (2017) Is the rate of mineralization of soil organic carbon under microbiological control? Soil Biol Biochem 112:127–139

Bronick CJ, Lal R (2005) Soil structure and management: a review. Geoderma 124:3–22

Burney JA, Davis SJ, Lobell DB (2010) Greenhouse gas mitigation by agricultural intensification. Proc Natl Acad Sci USA 107:12052–12057

Busch FA, Sage RF, Farquhar GD (2017) Plants increase CO_2 uptake by assimilating nitrogen via the photorespiratory pathway. Nat Plants. https://doi.org/10.1038/s41477-017-0065-x

Cailleau G, Braissant O, Verrecchia EP (2004) Biomineralization in plants as a long-term carbon sink. Naturwissenschaften 91:191–194

Cameron DD (2010) Arbuscular mycorrhizal fungi as (agro)ecosystem engineers. Plant Soil 333:1–5

Campbell EE, Paustian K (2015) Current developments in soil organic matter modeling and the expansion of model applications: a review. Environ Res Lett 10:123004. https://doi.org/10.1088/1748-9326/10/12/123004

Campbell JE, Berry JA, Seibt U et al (2017) Large historical growth in global terrestrial gross primary production. Nature 544:84–87

Canarini A, Kiær LP, Dijkstra FA (2017) Soil carbon loss regulated by drought intensity and available substrate: a meta-analysis. Soil Biol Biochem 112:90–99

Cardinale BJ, Srivastava DS, Duffy JE, Wright JP, Downing AL, Sankaran M, Jouseau C (2006) Effects of biodiversity on the functioning of trophic groups and ecosystems. Nature 443:989–992

Cerling T (1984) The stable isotopic composition of modern soil carbonate and its relation to climate. Earth Planet Lett 71:229–240

Chang J, Ciais P, Herrero M, Havlik P, Campioli M, Zhang X, Bai Y, Viovy N, Joiner J, Wang X, Peng S, Yue C, Piao S, Wang T, Hauglustaine DA, Soussana JF, Peregon A, Kosykh N, Mironycheva-Tokareva N (2016) Combining livestock production information in a pro-cess-based vegetation model to reconstruct the history of grassland management. Biogeo-sciences 13:3757–3776

Chantigny MH (2003) Dissolved and water-extractable organic matter in soils: a review on the influence of land use and management practices. Geoderma 113:357–380

Chapin FS III, Eviner VT (2014) Biogeochemical interactions governing terrestrial net primary production. In: Holland H, Turekian K (eds) Treatise on geochemistry. Elsevier, Amsterdam, Netherlands, pp 189–216

Chapin FS III, Matson PA, Mooney HA (2002) Principles of terrestrial ecosystem ecology. Springer, New York

Chappell A, Baldock J, Sanderman J (2016) The global significance of omitting soil erosion from soil organic carbon cycling schemes. Nat Clim Change 6:187–191. https://doi.org/10.1038/NCLIMATE2829

Chen M, Rafique R, Asrar GR et al (2017) Regional contribution to variability and trends of global gross primary productivity. Environ Res Lett 12:105005

Chen S, Huang Y, Zou J et al (2010) Modeling interannual variability of global soil respiration from climate and soil properties. Agric For Meteorol 150:590–605

Chen T, van der Werf GR, Gobron N, Moors EJ, Dolman AJ (2014) Global cropland monthly gross primary production in the year 2000. Biogeosciences 11:3871–3880

Chen Z, Pawluk S, Ng Juma (1997) Impact of variations in granular structures on carbon sequestration in two Alberta mollisols. In: Lal R, Kimble JM, Follett RF, Stewart BA (eds) Soil processes and the carbon cycle. CRC Press, Boca Raton, FL, pp 225–243

Cheng L, Zhang L, Wang YP et al (2017) Recent increases in terrestrial carbon uptake at little cost to the water cycle. Nat Commun 8:110. https://doi.org/10.1038/s41467-017-00114-5

Christensen BT (1996) Carbon in primary and secondary organomineral complexes. In: Carter MR, Stewart BA (eds) Structure and organic matter storage in agricultural soils. CRC Press, Boca Raton, FL, pp 97–165

Churkina G, Zaehle S, Hughes J, Viovy N, Chen Y, Jung M, Heumann BW, Ramankutty N, Heimann M, Jones C (2010) Interactions between nitrogen deposition, land cover conversion, and climate change determine the contemporary carbon balance of Europe. Biogeosciences 7:2749–2764

Ciais P, Friedlingstein P, Friend A, Schimel DS (2001) Integrating global models of terrestrial primary productivity. In: Roy A, Saugier B, Mooney HA (eds) Terrestrial global productivity. Academic Press, San Diego, CA, pp 449–478

Ciais P, Soussana J-F, Vuichard N, Luyssaert S, Don A, Janssens IA, Piao SL, Dechow R, Lathière J, Maignan F, Wattenbach M, Smith P, Ammann C, Freibauer A, Schulze E-D, Synthesis Team CARBOEUROPE (2010a) The greenhouse gas balance of European grasslands. Biogeosciences Discuss 7:5997–6050

Ciais P, Wattenbach M, Vuichard N, Smith P, Piao SL, Don A, Luyssaert S, Janssens IA, Bondeau A, Dechow R, Leip A, Smith PC, Beer C, Van DerWerf GR, Gervois S, Van Oost K, Tomelleri E, Freibauer A, Schulze E-D, Synthesis Team CARBOEUROPE (2010b) The European carbon balance. Part 2: croplands. Glob Change Biol 16:1409–1428

Cihacek LJ, Ulmer MG (2002) Effects of tillage on inorganic carbon storage in soils of the northern great plains of the U.S. In: Kimble JM, Lal R, Folett RF (eds) Agricultural practices and policies for carbon sequestration in soil. Lewis Publishers, CRC Press, Boca Raton, Fl, pp 63–69

Coleman K, Jenkinson DS (1995) ROTHC-26.3. A model for the turnover of carbon in soil. Model description and user's guide. Lawes Agricultural Trust, Harpenden, UK

Coops NC, Hilker T, Hall FG, Nichol CJ, Drolet GG (2010) Estimation of light-use efficiency f terrestrial ecosystems from space: a status report. Bioscience 60:788–797

Cornwell WK, Cornelissen JHC, Amatangelo K, Dorrepaal E, Eviner VT, Godoy O, Hobbie SE, Hoorens B, Kurokawa H, Pérez-Harguindeguy N, Quested HM, Santiago LS, Wardle DA, Wright IJ, Aerts R, Allison SD, van Bodegom P, Brovkin V, Chatain A, Callaghan TV, Díaz S, Garnier E, Gurvich DE, Kazakou E, Klein JA, Read J, Reich PB, Soudzilovskaia NA, Victoria Vaieretti M, Westoby M (2008) Plant species traits are the predominant control on litter decomposition rates within biomes worldwide. Ecol Lett 11:1065–1071

Cotrufo MF, Del Galdo I, Piermatteo D (2009) Litter decomposition: concepts, methods and future perspectives. In: Kutsch WL, Bahn M, Heinemeyer A (eds) Soil carbon dynamics: an integrated methodology. Cambridge University Press, Cambridge, UK, pp 76–90

Cramer W, Olson RJ, Prince SD, Scurlock JMO, Members of the Global Primary Production Data Initiative (2001) Determining present patterns of global productivity. In: Roy J, Saugier B, Mooney HA (eds) Terrestrial global productivity. Academic Press, San Diego, CA, pp 429–448

Crutzen PJ, Andreae MO (1990) Biomass burning in the tropics: impact on atmospheric chemistry and biogeochemical cycles. Science 250:1669–1678

Dai A (2011) Drought under global warming: a review. WIREs Clim Change 2:45–65

Davidson EA, Janssens IA (2006) Temperature sensitivity of soil carbon decomposition and feedbacks to climate change. Nature 440:165–173

de Graaff MA, Adkins J, Kardol P, Throop HL (2015) A meta-analysis of soil biodiversity impacts on the carbon cycle. Soil 1:257–271

de Graaff MA, van Groenigen K-J, Six J, Hungate B, van Kessel C (2006) Interactions between plant growth and soil nutrient cycling under elevated CO2: a meta-analysis. Glob Change Biol 12:2077–2091

Denef K, Six J (2006) Contributions of incorporated residue and living roots to aggregate-associated and microbial carbon in two soils with different clay mineralogy. Eur J Soil Sci 57:774–786

Denef K, Stewart CE, Brenner J, Paustian K (2008) Does long-term center-pivot irrigation increase soil carbon stocks in semi-arid agro-ecosystems? Geoderma 145:121–129

Denman KL, Brasseur G, Chidthaisong A, Ciais P, Cox PM, Dickinson RE, Hauglustaine D, Heinze C, Holland E, Jacob D, Lohmann U, Ramachandran S, da Silva Dias PL, Wofsy SC, Zhang X (2007) Couplings between changes in the climate system and biogeochemistry. In: Intergovernmental Panel on Climate Change (ed) Climate change 2007: the physical science basis, Chap. 7. Cambridge University Press, Cambridge

Dennis PG, Miller AJ, Hirsch PR (2010) Are root exudates more important than other sources of rhizodeposits in structuring rhizosphere bacterial communities? FEMS Microbial Ecol 72:313–327

Díaz-Hernández JL (2010) Is soil carbon storage underestimated? Chemosphere 80:346–349
Dietzel R, Liebman M, Archontoulis S (2017) A deeper look at the relationship between root carbon pools and the vertical distribution of the soil carbon pool. SOIL 3:139–152
Doetterl S, Berhe AA, Nadeu E, Wang Z, Sommer M, Fiener P (2016) Erosion, deposition and soil carbon: a review of process-level controls, experimental tools and models to address C cycling in dynamic landscapes. Earth Sci Rev 154:102–122
Don A, Böhme IH, Dohrmann AB, Poeplau C, Tebbe CC (2017) Microbial community composition affects soil organic carbon turnover in mineral soils. Biol Fertil Soils 53:445–456. https://doi.org/10.1007/s00374-017-1198-9
Don A, Schumacher J, Freibauer A (2011) Impact of tropical land-use change on soil organic carbon stocks—a meta-analysis. Glob Change Biol 17:1658–1670
Dong X, Hao Q, Li G, Lin Q, Zhao X (2017) Contrast effect of long-term fertilization on SOC and SIC stocks and distribution in different soil particle-size fractions. J Soils Sediments 17:1054–1063
Driscoll DA, Lindenmayer DB, Bennett AF, Bode M, Bradstock RA, Cary GJ, Clarke MF, Dexter N, Fensham R, Friend G, Gill M, James S, Kay G, Keith DA, MacGregor C, Russell-Smith J, Salt D, Watson JEM, Williams RJ, York A (2010) Fire management for biodiversity conservation: key research questions and our capacity to answer them. Biol Conserv 143:1928–1939
Dungait JA, Hopkins DW, Gregory AS, Whitmore AP (2012) Soil organic matter turnover is governed by accessibility not recalcitrance. Glob Change Biol 18:1781–1796
Duniway MC, Herrick JE, Monger HC (2007) The high water-holding capacity of petrocalcic horizons. Soil Sci Soc Am J 71:812–819
DuPont ST, Beniston J, Glover JD, Hodson A, Culman SW, Lal R, Ferris H (2014) Root traits and soil properties in harvested perennial grassland, annual wheat, and never-tilled annual wheat. Plant Soil 381:405–420
Easterling DR, Meehl GA, Parmesan C, Changnon SA, Karl TR, Mearns LO (2000) Climate extremes: observations, modeling, and impacts. Science 289:2068–2074
Edwards CA, Bohlen PJ (1996) Biology and ecology of earthworms. Chapman and Hall, London
Edwards DP, Lim F, James RH et al (2017) Climate change mitigation: potential benefits and pitfalls of enhanced rock weathering in tropical agriculture. Biol Lett 13:20160715. https://doi.org/10.1098/rsbl.2016.0715
Ekschmitt K, Kandeler E, Poll C, Brune A, Buscot F, Friedrich M, Gleixner G, Hartmann A, Kästner M, Marhan S, Miltner A, Scheu S, Wolters V (2008) Soil-carbon preservation through habitat constraints and biological limitations on decomposer activity. J Plant Nutr Soil Sci 171:27–35
Elbert W, Weber B, Büdel B, Andreae MO, Pöschl U (2009) Microbiotic crusts on soil, rock and plants: neglected major players in the global cycles of carbon and nitrogen? Biogeosciences Discuss 6:6983–7015
Emmerich WE (2003) Carbon dioxide fluxes in a semiarid environment with high carbonate soils. Agr For Meteorol 116:91–102
Erb KH, Kastner T, Plutzar C et al (2018) Unexpectedly large impact of forest management and grazing on global vegetation biomass. Nature 553:73–76
Eshel G, Fine P, Singer MJ (2007) Total soil carbon and water quality: an implication for carbon sequestration. Soil Sci Soc Am J 71:397–405
Eswaran H, Reich PF, Kimble JM (2000) Global carbon stocks. In: Lal R, Kimble JM, Eswaran H, Stewart BA (eds) Global climate change and pedogenic carbonates. CRC Press, Boca Raton, FL, pp 15–25
Ewing SA, Sandermann J, Baisden WT, Wang Y, Amundson R (2006) Role of large-scale soil structure in organic carbon turnover: evidence from California grassland soils. J Geophys Res 111:G03012. https://doi.org/10.1029/2006JG000174
Falloon P, Smith P (2003) Accounting for changes in soil carbon under the Kyoto Protocol: need for improved long-term data sets to reduce uncertainty in model projections. Soil Use Manage 19:265–269

Falloon P, Smith P (2009) Modelling soil carbon dynamics. In: Kutsch WL, Bahn M, Heinemeyer A (eds) Soil carbon dynamics: an integrated methodology. Cambridge University Press, Cambridge, UK, pp 221–244

Fan Y, Miguez-Macho G, Jobbágy EG, Jackson RB, Otero-Casal C (2017) Hydrologic regulation of plant rooting depth. Proc Natl Acad Sci USA 114:10572–10577

FAO Food and Agriculture Organization of the United Nations (1996) Rome declaration on world food security, FAO, Rome. http://www.fao.org/wfs/index_en.htm

Farquhar GD (1989) Models of integrated photosynthesis of cells and leaves. Phil Trans R Soc B 323:357–367

Fauria MM, Michaletz ST, Johnson EA (2011) Predicting climate change effects on wildfires requires linking processes across scales. WIREs Clim Change 2:99–112

Feng X, Hills KM, Simpson AJ, Whalen JK, Simpson MJ (2011) The role of biodegradation and photo-oxidation in the transformation of terrigenous organic matter. Org Geochem 42:262–274

Fierer N, Schimel JP, Holden PA (2003) Variations in microbial community composition through two soil depth profiles. Soil Biol Biochem 35:167–176

Fisher J, Diggle A, Bowden B (2003) Quantifying the acid balance for broad-range agricultural systems. In: Rengel Z (ed) Handbook of soil acidity. CRC Press, Boca Raton, FL, pp 117–133

Foereid B, Rivero MJ, Primo O, Ortiz I (2011) Modelling photodegradation in the global carbon cycle. Soil Biol Biochem 43:1383–1386

Fontaine S, Barot S, Barré P, Bdioui N, Mary B, Rumpel C (2007) Stability of organic carbon in deep soil layers controlled by fresh carbon supply. Nature 450:277–281

Fortunel C, Garnier E, Joffre R, Kazakou E, Quested H, Grigulis K, Lavorel S, Ansquer P, Castro H, Cruz P, Doležal J, Eriksson O, Freitas H, Golodets C, Jouany C, Kigel J, Kleyer M, Lehsten V, Leps J, Robson M, Sternberg M, Theau J-P, Thébault A, Zarovali M (2009) Leaf traits capture the effects of land use changes and climate on litter decomposability of grasslands across Europe. Ecology 90:598–611

Fox O, Vetter S, Ekschmitt K, Wolters V (2006) Soil fauna modifies the recalcitrance-persistence relationship of soil carbon pools. Soil Biol Biochem 38:1353–1363

Franzluebbers AJ (2010) Will we allow soil carbon to feed our needs? Carbon Manage 1:237–251

Fridley JD, Grime JP, Bilton M (2007) Genetic identity of interspecific neighbours mediates plant responses to competition and environmental variation in a species-rich grassland. J Ecol 95:908–915

Fujisaki K, Chapuis-Lardy L, Albrecht A et al (2018) Data synthesis of carbon distribution in particle size fractions of tropical soils: Implications for soil carbon storage potential in croplands. Geoderma 313:41–51

Garbaye J (1994) Helper bacteria—a new dimension to the mycorrhizal symbiosis. New Phytol 128:197–210

Garrity DP, Akinnifesi FK, Ajayi OC, Weldesemayat SG, Mowo JG, Kalinganire A, Larwanou M, Bayala J (2010) Evergreen agriculture: a robust approach to sustainable food security in Africa. Food Sec 2:197–214

Gartner TB, Cardon ZG (2004) Decomposition dynamics in mixed-species leaf litter. Oikos 104:230–246

Gaudinski JB, Trumbore SE, Davidson EA, Zheng S (2000) Soil carbon cycling in a temperate forest: radiocarbon-based estimates of residence times, sequestration rates and partitioning of fluxes. Biogeochemistry 51:33–69

Giardina CP, Ryan MG (2000) Evidence that decomposition rates of organic carbon in mineral soil do not vary with temperature. Nature 404:858–861

Gifford RM (2003) Plant respiration in productivity models: conceptualisation, representation and issues for global terrestrial carbon-cycle research. Funct Plant Biol 30:171–186

Gill RA, Jackson RB (2000) Global patterns of root turnover for terrestrial ecosystems. New Phytol 147:13–31

Gill RA, Kelly RH, Parton WJ, Day KA, Jackson RB, Morgan JA, Scurlock JMO, Tieszen LL, Castle JV, Ojima DS, Zhang XS (2002) Using simple environmental variables to estimate belowground productivity in grasslands. Glob Ecol Biogeogr 11:79–86

Gilmanov TG, Aires L, Barcza Z, Baron VS, Belelli L, Beringer J, Billesbach D, Bonal D, Bradford J, Ceschia E, Cook D, Corradi C, Frank A, Gianelle D, Gimeno C, Gruenwald T, Guo H, Hanan N, Haszpra L, Heilman J, Jacobs A, Jones MB, Johnson DA, Kiely G, Li S, Magliulo V, Moors E, Nagy Z, Nasyrov M, Owensby C, Pinter K, Pio C, Reichstein M, Sanz MJ, Scott R, Soussana J-F, Stoy PC, Svejcar T, Tuba Z, Zhou G (2010) Productivity, respiration, and light-response parameters of world grassland and agroecosystems derived from flux-tower measurements. Rangeland Ecol Manage 63:16–39

Glanz JT (1995) Saving our soil: solutions for sustaining earth's vital resource. Johnson Books, Boulder, CO, USA

Goddard MA, Mikhailova EA, Post CJ, Schlautman MA, Galbraith JM (2009) Continental United States atmospheric wet calcium deposition and soil inorganic carbon stocks. Soil Sci Soc Am J 73:989–994

Goebel M-O, Woche SK, Bachmann J (2009) Do soil aggregates really protect encapsulated organic matter against microbial decomposition? Biologia 64:443–448

Gombert P (2002) Role of karstic dissolution in global carbon cycle. Global Planet Change 33:177–184

Goudriaan J, Groot JJR, Uithol PWJ (2001) Productivity of agro-ecosystems. In: Roy J, Saugier B, Mooney HA (eds) Terrestrial global productivity. Academic Press, San Diego, CA, pp 301–313

Gray CM, Monson RK, Fierer N (2010) Emissions of volatile organic compounds during the decomposition of plant litter. J Geophys Res 115:G03015. https://doi.org/10.1029/2010JG001291

Greenhouse Gas Working Group (2010) Agriculture's role in greenhouse gas emissions & capture. Madison, WI, USA, Greenhouse Gas Working Group Rep. ASA, CSSA, and SSSA

Gregorich EG, Carter MR, Angers DA, Monreal CM, Ellert BH (1994) Towards a minimum data set to assess soil organic matter quality in agricultural soils. Can J Soil Sci 74:367–385

Grover SP, Butterly CR, Wang X, Tang C (2017) The short-term effects of liming on organic carbon mineralization in two acidic soils as affected by different rates and application depths of lime. Biol Fertil Soils 53:431–443. https://doi.org/10.1007/s00374-017-1196-y

Guanter L, Zhang Y, Jung M, Joiner J, Voigt M, Berry JA, Frankenberg C, Huete AR, Zarco-Tejada P, Lee JE, Moran MS, Ponce-Campos G, Beer C, Camps-Valls G, Buchmann N, Gianelle D, Klumpp K, Cescatti A, Baker JM, Griffis TJ (2014) Global and time-resolved monitoring of crop photosynthesis with chlorophyll fluorescence. Proc Natl Acad Sci USA 14: E1327–E1333

Guo LB, Gifford RM (2002) Soil carbon stocks and land use change: a meta analysis. Glob Change Biol 8:345–360

Guo D, Li H, Mitchell RJ, Han W, Hendricks JJ, Fahey TJ, Hendrick RL (2008) Fine root heterogeneity by branch order: exploring the discrepancy in root turnover estimates between minirhizotron and carbon isotopic methods. New Phytol 177:443–456

Haberl H, Erb K-H, Krausmann F, Gaube V, Bondeau A, Plutzar C, Gingrich S, Lucht W, Fischer-Kowalski M (2007) Quantifying and mapping the human appropriation of net primary production in earth's terrestrial ecosystems. Proc Natl Acad Sci USA 104:12942–12947

Hättenschwiler S, Tiunov AV, Scheu S (2005) Biodiversity and litter decomposition in terrestrial ecosystems. Annu Rev Ecol Evol Syst 36:191–218

Hamilton SK, Kurzman AL, Arango C, Jin L, Robertson GP (2007) Evidence for carbon sequestration by agricultural liming. Global Biogeochem Cy 21, GB2021. https://doi.org/10.1029/2006gb002738

Hammes K, Torn MS, Lapenas AG, Schmidt MWI (2008) Centennial black carbon turnover in a Russian steppe soil. Biogeosciences 5:1339–1350

Hanson DT, Stutz SS, Boyer JS (2016) Why small fluxes matter: the case and approaches for improving measurements of photosynthesis and (photo)respiration. J Exp Bot 67:3027–3039. https://doi.org/10.1093/jxb/erw139

Hanson PJ, Edwards NT, Garten CT, Andrews JA (2000) Separating root and soil microbial contributions to soil respiration: a review of methods and observations. Biogeochemistry 48:115–146

Harmon ME, Silver WL, Fasth B, Chen H, Burke IC, Parton WJ, Hart SC, Currie WS, LIDET (2009) Long-term patterns of mass loss during the decomposition of leaf and fine root litter: an intersite comparison. Glob Change Biol 15:1320–1338

Hartmann J, Jansen N, Dürr HH, Kempe S, Köhler P (2009) Global CO_2-consumption by chemical weathering: what is the contribution of highly active weathering regions? Glob Planet Change 69:185–194

Hashimoto S (2012) A new estimation of global soil greenhouse gas fluxes using a simple data-oriented model. Plos One 7:e41962

Hashimoto S, Carvalhais N, Ito A et al (2015) Global spatiotemporal distribution of soil respiration modeled using a global database. Biogeosciences 12:4121–4132

Hawkes CV, Waring BG, Rocca JD, Kivlin SN (2017) Historical climate controls soil respiration responses to current soil moisture. Proc Natl Acad Sci USA 114:6322–6327

He L, Chen JM, Croft H, et al (2017) Nitrogen availability dampens the positive impacts of CO_2 fertilization on terrestrial ecosystem carbon and water cycles. Geophys Res Lett 44:11,590–11,600. https://doi.org/10.1002/2017GL075981

Heitkötter J, Heinze S, Marschner B (2017) Relevance of substrate quality and nutrients for microbial C-turnover in top and subsoil of a Dystric Cambisol. Geoderma 302:89–99

Heldt H-W (2005) Plant biochemistry. Elsevier, San Diego

Herbst M, Welp G, Macdonald A et al (2018) Correspondence of measured soil carbon fractions and RothC pools for equilibrium and non-equilibrium states. Geoderma 314:37–46

Hernes PJ, Kaiser K, Dyda RY, Cerli C (2013) Molecular trickery in soil organic matter: hidden lignin. Environ Sci Technol 47:9077–9085

Herrick JE, Wander MM (1998) Relationships between soil organic carbon and soil quality in cropped and rangeland soils: the importance of distribution, composition and soil biological activity. In: Lal R, Kimble J, Follett J, Stewart BA (eds) Advances in soil science: soil processes and the carbon cycle. CRC Press, Boca Raton, FL, pp 405–425

Hinsinger P, Gobran GR, Gregory PJ, Wenzel WW (2005) Rhizosphere geometry and heterogeneity arising from root-mediated physical and chemical processes. New Phytol 168:293–303

Hodge A, Berta G, Doussan C, Merchan F, Crespi M (2009) Plant root growth, architecture and function. Plant Soil 321:153–187

Hofmann A, Heim A, Christensen BT, Miltner A, Gehre M, Schmidt MWI (2009) Lignin dynamics in two ^{13}C-labelled arable soils during 18 years. Eur J Soil Sci 60:205–257

Holland JE, Bennett AE, Newton AC et al (2017) Liming impacts on soils, crops and biodiversity in the UK: A review. Sci Tot Environ 610–611:316–332

Hooper DU, Chapin FSIII, Ewel JJ, Hector A, Inchausti P, Lavorel S, Lawton JH, Lodge DM, Loreau M, Naeem S, Schmid B, Setälä H, Symstad AJ, Vandermeer J, Wardle DA (2005) Effects of biodiversity on ecosystem functioning: a consensus of current knowledge. Ecol Monogr 75:3–35

Hopkins DW, Waite IS, McNicol JW, Poulton PR, Macdonald AJ, O'Donnell AG (2009) Soil organic carbon contents in long-term experimental grassland plots in the UK (Palace Leas and Park Grass) have not changed consistently in recent decades. Glob Change Biol 15:1739–1754

Horn R, Smucker A (2005) Structure formation and its consequences for gas and water transport in unsaturated arable and forest soils. Soil Till Res 82:5–14

Houghton RA (2007) Balancing the global carbon budget. Annu Rev Earth Planet Sci 35:313–347

Houghton RA (2010) How well do we know the flux of CO_2 from land-use change? Tellus 62B:337–351

Houghton RA, Nassikas AA (2017) Global and regional fluxes of carbon from land use and land cover change 1850–2015. Global Biogeochem Cycles 31:456–472. https://doi.org/10.1002/2016GB005546

House JI, Hall DO (2001) Productivity of tropical savannas and grasslands. In: Roy J, Saugier B, Mooney HA (eds) Terrestrial global productivity. Academic Press, San Diego, CA, pp 363–400

Huang J, Minasny B, McBratney AB, Padariana J, Triantafilis J (2018) The location- and scale-specific correlation between temperature and soil carbon sequestration across the globe. Sci Tot Environ 615:540–548

Hugelius G, Strauss J, Zubrzycki S, Harden JW, Schuur EAG, Ping CL, Schirrmeister L, Grosse G, Michaelson GJ, Koven CD, O'Donnell JA, Elberling B, Mishra U, Camill P, Yu Z, Palmtag J, Kuhry P (2014) Estimated stocks of circumpolar permafrost carbon with quantified uncertainty ranges and identified data gaps. Biogeosciences 11:6573–6593

Huo C, Luo Y, Cheng W (2017) Rhizosphere priming effect: a meta-analysis. Soil Biol Biochem 111:78–84

Hursh A, Ballantyne A, Cooper L et al (2017) The sensitivity of soil respiration to soil temperature, moisture, and carbon supply at the global scale. Glob Change Biol 23:2090–2103

Huston MA, Wolverton S (2009) The global distribution of net primary production: resolving the paradox. Ecol Monogr 79:343–377

Insam H, Seewald MSA (2010) Volatile organic compounds (VOCs) in soils. Biol Fertil Soils 46:199–213

Ito A, Nishina K, Reyer CPO et al (2017) Photosynthetic productivity and its efficiencies in ISIMIP2a biome models: benchmarking for impact assessment studies. Environ Res Lett 12:085001. https://doi.org/10.1088/1748-9326/aa7a19

Ito A, Oikawa T (2004) Global mapping of terrestrial primary productivity and light-use efficiency with a process-based model. In: Shiyomi M, Kawahata H, Koizumi H, Tsuda A, Awaya Y (eds) Global environmental change in the ocean and on land. Terrabup, Tokyo, Japan, pp 343–358

Jackson RB, Lajtha K, Crow SE et al (2017) The ecology of soil carbon: pools, vulnerabilities, and biotic and abiotic controls. Annu Rev Ecol Evol Syst 48:419–445

Jansen B, Wiesenberg GLB (2017) Opportunities and limitations related to the application of plant-derived lipid molecular proxies in soil science. SOIL 3:211–234

Jansen N, Hartmann J, Lauerwald R, Dürr HH, Kempe S, Loos S, Middelkoop H (2010) Dissolved silica mobilization in the conterminous USA. Chem Geol 270:90–109

Jansson C, Wullschleger SD, Kalluri UC, Tuskan GA (2010) Phytosequestration: carbon biosequestration by plants and the prospects of genetic engineering. BioScience 60:685–696

Janzen HH (2006) The soil carbon dilemma: shall we hoard it or use it? Soil Biol Biochem 38:419–425

Janzen HH (2015) Beyond carbon sequestration: soil as conduit of solar energy. Eur J Soil Sci 66:19–32

Jenkinson DS, Coleman K (2008) The turnover of organic carbon in subsoils. Part 2. Modelling carbon turnover. Eur J Soil Sci 59:400–413

Jiang GM, Noonan MJ, Buchan GD, Smith N (2005) Transport and deposition of *Bacillus subtilis* through an intact soil column. Aust J Soil Res 43:695–703

Jiang Y, Wang W, Xie Q et al (2017) Plants transfer lipids to sustain colonization by mutualistic mycorrhizal and parasitic fungi. Science 356:1172–1175

Jobbágy EG, Jackson RB (2000) The vertical distribution of soil organic carbon and its relation to climate and vegetation. Ecol Appl 10:423–436

Joergensen RG, Wichern F (2018) Alive and kicking: why dormant soil microorganisms matter. Soil Biol Biochem 116:419–430

Johnson JMF, Allmaras RR, Reicosky DC (2006) Estimating source carbon from crop residues, roots and rhizodeposits using the national grain-yield database. Agron J 98:622–636

Jones DL, Nguyen C, Finlay RD (2009) Carbon flow in the rhizosphere: carbon trading at the soil-root interface. Plant Soil 321:5–33

Kalbitz K, Solinger S, Park J-H, Michalzik B, Matzner E (2000) Controls on the dynamics of dissolved organic matter in soils: a review. Soil Sci 165:277–304

Kampichler C, Bruckner A (2009) The role of microarthropods in terrestrial decomposition: a meta-analysis of 40 years of litterbag studies. Biol Rev 84:375–389

Kantola IB, Masters MD, Beerling DJ, Long SP, DeLucia EH (2017) Potential of global croplands and bioenergy crops for climate change mitigation through deployment for enhanced weathering. Biol Lett 13:20160714. https://doi.org/10.1098/rsbl.2016.0714

Kaplan JO, Krumhardt KM, Ellis RC, Ruddiman WF, Lemmen C, Klein Goldewijk K (2011) Holocene carbon emissions as a result of anthropogenic land cover change. The Holocene in press

Karlen DL, Ditzler CA, Andrews SS (2003) Soil quality: why and how? Geoderma 114:145–156

Karlen DL, Mausbach MJ, Doran JW, Cline RG, Harris RF, Schuman GE (1997) Soil quality: a concept, definition, and framework for evaluation (a guest editorial). Soil Sci Soc Am J 61;4–10

Kassam A, Friedrich T, Shaxson F, Pretty J (2009) The spread of conservation agriculture: justification, sustainability and uptake. Int J Agric Sustain 7:293–320

Keeley JE, Rundel PW (2003) Evolution of CAM and C_4 carbon-concentrating mechanisms. Int J Plant Sci 164:S55–S77

Keiluweit M, Nico PS, Kleber M, Fendorf S (2016) Are oxygen limitations under recognized regulators of organic carbon turnover in upland soils? Biogeochemistry 127:157–171

Keiluweit M, Wanzek T, Kleber M, Nico P, Fendorf S (2017) Anaerobic microsites have an unaccounted role in soil carbon stabilization. Nat Commun 8:1771. https://doi.org/10.1038/s41467-017-01406-6

Kelleher BP, Simpson AJ (2006) Humic substances in soils: are they really chemically distinct? Environ Sci Technol 40:4605–4611

Kemmitt SJ, Lanyon CV, Waite IS, Wen Q, Addiscott TM, Bird NRA, O'Donnell T, Brookes PC (2008) Mineralization of native soil organic matter is not regulated by the size, activity or composition of the soil microbial biomass. Soil Biol Biochem 40:61–73

Kindler R, Siemens J, Kaiser K, Walmsley DC, Bernhofer C, Buchmann N, Cellier P, Eugster W, Gleixner G, Grûnwald T, Heim A, Ibrom A, Jones SK, Jones M, Klumpp K, Kutsch W, Larsen KS, Lehuger S, Loubet B, McKenzie R, Moor E, Osborne B, Pilegaard K, Rebmann C, Saunders M, Schmidt MWI, Schrumpf M, Seyfferth J, Skiba U, Soussana J-F, Sutton MA, Tefs C, Vowinckel B, Zeeman MJ, Kaupenjohann M (2011) Dissolved carbon leaching from soil is a crucial component of the net ecosystem carbon balance. Glob Change Biol 17:1167–1185

Kleber M (2010a) What is recalcitrant soil organic matter? Environ Chem 7:320–332

Kleber M (2010b) Response to the Opinion paper by Margit von Lützow and Ingrid Kögel-Knabner on 'What is recalcitrant soil organic matter?' by Markus Kleber. Environ Chem 7:336–337

Kleber M, Eusterhues K, Keiluweit M, Mikutta C, Mikutta R, Nico PS (2015) Mineral–organic associations: formation, properties, and relevance in soil environments. Adv Agron 130:1–140

Kleber M, Johnson MG (2010) Advances in understanding the molecular structure of soil organic matter: implications for interactions in the environment. Adv Agron 106:77–142

Kleber M, Nico PS, Plante A, Filley T, Kramer M, Swanston C, Sollins P (2011) Old and stable soil organic matter is not necessarily chemically recalcitrant: implications for modeling concepts and temperature sensitivity. Glob Change Biol 17:1097–1107

Kleber M, Sollins P, Sutton R (2007) A conceptual model of organo-mineral associations in soils: self-assembly of organic molecular fragments into zonal structures on mineral surfaces. Biogeochemistry 85:9–24

Klotzbücher T, Klotzbücher A, Kaiser K et al (2018) Variable silicon accumulation in plants affects terrestrial carbon cycling by controlling lignin synthesis. Glob Change Biol 24:e183–e189. https://doi.org/10.1111/gcb.13845

Knacker T, Förster B, Römbke J, Frampton GK (2003) Assessing the effects of plant protection products on organic matter breakdown in arable fields-litter decomposition test systems. Soil Biol Biochem 35:1269–1287

Köchy M, Hiederer R, Freibauer A (2015) Global distribution of soil organic carbon—Part 1: Masses and frequency distributions of SOC stocks for the tropics, permafrost regions, wetlands, and the world. SOIL 1:351–365

Kögel-Knabner I, Guggenberger G, Kleber M et al (2008) Organo-mineral associations in temperate soils: Integrating biology, mineralogy, and organic matter chemistry. J Plant Nutr Soil Sci 171:61–82. https://doi.org/10.1002/jpln.200700048

Kong AYY, Six J, Bryant DC, Denison RF, van Kessel C (2005) The relationship between carbon input, aggregation, and soil organic carbon stabilization in sustainable cropping systems. Soil Sci Soc Am J 69:1078–1085

Konikow LF (2011) Contribution of global groundwater depletion since 1900 to sea-level rise. Geophys Res Lett 38. https://doi.org/10.1029/2011gl048604

Kopittke PM, Dalal RC, Finn D, Menzies NW (2017) Global changes in soil stocks of carbon, nitrogen, phosphorus, and sulphur as influenced by long-term agricultural production. Glob Change Biol 23:2509–2519. https://doi.org/10.1111/gcb.13513

Kopp G, Lean JL (2011) A new, lower value of total solar irradiance: evidence and climate significance. Geophys Res Lett 38:L01706. https://doi.org/10.1029/2010GL045777

Koukoura Z (1998) Decomposition and nutrient release from C_3 and C_4 plant litters in a natural grassland. Acta Oecol 19:115–123

Koven CD, Hugelius G, Lawrence DM, Wieder WR (2017) Higher climatological temperature sensitivity of soil carbon in cold than warm climates. Nat Clim Change 7:817–824. https://doi.org/10.1038/NCLIMATE3421

Krull ES, Baldock JA, Skjemstad JO (2003) Importance of mechanisms and processes of the stabilisation of soil organic matter for modelling carbon turnover. Funct Plant Biol 30:207–222

Kunhikrishnan A, Thangarajan R, Bolan NS, Xu Y, Mandal S, Gleeson DB, Seshadri B, Zaman M, Barton L, Tang C, Luo J, Dalal R, Ding W, Kirkham MB, Naidu R (2016) Functional relationships of soil acidification, liming, and greenhouse gas flux. Adv Agron 139:1–71

Kutsch WL, Aubinet M, Buchmann N, Smith P, Osborne B, Eugster W, Wattenbach M, Schrumpf M, Schulze ED, Tomelleri E, Ceschia E, Bernhofer C, Béziat P, Carrara A, Di Tommasi P, Grünwald T, Jones M, Magliulo V, Marloie O, Moureaux C, Olioso A, Sanz MJ, Saunders M, Sogaard H, Ziegler W (2010) The net biome production of full crop rotations in Europe. Agric Ecosyst Environ 139:336–345

Kutsch WL, Bahn M, Heinemeyer A (2009a) Soil carbon relations: an overview. In: Kutsch WL, Bahn M, Heinemeyer A (eds) Soil carbon dynamics: an integrated methodology. Cambridge University Press, Cambridge, U.K., pp 1–15

Kutsch WL, Schimel J, Denef K (2009b) Measuring soil microbial parameters relevant for soil carbon fluxes. In: Kutsch WL, Bahn M, Heinemeyer A (eds) Soil carbon dynamics: an integrated methodology. Cambridge University Press, Cambridge, U.K., pp 169–186

Kuzyakov Y (2002) Review: factors affecting rhizosphere priming effects. J Plant Nutr Soil Sci 165:382–396

Kuzyakov Y (2006) Sources of CO_2 efflux from soil and review of partitioning methods. Soil Biol Biochem 38:425–448

Kuzyakov Y (2010) Priming effects: interactions between living and dead organic matter. Soil Biol Biochem 42:1363–1371

Kuzyakov Y, Blagodatskaya E (2015) Microbial hotspots and hot moments in soil: concept & review. Soil Biol Biochem 83:184–199

Kuzyakov Y, Shevtzova E, Pustovoytov K (2006) Carbonate re-crystallization in soil revealed by ^{14}C labeling: experiment, model and significance for paleo-environmental reconstructions. Geoderma 131:45–58

Lal R (2001) World cropland soils as a source or sink for atmospheric carbon. Adv Agron 71:145–191

Lal R (2003) Soil erosion and the global carbon budget. Environ Int 29:437–450

Lal R (2004a) Carbon sequestration in dryland ecosystems. Environ Manage 33:528–544

Lal R (2004b) Agricultural activities and the global carbon cycle. Nutr Cyc Agroecosyst 70:103–116

Lal R (2004c) Soil carbon sequestration impacts on global climate change and food security. Science 304:1623–1627

Lal R (2004d) Soil carbon sequestration to mitigate climate change. Geoderma 123:1–22

Lal R (2007) Farming carbon. Soil Till Res 96:1–5

Lal R (2009) Soils and food sufficiency. A review. Agron Sustain Dev 29:113–133

Lal R (2010a) Beyond Copenhagen: mitigating climate change and achieving food security through soil carbon sequestration. Food Sec 2:169–177

Lal R (2010b) Managing soils and ecosystems for mitigating anthropogenic carbon emissions and advancing global food security. BioScience 60:708–721

Lal R (2010c) Enhancing eco-efficiency in agro-ecosystems through soil carbon sequestration. Crop Sci 50:S120–S131

Lal R, Follett RF (2009a) Preface. In: Lal R, Follett RF (eds) Soil carbon sequestration and the greenhouse effect. SSSA Special Publication 57, 2nd ed. Soil Science Society of America, Madison, WI, pp xi–xii

Lal R, Follett RF (2009b) Soils and climate change. In: Lal R, Follett RF (eds) Soil carbon sequestration and the greenhouse effect. SSSA Special Publication 57, 2nd ed. Soil Science Society of America, Madison, WI, pp xxi–xxviii

Lal R, Kimble JM (2000) Pedogenic carbonates and the global carbon cycle. In: Lal R, Kimble JM, Eswaran H, Stewart BA (eds) Global climate change and pedogenic carbonates. CRC Press, Boca Raton, FL, pp 291–302

Lambers H, Robinson SA, Ribas-Carbo M (2005) Regulation of respiration in vivo. In: Lambers H, Ribas-Carbo M (eds) Plant respiration: from cell to ecosystem. Springer, Dordrecht, The Netherlands, pp 43–61

Lapenis AG, Lawrence GB, Bailey SW, Aparin BF, Shiklomanov AI, Speranskaya NA, Torn MS, Calef M (2008) Climatically driven loss of calcium in steppe soil as a sink for atmospheric carbon. Global Biogeochem Cy 22, GB2010. https://doi.org/10.1029/2007gb003077

Landi A, Mermut AR, Anderson DW (2003) Origin and rate of pedogenic carbonate accumulation in Saskatchewan soils, Canada. Geoderma 117:143–156

Laothawornkitkul J, Taylor JE, Paul ND, Hewitt CN (2009) Biogenic volatile organic compounds in the Earth system. New Phytol 183:27–51

Leake J, Johnson D, Donnelly D, Muckle G, Boddy L, Read D (2004) Networks of power and influence: the role of mycorrhizal mycelium in controlling plant communities and agroecosystem functioning. Can J Bot 82:1016–1045

Leff JW, Fierer N (2008) Volatile organic compound (VOC) emissions from soil and litter samples. Soil Biol Biochem 40:1629–1636

Lehmann J, Kleber M (2015) The contentious nature of soil organic matter. Nature 528:60–68

Lehmann J, Solomon D, Kinyangi J, Dathe L, Wirick S, Jacobsen C (2008) Spatial complexity of soil organic matter forms at nanometre scales. Nature Geosci 1:238–242

Lenton TM, Britton C (2006) Enhanced carbonate and silicate weathering accelerates recovery from fossil fuel CO_2 perturbations. Global Biogeochem Cy 20, GB3009. https://doi.org/10.1029/2005gb002678

Le Quéré C, Andrew RM, Friedlingstein P et al (2017) Global carbon budget 2017. Earth Syst Sci Data Discuss. https://doi.org/10.5194/essd-2017-123

Li Y, Zhang C, Wang N et al (2017) Substantial inorganic carbon sink in closed drainage basins globally. Nat Geosci 10:501–508

Liang C, Schimel JP, Jastrow JD (2017a) The importance of anabolism in microbial control over soil carbon storage. Nat Microbiol 2:17105. https://doi.org/10.1038/nmicrobiol.2017.105

Liang W, Lü Y, Zhang W et al (2017b) Grassland gross carbon dioxide uptake based on an im-proved model tree ensemble approach considering human interventions: global estimation and covariation with climate. Glob Change Biol 23:2720–2742. https://doi.org/10.1111/gcb.13592

Lin Y, Karlen SD, Ralph J, King JY (2018) Short-term facilitation of microbial litter decomposition by ultraviolet radiation. Sci Tot Environ 615:838–848
Lindeman RL (1942) The trophic-dynamic aspect of ecology. Ecology 23:399–417
Liu Z, Zhao J (2000) Contribution of carbonate rock weathering to the atmospheric CO_2 sink. Environ Geol 39:1053–1058
Louis BP, Maron PA, Viaud V, Leterme P, Menasseri-Aubry S (2016) Soil C and N models that integrate microbial diversity. Environ Chem Lett 14:331–344. https://doi.org/10.1007/s10311-016-0571-5
Loreau M, Naeem S, Inchausti P, Bengtsson J, Grime JP, Hector A, Hooper DU, Huston MA, Raffaelli D, Schmid B, Tilman D, Wardle DA (2001) Biodiversity and ecosystem functioning: current knowledge and future challenges. Science 294:804–808
Lorenz K, Lal R (2005) The depth distribution of soil organic carbon in relation to land use and management and the potential of carbon sequestration in subsoil horizons. Adv Agron 88:35–66
Lorenz K, Lal R, Preston CM, Nierop KGJ (2007) Strengthening the soil organic carbon pool by increasing contributions from recalcitrant aliphatic bio(macro)molecules. Geoderma 142:1–10
Loreto F, Kesselmeier J, Schnitzler JP (2008) Volatile organic compounds in the biosphere–atmosphere system: a preface. Plant Biol 10:2–7
Loveland P, Webb J (2003) Is there a critical level of organic matter in the agricultural soils of temperate regions: a review. Soil Till Res 70:1–18
Luginbuehl LH, Menard GN, Kurup S et al (2017) Fatty acids in arbuscular mycorrhizal fungi are synthesized by the host plant. Science 356:1175–1178
Luo Z, Feng W, Luo Y, Baldock J, Wang E (2017) Soil organic carbon dynamics jointly controlled by climate, carbon inputs, soil properties and soil carbon fractions. Glob Change Biol 23:4430–4439. https://doi.org/10.1111/gcb.13767
Lynch JP, Wojciechowski T (2015) Opportunities and challenges in the subsoil: pathways to deeper rooted crops. J Exp Bot 66:2199–2210
Macías F, Arbestain MC (2010) Soil carbon sequestration in a changing global environment. Mitig Adapt Strateg Glob Change 15:511–529
Mackey B, Prentice IC, Steffen W et al (2013) Untangling the confusion around land carbon science and climate change mitigation policy. Nat Clim Change 3:552–557. https://doi.org/10.1038/NCLIMATE1804
Manlay RJ, Feller C, Swift MJ (2007) Historical evolution of soil organic matter concepts and their relationships with the fertility and sustainability of cropping systems. Agric Ecosyst Environ 119:217–233. https://doi.org/10.1016/j.agee.2006.07.011
Manzoni S, Porporato A (2009) Soil carbon and nitrogen mineralization: theory and models across scales. Soil Biol Biochem 41:1355–1379
Marchant BP, Villanneau EJ, Arrouays D, Saby NPA, Rawlins BG (2015) Quantifying and mapping topsoil inorganic carbon concentrations and stocks: approaches tested in France. Soil Use Manage 31:29–38
Marion GM, Verburg PS, McDonald EV, Arnone JA (2008) Modeling salt movement through a Mojave Desert soil. J Arid Environ 72:1012–1033
Marlon JR, Bartlein PJ, Carcaillet C, Gavin DG, Harrison SP, Higuera PE, Joos F, Power MJ, Prentice IC (2008) Climate and human influences on global biomass burning over the past two millennia. Nature Geosci 1:697–702
Marschner B, Brodowski S, Dreves A, Gleixner G, Gude A, Grootes PM, Hamer U, Heim A, Jandl G, Ji R, Kaiser K, Kalbitz K, Kramer C, Leinweber P, Rethemeyer J, Schäffer A, Schmidt MWI, Schwark L, Wiesenberg GLB (2008) How relevant is recalcitrance for the stabilization of organic matter in soils? J Plant Nutr Soil Sci 171:91–110
Marschner B, Kalbitz K (2003) Controls of bioavailability and biodegradability of dissolved organic matter in soils. Geoderma 113:211–235
Marschner H (1995) Mineral nutrition of higher plants. Academic Press, San Diego, CA

Martin A (1991) Short- and long-term effects of the endogenic earthworm *Millsonia anomala* (Omodeo) (Megascolecidae: Oligochaeta) of tropical savanna, on soil organic matter. Biol Fertil Soils 11:234–238

Mason-Jones K, Kuzyakov Y (2017) "Non-metabolizable" glucose analogue shines new light on priming mechanisms: triggering of microbial metabolism. Soil Biol Biochem 107:68–76

Mathieu J, Hatté C, Balesdent J, Parent E (2015) Deep soil carbon dynamics are driven more by soil type than by climate: a worldwide meta-analysis of radiocarbon profiles. Glob Change Biolo 21:4278–4292. https://doi.org/10.1111/gcb.13012

Matson PA, Parton WJ, Power AG, Swift MJ (1997) Agricultural intensification and ecosystem properties. Science 277:504–509

Matthews HD, Weaver AJ (2010) Committed climate warming. Nature Geosci 3:142–143

McGuire KL, Treseder KK (2010) Microbial communities and their relevance for ecosystem models: decomposition as a case study. Soil Biol Biochem 42:529–535

McNally SR, Beare MH, Curtin D et al (2017) Soil carbon sequestration potential of permanent pasture and continuous cropping soils in New Zealand. Glob Change Biol 23:4544–4555. https://doi.org/10.1111/gcb.13720

MEA (2005) Millenium ecosystem assessment. In: Ecosystems and human well-being: biodiversity synthesis. World Resources Institute, Washington, DC

Melillo JM, Frey SD, DeAngelis KM et al (2017) Long-term pattern and magnitude of soil carbon feedback to the climate system in a warming world. Science 358:101–105

Mercado LM, Bellouin N, Sitch S, Boucher O, Huntingford C, Wild M, Cox PM (2009) Impact of changes in diffuse radiation on the global land carbon sink. Nature 458:1014–1018

Metting FB, Smith JL, Amthor JS, Izaurralde RC (2001) Science needs and new technology for increasing soil carbon sequestration. Clim Change 51:11–34

Mikhailova EA, Post CJ (2006) Effects of land use on soil inorganic carbon stocks in the Russian Chernozem. J Environ Qual 35:1384–1388

Mikhailova E, Post C, Cihacek L, Ulmer M (2009) Soil inorganic carbon sequestration as a result of cultivation in the Mollisols. In: McPherson BJ, Sundquist ET (eds) Carbon sequestration and its role in the global carbon cycle. Geophysical Monograph Series 183, American Geophysical Union, Washington, D.C., pp 129–133

Mikutta R, Mikutta C, Kalbitz K, Scheel T, Kaiser K, Jahn R (2007) Biodegradation of forest floor organic matter bound to minerals via different binding mechanisms. Geochim Cosmochim Ac 71:2569–2590

Milchunas DG (2009) Estimating root production: comparison of 11 methods in shortgrass steppe and review of biases. Ecosystems 12:1381–1402

Miller RM, Jastrow JD (1990) Hierarchy of root and mycorrhizal fungal interactions with soil aggregation. Soil Biol Biochem 22:579–584

Miltner A, Kopinke F-D, Kindler R, Selesi D, Hartmann A, Kästner M (2005) Non-phototrophic CO_2 fixation by soil microorganisms. Plant Soil 269:193–203

Ming DW (2006) Carbonates. In: Lal R (ed) Encyclopedia of soil science. Taylor & Francis, London, pp XXX–XXX

Moinet GYK, Hunt JE, Kirschbaum MUF et al (2018) The temperature sensitivity of soil organic matter decomposition is constrained by microbial access to substrates. Soil Biol Biochem 116:333–339

Mokany K, Raison RJ, Prokushkin AS (2006) Critical analysis of root: shoot ratios in terrestrial biomes. Glob Change Biol 12:84–96

Monfreda C, Ramankutty N, Foley JA (2008) Farming the planet: 2. Geographic distribution of crop areas, yields, physiological types, and net primary production in the year 2000. Global Biogeochem Cyc 22, GB1022. https://doi.org/10.1029/2007gb002947

Monger HC, Gallegos RA (2000) Biotic and abiotic processes and rates of pedogenic carbonate accumulation in the southwestern United States. Relationship to atmospheric CO_2 sequestration. In: Lal R, Kimble JM, Eswaran H, Stewart BA (eds) Global climate change and pedogenic carbonates. CRC/Lewis Publishers, Boca Raton, Florida, pp 273–290

Monger HC, Kraimer RA, Khresat S, Cole DR, Wang X, Wang J (2015) Sequestration of inorganic carbon in soil and groundwater. Geology 43:375–378
Moni C, Rumpel C, Virto I, Chabbi A, Chenu C (2010) Relative importance of sorption versus aggregation for organic matter storage in subsoil horizons of two contrasting soils. Eur J Soil Sci 61:958–969
Monreal CM, Dinel H, Schnitzer M, Gamble DS, Biederbeck VO (1998) Impact of carbon sequestration on functional indicators of soil quality as influenced by management in sustainable agriculture. In: Lal R, Kimble JM, Follett RF, Stewart BA (eds) Soil processes and the carbon cycle. CRC Press, Boca Raton, FL, pp 435–447
Moore TR, Trofymow JA, Prescott CE, Titus BD, CIDET Working Group (2011) Nature and nurture in the dynamics of C, N and P during litter decomposition in Canadian forests. Plant Soil 339:163–175
Moorhead DL, Sinsabaugh RL (2006) A theoretical model of litter decay and microbial interaction. Ecol Monogr 76:151–174
Moosdorf N, Renforth P, Hartmann J (2014) Carbon dioxide efficiency of terrestrial enhanced weathering. Environ Sci Technol 48:4809–4816
Moran KK, Jastrow JD (2010) Elevated carbon dioxide does not offset loss of soil carbon from a corn–soybean agroecosystem. Environ Pollut 158:1088–1094
Morriën E, Hannula SE, Snoek LB et al (2017) Soil networks become more connected and take up more carbon as nature restoration progresses. Nat Commun 8:14349. https://doi.org/10.1038/ncomms14349
Mueller L, Schindler U, Mirschel W, Sheperd G, Ball BC, Helming K, Rogaski J, Eulenstein F, Wiggering H (2010) Assessing the productivity function of soils. A review. Agron Sust Dev 30:601–614
Nair PKR, Nair VD, Kumar BM, Showalter JM (2010) Carbon sequestration in agroforestry systems. Adv Agron 108:237–307
Newcomb CJ, Qafoku NP, Grate JW, Bailey VL, De Yoreo JJ (2017) Developing a molecular picture of soil organic matter–mineral interactions by quantifying organo–mineral binding. Nat Commun 8:396. https://doi.org/10.1038/s41467-017-00407-9
Nguyen C (2003) Rhizodeposition of organic C by plants: mechanisms and controls. Agronomie 23:375–396
Nieder R, Benbi DK (2008) Carbon and nitrogen in the terrestrial environment. Springer, Dordrecht, The Netherlands
Nordt LC, Wilding LP, Drees LR (2000) Pedogenic carbonate transformations in leaching soil systems: implications for the global C cycle. In: Lal R, Kimble JM, Eswaran H, Stewart BA (eds) Global climate change and pedogenic carbonates. CRC Press, Boca Raton, FL, pp 43–64
Oades JM (1984) Soil organic matter and structural stability, mechanisms and implications for management. Plant Soil 76:319–337
Oh N-H, Raymond PA (2006) Contribution of agricultural liming to riverine bicarbonate export and CO_2 sequestration in the Ohio River basin. Global Biogeochem Cy 20, GB3012. https://doi.org/10.1029/2005gb002565
Oldeman LR (1994) The global extent of soil degradation. In: Greenland DJ, Szabolcs I (eds) Soil resilience and sustainable land use. CAB International, Wallingford, Oxon, UK, pp 99–118
Olmstead AL, Rhode PW (2011) Adapting North American wheat production to climatic challenges, 1839–2009. Proc Natl Acad Sci USA 108:480–485
Osborne CP, Beerling DJ (2006) Review. Nature's green revolution: the remarkable evolutionary rise of C_4 plants. Phil Trans Royal Soc B 361:173–194
Osborne CP, Freckleton RP (2009) Ecological selection pressures for C4 photosynthesis in the grasses. Proc R Soc B. https://doi.org/10.1098/rspb.2008.1762
Osler GHR, Sommerkorn M (2007) Toward a complete soil C and N cycle: incorporating the soil fauna. Ecology 88:1611–1621
Owen NA, Fahy KF, Griffiths H (2016) Crassulacean acid metabolism (CAM) offers sustainable bioenergy production and resilience to climate change. GCB Bioenerg 8:737–749

Page KL, Allen DE, Dalal RC, Slattery W (2009) Processes and magnitude of CO_2, CH_4, and N_2O fluxes from liming of Australian acidic soils: a review. Austr J Soil Res 47:747–762
Papiernik SK, Lindstrom MJ, Schumacher TE, Schumacher JA, Malo DD, Lobb DA (2007) Characterization of soil profiles in a landscape affected by long-term tillage. Soil Till Res 93:335–345
Paradelo R, Virto I, Chenu C (2015) Net effect of liming on soil organic carbon stocks: a review. Agric Ecosyst Environ 202:98–107
Parton WJ, Schimel DS, Cole CV, Ojima DS (1987) Analysis of factors controlling soil organic matter levels in Great Plains grasslands. Soil Sci Soc Am J 51:1173–1179
Paul EA (2016) The nature and dynamics of soil organic matter: plant inputs, microbial transformations, and organic matter stabilization. Soil Biol Biochem 98:109–126
Pausch J, Kuzyakov Y (2018) Carbon input by roots into the soil: quantification of rhizodeposition from root to ecosystem scale. Glob Change Biol 24:1–12. https://doi.org/10.1111/gcb.13850
Pechony O, Shindell DT (2010) Driving forces of global wildfires over the past millennium and the forthcoming century. Proc Natl Acad Sci USA 107:19167–19170
Perrin AS, Probst A, Probst JL (2008) Impact of nitrogenous fertilizers on carbonate dissolution in small agricultural catchments: Implications for weathering CO_2 uptake at regional and global scales. Geochim Comochim Ac 72:3105–3123
Peyton Smith A, Bond-Lamberty B, Benscoter BW et al (2017) Shifts in pore connectivity from precipitation versus groundwater rewetting increases soil carbon loss after drought. Nat Commun 8:1335. https://doi.org/10.1038/s41467-017-01320-x
Phillips CL, Bond-Lamberty B, Desai AR et al (2017) The value of soil respiration measurements for interpreting and modeling terrestrial carbon cycling. Plant Soil 413:1–25
Pinstrup-Andersen P (2009) Food security: definition and measurement. Food Sec 1:5–7
Poeplau C, Don A, Vesterdal L, Leifeld J, Van Wesemael B, Schumacher J, Gensior A (2011) Temporal dynamics of soil organic carbon after land-use change in the temperate zone—carbon response functions as a model approach. Glob Change Biol 17:2415–2427
Polyakov V, Lal R (2004) Modeling soil organic matter dynamics as affected by soil water erosion. Environ Int 30:547–556
Poole P (2017) Shining a light on the dark world of plant root–microbe interactions. Proc Natl Acad Sci USA 114:4281–4283
Porras RC, Hicks Pries CE, Torn MS, Nico PS (2018) Synthetic iron (hydr)oxide-glucose associations in subsurface soil: effects on decomposability of mineral associated carbon. Sci Tot Environ 613–614:342–351
Portmann FT, Siebert S, Döll P (2010) MIRCA2000—Global monthly irrigated and rainfed crop areas around the year 2000: a new high-resolution data set for agricultural and hydrological modeling. Global Biogeochem Cy 24, GB101. https://doi.org/10.1029/2008gb003435
Power AG (2010) Ecosystem services and agriculture: tradeoffs and synergies. Phil Trans R Soc B 365:2959–2971
Powlson DS, Whitmore AP, Goulding KWT (2011) Soil carbon sequestration to mitigate climate change: a critical re-examination to identify the true and the false. Eur J Soil Sci 62:42–55
Prescott CE (2005) Do rates of litter decomposition tell us anything we really need to know? For Ecol Manage 220:66–74
Prescott CE (2010) Litter decomposition: what controls it and how can we alter it to sequester more carbon in forest soils? Biogeochemistry 101:133–149
Preston CM, Trofymow JA, Sayer BG, Niu J (1997) ^{13}C nuclear magnetic resonance spectroscopy with cross-polarization and magic-angle spinning investigation of the proximate-analysis fractions used to assess litter quality in decomposition studies. Can J Bot 75:1601–1613
Pribyl DW (2010) A critical review of the conventional SOC to SOM conversion factor. Geoderma 156:75–83
Pritchard SG, Strand AE (2008) Can you believe what you see? Reconciling minirhizotron and isotopically derived estimates of fine root longevity. Plant Phytol 177:287–291
Purin S, Rillig MC (2007) The arbuscular mycorrhizal fungal protein glomalin: limitations, progress, and a new hypothesis for its function. Pedobiologia 51:123–130

Qian B, Gregorich EG, Gameda S, Hopkins DW, Wang XL (2011) Observed soil temperature trends associated with climate change in Canada. J Geophys Res 116:D02106. https://doi.org/10.1029/2010JD015012

Quinton JN, Govers G, Van Oost K, Bardgett RD (2010) The impact of agricultural soil erosion on biogeochemical cycling. Nature Geosci 3:311–314

Raich JW, Lambers H, Oliver DJ (2014) Respiration in terrestrial ecosystems. In: Holland H, Turekian K (eds) Treatise on geochemistry, vol 10. Biogeochemistry. Elsevier, Amsterdam, Netherlands, pp 613–648

Raich JW, Potter CS, Bhagawati D (2002) Interannual variability in global soil respiration, 1980–94. Glob Change Biol 8:800–812

Rasmussen J (2011) Why we need to restrict the use of "rhizodeposition" and the Janzen and Bruinsma equation. Soil Biol Biochem 43:2213–2214

Rasse DP, Rumpel C, Dignac M-F (2005) Is soil carbon mostly root carbon? Mechanisms for a specific stabilization. Plant Soil 269:341–356

Raven JA, Edwards D (2001) Roots: evolutionary origins and biogeochemical significance. J Exp Bot 52:381–401

Reay DS (2003) Sinking methane. Biologist 50:15–19

Rees RM, Bingham IJ, Baddeley JA, Watson CA (2005) The role of plants and land management in sequestering soil carbon in temperate arable and grassland ecosystems. Geoderma 128:130–154

Reichstein M, Falge E, Baldocchi DS et al (2005) On the separation of net ecosystem exchange into assimilation and ecosystem respiration: review and improved algorithm Glob Change Biol 11:1424–1439

Richey JE (2004) Pathways of atmospheric CO_2 through fluvial systems. In: Field CB, Raupach MR (eds) The global carbon cycle: integrating humans, climate, and the natural world. Island, Washington, D.C., pp 329–340

Rillig MC, Ramsey PW, Morris S, Paul EA (2003) Glomalin, an arbuscular-mycorrhizal fungal soil protein, responds to land-use change. Plant Soil 253:293–299

Ritz K (1995) Growth responses of some fungi to spatially heterogeneous nutrients. FEMS Microbiol Ecol 16:269–280

Robinson D (2007) Implications of a large global root biomass for carbon sink estimates and for soil carbon dynamics. Proc R Soc B 274:2753–2759

Rosier CL, Hoye AT, Rillig MC (2006) Glomalin-related soil protein: assessment of current detection and quantification tools. Soil Biol Biochem 38:2205–2211

Rounsevell MDA, Evans SP, Bullock P (1999) Climate change and agricultural soils: impacts and adaptation. Clim Change 43:683–709

Rovira AD, Greacen EL (1957) The effect of aggregate disruption on the activity of microorganisms in soil. Austr J Agr Res 8:659–673

Roy J, Saugier B, Mooney HA (2001) Terrestrial global productivity. Academic Press, San Diego, CA

Ruddiman WF (2003) The anthropogenic greenhouse area began thousands of years ago. Clim Change 61:261–293

Ruddiman WF (2005) The early anthropogenic hypothesis a year later. Clim Change 69:427–434

Ruimy A, Jarvis PG, Baldocchi DD, Saugier B (1996) CO_2 fluxes over plant canopies and solar radiation: a review. Adv Ecol Res 26:1–68

Ruimy A, Saugier B, Dedieu G (1994) Methodology for the estimation of terrestrial net primary production from remotely sensed data. J Geophys Res 99:5263–5283

Rumpel C, Kögel-Knabner I (2011) Deep soil organic matter—a key but poorly understood component of terrestrial C cycle. Plant Soil 338:143–158

Rutledge S, Campbell DI, Baldocchi D, Schipper LA (2010) Photodegradation leads to increased carbon dioxide losses from terrestrial organic matter. Glob Change Biol 16:3065–3074

Sala OE (2001) Productivity of temperate grasslands. In: Roy J, Saugier B, Mooney HA (eds) Terrestrial global productivity. Academic Press, San Diego, CA, pp 285–300

Salomé C, Nunan N, Pouteau V, Lerch TZ, Chenu C (2010) Carbon dynamics in topsoil and in subsoil may be controlled by different regulatory mechanisms. Glob Change Biol 16:416–426

Sanaullah M, Chabbi A, Leifeld J, Bardoux G, Billou D, Rumpel C (2011) Decomposition and stabilization of root litter in top- and subsoil horizons: what is the difference? Plant Soil 338:127–141

Sanderman J (2012) Can management induced changes in the carbonate system drive soil carbon sequestration? A review with particular focus on Australia. Agric Ecosyst Environ 155:70–77

Sanderman J, Amundson R (2008) A comparative study of dissolved organic carbon transport and stabilization in California forest and grassland soils. Biogeochemistry 89:309–327

Sanderman J, Baldock JA (2010) Accounting for soil carbon sequestration in national inventories: a soil scientist's perspective. Environ Res Lett 5:034003

Sanderman J, Berhe AA (2017) The soil carbon erosion paradox. Nat Clim Change 7:317–319

Sanderman J, Farquharson R, Baldock J (2010) Soil carbon sequestration potential: a review for Australian agriculture—a report prepared for Department of Climate Change and Energy Efficiency, CSIRO Land and Water. http://www.csiro.au/resources/Soil-Carbon-Sequestration-Potential-Report.html

Sanderman J, Hengl T, Fiske GJ (2017) Soil carbon debt of 12,000 years of human land use. Proc Natl Acad Sci USA Early Edition

Sarker TC, Incerti G, Spaccini R et al (2018) Linking organic matter chemistry with soil aggregate stability: Insight from ^{13}C NMR spectroscopy. Soil Biol Biochem 117:175–184

Sartori F, Lal R, Ebinger MH, Eaton JA (2007) Changes in soil carbon and nutrient pools along a chronosequence of poplar plantations in the Columbia Plateau, Oregon, USA. Agr Ecosyst Environ 122:325–339

Schenk HJ, Jackson RB (2005) Mapping the global distribution of deep roots in relation to climate and soil characteristics. Geoderma 126:129–140

Schimel JP, Schaeffer SM (2012) Microbial control over carbon cycling in soil. Front Microbiol 3:348. https://doi.org/10.3389/fmicb.2012.00348

Schindler FV, Mercer EJ, Rice JA (2007) Chemical characteristics of glomalin-related soil protein (GRSP) extracted from soils of varying organic matter content. Soil Biol Biochem 39:320–329

Schipper LA, Baisden WT, Parfitt RL, Ross W, Claydon JJ, Arnold G (2007) Large losses of soil C and N from soil profiles under pasture in New Zealand during the past 20 years. Glob Change Biol 13:1138–1144

Schipper LA, Parfitt RL, Ross W, Baisden WT, Claydon JJ, Fraser S (2010) Gains and losses in C and N stocks of New Zealand pasture soils depend on land use. Agric Ecosyst Environ 139:611–617

Schlesinger WH (1997) Biogeochemistry: an analysis of global change. Academic Press, San Diego, CA

Schlesinger WH (2000) Carbon sequestration in soils: some cautions amidst optimism. Agric Ecosys Environ 82:121–127

Schlesinger WH (2006) Inorganic carbon and the global C cycle. In: Lal R (ed) Encyclopedia of soil science. Taylor & Francis, London, pp 879–881

Schlesinger WH (2017) An evaluation of abiotic carbon sinks in deserts. Glob Change Biol 23:25–27

Schmidt MWI, Torn MS, Abiven S, Dittmar T, Guggenberger G, Janssens IA, Kleber M, Kögel-Knabner I, Lehmann J, Manning DAC, Nannipieri P, Rasse DP, Weiner S, Trumbore SE (2011) Persistence of soil organic matter as an ecosystem property. Nature 478:49–56

Schulze E-D, Beck E, Müller-Hohenstein K (2005) Plant ecology. Springer, Berlin

Schulze E-D, Ciais P, Luyssaert S, Schrumpf M, Janssens IA, Thiruchittampalam B, Theloke J, Saurat M, Bringezu S, Lelieveld J, Lohila A, Rebmann C, Jung M, Bastviken D, Abril G, Grassi G, Leip A, Freibauer A, Kutsch W, Don A, Nieschulze J, Börner A, Gash JH, Dolman AJ (2010) The European carbon balance. Part 4: integration of carbon and other trace-gas fluxes. Global Change Biol 16:1451–1469

Schulze E-D, Luyssaert S, Ciais P, Freibauer A, Janssens IA et al (2009) Importance of methane and nitrous oxide for Europe's terrestrial greenhouse-gas balance. Nature Geosci 2:842–850
Selesi D, Schmid M, Hartmann A (2005) Diversity of green-like and red-like ribulose-1,5-bisphosphate carboxylase/oxygenase large-subunit genes (cbbL) in differently managed agricultural soils. Appl Environ Microb 71:175–184
Senthilkumar S, Basso B, Kravchenko AN, Robertson GP (2009) Contemporary evidence of soil carbon loss in the U.S. corn belt. Soil Sci Soc Am J 73:2078–2086
Séquaris J-M, Guisado G, Magarinos M, Moreno C, Burauel P, Narres H-D, Vereecken H (2010) Organic-carbon fractions in an agricultural topsoil assessed by the determination of the soil mineral surface area. J Plant Nutr Soil Sci 173:699–705
Serna-Pérez A, Monger HC, Herrick JE, Murray L (2006) Carbon dioxide emissions from exhumed petrocalcic horizons. Soil Sci Soc Am J 70:795–805
Serrano-Ortiz P, Roland M, Sanchez-Moral S, Janssens IA, Domingo F, Goddéris Y, Kowalski AS (2010) Hidden, abiotic CO_2 flows and gaseous reservoirs in the terrestrial carbon cycle: review and perspectives. Agr For Meteorol 150:321–329
Shibu ME, Leffelaar PA, Van Keulen H, Aggarwal PK (2006) Quantitative description of soil organic matter dynamics-A review of approaches with reference to rice-based cropping systems. Geoderma 137:1–18
Siletti CE, Zeiner CA, Bhatnagar JM (2017) Distributions of fungal melanin across species and soils. Soil Biol Biochem 113:285–293
Silver WL, Miya RK (2001) Global patterns in root decomposition: comparisons of climate and litter quality effects. Oecologia 129:407–419
Sindelarova K, Granier C, Bouarar I et al (2014) Global data set of biogenic VOC emissions calculated by the MEGAN model over the last 30 years. Atmos Chem Phys 14:9317–9341
Singh BK, Bardgett RD, Smith P, Reay DS (2010) Microorganisms and climate change: terrestrial feedbacks and mitigation options. Nat Rev Microbiol 8:779–790
Singh M, Sarkar B, Sarkar S et al (2018) Stabilization of soil organic carbon as influenced by clay mineralogy. Adv Agron. https://doi.org/10.1016/bs.agron.2017.11.001
Six J, Elliott ET, Paustian K (2000) Soil macroaggregate turnover and microaggregate formation: a mechanism for C sequestration under no-tillage agriculture. Soil Biol Biochem 32:2099–2103
Six J, Feller C, Denef K, Ogle SM, de Moraes Sa JC, Albrecht A (2002) Soil organic matter, biota and aggregation in temperate and tropical soils—effects of no-tillage. Agronomie 22:755–775
Six J, Frey SD, Thiet RK, Batten KM (2006) Bacterial and fungal contributions to carbon sequestration in agroecosystems. Soil Sci Soc Am J 70:555–569
Smith KA, Conen F (2004) Impacts of land management on fuxes of trace greenhouse gases. Soil Use Manage 20:255–263
Smith P, Fang C (2010) A warm response by soils. Nature 464:499–500
Smith P, Lutfalla S, Riley WJ et al (2018) The changing faces of soil organic matter research. Eur J Soil Sci 69:23–30. https://doi.org/10.1111/ejss.12500
Smith P, Martino D, Cai Z, Gwary D, Janzen H, Kumar P, McCarl B, Ogle S, O'Mara F, Rice C, Scholes B, Sirotenko O, Howden M, McAllister T, Pan G, Romanenkov V, Schneider U, Towprayoon S, Wattenbach M, Smith J (2008) Greenhouse gas mitigation in agriculture. Phil Trans R Soc B 363:789–813
Smith P, Olesen JE (2010) Synergies between the mitigation of, and adaptation to, climate change in agriculture. J Agric Sci 148:543–552
Smith SE, Read DJ (2008) Mycorrhizal symbiosis. Academic Press, London
Solomon S, Plattner G-K, Knutti R, Friedlingstein P (2009) Irreversible climate change due to carbon dioxide emissions. Proc Natl Acad Sci USA 105:14239–14240
Soper FM, McCalley CK, Sparks K, Sparks JP (2016) Soil carbon dioxide emissions from the Mojave Desert: isotopic evidence for a carbonate source. Geophys Res Lett 44:245–251
Sposito G (1989) The chemistry of soils. Oxford Univ Press, NewYork
Stanbery CA, Pierce JL, Benner SG, Lohse K (2017) On the rocks: Quantifying storage of inorganic soil carbon on gravels and determining pedon-scale variability. Catena 157:436–442

Steffens M, Rogge DM, Mueller CW et al (2017) Identification of distinct functional microstructural domains controlling C storage in soil. Environ Sci Technol 51:12182–12189. https://doi.org/10.1021/acs.est.7b03715

Stevenson FJ (1994) Humus chemistry: genesis, composition, reactions. Wiley, New York, NY

Stewart CE, Plante AF, Paustian K, Conant RT, Six J (2008) Soil carbon saturation: linking concept and measurable carbon pools. Soil Sci Soc Am J 72:379–392

Stewart CE, Paustian K, Conant RT, Plante AF, Six J (2007) Soil carbon saturation: concept, evidence and evaluation. Biogeochemistry 86:19–31

Stewart CE, Paustian K, Conant RT, Plante AF, Six J (2009) Soil carbon saturation: implications for measurable carbon pool dynamics in long-term incubations. Soil Biol Biochem 41:357–366

Stockmann U, Adams MA, Crawford JW et al (2013) The knowns, known unknowns and unknowns of sequestration of soil organic carbon. Agric Ecosyst Environ 164:80–99

Suarez DL (2000) Impact of agriculture on CO_2 as affected by changes in inorganic carbon. In: Lal R, Kimble JM, Eswaran H, Stewart BA (eds) Global climate change and pedogenic carbonates. CRC Press, Boca Raton, FL, pp 257–272

Sun Y, Frankenberg C, Wood JD, et al (2017) OCO-2 advances photosynthesis observation from space via solar-induced chlorophyll fluorescence. Science 358:eaam5747. https://doi.org/10.1126/science.aam5747

Swift MJ, Heal OW, Anderson JM (1979) Decomposition in terrestrial ecosystems. University of California Press, Berkeley

Talbot JM, Allison SD, Treseder KK (2008) Decomposers in disguise: mycorrhizal fungi as regulators of soil C dynamics in ecosystems under global change. Funct Ecol 22:955–963. https://doi.org/10.1111/j.1365-2435.2008.01402.x

Tao F, Zhang Z (2010) Adaptation of maize production to climate change in North China Plain: quantify the relative contributions of adaptation options. Eur J Agron 33:103–116

Taylor LT, Quirk J, Thorley RMS et al (2016) Enhanced weathering strategies for stabilizing climate and averting ocean acidification. Nat Clim Change 6:402–408. https://doi.org/10.1038/NCLIMATE2882

Terrer C, Vicca S, Stocker BD et al (2018) Ecosystem responses to elevated CO_2 governed by plant–soil interactions and the cost of nitrogen acquisition. New Phytol 217:507–522. https://doi.org/10.1111/nph.14872

Thauer RK (2007) A fifth pathway of carbon fixation. Science 318:1732–1733

Thevenot M, Dignac M-F, Rumpel C (2010) Fate of lignins in soils: a review. Soil Biol Biochem 42:1200–1211

Torn MS, Swanston CW, Castanha C, Trumbore SE (2009) Storage and turnover of natural organic matter in soil. In: Senesi N, Xing B, Huang PM (eds) Biophysico-chemical processes involving natural nonliving organic matter in environmental systems. Wiley, Hoboken, New Jersey, USA, pp 219–272

Torn MS, Trumbore SE, Chadwick OA, Vitousek PM, Hendricks DM (1997) Mineral control of soil organic carbon storage and turnover. Nature 389:170–173

Treseder KK, Cross A (2006) Global distributions of arbuscular mycorrhizal fungi. Ecosystems 9:305–316

Treseder KK, Turner KM (2007) Glomalin in ecosystems. Soil Sci Soc Am J 71:1257–1266

Trumbore SE (1997) Potential responses of soil organic carbon to global environmental change. Proc Natl Acad Sci USA 94:8284–8291

Turner DP, Urbanski S, Bremer D, Wofsy SC, Meyers T, Gower ST, Gregory M (2003) A cross-biome comparison of daily light use efficiency for gross primary production. Global Change Biol 9:383–395

U.S. Department of Energy (2008) Carbon cycling and biosequestration: integrating biology and climate through systems science. Report from the March 2008 Workshop, DOE/SC-108, U.S. Department of Energy Office of Science, http://genomicsgtl.energy.gov/carboncycle/

U.S. Environmental Protection Agency (2006) Global anthropogenic non-CO_2 greenhouse gas emissions, 1990–2020. EPA Report 430-R-06-003, U.S. Environmental Protection Agency, Washington, DC

Vancampenhout K, Wouters K, De Vos B, Buurman P, Swennen R, Deckers J (2009) Differences in chemical composition of soil organic matter in natural ecosystems from different climatic regions - a pyrolysis-GC/MS study. Soil Biol Biochem 41:568–579

van der Heijden MGA, Martin FM, Selosse MA, Sanders IR (2015) Mycorrhizal ecology and evolution: the past, the present, and the future. New Phytol 205:1406–1423

van der Voort TS, Zell CI, Hagedorn F, et al (2017) Diverse soil carbon dynamics expressed at the molecular level. Geophys Res Lett 44:11,840–11,850. https://doi.org/10.1002/2017GL076188

van der Wal A, de Boer W (2017) Dinner in the dark: illuminating drivers of soil organic matter decomposition. Soil Biol Biochem 105:45–48

van Groenigen JW, Lubbers IM, Vos HMJ et al (2014) Earthworms increase plant production: a meta-analysis. Sci Rep 4:6365. https://doi.org/10.1038/srep06365

van Groenigen JW, van Kessel C, Hungate BA et al (2017) Sequestering soil organic carbon: a nitrogen dilemma. Environ Sci Technol 51:4738–4739. https://doi.org/10.1021/acs.est.7b01427

Van Hemelryck H, Fiener P, Van Oost K, Govers G (2009) The effect of soil redistribution on soil organic carbon: an experimental study. Biogeosci Discuss 6:5031–5071

Van Oost K, Quine TA, Govers G, De Gryze S, Six J, Harden JW, Ritchie JC, McCarty GW, Heckrath G, Kosmas C, Giraldez JV, Marques Da Silva JR, Merckx R (2007) The impact of agricultural soil erosion on the global carbon cycle. Science 318:626–629

Van Pelt RS, Zobeck TM (2009) Carbonaceous materials in soil-derived dusts. In: Lal R, Follett RF (eds) Soil carbon sequestration and the greenhouse effect. SSSA Special Publication 57, 2nd ed. Soil Science Society of America, Madison, WI, pp 365–391

Verbruggen E, Kiers ET (2010) Evolutionary ecology of mycorrhizal functional diversity in agricultural systems. Evol Appl 3:547–560

Vidal A, Quenea K, Alexis M, Derenne S (2016) Molecular fate of root and shoot litter on incorporation and decomposition in earthworm casts. Org Geochem 101:1–10

Virto I, Antón R, Apesteguía M, Plante A (2018) Role of carbonates in the physical stabilization of soil organic matter in agricultural Mediterranean soils. In: Muñoz MÁ, Zornoza R (eds) Soil management and climate change. Academic Press, London, UK, pp 121–136

Von Lützow M, Kögel-Knabner I (2010) Response to the concept paper: 'What is recalcitrant soil organic matter?' by Markus Kleber. Environ Chem 7:333–335

Von Lützow M, Kögel-Knabner I, Ekschmitt K, Matzner E, Guggenberger G, Marschner B, Flessa H (2006) Stabilization of organic matter in temperate soils: mechanisms and their relevance under different soil conditions—a review. Eur J Soil Sci 57:426–445

Vossbrinck CR, Coleman DC, Wooley TA (1979) Abiotic and biotic factors in litter decomposition in a semiarid grassland. Ecology 60:265–271

Wada Y (2016) Modeling groundwater depletion at regional and global scales: present state and future prospects. Surv Geophys 37:419–451. https://doi.org/10.1007/s10712-015-9347-x

Walker AP, Quaife T, van Bodegom PM et al (2017) The impact of alternative trait-scaling hypotheses for the maximum photosynthetic carboxylation rate (V_{cmax}) on global gross primary production. New Phytol 215:1370–1386. https://doi.org/10.1111/nph.14623

Walker JCG (1980) The oxygen cycle. In: Hutzinger O (ed) The natural environment and the biogeochemical cycles. Springer, Berlin, pp 87–104

Wander MM, Drinkwater LE (2000) Fostering soil stewardship through soil quality assessment. Appl Soil Ecol 15:61–73

Wang B, Lerdau M, He Y (2017a) Widespread production of nonmicrobial greenhouse gases in soils. Glob Change Biol 23:4472–4482. https://doi.org/10.1111/gcb.13753

Wang Z, Hoffmann T, Six J et al (2017b) Human-induced erosion has off set one-third of carbon emissions from land cover change. Nat Clim Change 7:345–350

Wang Z, Van Oost K, Govers G (2015) Predicting the long-term fate of buried organic carbon in colluvial soils. Global Biogeochem Cycles 29:65–79. https://doi.org/10.1002/2014GB004912

Watanabe Y, Stewart BW, Ohmoto H (2004) Organic- and carbonate-rich soil formation 2.6 billion years ago at Schagen, East Transvaal district, South Africa. Geochim Cosmochim Ac 68:2129–2151

Weedon JT, Cornwell WK, Cornelissen JHC, Zanne AE, Wirth C, Coomes DA (2009) Global meta-analysis of wood decomposition rates: a role for trait variation among tree species? Ecol Lett 12:45–56

Wei S, Yi C, Fang W, Hendrey G (2017) A global study of GPP focusing on light-use efficiency in a random forest regression model. Ecosphere 8(5):e01724. https://doi.org/10.1002/ecs2.1724

West TO, McBride AC (2005) The contribution of agricultural lime to carbon dioxide emissions in the United States: dissolution, transport, and net emissions. Agr Ecosyst Environ 108:145–154

Whalen JK, Sampedro L (2009) Primary production. In: Whalen JK, Sampedro L (ed) Soil ecology and management. CAB International, Wallingford, UK, pp 109–133

White KE, Coale FJ, Reeves JB III (2018) Degradation changes in plant root cell wall structural molecules during extended decomposition of important agricultural crop and forage species. Org Geochem 115:233–245

Whitham TG, Bailey JK, Schweitzer JA, Shuster SM, Bangert RK, LeRoy CJ, Lonsdorf EV, Allan GJ, DiFazio SP, Potts BM, Fischer DG, Gehring CA, Lindroth RL, Marks JC, Hart SC, Wimp GM, Wooley SC (2006) A framework for community and ecosystem genetics: from genes to ecosystems. Nat Rev Genet 7:510–523

Wilkinson MT, Richards PJ, Humphreys GS (2009) Breaking ground: pedological, geological, and ecological implications of soil bioturbation. Earth Sci Rev 97:254–269

Wilson GWT, Rice CW, Rillig MC, Springer A, Hartnett DC (2004) Soil aggregation and carbon sequestration are tightly correlated with the abundance of arbuscular mycorrhizal fungi: results from long-term field experiments. Ecol Lett 12:452–461

Wolf J, West TO, Le Page Y et al (2015) Biogenic carbon fluxes from global agricultural production and consumption. Global Biogeochem Cycles 29:1617–1639. https://doi.org/10.1002/2015GB005119

Wolters V (2000) Invertebrate control of soil organic matter stability. Biol Fert Soils 31:1–19

Wood WW, Hyndman DW (2017) Groundwater depletion: a significant unreported source of atmospheric carbon dioxide. Earth's Future 5. https://doi.org/10.1002/2017EF000586

Wright SF, Upadhyaya A (1996) Extraction of an abundant and unusual protein from soil and comparison with hyphal protein of arbuscular mycorrhizal fungi. Soil Sci 161:575–586

Wu H, Guo Z, Gao Q, Peng C (2009) Distribution of soil inorganic carbon storage and its changes due to agricultural land use activity in China. Agr Ecosyst Environ 129:413–421

Wutzler T, Reichstein M (2007) Soils apart from equilibrium—consequences for soil carbon balance modelling. Biogeosciences 4:125–136

Wutzler T, Zaehle S, Schrumpf M, Ahrens B, Reichstein M (2017) Adaptation of microbial resource allocation affects modelled long term soil organic matter and nutrient cycling. Soil Biol Biochem 115:322–336

Xia L, Lam SK, Yan X, Chen D (2017) How does recycling of livestock manure in agroecosystems affect crop productivity, reactive nitrogen losses, and soil carbon balance? Environ Sci Technol 51:7450–7457

Yan Y, Zhou X, Jiang L, Luo Y (2017) Effects of carbon turnover time on terrestrial ecosystem carbon storage. Biogeosciences 14:5441–5454

Young IM, Crawford JW, Nunan N, Otten W, Spiers A (2008) Microbial distribution in soils: physics and scaling. Adv Agron 100:81–121

Yu Z, Loisel J, Brosseau DP, Beilman DW, Hunt SJ (2010) Global peatland dynamics since the last glacial maximum. Geophys Res Lett 37:L13402. https://doi.org/10.1029/2010GL043584

Yu G, Xiao J, Hu S et al (2017) Mineral availability as a key regulator of soil carbon storage. Environ Sci Technol 51:4960–4969

Zamanian K, Pustovoytov K, Kuzyakov Y (2016) Pedogenic carbonates: forms and formation processes. Earth Sci Rev 157:1–17

Zhang D, Hui D, Luo Y, Zhou G (2008) Rates of litter decomposition in terrestrial ecosystems: global patterns and controlling factors. J Plant Ecol 1:85–93

Zhang Y, Xu M, Chen H, Adams J (2009) Global pattern of NPP to GPP ratio derived from MODIS data: effects of ecosystem type, geographical location and climate. Global Ecol Biogeogr 18:280–290

Zhao Z, Peng C, Yang Q et al (2017) Model prediction of biome-specific global soil respiration from 1960 to 2012. Earth's Future 5:715–729. https://doi.org/10.1002/2016EF000480
Zhu X-G, Long SP, Ort DR (2010) Improving photosynthetic efficiency for greater yield. Annu Rev Plant Biol 61:235–261
Zimmerman AR, Chorover J, Goyne KW, Brantley SL (2004) Protection of mesopore-adsorbed organic matter from enzymatic degradation. Environ Sci Technol 38:4542–4548
Zsolnay Á (2003) Dissolved organic matter: artefacts, definitions, and functions. Geoderma 113:187–209

Chapter 3
Carbon Sequestration in Cropland Soils

Contents

Abstract Humans began to cultivate land thousands of years ago, for growing crops after clearing the previous vegetation cover and plowing the soil. The soil disturbance altered, in particular, soil C dynamics which has recently been exacerbated by the increase in crop intensification (i.e., fertilization, irrigation, liming, mechanization). This has resulted in major losses of terrestrial carbon (C); i.e., global emissions by the expansion of agriculture for croplands were estimated at 98.4 Pg C for the period 1850–2015. Cropland management can alter soil inorganic carbon (SIC) stocks which, in arid and semiarid climates, can be similar to or more than the amount of soil organic carbon (SOC) stock. However, whether a specific practice leads to a net sequestration of SIC in cropland soils needs additional research. Otherwise, the conversion of land to cropland in temperate regions may release up to 36% of the SOC stock to 27 cm depth, and up to 30% of the SOC stock to 48 cm depth in tropical regions. In 2000, about 12% of Earth's ice-free land surface or 15 million km^2 was used for croplands. Climate, geology, and land and crop management practices control the magnitude of the cropland SOC stock. A major fraction (25–70%) of the carbon dioxide (CO_2) fixed during crop photosynthesis by gross primary production (GPP) is respired autotrophically (R_a) back to the atmosphere. Estimates of global cropland GPP vary between 8.2 and 20.0 Pg C yr^{-1} (1 Pg = 10^{15} g). The remaining net primary production (NPP = GPP − R_a) is the main C input into cropland soils. Cropland NPP includes

K. Lorenz and R. Lal, *Carbon Sequestration in Agricultural Ecosystems*,
https://doi.org/10.1007/978-3-319-92318-5_3

the production of biomass in foliage, shoots and roots, weed and seed production, root exudation, the carbon (C) transfer to microorganisms that are symbiotically associated with roots, and the volatile organic carbon (VOC) emissions that are lost from leaves to the atmosphere. However, not all components of cropland NPP have yet been measured in a single study. NPP enters soil by rhizodeposition and decomposition of plant litter, but the major fraction is heterotrophically converted back to CO_2 by soil respiration and some is lost as methane (CH_4). Aside decomposition, C losses from croplands soils occur also by fire, erosion, leaching, and harvest. Thus, a small amount of fixed C remains in cropland soils and accumulates as SOC due to a combination of short- and long-term stabilization processes. Stabilization processes include physical protection of organic matter (OM) against decomposers and their enzymes, stabilization by organo-mineral complexes and organo-metal interactions, and some as biochemically recalcitrant black carbon (BC). Soil aggregation, in particular, may be the most important stabilization process in the surface layers of cropland soils. Site-specific factors including climate, physicochemical characteristics, soil and vegetation management determine the balance between C input and losses. Cropland soils can be recarbonized through the adoption of recommended management practices (RMPs) such as conservation tillage, residue mulching and use of cover crops, practices which all contribute to SOC accumulation and sequestration by an additional transfer of C from the atmosphere to the soil. Translating science of cropland soil management into a global restorative program is a high priority for feeding the world, mitigating climate change and improving the environment.

Keywords Cropland types · Plant roots · Rhizodeposition · Manure
Anthropogenic drivers · Recarbonization

3.1 Introduction

There is no single definition for the term cropland. For example, FAO does not clearly define 'cropland' but defines 'arable lands' as 'land under temporary crops (double-cropped areas are counted only once), temporary meadows for mowing or pasture, land under market and kitchen gardens and land temporarily fallow (<5 years)' (http://faostat.fao.org/site/375/default.aspx). 'Permanent crops' are defined by FAO as 'land cultivated with crops that occupy the land for long periods and need not be replanted after each harvest, such as cocoa (*Theobroma cacao* L.), coffee (*Coffea arabica* L.), and rubber (*Hevea brasiliensis* Müll.Arg.).' This category includes land under flowering shrubs, fruit trees, nut trees, and vines, but excludes land under trees grown for wood or timber." There remains much misunderstanding and confusion regarding the definition of cropland (Ramankutty et al. 2008). See et al. (2015) proposed to use the definition of the International Food Policy Research Institute (IFPRI), i.e., cropland is arable land and land under permanent crops. This excludes hay and fodder crops or managed grasslands.

Historically, the global cropland area has increased drastically by 110% from 1850 to 2015 (Houghton and Nassikas 2017). Crops have been cultivated by humans for thousands of years on land reclaimed by converting forests, savannas, and grasslands causing the largest emissions of carbon (C) from land-use change (Houghton 2010; Pielke et al. 2011). Specifically, expansion of agriculture for croplands resulted in cumulative land-use and land-cover change (LULCC) emissions of 98.4 Pg C from 1850 to 2015, while total net emissions attributable to terrestrial land use were 145.5 Pg C for the same time period (Houghton and Nassikas 2017). Predominantly in Asia, forests have been cleared for croplands, and a substantial amount of savanna and grasslands have been converted to croplands in North America, Africa, and the Former Soviet Union (Ramankutty et al. 2008). About 80% of new croplands are currently replacing forests in the tropics (Gibbs et al. 2010). Between 1985 and 2005, global cropland area increased by only about 2.4% but harvested area by about 7% (Foley et al. 2011). This cropland intensification is associated with large increases in energy, water, and fertilizer consumption, along with considerable losses in biodiversity (Foley et al. 2005).

In 2000, about 12% of Earth's ice-free land surface or 15 million km^2 were covered by croplands (Ramankutty et al. 2008). However, uncertainties are high, i.e., current global estimates of the amount of land under crop production vary by 3 million km^2 (Fritz et al. 2011). Thus, there is a need for an improved global spatially explicit cropland map (See et al. 2015). For example, Ramankutty et al. (2008) reported that in the year 2000 the largest cropland areas were located in South Asia, the Former Soviet Union and Tropical Africa (Table 3.1). Based on a

Table 3.1 Cropland area by different regions and anthropogenic biomes (Ellis and Ramankutty 2008)

Region/anthropogenic cropland biome	Million km^2
Canada	0.41
USA	1.79
Mexico and Central America	0.50
South America, northern portions	0.74
Argentina, Uruguay, and Chile	0.37
Europe	1.25
North Africa and Middle East	0.84
Tropical Africa	1.94
Former Soviet Union	2.07
China	1.40
South Asia	2.22
Southeast Asia	0.97
Pacific developed countries	0.40
Residential irrigated	0.97
Residential rainfed mosaic	4.02
Populated irrigated	0.25
Populated rainfed	2.20
Remote	0.51

more detailed understanding of large areas that were never mapped before or were inaccurately mapped as non-croplands, USGS (2017) estimated that there were 18.7 million km^2 of croplands in 2015. Cropland area of India, USA, China, and Russia were estimated at 1.798, 1.678, 1.652, and 1.558 million km^2. For the same year, Houghton and Nassikas (2017) estimated the cropland areas (million km^2) of 3.33, 2.41, 2.00, and 1.98 million km^2 for South and Southeast Asia, Tropical Africa, North America, and Latin America, respectively. Cropland area can also be distinguished by biomes such as residential irrigated (i.e., irrigated cropland with substantial human populations), residential rainfed mosaic (i.e., mix of trees and rainfed cropland with substantial human populations), populated irrigated, populated rainfed and remote croplands (Ellis and Ramankutty 2008). In 2008, residential rainfed mosaic was by far the most extensive cropland biome and was the most abundant in Africa and Asia (Table 3.1; Ellis and Ramankutty 2008).

Global annual harvested area increased by 27% from 9.92 million km^2 in 1961 to 12.57 million km^2 in 2011 (Wolf et al. 2015). In 2000, the globally harvested areas (million km^2) for cereals, oil crops, and forage were about 6.6, 1.8, and 1.4, respectively (Monfreda et al. 2008) Annual grains were planted on 75% of the global cropland area. The largest cereal-harvested areas were located in Asia (2.7 million km^2), and Europe and the Former Soviet Union (in total 1.3 million km^2). Oil crops were harvested on about 0.7 million km^2 in Asia, and forages were grown on similar-sized area in combined Europe and the Former Soviet Union. Perennial crops can be distinguished from annual crops based on longevity. Specifically, crops that grow for more than 1 year under typical cultivation practices can be classified as perennials. Distinguishing between the area under annual and perennial crops is important with regard to soil organic carbon (SOC) sequestration. Specifically, perennial crops allocate a higher proportion of photosynthetically fixed C belowground compared to annual crops by maintaining a permanent vegetation cover and a high root turnover with a high SOC input (Don et al. 2011). Further, compared with annual counterparts, perennial crops tend to have longer growing seasons and deeper rooting depths, and intercept, retain, and utilize more precipitation (Glover et al. 2010). About 1.8 million km^2 of harvested crop area in 2000 was under perennial orchards, grasses and pastures (Monfreda et al. 2008). In general, high proportions of perennial crops are located in the tropics, particularly in Indonesia, Thailand, Malaysia, and Papua New Guinea. The perennial crops produce often high-value products including coffee, cocoa, fruit, oil palm (*Elaeis guineensis* or *E. oleifera*), and nuts for export.

For the assessment of SOC sequestration potential in cropland, it is useful to classify the cropland area on the basis of other plant functional types (PFTs) aside annual/perennial PFTs (Smith et al. 1993). For example, the crop life form or physiognomy classification herbaceous (forbs and graminoids) dominated 91% of all harvested crop area in 2000 (12.3 million km^2; Monfreda et al. 2008). Annual crops, in particular, are exclusively herbaceous. The remaining 9% of the area under non-herbaceous crops was under the life forms wood perennial shrubs and trees (0.4 and 0.8 million km^2, respectively). Shrubs were predominant in the coffee and cocoa growing regions of western and eastern Africa, the Central American countries south of Mexico, the eastern coast of Brazil, northern South America (e.g., Ecuador, Colombia, and Venezuela) and in pockets of Southeast Asia. Grapes (*Vitis* spp.) and berry-bearing shrubs covered small areas in temperate countries. Further, herbaceous crops occupied much more area than either shrub or tree crops throughout the temperate zones. Although fruit-bearing orchards were present in the humid tropics, coconuts (*Cocos nucifera* L.) and oil palm were the predominate perennials. Specifically, Indonesia accounted for about one-quarter of the global coconuts area, and Indonesia and Malaysia together grew one-half of all oil palm trees in 2000 (Monfreda et al. 2008).

The proportion of C_3 and C_4 vegetation in a region affects the flux of carbon dioxide (CO_2) between plants and the atmosphere (Still et al. 2003). Thus, classification of croplands with respect to SOC sequestration is also possible by distinguishing crop functional types C_3 and C_4 based on photosynthesis pathways. In 2000, the C_4 crops {i.e., corn or maize (*Zea mays* L.), sorghum [*Sorghum bicolor* (L.) Moench], millet, sugarcane (*Saccharum* spp.)} and some grasses comprised 3.2 million km^2 or a disproportionate 24% of all harvested area (Monfreda et al. 2008). The greatest proportion of C_4 crops was located in central Mexico through the Yucatan Peninsula, the eastern portion of southern Africa extending from South Africa to Mozambique and Zambia, and a huge swath of the Sahel running from Mauritania in the west to Ethiopia in the east (Monfreda et al. 2008). Further, at least 50% of all crops planted in the US Corn Belt and of the crops grown in northern Argentina and southern Brazil were C_4 crops. In these regions, corn is grown in rotation with soybeans [*Glycine max* (L.) Merr.] and fodder grass to produce animal feedstock. The C_4 crops are, in particular, important in arid, agriculturally marginal regions (Monfreda et al. 2008).

Legumes are the second most important source for nitrogen (N) in the terrestrial biosphere contributing about 40 Tg N yr^{-1} (1 Tg = 10^{12} g), or about half of the amount of fertilizer N applied to croplands (Monfreda et al. 2008; Conant et al. 2013). In 2000, leguminous crops occupied 2.4 million km^2 or 18% of the global harvested cropland area which included all pulses plus groundnuts, soybeans, alfalfa (*Medicago sativa* L.), and other leguminous forage crops, and half of the area of mixed grasses and legumes. Specifically, soybeans and alfalfa covered much of the central USA, the region near the border of Argentina and Brazil, and southern Australia. Soybeans and alfalfa were the primary protein source for livestock in these regions and were grown in rotation with corn, which supply livestock with the bulk of their energy requirements (Monfreda et al. 2008).

3.2 Cropland Soil Inorganic Carbon

Cropland practices may have a large impact on the inorganic C cycle of soils. Irrigation, fertilization and liming, in particular, have the potential to greatly alter soil inorganic carbon (SIC) and, thus, affect cropland soil C sequestration (Lal and Kimble 2000). SIC can accumulate or be lost at similar or even greater rates (up to

1 Mg C ha^{-1} yr^{-1}) as SOC but the net effects of cropland management practices on the SIC stock are uncertain (Sanderman 2012). The accumulation of pedogenic inorganic C (PIC) in croplands depends, in particular, on the Ca^{2+} and Mg^{2+} availability. For example, various sources of Ca and/or Mg include fertilizers, dust, irrigation water and weathering of calcium/magnesium silicate minerals, which may have contributed to the high rates of PIC accumulation (>20 g C m^{-2} yr^{-1}) in arid croplands of northwest China (Wang et al. 2015). In contrast, croplands in Canada, USA, and New Zealand accumulated PIC at lower rates of <3 g C m^{-2} yr^{-1}.

Agricultural practices in cropland may affect SIC by interacting with the weathering of silicate minerals, carbonate dissolution and pedogenic carbonate formation (Sanderman 2012). The acceleration of silicate mineral weathering by cropland management has negligible effects on C sequestration as only 0.001–0.01 Mg CO_2–C ha^{-1} yr^{-1} may be consumed by this process (Chadwick et al. 1994). However, carbonate dissolution due to cropland management can be a major C sequestration mechanism especially when (i) CO_2 is derived from the recent atmosphere (i.e., root respiration or decomposition of soil organic matter [SOM]), and (ii) the HCO_3^- is leached into groundwater systems or the oceanic pool where its mean residence time (MRT) is of the order of 1000s of years or longer (Sanderman 2012).

The rate of carbonate dissolution may be altered by enhanced biological activity under cropland which results in increased soil CO_2 levels. Various agricultural practices in croplands (i.e., application of irrigation water, installation of tile drainage and the slow acidification of many cropland soils due to the application of ammonium-based fertilizers) may also accelerate carbonate dissolution and loss rates. However, due consideration must be given for both the temporal (i.e., MRTs of atmospheric CO_2, HCO_3^-, and soil carbonates are 5 , 100–1,000 year, and 1000–10,000 year, respectively), and spatial scales (i.e., farm, groundwater, ocean) over which the C accounting is made (Sanderman 2012). The formation of pedogenic carbonates in cropland soils leads to net sequestration of CO_2 only if (i) the HCO_3^- is derived from active biological respiration (i.e., root respiration or microbial breakdown of recently formed SOM), and/or (ii) the source of Ca is derived from the weathering of Ca-/Mg-bearing silicate minerals, from parent material or eolian deposition (Lal and Kimble 2000).

It has been hypothesized that irrigation, especially when combined with tile drainage, may cause large changes in cropland SIC stocks due to the acceleration of mineral dissolution rates, and by supplying dissolved Ca and HCO_3^- (Sanderman 2012). For example, Entry et al. (2004) found that after >30 years of irrigated cropping in the western USA, SIC stocks to 1 m depth were 51 Mg C ha^{-1} greater than those under native vegetation. In contrast, Wu et al. (2008) reported that depending on the source of the irrigation water, SIC stocks increased, decreased or remained essentially unchanged to 1 m depth. However, SIC changes in the entire cropland soil profile must be studied as redistribution of SIC within the profile and HCO_3^- losses from the soil profile may occur (Eshel et al. 2007). Thus, irrigated agriculture can increase or decrease cropland SIC stocks or have negligible effects. Most importantly, the movement of HCO_3^- is a core question that needs to be addressed for evaluating irrigation-induced changes in SIC stock (Ahmad et al. 2015). The fate of HCO_3^- must also be studied for evaluating the effects of ubiquitous acidification of cropland soils on SIC. Acidification is caused by the removal of alkalinity by harvest, nitrate leaching, accelerated N-fixation by leguminous crops and the application of ammonium-based fertilizers. Further, a net atmospheric sink is established when HCO_3^- is leached from the soil profile after lime application to croplands. In conclusion, more research is needed to ascertain the importance of crop management-induced changes in SIC stocks on the net CO_2 emissions (Sanderman 2012). Specifically, all reactants and products must be fully accounted for to elucidate whether or not specific processes in croplands lead to a net sequestration of SIC (Ahmad et al. 2015).

3.3 Cropland Soil Organic Carbon

The fate of C recently fixed during plant photosynthesis (i.e., gross primary production or GPP) is the major determinant of the cropland SOC stock. Only the fraction remaining after accounting for C losses by autotrophic respiration R_a (i.e., the net primary production or NPP = GPP − R_a) is stored in new plant biomass. While 10 Pg C are currently stored in cropland biomass, potential biomass stock would be 139–141 Pg C under the hypothetical absence of land use and current climate conditions (Erb et al. 2018). Specifically, crop NPP is lost by biomass removal during harvest (i.e., grains, pulses, and other harvestable products) before reaching the soil. About 42% of global crop biomass is harvested, transported, and respired offsite (Wolf et al. 2015). For example, 30–50% of the aboveground dry mass is harvested from cereal croplands (Johnson et al. 2006). In the 1990 s, global crop harvest was about 2.2 Pg C yr^{-1} (1 Pg = 10^{15} g) (Bondeau et al. 2007). Global mean harvested biomass per area increased from 67 Mg C km^{-2} in 1961 to 163 Mg C km^{-2} in 2011 (Wolf et al. 2015). Further, 0.54 Pg C yr^{-1} was collected from crop residues for livestock fodder. The cropland net C exchange can be estimated as the sum of C uptake by crops, decomposition of in situ crop biomass, and the release of C occurring from the consumption of crops by livestock and humans. Due to production and export of large quantities of crop C, North and South America act as apparent regional C sinks (Wolf et al. 2015).

Herbivory by insects and mammals, and emissions of volatile organic compounds (VOCs) also contribute to biomass loss from croplands (Ciais et al. 2010). Some cropland C is lost by fire, erosion, and leaching (Torn et al. 2009). For example, about 1.6 kg CO_2 is emitted per kg of crop residues burned (Andreae 2004). Globally, about 1,200 Tg of crop residues was burnt in the late 1990s. Especially in the tropics, the ignition of fires is a common human activity to clear land for shifting cultivation by converting forests into croplands (Crutzen and Andreae 1990). Dry vegetation is also removed by human-induced fires to promote crop productivity. However, burning of crop residues has been greatly reduced in developed countries, especially during the industrial era (Marlon et al. 2008). For example, the regulation to improve air quality has significantly reduced cropland burning in the western USA in the last decades (Andela et al. 2017). In contrast, fire activity has increased in some densely populated agricultural regions of India and China. Thus, agricultural intensification may increase fire activity in regions where crop residue burning is the dominant fire type (Andela et al. 2017).

Some of the fixed C remaining in non-erosional croplands, after accounting for the losses by fire, harvest, herbivory, and VOCs, is deposited as above- and belowground residues. Aboveground, but more importantly belowground residues, are sources of SOC (Rasse et al. 2005). Other belowground input aside residues is GPP transferred into the soil as root and mycorrhizal exudates including turnover of fine roots (i.e., <2 mm in diameter) and fungal hyphae, and as other dead microbial cells. Transfer of crop C into soils occurs also by the activity of the mesofauna (e.g., earthworms), hydrological flows of soluble C, and root senescence and mortality. About 50–70% of C fixed in croplands remains aboveground, and 20–30% is transferred belowground (Johnson et al. 2006). Aside by natural processes, organic C is also entering cropland soils by plant and soil management (i.e., addition of manure, and of non-harvested and non-burned residues).

The inputs of organic matter (OM) are substrate for decomposition and subject to losses by heterotrophic, primarily microbial respiration. The remaining SOC stock consists of organic C compounds in plant, microbial and faunal residues at various stages of decomposition with MRTs ranging from days to millennia (Sandermann et al. 2010). Natural processes, environmental factors (i.e., climate, soil parent material), and vegetation and soil management (i.e., crop type, fertilization, irrigation, harvest, residue management, tillage) determine amount and composition of the SOC stock. The persistence of SOC is largely due to complex interactions between OM and its environments, such as the interdependence of compound chemistry, reactive mineral surfaces, climate, water availability, soil acidity, soil redox state, and the presence of potential degraders in the immediate microenvironment (Schmidt et al. 2011). Among natural processes affecting the SOC stock in croplands are (i) C allocation and partitioning among plant organs (i.e., leaves, stems, roots), (ii) C fluxes within plant organs (i.e., to respiration, storage compounds, defensive compounds, structural components), and (iii) C fluxes among soil pools (US DOE 2008). The fraction of organic C accumulating in croplands (i.e., net biome production or NBP) which enters the soil (NBP_{soil}) primarily determines the amount of SOC sequestered (Schulze et al. 2010).

3.3.1 Cropland Soil Carbon Input

The cropland SOC stock is primarily derived from photosynthetically fixed C. Thus, cropland GPP or the annual photosynthetic C uptake of all plant tissues (i.e., crops, weeds) over a specified cropland area determines the potential C input into cropland soils (US DOE 2008). Globally, cropland GPP was estimated at between 8.2 and 20.0 Pg C yr^{-1} (Saugier et al. 2001; Beer et al. 2010; Guanter et al. 2014). From a crop production perspective, photosynthesis includes all the events from light interception to the export of photosynthate for biomass accumulation and grain production (Murchie et al. 2009). Thus, GPP is the basis for food, fiber and biofuel production, and one of the major processes providing the capacity of croplands to partly offset anthropogenic CO_2 emissions (Beer et al. 2010). The net C stored in croplands as new plant material before harvest and other losses is the NPP. The portion of crop NPP at non-erosional croplands remaining after accounting for losses by harvest, herbivory, and VOCs is deposited as above- and belowground residues. Decomposition of aboveground residues, but even more importantly, of the belowground residues together with root exudates, and biotic and hydrologic belowground C transfer, are the major natural cropland soil C input processes (Lorenz and Lal 2005). Addition of manure and residues during land use and soil management (e.g., tillage) may result in additional direct C input into cropland soils.

Natural Input Processes

The input of crop residue C depends on the biomass that grows from a unit input of solar radiation (Amthor 2010). The crop functional types (i.e., C_3 and C_4) differ in their efficiency in converting solar energy into biomass. Specifically, crops using the C_4 photosynthetic pathway are generally more efficient than C_3 crops, but this difference varies with solar radiation, temperature and water supply (Jansson et al. 2010; Sanderman et al. 2010). The reduced photosynthetic efficiency of C_3 crops arises, in particular, from reduced performance (Murchie et al. 2009). However, in spite of the greater capacity and water use efficiency (WUE) of the C_4 photosynthetic pathway C_4, crops may be equally or even more sensitive to water stress than C_3 crops (Ghannoum 2009). Many economically important agricultural crops (e.g., wheat (*Triticum* L.), rice (*Oryza sativa*) and soybean) are C_3 crops (Schulze et al. 2005). Otherwise, the C_4 photosynthetic pathway occurs in economically important crops such as corn, sugarcane, millet, and sorghum.

Gross Primary Production

The GPP is controlled by leaf area, nitrogen (N) supply, season, temperature, light, and atmospheric CO_2 concentration (Chapin et al. 2002). Differences in annual GPP among croplands depend primarily on the quantity of leaf area and the length of time this leaf area is photosynthetically active which are both ultimately determined by the interacting effects of soil resources, climate, vegetation, and disturbance regime (Falge et al. 2002).

Previously, plant- and stand-level GPP was calculated as two times biomass production with considerable variation among biomes and sites (Beer et al. 2010). However, the assumption that the NPP/GPP ratio is consistent regardless of ecosystem type has been challenged (Zhang et al. 2009). For example, under the assumption that NPP/GPP equals 2, global cropland GPP is estimated to be about 8.2 Pg C yr^{-1}, but may range between 11.1 and 20.0 Pg C yr^{-1} based on eddy covariance flux data, diagnostic models, light-use efficiency models and chlorophyll fluorescence (Saugier et al. 2001; Beer et al. 2010; Chen et al. 2014; Guanter et al. 2014). For the year 2000, GPP estimates for maize, rice, and wheat were 1.55, 1.51, and 1.38 Pg C yr^{-1}, respectively (Chen et al. 2014). Cropland GPP can be measured but is generally estimated as it is impossible to measure the net C exchange of all leaves of a cropland in isolation from other components (e.g., soil respiration; Chapin et al. 2002). Thus, estimations of GPP are based on (i) observations, (ii) diagnostic modeling approaches, (iii) process-oriented models, and (iv) satellite-data-based models (Yang et al. 2007; Beer et al. 2010). Field studies, for example, are based on tower eddy covariance systems to calculate seasonal and inter-annual dynamics of cropland GPP (Peng and Gitelson 2011). Some examples of cropland GPP estimated for different time periods, regions and crop species are given in the following section.

Global mean annual GPP of the period 2000–2005 was estimated to be 765 g C m^{-2} yr^{-1} based on monitoring vegetation productivity (Zhao et al. 2010). For the period 1982–2004, estimates of the simulated global crop GPP ranged from 340 to 788 g C m^{-2} yr^{-1} (Bonan et al. 2011). For EU-25, the mean cropland GPP was estimated at 1,120 g C m^{-2} yr^{-1} for the period 2000–2005 based on a combination of top-down estimates of atmospheric observations and bottom-up estimates derived from ground-based measurements (Schulze et al. 2009). Otherwise, based on process-oriented and remote sensing models, cropland GPP for EU-25 was 1,360 g C m^{-2} yr^{-1} and 879 g C m^{-2} yr^{-1} over the period 1990–1999, respectively (Ciais et al. 2010). By applying a NPP/GPP ratio of 0.55, cropland GPP of Europe (but excluding Russia, Belarus, and Ukraine) was estimated at 1,591 g C m^{-2} yr^{-1} (calculated based on Beer et al. 2007).

Seasonal changes in crop GPP can be characterized by CO_2 flux measurements. For mixed croplands, CO_2 fluxes within a season were as low as 0.2 μmol CO_2 m^{-2} s^{-1} and as high as 33.5 μmol CO_2 m^{-2} s^{-1} (Falge et al. 2002). The maximum flux was higher for C_4 than that for C_3 crops (60.5 μmol CO_2 m^{-2} s^{-1} vs. 27.3 μmol CO_2 m^{-2} s^{-1}). Estimated annual GPP based on these measurements were 599 g C m^{-2} yr^{-1} for C_3 crops, 1,101–1,396 g C m^{-2} yr^{-1} for mixed croplands, and 1,471 g C m^{-2} yr^{-1} for C_4 crops. Often, a second maximum of CO_2 emissions developed after harvest due to photosynthetic activity of intercrops or weeds (Falge et al. 2002).

Over 6 years, GPP of irrigated and rainfed corn–soybean cropping systems were estimated by continuous CO_2 measurements in eastern Nebraska, USA (Suyker and Verma 2010). Peak daily GPP values for irrigated and rainfed corn, and for irrigated and rainfed soybean were 24.9 g C m^{-2} d^{-1} and 22.9 g C m^{-2} d^{-1}, and 15.4 g C m^{-2} d^{-1} and 14.4 g C m^{-2} d^{-1}, respectively. Mean growing season totals of GPP for irrigated corn and soybean were 1,738 g C m^{-2} and 996 g C m^{-2}, respectively. Corresponding values for rainfed corn and soybean were 1,553 g C m^{-2} and 895 g C m^{-2}, respectively (Suyker and Verma 2010). The mean annual GPP based on flux-tower sites in North America and Europe was 1,221, 652, and 833 g C m^{-2} yr^{-1} for maize, wheat, and legume fields, respectively (calculated from Gilmanov et al. 2014). However, the CO_2 exchange of leguminous crops is more uncertain than that of non-leguminous crops.

Wagle et al. (2016) used three GPP modeling approaches (space-borne solar-induced chlorophyll fluorescence retrievals, a LUE-based vegetation photosynthesis model, and a process-based soil canopy observation of photochemistry and energy flux model constrained by solar-induced chlorophyll fluorescence) at a maize site in Mead, Nebraska, USA. Predicted annual GPP for the period 2007–2011 ranged from 1,343.8 to 2,014.6 g C m^{-2} yr^{-1}. In contrast, tower-based GPP ranged from 1,566.0 to 1,840.0 g C m^{-2} yr^{-1}. It was concluded that all three modeling approaches captured seasonal dynamics and magnitude of maize GPP (Wagle et al. 2016).

In the North China Plain, C fluxes were measured during four seasons in a typical irrigated cropland under wheat/corn rotation (Lei and Yang 2010). The seasonal GPP ranged between 782 and 1,114 g C m^{-2} for wheat, and between 872 and 880 g C m^{-2} for corn, respectively. The annual GPP values for this rotational cropland were 2,008 g C m^{-2} yr^{-1} for the period 2006–2007, and 1,668 g C m^{-2} yr^{-1} for the period 2007–2008 (Lei and Yang 2010). In a corn cropland in Northeast China, GPP was simulated for a three years period using (i) a photosynthesis model and (ii) estimations based on tower CO_2 flux data (Wang et al. 2010). Annual simulated GPP in 2004, 2005, and 2006 were 310, 464, and 360 g C m^{-2} yr^{-1}, respectively. In contrast, annual estimated GPP for the same years were 392, 504, and 437 g C m^{-2} yr^{-1}, respectively (Wang et al. 2010).

For a winter wheat/summer maize cropping system in China, GPP ranged between 1,370 and 1,531 g C m^{-2} yr^{-1} during two years (Yu et al. 2006). Depending on crop development stage, estimates of daily GPP for winter wheat at a site in Belgium ranged between 0.92 and 18.72 g C m^{-2} d^{-1} during three growing seasons (Dufranne et al. 2011). Growing season GPP at this site was as low as 1,568 g C m^{-2} and as high as 1,716 g C m^{-2}. Satellite-based remote sensing technologies can also be used to characterize GPP of croplands during the growing season. For example, Peng et al. (2011) used chlorophyll-related vegetation indices to estimate corn GPP over a period of eight years at eight rainfed and irrigated sites in Nebraska, USA. GPP followed seasonal changes in chlorophyll content. Daytime growing season GPP varied strongly between 0 and 35 g C m^{-2} d^{-1} depending on weather conditions and growth stage with mean values between 10.4 and 15.1 g C m^{-2} d^{-1} (Peng et al. 2011).

Net Primary Production

The bulk of NPP in croplands is allocated to the production of biomass in foliage, shoots, and roots. Cropland NPP includes also weed and seed production, root exudation, and C transfer to microorganisms that are symbiotically associated with roots (e.g., mycorrhizae and N-fixing bacteria), and the VOC emissions that are lost from leaves to the atmosphere. Further, NPP also includes the biomass removed by herbivory often accounting for 5–10% of NPP. Direct measurements of cropland NPP are not possible as not all of the biomass produced remains in a cropland (Ciais et al. 2010). In addition, other components are rarely measured such as weed and seed production, emission of VOCs, exudation from roots and C transfer to root symbionts. Specifically, all components of cropland NPP have rarely if ever been measured.

Global cropland NPP may be estimated based on inventory, empirical model simulation, biogeochemical model simulation, dynamic global vegetation model simulation, and remote sensing estimation (Ito 2011). At the regional level, crop inventory, biogeochemical model simulation, and remote sensing estimation using a satellite-based model may be used. Often, cropland NPP estimates rely on census and survey data but these data measure agricultural production and not NPP (Ciais et al. 2010). Further, estimates of cropland NPP from yield inventory statistics are biased due to uncertainties in NPP definition, allometry, cropland area and input yield data themselves (Ciais et al. 2010). Recently, multiple satellite data spanning a diverse spectral range (from visible, near-infrared, thermal to microwave) have been synthesized into a coherent framework to estimate crop yield over the US Corn Belt (Guan et al. 2017). Some examples of cropland NPP for different regions and species are given in the following section.

The average EU-25 cropland NPP for the decade 1990–1999 ranged from 586 to 646 g C m^{-2} yr^{-1} based on yield data, from 482 to 585 g C m^{-2} yr^{-1} based on process-oriented and from 419 to 510 g C m^{-2} yr^{-1} based on remote sensing models, respectively (Ciais et al. 2010). The inter-annual variability of NPP in EU-25 is determined by the fluctuating climate conditions but the long-term trend is driven by technological changes.

For the conterminous USA, agricultural statistics (yields and cropped area) have been used to estimate cropland NPP (Hicke et al. 2004). The NPP increased from about 350 g C m^{-2} yr^{-1} in 1972 to about 490 g C m^{-2} yr^{-1} in 2001. This gain was explained by more effective fertilization and pest management, higher yielding cultivars, more favorable climate, shifts to productive crop types (e.g., wheat to corn), and economic factors. Further, a substantial inter-annual variation in cropland NPP in the conterminous USA may have been driven by changes in climate, but economic and management decisions (e.g., increases in irrigation) may have also contributed (Hicke et al. 2004). A 41% increase in cropland NPP in the southern USA was reported by Tian et al. (2010) for the period 1895–2007. Specifically, the annual mean cropland NPP based on a process-based model increased from 369 g C m^{-2} yr^{-1} for the period 1895–1950 to 520 g C m^{-2} yr^{-1} for the period 1951–2007. Drought events were identified as major threats to cropland NPP in the southern USA (Tian et al. 2010). Mean cropland NPP in the Midwestern USA for 2007 and 2008 was 469–687 g C m^{-2} yr^{-1} estimated based on inventory, remote sensing and a process-based model (Li et al. 2014).

The NPP of croplands in China was estimated at 495 g C m^{-2} yr^{-1} during 1901–2005 based on published literature and the Multi-scale Synthesis and Terrestrial Model Intercomparison Project (Shao et al. 2016). Field surveys in China indicated that cropland NPP ranges between 239 and 760 g C m^{-2} yr^{-1} (Zhang et al. 2014).

Using plant C allocation coefficients obtained from studies published after 1970, Bolinder et al. (2007) estimated a NPP of 537 g C m^{-2} yr^{-1} for small-grain cereals in Canada. However, estimating belowground NPP in this study was associated with considerable uncertainty.

Among species, wheat and rice were the most dominant global crops over the period from 1961 to 2011, each accounting for more than 15% of global crop NPP (Wolf et al. 2015). In recent decades, the percentages of total crop NPP contributed by maize grain, soybean, and other oil crops have increased, while the percentages contributed by barley (*Hordeum vulgare* L.), millet, oats (*Avena sativa* L.), rye

(*Secale cereale* L.), and grain sorghum, and potatoes (*Solanum tuberosum* L.) have decreased (Wolf et al. 2015). The NPP in 1991 was as low as 181 g C m^{-2} yr^{-1} for pulses and as high as 801 g C m^{-2} yr^{-1} for sugar crops globally based on agricultural production data (Goudriaan et al. 2001). The NPP of corn, soybean, spring wheat and winter wheat in the Midwestern USA in 2007 and 2008 was 952–1047 g C m^{-2} yr^{-1}, 334–375 g C m^{-2} yr^{-1}, 366–457 g C m^{-2} yr^{-1}, and 370–579 g C m^{-2} yr^{-1}, respectively (Li et al. 2014). Globally, paddy rice had a higher NPP per unit land in 1991 than wheat (376 vs. 271 g C m^{-2} yr^{-1}) as rice is produced under more intensified growing conditions than wheat (Goudriaan et al. 2001). Further, small and coarse grains together accounted for 60% of global crop NPP, whereas oil and sugar crops each accounted for 9%. Globally, the highest cultivated crop NPP has been estimated for western Europe, eastern Asia including Japan, South Korea, and China, and the central USA (Monfreda et al. 2008). High crop NPP has also been estimated for southern Brazil and northern Argentina. These are the only large cultivated areas with NPP > 1,000 g C m^{-2}. The highest rates of all exceed 2,000 g C m^{-2} in the countries of northern Europe bordering the North Sea. Especially intensive agriculture, usually associated with irrigation, attains comparable rates in smaller areas scattered throughout the world, including New Zealand, Israel, Egypt, Saudi Arabia, California, Oregon, and Washington in the western USA, Java, and pockets of Columbia, and India, and Mexico (Monfreda et al. 2008).

Using a method based on production and crop-specific factors, Prince et al. (2001) estimated that county-level NPP in 1992 in the USA ranged from <200 g C m^{-2} yr^{-1} in North Dakota, Wisconsin, and Minnesota to over 850 g C m^{-2} yr^{-1} in central Iowa, Illinois, and Ohio (assuming 50% of dry matter is C). Further, corn grain NPP reached the highest values (>850 g C m^{-2} yr^{-1}) over extensive areas but that for corn silage was lower, and the lowest values were for wheat and barley. Large inter-annual variability in NPP (<350 to >700 g C m^{-2} yr^{-1}) was observed in Iowa from 1982 to 1996, with the lowest NPP values in the years 1983 (which had an unusually wet spring), 1988 (which was a drought year), and 1993 (which experienced floods). In 1992, NPP in Iowa ranged from 232 g C m^{-2} yr^{-1} for soybean to 547 g C m^{-2} yr^{-1} for barley (Prince et al. 2001). The increases in total NPP for croplands in the Great Plains from 1991 to 1996 were largely attributed to high NPP for corn (Bradford et al. 2005).

Belowground Carbon Input

Surface plant residues are incorporated into the mineral soil by physical mixing and solubilization, transport and subsequent adsorption. The relative importance of these processes versus root litter and rhizodeposition for profile SOC distribution and dynamics depend on climate, soil, and vegetation types. DOC represents only a small proportion of SOC but the most mobile and actively cycling SOC fraction. However, there are only a few studies on DOC in cropland soils.

Tillage mixes surface crop residues into the mineral soil which may result in subsurface SOC formation (Wright et al. 2007). For example, stubble incorporated into the soil makes a significant direct contribution to SOC sequestration (Sanderman et al. 2010). Further, repeated erosion events over time can lead to successive layers of sediments and to large amounts of SOC buried at depth in croplands (Chaopricha and Marín-Spiotta 2014). Over and above the effects of crop management and translocation of DOC, crop rooting and belowground biomass also impact SOC formation in the subsoil. Further, in extensively managed croplands and, in particular, those with low or no tillage activity, there is a myriad of soil biota that affect directly and indirectly the inputs of SOC into subsoil (Wilkinson et al. 2009). At European multiyear field experiments, both non-inversion tillage practices and applying organic amendments resulted in larger earthworm populations, a higher microbial biomass content and a marked increase in bacterivorous nematodes (D'Hose et al. 2018). The biotic activity of earthworms, termites, ants, arthropods but also roots efficiently buries SOC. Indirect inputs of SOC into subsoils may occur by infilling of biogenic voids with litter, redistribution of SOC, and subsurface mixing and burial (Rumpel and Kögel-Knabner 2011).

Plant Roots

Roots of crop plants can grow deep into the soil but important knowledge gaps remain (Fan et al. 2017). For the better-studied maize, a mean rooting depth of 0.91 m has been reported. About 20% of the C from photosynthates is transferred to roots in annual crops during the growing season (Whalen and Sampedro 2009). However, not all fluxes and the entire crop root biomass can be directly measured. For example, only 64% of wheat roots were recovered by direct sampling after ^{13}C-labeling (Subedi et al. 2006). Also, there is a lack of spatially representative and reliable data on R_h for croplands which represents a key flux in determining the rate of SOC sequestration in croplands (Schulze et al. 2010). About 65% of the total NPP in EU-25 croplands is consumed in soil respiration, NBP is a sink or source for 3% of NPP, but only 0.06–0.25% of NPP is available for sequestration in cropland soils (Ciais et al. 2010).

About 50% of the C translocated belowground in croplands is used for root growth (Johnson et al. 2006). The root biomass ranges between 6% of the shoot biomass for sunflower (*Helianthus annuus* L.) and 50% for barley (Whalen and Sampedro 2009). Root biomass C of annual cereal crops (wheat, barley (*Hordeum vulgare* L.), oat (*Avena sativa* L.), triticale (x *Triticosecale* Wittm. ex A. Camus)) is 36–67.5 g C m^{-2} (Pausch and Kuzyakov 2017). About 30% of the C translocated belowground moves into the rhizosphere or is released during root respiration (Johnson et al. 2006). Further, as much as 30% of NPP may be transferred to mycorrhizal fungi. In soybean, about 20–30% of photosynthates are allocated to support rhizobia in nodules (Whalen and Sampedro 2009).

The root C is preferentially stabilized as SOC relative to that from surface residue. For example, root-derived C from hairy vetch (*Vicia dasycarpa* Ten.) is stabilized three times more readily in soils under corn-based cropping systems than is surface residue-derived C (Kong and Six 2010). Similarly, Bolinder et al. (1999) and Rasse et al. (2006) estimated that the humification ratio (i.e., SOC/input ratio) of corn residues is about 0.12 for shoots and 0.19 for roots. Further, the humification ratio for barley residues is 0.12 for shoots and 0.16 for roots (Broadbent and Nakashima 1974). For different crops, the contribution factor of roots versus shoots to total SOC (i.e., [root-derived C/total root C input]/[shoot derived C/total shoot C input]) is 2.4 (Rasse et al. 2005). Thus, roots are a more important SOC source, in particular in subsoils, than are shoots. However, the flux of root C into soil is poorly understood because accurate measurements of root decomposition, exudation and root cell sloughing are difficult. Further, it is not known what specific properties of roots influence SOC stability or rhizodeposition (Mendez-Millan et al. 2010)? While inputs of soil C in annual crops can be modeled on the basis of root-to-shoot ratios, the above- and belowground biomass in perennial species may not be representative of C inputs into the soil (Gill et al. 2002).

The inputs of belowground C in croplands are often estimated from root-to-shoot ratios but much current knowledge about root–shoot biomass relations is based on scanty and often unreliable data. By omitting unreliable data, higher estimates for median root-to-shoot ratios are calculated than previously obtained. Accordingly, vegetation-specific root-to-shoot ratios must be used to predict the root biomass (Mokany et al. 2006). Static measurements of live root biomass are also biased as the non-woody roots of many crop plants, which lack secondary growth and cell wall thickening, are short-lived (Hodge et al. 2009).

Not all net fixed C allocated belowground is used for root growth. For example, when the total belowground C allocation is considered, the estimated root-to-shoot ratios of 0.50, 0.30, and 0.33 for wheat, corn, and soybean increase to 0.82, 0.55, and 0.62, respectively (Johnson et al. 2006). However, most studies have focused on young plants at a vegetative stage but partitioning is strongly affected by plant age. For example, the partitioning of C to the rhizosphere decrease by 43, 28, and 20% for roots, rhizosphere respiration and soil residues, respectively, for annual plants of ages ranging from 28–600 days (Nguyen 2003). The root-to-shoot ratios for wheat, corn, soybean, and lupin (*Lupinus* L.) decrease from about 0.4 to 0.6 during vegetative growth phases to as low as 0.1 at flowering (Gregory et al. 1997). Further, almost half of the published data on rhizodeposition are for wheat and ryegrass, and 76% of the studies are related to only five crop/grassland species. Thus, the knowledge of C rhizodeposition, in particular of the mixed crop plant communities, is scanty (Jones et al. 2009).

Rhizodeposition

Rhizodeposition has direct and indirect effects on SOC sequestration in croplands as root exudation, in particular, may represent a significant C loss for crop plants. The magnitude of photosynthates secreted varies with soil type, age, and physiological state of the plant, and nutrient availability (Bais et al. 2006). However, many of the published data on C flow in both soil and roots are severely biased (Rees et al. 2005; Kuzyakov 2006). For annual cereal crops, annual net C input via rhizodeposition may be 18–34 g C m^{-2} but estimates are uncertain as those are based on root sampling and did not consider root turnover during the vegetation period (Pausch and Kuzyakov 2017). Of the 1.5–2.2 Mg C ha^{-1} allocated belowground by cereals during the vegetation period, ~0.4 to 0.6 Mg C ha^{-1} enters the soil in the form of rhizodeposition (Jones et al. 2009; Kuzyakov and Domanski 2000). However, these estimates are highly uncertain as the partitioning of rhizosphere respiration from mycorrhizal respiration is almost impossible. Also, whether root exudates directly contribute to the SOC stock is under discussion (Sanderman et al. 2010). Complex compounds derived from root turnover may contribute indirectly to the SOC stock by enhancing aggregation (Rees et al. 2005).

Mycorrhizal Fungi

Mycorrhizal fungi live in symbiosis with the roots of plants. Almost all agricultural crops are mycorrhizal plants with the exception of the Brassicaceae family (cauliflower [*Brassica oleracea* L. Botrytis group], cabbage [*Brassica oleracea* L. Capitata group], canola [*Brassica campestris* L.], and others; Whalen and Sampedro 2009). In addition, soil yeasts may affect the colonization of crop plant roots by arbuscular mycorrhizal (AM) fungi (Botha 2011). AM fungi are important for sustainable crop production (Sen 2003). For example, AM fungi may facilitate uptake of up to 80% of the P, 25% of the N, 10% of the K, 25% of the Zn and 60% of the Cu requirements of the host plant (Marschner and Dell 1994). Other functional benefits provided by AM fungi are pathogen protection and facilitation of water uptake by the host plant. AM fungi receive up to 20% of photosynthate from their host plant, and the hyphae outside of the plant root play a major role in C translocation into the soil (Treseder and Cross 2006), which may enhance SOC sequestration as C is translocated away from the zone of high respiratory activity around the root. The large network of hyphae outside of the root may represent up to 15% of the SOC stock (Leake et al. 2004).

Manure application

Manure applied to croplands may increase soil C inputs directly through its C content, and/or indirectly through increases in NPP including roots and crop residues (Maillard and Angers 2014). Compared to unfertilized controls, manure applications increased SOC stocks on average by 9.4 Mg C ha^{-1} for annual crops, perennial crops, and rice paddies averaged to a soil depth of 26 cm and a study duration of eighteen years. Above animal species, land use or soil properties, the direct C input by manure was proposed as the major mechanism for SOC stock enhancement with a significant linear relationship between the SOC stock difference and cumulative manure C input. The global manure C retention coefficient or average proportion of manure C remaining in soil was 12%. Despite apparently faster early decomposition of crop residues compared to manure, longer-term stabilization of crop C and manure C may be relatively similar for both. However, there were obvious gaps in the data especially with regards to animal species (i.e., pig (*Sus scrofa domesticus* Erxleben, 1777) and poultry manure), manure management system (e.g., liquid vs. solid storage), and underrepresentation of some geographical regions. For example, at long-term European field experiments, bovine farmyard manure alone and in combination with N fertilizer increased SOC concentrations by 32.4 and 33.4% compared to N fertilizer alone treatments, respectively (Zavattaro et al. 2017). In contrast, bovine liquid slurry alone or combined with N fertilizers increased SOC concentrations on average by 17.4%. Long-term field studies covering important knowledge gaps should either be initiated or continued, and these studies should document important explanatory factors such as manure application rate and C concentration which were very often absent from research articles (Maillard and Angers 2014).

3.3.2 *Cropland Soil Organic Carbon Sequestration*

The net balance between soil C inputs and losses determines the capacity of a crop soil to sequester C. Aside biotic parameters, abiotic processes such as photodegradation of surface OM may contribute to decomposition. Photodegradation may occur in croplands of arid and semiarid areas, bare burnt areas, sparsely vegetated croplands like shrublands, croplands after cultivation or harvest especially when crop residues are left on the surface, during prolonged drought and croplands with large amount of exposed standing dead material like croplands under no-till (NT; Rutledge et al. 2010). Photodegradation of SOM may also be an important process after crop residues have been removed or where erosion exposes SOM to solar radiation (Feng et al. 2011). For example, photooxidation increased the solubility of SOM but did not substantially affect the organo-chemical composition of corn litter. Thus, photooxidation may contribute to SOC loss through leaching and oxidation (Feng et al. 2011).

Net C losses from croplands may also occur by soil erosion. Historically, crop soils have lost significant amounts of SOC by this process (Lal 2003). However, erosion can induce also a net C sink depending on crop management practices, the extent to which SOC oxidation and production characteristics change with erosion, and the fate of eroded SOC (Billings et al. 2010). While lateral export of C from cropland may have approximately the same magnitude as additional SOC sequestration in SOC-depleted eroded soils, the ultimate impact of erosion on the landscape SOC balance is also determined by the fate of the exported SOC (Doetterl et al. 2016).

Some cropland C may also be lost by the emission of CH_4, e.g., by anaerobic decomposition of OM in flooded soils under rice cultivation emitting 31–112 Tg CH_4 yr^{-1}, and of crop residues under anaerobic or wet field conditions (Greenhouse Gas Working Group 2010). The impact of CO_2, CH_4, and N_2O on radiative forcing in Earth's atmosphere is quantitatively estimated by calculating net global warming potential (GWP) which accounts for all sources and sinks of CO_2 equivalents from farm operations, chemical inputs, SOC sequestration, and N_2O and CH_4 emissions (Sainju 2016). Another measure of cropland GHGs' impact is net greenhouse gas intensity (GHGI) which is expressed as net GWP per unit crop yield. Both GWP and GHGI are typically controlled by the balance between SOC sequestration rate, N_2O and CH_4 emissions, and crop yields (Sainju 2016).

Cropland soils may also lose small amounts of C by emitting VOCs. Carbon losses through leaching are important for the C balance of croplands. For example, leaching losses of biogenic C (DOC plus biogenic dissolved inorganic carbon or DIC) increase the net C loss from European cropland soils by 25%. Leaching of DOC from European croplands can be up to 4 g C m^{-2} yr^{-1} (Kindler et al. 2011).

Recently deposited crop residues and application of organic amendments such as biosolids and manures are the most important sources of DOC in arable soils (Bolan et al. 2011). Minor sources are root decay, exudates, and microbial metabolites.

3.3.2.1 Stable Soil Organic Matter and Humic Substances

All crop residues decay rather rapidly in aerobic soils with adequate moisture and temperature. Some fresh plant C is reworked by the biotic community, and the remnants of soil biota attach to mineral surfaces and can be protected against decomposition for centennial to millennial timescales as SOM (Kleber et al. 2015). Thus, previous assumptions about the supramolecular structure of SOM and its recalcitrance have been challenged (Lehmann and Kleber 2015). In fact, simple relatively fresh OM inputs can directly contribute to the stable SOM stock (Sanderman et al. 2010).

3.3.2.2 Decomposition

The unharvested remnants of crops such as leaf, stem, and root tissue not removed from the field and the dieback of legumes such as leaves and stems are subject to decomposition (Whalen and Sampedro 2009). Aside the primary resource crop OM, biomass of microorganisms and microfauna (Ø < 0.2 mm) are themselves secondary resources for decomposition. Depending on tillage practices, the residues may be decomposed rapidly or slowly as litter placement can strongly influence the decomposer community composition and decomposition rates (Beare et al. 1992). In particular in no-till (NT) croplands, macrofauna such as earthworms fragment litter and redistribute it in the soil profile.

The main initial decomposers fungi are mostly concentrated closer to the soil surface. Fungal C is about 75% of total microbial residue C in soils of croplands, and fungal respiration is about 61% of microbial respiration in cropland soils (Joergensen and Wichern 2008). Fungal enzymes can break down virtually all classes of plant compounds. SOM in croplands favoring a fungal-dominated community is improved quantitatively as fungi have higher C assimilation efficiencies (Six et al. 2006).

However, the long-term fate of leaf and root residues in cropland soil is less well studied and the widely used litter bag method to study decomposition has its limitations. For example, up to 65% of the initial *Triticum aestivum* L. leaf mass and up to 55% of the initial *Andropogon gerardii* Vitman root mass may remain after 10 years of decomposition in a soil (Harmon et al. 2009).

3.3.2.3 Organic Matter Stabilization

Inputs of OM to soils can increase the SOC stock when they are stabilized and not completely mineralized to CO_2. However, the molecular structure of plant inputs and OM plays only a secondary role in determining C residence times over decades to millennia. The persistence of SOM is largely due to complex interactions between OM and its environments, such as the interdependence of compound chemistry, reactive mineral surfaces, climate, water availability, soil acidity, soil redox state and the presence of potential degraders in the immediate microenvironment (Schmidt et al. 2011). In the surface layer of cropland soils, aggregation may be the most important factor influencing the spatial separation between decomposers or their extracellular enzymes and OM. Stabilization of OM within macroaggregates may be restricted to surface horizons but stabilization within silt-size aggregates is also important in subsurface horizons (Moni et al. 2010). Protection of OM by aggregation against microbial decomposition is effective primarily for soils with a large pool of labile OM (Goebel et al. 2009). Most of the SOM in soils of croplands, with poor structure following cultivation for many years, is often associated with mineral surfaces and stabilized in organo-mineral complexes (Basile-Doelsch et al. 2009).

3.3.2.4 Controls on Decomposition and Stabilization

Litter decomposition is a biological process and mainly controlled by variations in litter 'quality,' microclimate, soil properties and microbial community composition. Primary controls of decomposition are microbial activity and ultimately enzymatic activity, except for the abiotic process of photodegradation, and initial litter communition and mixing by soil fauna. Under the same environmental conditions, litter 'quality' controls the decomposition rate. Litter of high 'quality' decomposes faster than that of lower 'quality.' For example, alfalfa (*Medicago sativa* L.) residues decompose faster than those of grain sorghum (*Sorghum bicolor* [L.] Moench) and winter wheat (*Triticum aestivum* L. emend. Thell.) under the same environmental conditions (Schomberg et al. 1994). Thus, alfalfa litter supposedly has a higher 'quality' than those of sorghum and wheat. However, there is neither a common definition nor a quantitative index of 'quality' (Cotrufo et al. 2009). Soil fauna mediate decomposition and typically increase rates of mass loss from litter and SOM. For example, earthworms may increase the amount of OM in soil aggregates and, thus, stabilize SOM (Bossuyt et al. 2005). Further, the oribatid mite *Scheloribates moestus* (Acari: Oribatida) which is abundant in many US ecosystems, stimulates extracellular enzyme activity, enhances microbial respiration rates and increases water-extractable organic C during decomposition of corn litter (Wickings and Grandy 2011). The mites decrease the relative abundance of polysaccharides in decomposing litter. Further, the feces have a higher relative abundance of polysaccharides and phenols and a lower relative abundance of lignin compared to unprocessed corn litter. Thus, *S. moestus* may play an important role in SOC cycling dynamics (Wickings and Grandy 2011).

Climate has an overriding control on OM decomposition and stabilization. Water addition through irrigation, for example, may increase decomposition rates of crop residues (Schomberg et al. 1994). In general, faster litter decomposition rates are measured under warmer and wetter conditions when soil temperature and moisture conditions are ideal for microbial activity. Thus, over large geographical areas climate (i.e., temperature and soil moisture) exerts the strongest controls on losses of OM from soil (Sanderman et al. 2010).

Temperature and soil moisture regimes affect OM breakdown in aggregates through their effect on microbial activity. The aggregate turnover is slower in cold or dry climates but faster in moist and/or warm climates. However, very wet climates may reduce OM breakdown under anaerobic soil conditions. Plant species in croplands can also affect aggregation. For example, N-fixing species and mycorrhizal associations may boost microbial populations and increase aggregate stability. Further, molecular-level properties of SOM under different plant species may impact the resistance to degradation in microaggregates (Bachmann et al. 2008).

The decomposition of a substrate and destabilization of SOM are accelerated when (i) conditions are suitable for microbial activity, enzyme production and diffusion, (ii) OM is vulnerable to enzymatic degradation, (iii) microbial transformations that lead to the creation of recalcitrant products are constrained, (iv) chemical reactions that lead to the formation of stabilized SOM are limited, (v) physical protection of OM in soil aggregates is weak and, (vi) chemical protection of OM in soil is also weak (Prescott 2010).

3.3.3 The Net Balance of Cropland Soil Organic Carbon

At steady state and under similar soil and vegetation management, the C inputs to cropland soil and losses from it are approximately balanced depending on site-specific factors (Sanderman et al. 2010). Quantitative descriptions of SOM dynamics have been historically based on the assumption that SOM consists of a single homogenous pool decomposing at varying relative rates (Shibu et al. 2006). Others consider SOM as comprising of heterogeneous components and decomposition of the components occurs at different relative rates. The components are represented by arbitrary SOM pools. For example, the 'active pool' decomposes on timescale of hours to months to years and may consist of root exudates, microbial cell contents and some fresh litter compounds (Torn et al. 2009). The 'intermediate' or 'slow' SOM pool has turnover times in the range of decades to centuries and may consist of structural components of plants more resistant to decay or OM stabilized by association with soil minerals or aggregate structures. The 'passive' or 'millennial cycling' SOM pool persists in soils for thousands of years and consists of highly stabilized OM, typically associated with soil minerals or stable aggregates (Torn et al. 2009). However, the molecular model for stable SOM is under discussion but the specific structure of OM may be responsible for long-term preservation of OM in soil mineral fractions (Clemente et al. 2011; Kleber et al. 2011; Lehmann and Kleber 2015). Some SOM models and their limitations are discussed in Chap. 2. However, recent advances in mechanistic understanding of soils have not yet been incorporated into widely used models of SOM cycling

(Schmidt et al. 2011). Current models assume a pool of OM that will have an intrinsic decay rate. These models rely on simple proxies such as soil texture as a surrogate for sorption and other organo-mineral interactions. Further, litter quality (e.g., lignin:N ratios or structural C groupings) is used a proxy of partitioning plant inputs into pools of different turnover times. However, these parameters are not consistent with the observations that emerge. Also, global models largely ignore deep mineral soils (Schmidt et al. 2011).

3.3.4 Anthropogenic Drivers of Soil Organic Carbon in Croplands

The SOC balance of croplands is affected by past and present agricultural technology, and farming practices such as manure and tillage management (Ciais et al. 2011). Technology may impact the soil C input directly by affecting the management of harvest residues, and indirectly via effects on yield and NPP. Anything that increases crop biomass production (e.g., rotations, nutrients, improved cultivars and irrigation or, agricultural intensification) affects the soil C input (Robbins 2011). Further, solar brightening (decadal-scale increases in incident solar radiation) may also contribute to yield increases as was shown for the trends in yield of the US Corn Belt from 1984 to 2013 (Tollenaar et al. 2017). As croplands are often intensively managed, they offer opportunities to deliberately alter the SOC dynamics (Smith et al. 2008). Practices include tillage management (in some cases, Powlson et al. 2014); crop rotations and cover crops (Poeplau and Don 2015); improving crop production through fertilization and irrigation management; selection of high residue yielding crops; crop intensification by removing bare fallow management in a cropping system; and application of organic amendments with manure or biochar (Harden et al. 2017). However, C budgeting studies over croplands are scarce in comparison to those on crop yields and on processes controlling plant and soil fertility (Ciais et al. 2011).

Improving cropland management, restoring degraded croplands and cultivated organic soils greatly affect SOC dynamics (Smith et al. 2008). For example, improved agronomic practices that increase yields and generate higher inputs of residue C can lead to increased SOC stocks. Relevant practices include (i) using improved crop varieties, (ii) extending crop rotations, notably those with perennial crops which allocate more C belowground, and (iii) avoiding or reducing the use of bare fallow (Lal 2004b; Smith and Conen 2004). Adding more nutrients, when deficient, can also promote SOC gains (Alvarez 2005). By providing temporary vegetative cover between agricultural crops, 'catch' or 'cover' crops also add C to cropland soils (Freibauer et al. 2004). Since soil disturbance tends to stimulate SOC losses through enhanced decomposition and erosion, reduced- or NT agriculture (conservation agriculture) often results in SOC gain, though not always (Govaerts et al. 2009). Cropland systems that retain crop residues (e.g., avoiding the burning of residues) also tend to increase SOC stocks because these residues are the precursors for SOM, the main store of C in the soil (Smith et al. 2008).

The management of croplands is an important factor in altering the SOC dynamics (Franzluebbers 2010). Conservation practices, in particular, have a great potential to increase the SOC stock. Some of the guiding conservation principles that can be globally applied are to (i) minimize soil disturbance, (ii) maximize surface cover by managing crops and crop residues, and, (iii) stimulate biological activity through crop rotations, cover crops and integrated nutrient and pest management (Franzluebbers 2010). However, the mechanisms that govern changes in SOC after reducing tillage operations are less clear (Govaerts et al. 2009). Factors that may play a role include root development and rhizodeposits, baseline SOC content, bulk density and porosity, climate, landscape position, and erosion/deposition history. Further, altering crop rotation may influence SOC by changing quantity and quality of OM input (Govaerts et al. 2009).

Cropland irrigation may affect SOC dynamics by altering crop yields and residue returns (Lal 2004b). About 25% of the global harvested crop area (i.e., area of all major food crops and cotton (*Gossypium* L.) as well as those of perennial, annual and fodder grasses) was irrigated in 2000 (Portmann et al. 2010). Further, 3.46 million km^2 of irrigated crops were harvested in 2011 on 2.61 million km^2 actually irrigated (AQUASTAT 2014). Cereals and vegetables covered 61% and 10% of the irrigated harvested crop area, respectively. Rice is the world's largest irrigated cereal, covering 29% of the total irrigated crop area and almost half of the irrigated cereals area (AQUASTAT 2014). The average crop yield of irrigated cereals was 442 Mg km^{-2}, while average yield of rainfed cereals was 266 Mg km^{-2} (Siebert and Döll 2010). Crop yields and residue returns and, thus, soil C input may decrease when crop irrigation is discontinued. For example, the global production of dates (*Phoenix dactylifera* L.), rice, cotton, citrus (*Citrus*), and sugarcane would decrease by 60, 39, 38, 32, and 31%, respectively, if currently irrigated crops were not irrigated. Further, cereal production on irrigated land would decrease by 47% without irrigation (Siebert and Döll 2010). In contrast, drainage of agricultural lands in humid regions can promote productivity and hence SOC accrual (Smith et al. 2008).

Agroforestry refers to the practice of purposeful growing of trees and crops, and/or animals, in interacting combinations, for a variety of benefits and services such as increasing crop yields, reducing food insecurity, enhancing environmental services and resilience of agroecosystems (Lal 1989 a, b, c, d, e; Ajayi et al. 2011; Nair et al. 2008). In the tropics, agroforestry systems include alley cropping, homegardens, improved fallows, multipurpose trees on farms and rangelands, silvopastoral grazing systems, shaded perennial-crop systems, shelterbelts, windbreaks, and taungya (i.e., growing agricultural crops during early stages of the establishment of forestry plantations). In temperate regions, agroforestry practices include alley cropping, forest farming, riparian buffer strips, silvopasture and windbreaks (Nair et al. 2009). In particular, planting trees may affect SOC dynamics and the SOC stock may be higher compared to that of croplands, pastures or natural grasslands replaced by the agroforestry system (Nair et al. 2010; Lorenz and Lal 2014).

Organic soils contain high SOC densities because of restricted decomposition under flooded conditions. When converted to croplands, organic soils must be drained (Freibauer et al. 2004). Thus, SOC is lost from organic soils, specifically by deep drainage and intensive mechanical disturbance such as deep plowing. Potential alternative uses for organic soils are maintaining a more shallow water table and avoiding deep plowing as well as avoiding cultivation with potatoes and sugar beets (*Beta vulgaris* L.) and instead cropping to permanent cultures (Freibauer et al. 2004).

3.4 Recarbonization of Cropland Soils

The depletion of the SOC stock by arable land use can be partially reversed by recarbonization of the soil through the adoption of practices which create a positive soil and ecosystem C budget. SOC losses may occur when land under perennial uses is converted for the cultivation of crops. For example, about 25–30% of the SOC stored in the top meter of soil is released by conversion of native soils (whether under forest or prairie vegetation) to agroecosystems (Houghton 2010). Specifically, in temperate regions, conversion from forest to cropland and from grassland to cropland may cause the loss of 31 ± 20% SOC to 28.5 ± 13.5 cm depth and of 36 ± 5% SOC to 27.1 ± 11.1 cm depth, respectively (Poeplau et al. 2011). New SOC equilibrium under cropland may be reached, if at all, about 23 years after conversion from forest and 17 years following conversion from grassland (Bell et al. 2011). In tropical regions, SOC losses of 25% to 36 ± 4 cm depth and of 30% to 48 ± 8 cm depth occurred by conversion of primary forest to cropland or perennial crops, respectively (Don et al. 2011). Losses of 21% SOC to 39 ± 5 cm depth occurred when secondary tropical forest was converted to cropland but no changes were observed to 51 ± 9 cm depth when converted to perennial crops. When tropical grassland was converted to cropland, 10.4 ± 6.1% of SOC were lost to 38 ± 11 cm depth (Don et al. 2011). However, cropping practices, irrigation, use of fertilizers and different types of tillage affect changes in SOC stocks. Thus, both conversions of native soils to crop soils and cultivation of croplands may decrease SOC stocks. Croplands can be partially recarbonized through adoption of recommended management practices (RMPs) such as conservation agriculture, residue mulching and use of cover crops, practices which all contribute to SOC accumulation and sequestration by an additional transfer of C from the atmosphere to the soil (Lal 2007, 2016; Powlson et al. 2011).

Deep tillage may also contribute to SOC sequestration by enlarging the storage space for SOC-rich material (Alcántara et al. 2016). For example, after 25–48 years, deep-tilled subsoil contained 67% more SOC than reference subsoil (Alcántara et al. 2017). Buried SOC was 32% more stable than reference SOC and had higher apparent radiocarbon ages indicating that it was largely isolated from exchange with atmospheric CO_2. Deep tillage increased subsoil SOC storage, and the higher subsoil SOC stability was not only a result of selective preservation of more stable SOC fractions (Alcántara et al. 2017). Similarly, after 17 years full inversion tillage resulted in higher SOC stocks in 0–90 cm depth compared to those under reduced tillage treatments, in particular, due to higher SOC stocks in 30–60 cm depth (Knebl et al. 2017). Based on a meta-analysis, Angers and Eriksen-Hamel (2008) reported that at the 21 to 25 cm soil depth, which corresponded to the mean tillage depth for the data set, the SOC content was higher under full inversion tillage than NT. Under full inversion tillage, higher SOC content was observed just below the average depth of tillage (26–35 cm). Deep tillage may lead to increased SOC storage by (i) greater stability of buried SOC and, (ii) additional SOC accumulation in the 'newly established' topsoil (Alcántara et al. 2017).

Agricultural systems can be managed to enhance SOC sequestration (Power 2010). Practices that increase C inputs can result in an increase in SOC stocks (Paustian et al. 2016). Among cropland practices are (i) improved crop varieties or species with greater root-derived C input into deeper layers where turnover is slower, (ii) adopting crop rotations that provide greater C inputs, (iii) more residue retention, and (iv) cover crops during fallow periods to provide year-round C inputs. For example, increasing C inputs through straw amendments contributed to the improvement of SOC stocks to 50 cm depth in China's cropland during the last two decades of the twentieth century (Zhang et al. 2017). Other practices for increasing C inputs into cropland soils include irrigation in water-limited systems, and additional fertilizer input to increase productivity in low-yielding, nutrient-deficient systems. For example, SOC stocks in China's cropland decreased without but increased with irrigation (Zhang et al. 2017). Further, applying chemical fertilizers, and chemical fertilizers with straw or manure application may result in increases in topsoil SOC contents (Han et al. 2016).

Croplands can also sequester SOC through less intensive tillage, particularly NT. For example, the C sequestration rate under NT in 0–25 cm depth was 864 kg C ha^{-1} yr^{-1} in tropical regions and 173 kg C ha^{-1} yr^{-1} in temperate regions, respectively (Mangalassery et al. 2015). However, the time required to reach a 'steady state' in C sequestration varies with respect to climate, soil type, and management practices. In temperate soils, the time period to attain sink saturation is $\sim$100 years, with lower values for tropical soils (20–50 years; Mangalassery et al. 2015). A change from annual to perennial crops typically increases belowground C inputs and, may also lead to SOC sequestration. Based on a meta-analysis, the most effective crop rotations for increasing SOC concentrations in 0-30 cm depth relative to grain-only rotations were those that included a perennial crop or a cover crop (King and Blesh 2018). Thus, among the options for crop rotations to increase SOC are (i) increasing functional diversity with cover crops and perennials; (ii) increasing C input, especially root C input; and (iii) increasing perenniality, the time of living soil cover (King and Blesh 2018).

Addition of plant-derived C from external (offsite) sources such as composts or biochars can potentially also increase SOC stocks. However, a broader life-cycle assessment (LCA) approach is needed as the origin of inputs may be outside the cropland ecosystem boundary (Paustian et al. 2016). Crop rotations and crop residues are not adequately represented in current LCA and product carbon footprint modeling practice (Brankatschk and Finkbeiner 2017). Crop rotations and straw harvest should be considered, in particular, for the product carbon footprints of bread, milk, and first- and second-generation biofuels. Further, net GWP and GHGI were 70–87% lower for improved cropland management that included NT, crop rotation/perennial crop, and reduced N rate compared to those for the traditional combined management that included conventional tillage, monocropping/ annual crop, and recommended N rate (Sainju 2016).

In boreo-temperate regions, soils under NT had higher SOC stocks to 30 cm depth than soils under high or intermediate tillage intensity based on a meta-analysis of data from 351 studies (Haddaway et al. 2017). The SOC stock increase to 30 cm depth under NT compared to both more intensive tillage practices was about 4.6 Mg C ha^{-1} over ≥ 10 year, while no effect was found for the full soil profile. Both NT and intermediate tillage increased SOC in the topsoil of boreo-temperate regions, and implementation of those practices should be supported by policy recommendations (Haddaway et al. 2017). Based on the database, Meurer et al. (2018) extracted those studies (101 long-term field experiments) allowing for calculation of the equivalent soil mass (ESM). Because the mass of soil over a pre-defined depth differs between treatments with different bulk densities as a result of loosening or compaction, depth must be adjusted when comparing treatment effects on SOC stocks using an ESM approach (Ellert and Bettany 1995). Calculating SOC stocks based on fixed depth layers and without consideration of ESM, respectively, resulted in an overestimation of the increase with 15% for high versus intermediate tillage, and 47% for both more intensive tillage practices versus NT. The positive effect of NT over high-intensity tillage decreased with increase in soil depth, and annual changes were below 0.10 Mg ha^{-1} at 0–60 cm depth. Meurer et al. (2018) concluded that the NT sequestration potential with respect to mitigating climate change is likely to be overoptimistic.

A net gain of the SOC stock can be achieved by conversion of plow tillage (PT) to NT and other conservation agriculture practices, along with crop residue mulch or cover crops (Lal 2009, 2016). The transfer of C from the atmosphere into both the SIC and SOC stock for enhancement of soil C sinks can be accelerated (Macías and Arbestain 2010). Such a positive balance may be achieved by (i) favouring growth of crop biomass which is the major source for SOC, (ii) promoting and facilitating carbonation processes to increase the SIC stock, (iii) reducing erosional C loss from croplands and favoring pedogenesis for build-up of the soil profile SOC stock, (iv) developing OM-rich horizons, and/or (v) recovering degraded or contaminated crop soils to restore the soil C sink. Some of the C lost in the past from cropland soils by changes in land use and cultivation can be recovered through improved management, thereby withdrawing atmospheric CO_2 (Smith et al. 2008), and creating net negative emissions. Most agricultural soils contain 30–75% less C than their potential capacity as determined by soil, climate, terrain, drainage, land use, and soil and crop management practices (Lal and Follett 2009b). The term 'soil C sequestration' implies that the total C stock in the soil profile is increasing through managerial interventions aimed at transferring atmospheric CO_2 to the soil C stock by moderating either organic and/or inorganic transformations (Lal and Follett 2009b; Olson et al. 2014). Thus, practices that retain or return more of the C captured by growing plants increase the soil C stock (Sanderman and Baldock 2010). Carbon sequestration in crop soil implies an additional transfer of C from the atmosphere to the soil and, thus, a genuine contribution to climate change mitigation (Powlson et al. 2011). NT and reduced tillage systems are able to stock more SOC than PT to 40 cm depth at Mediterranean cereal systems under future climate scenarios and are, thus, proposed for climate change mitigation (Iocola et al. 2017).

The rate of SOC sequestration in croplands with the adoption of RMPs depends on soil texture and structure, rainfall, temperature, farming system, and soil management (Lal 2004a, b). The SOC stock in croplands can be enhanced by increasing the use efficiency of input, decreasing losses by erosion and leaching, and improving soil structure (Lal and Follett 2009b). Further, the SIC stock can be enhanced by the application of biosolids, liming/application of cations, and conserving water in the root zone. However, the potential of SIC sequestration by pedogenic carbonate formation is less well known. Soils of irrigated croplands may sequester both SIC and SOC. The rate of soil C sequestration ranges from about 100 to 1,000 kg ha^{-1} yr^{-1} for SOC and 5 to 15 kg ha^{-1} yr^{-1} for SIC (Lal and Follett 2009a). However, some agricultural field trials indicate that the relative increase in soil C stock with the adoption of RMPs is not an actual increase but rather due to a reduction or cessation of soil C losses (Sanderman and Baldock 2010). The global C sink capacity of agricultural pools is estimated to be up to 78 Pg C (Lal 1999) and can be filled at the potential maximum rate of about 1–3 Pg C yr^{-1} (Lal 2010). However, the attainable and actual cumulative global rate of soil C sequestration may be lower because of managerial, economic, and policy constraints (Lal and Follett 2009a). The duration of soil C sequestration may be 25–50 years (Lal 2004b). With increasing saturation of the soil C stock, the sink activity diminishes but sequestration may continue due to climate change and land-management change (Bell et al. 2011).

Common RMPs for SOC sequestration and those which create a positive soil C budget are mulch farming, conservation agriculture, agroforestry and diverse cropping systems, cover crops, and integrated nutrient management, including the use of manure, compost, and biosolids (Lal 2004b). Irrigation can be used to buffer against soil moisture deficits and to sustain crop productivity. Where irrigation is constrained, the selection of drought-resistant crops promotes SOC sequestration. Further, crop production can also be adapted to temperature. For example, wheat in North America is now cultivated in environments once considered too arid, too variable, and too harsh to cultivate (Olmstead and Rhode 2011). Rising atmospheric CO_2 concentrations may directly alter crop yield and the SOC stock (Ainsworth and McGrath 2010). Free-air CO_2 enrichment (FACE) experiments indicate that soybean and rice grain yield may increase by 13% at 550 ppm CO_2. In contrast, grain yield of sorghum and corn are not expected to increase at elevated CO_2 when water supply is adequate. However, in the long-term responses of row crop agroecosystems to management practices may be more important than responses to elevated CO_2 (Moran and Jastrow 2010).

3.5 Conclusions

By converting land under other uses to cropland, by cropland management and its intensification (i.e., fertilization, irrigation, mechanization), humans have altered SOC dynamics on 12% of Earth's ice-free land area. Up to 36% of SOC stock may have been lost from cropland topsoils. Thus, most cropland soils contain less SOC than their potential capacity as determined by soil, climate, terrain, drainage, land use, and soil and crop management practices. This provides a tremendous opportunity for the recarbonization of global cropland soils. Photosynthesis is the major natural C input into cropland soils, while direct input occurs by the addition of manure and organic residues. Harvest removes a major proportion of cropland NPP before it enters the soil. Natural crop SOC losses occur by decomposition, erosion, and leaching. Cropland SOC can be increased by mulch farming, conservation tillage, agroforestry and diverse cropping systems, cover crops, and integrated nutrient management, including the use of manure, compost, and biosolids. Recent advances have improved our understanding of SOC dynamics and SOC persistence. However, the deep mineral cropland soils have been largely ignored. Further, assessing the potential of SOC-accreting crops (e.g., crop plants with a bushy and deep root system, C_4 crops, perennial crops) to recarbonize cropland soils is among the top researchable priorities.

3.6 Review Questions

1. What processes affect loss of SOC upon conversion of natural to cropland ecosystems?
2. How can protection and stabilization of SOC stocks in croplands be managed?
3. What is the soil C sink capacity and what are its determinants?
4. How can SIC be managed in croplands to improve formation of PIC and increase leaching of bicarbonates?
5. Why is SOM considered an agroecosystem property and what are the implications of this concept to soil management for C sequestration?
6. What determines the MRT of soil C?
7. Can soil C sequestration offset anthropogenic emissions through programs such as '4 per Thousand'?
8. How can management of cropland soils contribute to storing C for millennia?
9. Why is SOC sequestration sometimes described as a win-win-win option?
10. Discuss potential negative environmental effects of increasing SOC stocks in previously SOC-depleted soils of croplands and how those effects can be minimized.
11. What soil factors must be considered in improved dynamic SOC models?

References

Ahmad W, Singh B, Dalal RC, Dijkstra FA (2015) Carbon dynamics from carbonate dissolution in Australian agricultural soils. Soil Res 53:144–153

Ainsworth EA, McGrath JM (2010) Direct effects of rising atmospheric carbon dioxide and ozone on crop yields. In: Lobell D, Burke M (eds) Climate change and food security. Springer, Dordrecht, The Netherlands, pp 109–130

Ajayi OC, Place F, Akinnifesi FK, Sileshi GW (2011) Agricultural success from Africa: the case of fertilizer tree systems in southern Africa (Malawi, Tanzania, Mozambique, Zambia and Zimbabwe). Int J Agric Sustain 9:129–136

Alcántara V, Don A, Vesterdal L, Well R, Nieder R (2017) Stability of buried carbon in deepploughed forest and cropland soils—implications for carbon stocks. Sci Rep 7:5511. https://doi.org/10.1038/s41598-017-05501-y

Alcántara V, Don A, Well R, Nieder R (2016) Deep ploughing increases agricultural soil organic matter stocks. Soil Biol Biochem 22:2939–2956. https://doi.org/10.1111/gcb.13289

Alvarez R (2005) A review of nitrogen fertilizer and conservative tillage effects on soil organic storage. Soil Use Manage 21:38–52

Amthor JS (2010) From sunlight to phytomass: on the potential efficiency of converting solar radiation to phyto-energy. New Phytol 188:939–959

Andela N, Morton DC, Giglio L et al (2017) A human-driven decline in global burned area. Science 356:1356–1362

Andreae MO (2004) Assessment of global emissions from vegetation fires. International Forest Fire News 31:112–121

Angers DA, Eriksen-Hamel NS (2008) Full-inversion tillage and organic carbon distribution in soil profiles: a meta-analysis. Soil Sci Soc Am J 72:1370–1374. https://doi.org/10.2136/sssaj2007.0342

AQUASTAT, FAO's global water information system (2014). http://www.fao.org/nr/aquastat

Bachmann J, Guggenberger G, Baumgartl Th, Ellerbrock R, Urbanek E, Goebel M-O, Kaiser K, Horn R, Fischer WR (2008) Physical carbon-sequestration mechanisms under special consideration of soil wettability. J Plant Nutr Soil Sci 171:14–26

Bais HP, Weir TL, Perry LG, Gilroy S, Vivanco JM (2006) The role of root exudates in rhizosphere interactions with plants and other organisms. Annu Rev Plant Biol 57:233–266

Basile-Doelsch I, Brun T, Borschneck D, Masion A, Marol C, Balesdent J (2009) Effect of landuse on organic matter stabilized in organomineral complexes: a study combining density fractionation, mineralogy and $\delta^{13}C$. Geoderma 151:77–86

Beare MH, Parmelee RW, Hendrix PF, Cheng W, Coleman DC, Crossley DA (1992) Microbial and faunal interactions and effects on litter nitrogen and decomposition in agroecosystems. Ecol Monogr 62:569–591

Beer C, Reichstein M, Ciais P, Farquhar GD, Papale D (2007) Mean annual GPP of Europe derived from its water balance. Geophys Res Lett 34:L05401. https://doi.org/10.1029/2006GL029006

Beer C, Reichstein M, Tomelleri E, Ciais P, Jung M, Carvalhais N, Rödenbeck C, Altaf Arain M, Baldocchi D, Bonan GB, Bondeau A, Cescatti A, Lasslop G, Lindroth A, Lomas M, Luyssaert S, Margolis H, Oleson KW, Roupsard O, Veenendal E, Viovy N, Williams C, Ian Woodward F, Papale D (2010) Terrestrial gross carbon dioxide uptake: global distribution and covariation with climate. Science 329:834–838

Bell MJ, Worrall F, Smith P, Bhogal A, Black H, Lilly A, Barraclough D, Merrington G (2011) UK land-use change and its impact on SOC: 1925–2007. Global Biogeochem Cycles, 25, GB4015. https://doi.org/10.1029/2010gb0038

Billings SA, Buddemeier RW, deB Richter D, Van Oost K, Bohling G (2010) A simple method for estimating the influence of eroding soil profiles on atmospheric CO_2. Global Biogeochem Cyc 24, GB2001. https://doi.org/10.1029/2009gb003560

Bolan NS, Adriano DC, Kunhikrishnan A, James T, McDowell R, Senesi N (2011) Dissolved organic matter: biogeochemistry, dynamics, and environmental significance in soils. Adv Agron 110:1–75

Bolinder MA, Angers DA, Giroux M, Laverdiere MR (1999) Estimating C inputs retained as soil organic matter from corn (*Zea mays* L.). Plant Soil 215:85–91

Bolinder MA, Janzen HH, Gregorich EG, Angers DA, VandenBygaart AJ (2007) An approach for estimating net primary productivity and annual carbon inputs to soil for common agricultural crops in Canada. Agric Ecosyst Environ 118:29–42

Bonan GB, Lawrence PJ, Oleson KW, Levis S, Jung M, Reichstein M, Lawrence DM, Swenson SC (2011) Improving canopy processes in the Community Land Model version 4 (CLM4) using global flux fields empirically inferred from FLUXNET data. J Geophys Res 116:G02014. https://doi.org/10.1029/2010JG001593

Bondeau A, Smith P, Zaehle S, Schaphoff S, Lucht W, Cramer W, Gerten D, Lotze-Campen H, Müller C, Reichstein M, Smith B (2007) Modelling the role of agriculture for the 20th century global terrestrial carbon balance. Glob Change Biol 13:679–706

Bossuyt H, Six J, Hendrix PF (2005) Protection of soil carbon by microaggregates within earthworm casts. Soil Biol Biochem 37:251–258

Botha A (2011) The importance and ecology of yeasts in soil. Soil Biol Biochem 43:1–8

Bradford JB, Lauenroth WK, Burke IC (2005) The impact of cropping on primary production in the U.S. Great Plains. Ecology 86:1863–1872

Brankatschk G, Finkbeiner M (2017) Crop rotations and crop residues are relevant parameters for agricultural carbon footprints. Agron Sustain Dev 37:58. https://doi.org/10.1007/s13593-017-0464-4

Broadbent FE, Nakashima T (1974) Mineralization of carbon and nitrogen in soil amended with carbon-13 and nitrogen-15 labeled plant material. Soil Sci Soc Am Proc 38:313–315

Chadwick OA, Kelly EF, Merritts DM, Amundson RG (1994) Carbon-dioxide consumption during soil development. Biogeochemistry 24:115–127

Chaopricha NT, Marín-Spiotta E (2014) Soil burial contributes to deep soil organic carbon storage. Soil Biol Biochem 69:251–264

Chapin FS III, Matson PA, Mooney HA (2002) Principles of terrestrial ecosystem ecology. Springer, New York

Chen T, van der Werf GR, Gobron N, Moors EJ, Dolman AJ (2014) Global cropland monthly gross primary production in the year 2000. Biogeosciences 11:3871–3880

Ciais P, Gervois S, Vuichard N, Piao SL, Viovy N (2011) Effects of land use change and management on the European cropland carbon balance. Glob Change Biol 17:320–338

Ciais P, Wattenbach M, Vuichard N, Smith P, Piao SL, Don A, Luyssaert S, Janssens IA, Bondeau A, Dechow R, Leip A, Smith PC, Beer C, Van DerWerf GR, Gervois S, Van Oost K, Tomelleri E, Freibauer A, Schulze E-D, Synthesis Team CARBOEUROPE (2010) The European carbon balance. Part 2: croplands. Glob Change Biol 16:1409–1428

Clemente JS, Simpson AJ, Simpson MJ (2011) Association of specific organic matter compounds in size fractions of soils under different environmental controls. Org Geochem 42:1169–1180

Conant RT, Berdanier AB, Grace PR (2013) Patterns and trends in nitrogen use and nitrogen recovery efficiency in world agriculture. 27:1–9. https://doi.org/10.1002/gbc.20053

Cotrufo MF, Del Galdo I, Piermatteo D (2009) Litter decomposition: concepts, methods and future perspectives. In: Kutsch WL, Bahn M, Heinemeyer A (eds) Soil carbon dynamics: an integrated methodology. Cambridge University Press, Cambridge, UK, pp 76–90

Crutzen PJ, Andreae MO (1990) Biomass burning in the tropics: impact on atmospheric chemistry and biogeochemical cycles. Science 250:1669–1678

D'Hose T, Molendijk L, Van Vooren L et al (2018) Responses of soil biota to non-inversion tillage and organic amendments: An analysis on European multiyear field experiments. Pedobiologia 66:18–28

Doetterl S, Berhe AA, Nadeu E, Wang Z, Sommer M, Fiener P (2016) Erosion, deposition and soil carbon: A review of process-level controls, experimental tools and models to address C cycling in dynamic landscapes. Earth Sci Rev 154:102–122

Don A, Schumacher J, Freibauer A (2011) Impact of tropical land-use change on soil organic carbon stocks—a meta-analysis. Glob Change Biol 17:1658–1670

Dufranne D, Moureaux C, Vancutsem F, Bodson B, Aubinet M (2011) Comparison of carbon fluxes, growth and productivity of a winter wheat crop in three contrasting growing seasons. Agric Ecosyst Environ 141:133–142

Ellert BH, Bettany JR (1995) Calculation of organic matter and nutrients stored in soils under contrasting management regimes. Can J Soil Sci 75:529–538

Ellis EC, Ramankutty N (2008) Putting people in the map: anthropogenic biomes of the world. Front Ecol Environ 6:439–447

Entry JA, Sojka RE, Shewmaker GE (2004) Irrigation increases inorganic carbon in agricultural soils. Environ Manage 33:S309–S317

Erb KH, Kastner T, Plutzar C et al (2018) Unexpectedly large impact of forest management and grazing on global vegetation biomass. Nature 553:73–76

Eshel G, Fine P, Singer MJ (2007) Total soil carbon and water quality: an implication for carbon sequestration. Soil Sci Soc Am J 71:397–405

Falge E, Baldocchi D, Tenhunen J, Aubinet M, Bakwin P, Berbigier P, Bernhofer C, Burba G, Clement R, Davis KJ, Elbers JA, Goldstein AH, Grelle A, Granier A, Guomundsson J, Hollinger D, Kowalski AS, Katul G, Law BE, Malhi Y, Meyers T, Monson RK, Munger JW, Oechel W, Paw UKT, Pilegaard K, Rannik U, Rebmann C, Suyker AE, Valentini R, Wilson K, Wofsy S (2002) Seasonality of ecosystem respiration and gross primary production as derived from FLUXNET measurements. Agric For Meteorol 113:53–74

Fan Y, Miguez-Macho G, Jobbágy EG, Jackson RB, Otero-Casal C (2017) Hydrologic regula-tion of plant rooting depth. Proc Natl Acad Sci USA 114:10572–10577

Feng X, Hills KM, Simpson AJ, Whalen JK, Simpson MJ (2011) The role of biodegradation and photo-oxidation in the transformation of terrigenous organic matter. Org Geochem 42:262–274

Foley JA, DeFries R, Asner GP, Barford C, Bonan G, Carpenter SR, Chapin FS, Coe MT, Daily GC, Gibbs HK, Helkowski JH, Holloway T, Howard EA, Kucharik CJ, Monfreda C, Patz JA, Prentice IC, Ramankutty N, Snyder PK (2005) Global consequences of land use. Science 309:570–574

Foley JA, Ramankutty N, Brauman KA, Cassidy ES, Gerber JS, Johnston M, Mueller ND, O'Connell C, Ray DK, West PC, Balzer C, Bennett EM, Carpenter SR, Hill J, Monfreda C, Polasky S, Rockström J, Sheehan J, Siebert S, Tilman D, Zaks DPM (2011) Solutions for a cultivated planet. Nature 478:337–342

Franzluebbers AJ (2010) Will we allow soil carbon to feed our needs? Carbon Manage 1:237–251

Freibauer A, Rounsevell M, Smith P, Verhagen A (2004) Carbon sequestration in the agricultural soils of Europe. Geoderma 122:1–23

Fritz S, See L, McCallum I, Schill C, Obersteiner M, Velde M, van der Boettcher H, Havlík P, Achard F (2011) Highlighting continued uncertainty in global land cover maps for the user community. Environ Res Lett 6:044005

Ghannoum O (2009) C_4 photosynthesis and water stress. Ann Bot 103:635–644

Gibbs KH, Ruesch AS, Achard F, Clayton MK, Holmgren P, Ramankutty N, Foley JA (2010) Tropical forests were the primary sources of new agricultural land in the 1980s and 1990s. Proc Natl Acad Sci USA 107:16732–16737

Gill RA, Kelly RH, Parton WJ, Day KA, Jackson RB, Morgan JA, Scurlock JMO, Tieszen LL, Castle JV, Ojima DS, Zhang XS (2002) Using simple environmental variables to estimate belowground productivity in grasslands. Glob Ecol Biogeogr 11:79–86

Gilmanov TG, Baker JM, Bernacchi CJ, Billesbach DP, Burba GG, Castro S, Chen J, Eugster W, Fischer ML, Gamon JA, Gebremedhin MT, Glenn AJ, Griffis TJ, Hatfield JL, Heuer MW, Howard DM, Leclerc MY, Loescher HW, Marloie O, Meyers TP, Olioso A, Phillips RL, Prueger JH, Skinner RH, Suyker AE, Tenuta M, Wylie BK (2014) Productivity and carbon dioxide exchange of leguminous crops: estimates from flux tower measurements. Agron J 106:545–559

Glover JD, Reganold JP, Bell LW, Borevitz J, Brummer EC, Buckler ES, Cox CM, Cox TS, Crews TE, Culman SW, DeHaan LR, Eriksson D, Gill BS, Holland J, Hu F, Hulke BS,

Ibrahim AMH, Jackson W, Jones SS, Murray SC, Paterson AH, Ploschuk E, Sacks EJ, Snapp S, Tao D, Van Tassel DL, Wade LJ, Wyse DL, Xu Y (2010) Increased food and ecosystem security via perennial grains. Science 328:1638–1639
Goebel MO, Woche SK, Bachmann J (2009) Do soil aggregates really protect encapsulated organic matter against microbial decomposition? Biologia 64:443–448. https://doi.org/10.2478/s11756-009-0065-z
Goudriaan J, Groot JJR, Uithol PWJ (2001) Productivity of agro-ecosystems. In: Roy J, Saugier B, Mooney HA (eds) Terrestrial global productivity. Academic Press, San Diego, CA, pp 301–313
Govaerts B, Verhulst N, Castellanos-Navarrete A, Sayre KD, Dixon J, Dendooven L (2009) Conservation agriculture and soil carbon sequestration: between myth and farmer reality. Crit Rev Plant Sci 28:97–122
Greenhouse Gas Working Group (2010) Agriculture's role in greenhouse gas emissions & capture. Greenhouse Gas Working Group Rep. ASA, CSSA, and SSSA, Madison, WI, USA
Gregory PJ, Palta JA, Batts GR (1997) Root systems and root:mass ratio—carbon allocation under current and projected atmospheric conditions in arable crops. Plant Soil 187:221–228
Guan K, Wu J, Kimball JS et al (2017) The shared and unique values of optical, fluorescence, thermal and microwave satellite data for estimating large-scale crop yields. Remote Sens Environ 199:333–349
Guanter L, Zhang Y, Jung M, Joiner J, Voigt M, Berry JA, Frankenberg C, Huete AR, Zarco-Tejada P, Lee JE, Moran MS, Ponce-Campos G, Beer C, Camps-Valls G, Buchmann N, Gianelle D, Klumpp K, Cescatti A, Baker JM, Griffis TJ (2014) Global and time-resolved monitoring of crop photosynthesis with chlorophyll fluorescence. Proc Natl Acad Sci USA 14: E1327–E1333
Haddaway NR, Hedlund K, Jackson LE et al (2017) How does tillage intensity affect soil organic carbon? A systematic review. Environ Evid 6:30. https://doi.org/10.1186/s13750-017-0108-9
Han P, Zhang W, Wang G, Sun W, Huang Y (2016) Changes in soil organic carbon in croplands subjected to fertilizer management: a global meta-analysis. Sci Rep 6:27199. https://doi.org/10.1038/srep27199
Harden JW, Hugelius G, Ahlström A et al (2017) Networking our science to characterize the state, vulnerabilities, and management opportunities of soil organic matter. Glob Change Biol. https://doi.org/10.1111/gcb.13896
Harmon ME, Silver WL, Fasth B, Chen H, Burke IC, Parton WJ, Hart SC, Currie WS, LIDET (2009) Long-term patterns of mass loss during the decomposition of leaf and fine root litter: an intersite comparison. Glob Change Biol 15:1320–1338
Hicke JA, Lobell DB, Asner GP (2004) Cropland area and net primary production computed from 30 Years of USDA agricultural harvest data. Earth Interact 8:1–21
Hodge A, Berta G, Doussan C, Merchan F, Crespi M (2009) Plant root growth, architecture and function. Plant Soil 321:153–187
Houghton RA (2010) How well do we know the flux of CO_2 from land-use change? Tellus 62B:337–351
Houghton RA, Nassikas AA (2017) Global and regional fluxes of carbon from land use and land cover change 1850–2015. Glob Biogeochem Cycles 31:456–472
Iocola I, Bassu S, Farina R et al (2017) Can conservation tillage mitigate climate change impacts in Mediterranean cereal systems? A soil organic carbon assessment using long term experiments. Europ J Agron 90:96–107
Ito A (2011) A historical meta-analysis of global terrestrial net primary productivity: are estimates converging? Glob Change Biol 17:3161–3175
Jansson C, Wullschleger SD, Kalluri UC, Tuskan GA (2010) Phytosequestration: carbon biosequestration by plants and the prospects of genetic engineering. Bioscience 60:685–696
Joergensen RG, Wichern F (2008) Quantiative assessment of the fungal contribution to microbial tissue in soil. Soil Biol Biochem 40:2977–2991

Johnson JMF, Allmaras RR, Reicosky DC (2006) Estimating source carbon from crop residues, roots and rhizodeposits using the national grain-yield database. Agron J 98:622–636
Jones DL, Nguyen C, Finlay RD (2009) Carbon flow in the rhizosphere: carbon trading at the soil-root interface. Plant Soil 321:5–33
Kindler R, Siemens J, Kaiser K, Walmsley DC, Bernhofer C, Buchmann N, Cellier P, Eugster W, Gleixner G, Grûnwald T, Heim A, Ibrom A, Jones SK, Jones M, Klumpp K, Kutsch W, Larsen KS, Lehuger S, Loubet B, McKenzie R, Moor E, Osborne B, Pilegaard K, Rebmann C, Saunders M, Schmidt MWI, Schrumpf M, Seyfferth J, Skiba U, Soussana J-F, Sutton MA, Tefs C, Vowinckel B, Zeeman MJ, Kaupenjohann M (2011) Dissolved carbon leaching from soil is a crucial component of the net ecosystem carbon balance. Glob Change Biol 17:1167–1185
King AE, Blesh J (2018) Crop rotations for increased soil carbon: perenniality as a guiding principle. Ecol Appl 28:249–261
Kleber M, Eusterhues K, Keiluweit M, Mikutta C, Mikutta R, Nico PS (2015) Mineral–organic associations: formation, properties, and relevance in soil environments. Adv Agron 130:1–140
Kleber M, Nico PS, Plante A, Filley T, Kramer M, Swanston C, Sollins P (2011) Old and stable soil organic matter is not necessarily chemically recalcitrant: implications for modeling concepts and temperature sensitivity. Glob Change Biol 17:1097–1107
Knebl L, Leithold G, Schulz F, Brock C (2017) The role of soil depth in the evaluation of management-induced effects on soil organic matter. Europ J Soil Sci 68:979–987. https://doi.org/10.1111/ejss.12492
Kong AYY, Six J (2010) Tracing root vs. residue carbon into soils from conventional and alternative cropping systems. Soil Sci Soc Am J 74:1201–1210
Kuzyakov Y (2006) Sources of CO_2 efflux from soil and review of partitioning methods. Soil Biol Biochem 38:425–448
Kuzyakov Y, Domanski G (2000) Carbon input by plants into the soil. Review J Plant Nutr Soil Sci 163:421–431
Lal R (1989a) Agroforestry systems and soil surface management of a tropical Alfisol: Introduction. Agrofor Syst 8:1–6
Lal R (1989b) Agroforestry systems and soil surface management of a tropical Alfisol: I. Soil moisture and crop yields. Agrofor Syst 8:7–29
Lal R (1989c) Agroforestry systems and soil surface management of a tropical Alfisol. II. Soil erosion and nutrient loss. Agrofor Syst 8:97–111
Lal R (1989d) Agroforestry systems and soil surface management of a tropical Alfisol III. Soil chemical properties. Agrofor Syst 8:113–132
Lal R (1989e) Agroforestry systems and soil surface management of a tropical Alfisol. IV. Effects on soil physical and mechanical properties. Agrofor Syst 8:197–215
Lal R (1999) Soil management and restoration for C sequestration to mitigate the greenhouse effect. Prog Env Sci 1:307–326
Lal R (2003) Soil erosion and the global carbon budget. Environ Int 29:437–450
Lal R (2004a) Soil carbon sequestration impacts on global climate change and food security. Science 304:1623–1627
Lal R (2004b) Soil carbon sequestration to mitigate climate change. Geoderma 123:1–22
Lal R (2007) Farming carbon. Soil Till Res 96:1–5
Lal R (2009) Soils and food sufficiency. A review. Agron Sustain Dev 29:113–133
Lal R (2010) Beyond Copenhagen: mitigating climate change and achieving food security through soil carbon sequestration. Food Security 2:169–177
Lal R (2016) Potential and challenges of conservation agriculture in sequestration of atmospheric CO_2 for enhancing climate-resilience and improving productivity of soil of small landholder farms. CAB Reviews 11:009
Lal R, Follett RF (2009a) Preface. In: Lal R, Follett RF (eds) Soil carbon sequestration and the greenhouse effect. SSSA Special Publication 57, second edition, Soil Science Society of America, Madison, WI, pp xi–xii

Lal R, Follett RF (2009b) Soils and climate change. In: Lal R, Follett RF (eds) Soil carbon sequestration and the greenhouse effect. SSSA Special Publication 57, second edition, Soil Science Society of America, Madison, WI, pp xxi–xxviii

Lal R, Kimble JM (2000) Pedogenic carbonates and the global carbon cycle. In: Lal R, Kimble JM, Eswaran H, Stewart BA (eds) Global climate change and pedogenic carbonates. CRC Press, Boca Raton, FL, pp 291–302

Leake J, Johnson D, Donnelly D, Muckle G, Boddy L, Read D (2004) Networks of power and influence: the role of mycorrhizal mycelium in controlling plant communities and agroecosystem functioning. Can J Bot 82:1016–1045

Lehmann J, Kleber M (2015) The contentious nature of soil organic matter. Nature 528:60–68

Lei H-M, Yang D-W (2011) Seasonal and interannual variations in carbon dioxide exchange over a cropland in the North China Plain. Global Change Biol 16:2944–2957

Li Z, Liu S, Tan Z, Bliss NB, Young CJ, West TO, Ogle SM (2014) Comparing cropland net primary production estimates from inventory, a satellite-based model, and a process-based model in the Midwest of the United States. Ecol Model 277:1–12

Lorenz K, Lal R (2005) The depth distribution of soil organic carbon in relation to land use and management and the potential of carbon sequestration in subsoil horizons. Adv Agron 88:35–66

Lorenz K, Lal R (2014) Soil organic carbon sequestration in agroforestry systems. A review. Agron Sustain Dev 34:443–454

Macías F, Arbestain MC (2010) Soil carbon sequestration in a changing global environment. Mitig Adapt Strateg Glob Change 15:511–529

Maillard É, Angers DA (2014) Animal manure application and soil organic carbon stocks: a meta-analysis. Glob Change Biol 20:666–679

Mangalassery S, Sjögersten S, Sparkes DL, Mooney SJ (2015) Examining the potential for climate change mitigation from zero tillage. J Agric Sci 153:1151–1173. https://doi.org/10.1017/S0021859614001002

Marlon JR, Bartlein PJ, Carcaillet C, Gavin DG, Harrison SP, Higuera PE, Joos F, Power MJ, Prentice IC (2008) Climate and human influences on global biomass burning over the past two millennia. Nature Geosci 1:697–702

Marschner H, Dell B (1994) Nutrient uptake in mycorrhizal symbiosis. Plant Soil 159:89–102

Mendez-Millan M, Dignac MF, Rumpel C, Rasse DP, Derenne S (2010) Molecular dynamics of shoot vs. root biomarkers in an agricultural soil estimated by natural abundance ^{13}C labelling. Soil Biol Biochem 42:169–177

Meurer KHE, Haddaway NR, Bolinder MA, Kätterer T (2018) Tillage intensity affects total SOC stocks in boreo-temperate regions only in the topsoil—a systematic review using an ESM approach. Earth Sci Rev 177:613–622

Mokany K, Raison RJ, Prokushkin AS (2006) Critical analysis of root:shoot ratios in terrestrial biomes. Glob Change Biol 12:84–96

Monfreda C, Ramankutty N, Foley JA (2008) Farming the planet: 2. Geographic distribution of crop areas, yields, physiological types, and net primary production in the year 2000. Global Biogeochem Cyc 22, GB1022. https://doi.org/10.1029/2007gb002947

Moni C, Rumpel C, Virto I, Chabbi A, Chenu C (2010) Relative importance of sorption versus aggregation for organic matter storage in subsoil horizons of two contrasting soils. Eur J Soil Sci 61:958–969

Moran KK, Jastrow JD (2010) Elevated carbon dioxide does not offset loss of soil carbon from a corn–soybean agroecosystem. Environ Pollut 158:1088–1094

Murchie EH, Pinto M, Horton P (2009) Agriculture and the new challenges for photosynthesis research. New Phytol 181:532–552

Nair PKR, Gordon AM, Mosquera-Losada M-R (2008) Agroforestry. In: Jorgensen SE, Fath BD (eds) Ecological engineering, encyclopedia of ecology, vol 1. Elsevier. Oxford, UK, pp 101–110

Nair PKR, Kumar BM, Nair VD (2009) Agroforestry as a strategy for carbon sequestration. J Plant Nutr Soil Sci 172:10–23

Nair PKR, Nair VD, Kumar BM, Showalter JM (2010) Carbon sequestration in agroforestry systems. Adv Agron 108:237–307
Nguyen C (2003) Rhizodeposition of organic C by plants: mechanisms and controls. Agronomie 23:375–396
Olmstead AL, Rhode PW (2011) Adapting North American wheat production to climatic challenges, 1839–2009. Proc Natl Acad Sci USA 108:480–485
Olson KR, Al-Kaisi MM, Lal R, Lowery B (2014) Experimental consideration, treatments, and methods in determining soil organic carbon sequestration rates. Soil Sci Soc Am J 78:348–360
Pausch J, Kuzyakov Y (2017) Carbon input by roots into the soil: Quantification of rhizodeposition from root to ecosystem scale. Glob Change Biol in press. https://doi.org/10.1111/gcb.13850
Paustian K, Lehmann J, Ogle S et al (2016) Climate-smart soils. Nature 532:49–57
Peng Y, Gitelson AA (2011) Application of chlorophyll-related vegetation indices for remote estimation of maize productivity. Agric For Meteorol 151:1267–1276
Peng Y, Gitelson AA, Keydan G, Rundquist DC, Moses W (2011) Remote estimation of gross primary production in maize and support for a new paradigm based on total crop chlorophyll content. Remote Sens Environ 115:978–989
Pielke RA, Pitman A, Niyogi D, Mahmood R, McAlpine C, Hossain F, Goldewijk KK, Nair U, Betts R, Fall S, Reichstein M, Kabat P, de Noblet N (2011) Land use/land cover changes and climate: modeling analysis and observational evidence. WIREs Clim Change 2:828–850
Poeplau C, Don A (2015) Carbon sequestration in agricultural soils via cultivation of cover crops—a meta-analysis. Agric Ecosyst Environ 200:33–41
Poeplau C, Don A, Vesterdal L, Leifeld J, Van Wesemael B, Schumacher J, Gensior A (2011) Temporal dynamics of soil organic carbon after land-use change in the temperate zone—carbon response functions as a model approach. Glob Change Biol 17:2415–2427
Portmann FT, Siebert S, Döll P (2010) MIRCA2000—Global monthly irrigated and rainfed crop areas around the year 2000: a new high-resolution data set for agricultural and hydrological modeling. Global Biogeochem Cycles 24, GB1011. https://doi.org/10.1029/2008gb003435
Power AG (2010) Ecosystem services and agriculture: tradeoffs and synergies. Phil Trans R Soc B 365:2959–2971
Powlson DS, Stirling CM, Jat ML et al (2014) Limited potential of no-till agriculture for climate change mitigation. Nature Clime Change 4:678–683
Powlson DS, Whitmore AP, Goulding KWT (2011) Soil carbon sequestration to mitigate climate change: a critical re-examination to identify the true and the false. Eur J Soil Sci 62:42–55
Prescott CE (2010) Litter decomposition: what controls it and how can we alter it to sequester more carbon in forest soils? Biogeochemistry 101:133–149
Prince SD, Haskett J, Steininger M, Strand H, Wright R (2001) Net primary production of U.S. Midwest croplands from agricultural harvest yield data. Ecol Appl 11:1194–1205
Ramankutty N, Evan AT, Monfreda C, Foley JA (2008) Farming the planet: 1. Geographic distribution of global agricultural lands in the year 2000. Global Biogeochem Cyc 22, GB1003. https://doi.org/10.1029/2007gb002952
Rasse DP, Mulder J, Moni C, Chenu C (2006) Carbon turnover kinetics with depth in a French loamy soil. Soil Sci Soc Am J 70:2097–2105
Rasse DP, Rumpel C, Dignac M-F (2005) Is soil carbon mostly root carbon? Mechanisms for a specific stabilization. Plant Soil 269:341–356
Rees RM, Bingham IJ, Baddeley JA, Watson CA (2005) The role of plants and land management in sequestering soil carbon in temperate arable and grassland ecosystems. Geoderma 128:130–154
Robbins M (2011) Crops and carbon. Earthscan, Oxon, U.K.
Rumpel C, Kögel-Knabner I (2011) Deep soil organic matter—a key but poorly understood component of terrestrial C cycle. Plant Soil 338:143–158
Rutledge S, Campbell DI, Baldocchi D, Schipper LA (2010) Photodegradation leads to increased carbon dioxide losses from terrestrial organic matter. Glob Change Biol 16:3065–3074

Sainju UM (2016) A global meta-analysis on the impact of management practices on net global warming potential and greenhouse gas intensity from cropland soils. PLoS ONE 11:e0148527
Sanderman J (2012) Can management induced changes in the carbonate system drive soil carbon sequestration? A review with particular focus on Australia. Agric Ecosyst Environ 155:70–77
Sanderman J, Baldock JA (2010) Accounting for soil carbon sequestration in national inventories: a soil scientist's perspective. Environ Res Lett 5:034003
Sanderman J, Farquharson R, Baldock J (2010) Soil carbon sequestration potential: a review for Australian agriculture—a report prepared for Department of Climate Change and Energy Efficiency, CSIRO Land and Water. http://www.csiro.au/resources/Soil-Carbon-Sequestration-Potential-Report.html
Saugier B, Roy J, Mooney HA (2001) Estimations of global terrestrial productivity: converging toward a single number? In: Roy J, Saugier B, Mooney HA (eds) Terrestrial global productivity. Academic Press, San Diego, pp 543–557
Schmidt MWI, Torn MS, Abiven S, Dittmar T, Guggenberger G, Janssens IA, Kleber M, Kögel-Knabner I, Lehmann J, Manning DAC, Nannipieri P, Rasse DP, Weiner S, Trumbore SE (2011) Persistence of soil organic matter as an ecosystem property. Nature 478:49–56
Schomberg HH, Steiner JL, Unger PW (1994) Decomposition and nitrogen dynamics of crop residues: residue quality and water effects. Soil Sci Soc Am J 58:372–381
Schulze E-D, Beck E, Müller-Hohenstein K (2005) Plant ecology. Springer, Berlin
Schulze E-D, Ciais P, Luyssaert S, Schrumpf M, Janssens IA, Thiruchittampalam B, Theloke J, Saurat M, Bringezu S, Lelieveld J, Lohila A, Rebmann C, Jung M, Bastviken D, Abril G, Grassi G, Leip A, Freibauer A, Kutsch W, Don A, Nieschulze J, Börner A, Gash JH, Dolman AJ (2010) The European carbon balance. Part 4: integration of carbon and other trace-gas fluxes. Global Change Biol 16:1451–1469
Schulze E-D, Luyssaert S, Ciais P, Freibauer A, Janssens IA et al (2009) Importance of methane and nitrous oxide for Europe's terrestrial greenhouse-gas balance. Nature Geosci 2:842–850
See L, Fritz S, You L, Ramankutty N, Herrero M, Justice C, Becker-Reshef I, Thornton P, Erb K, Gong P, Tang H, van der Velde M, Ericksen P, McCallum I, Kraxner F, Obersteiner M (2015) Improved global crop land data as an essential ingredient for food security. Global Food Sec 4:37–45
Sen R (2003) The root–microbe–soil interface: new tools for sustainable plant production. New Phytol 157:391–394
Shao J, Zhou X, Luo Y et al (2016) Uncertainty analysis of terrestrial net primary productivity and net biome productivity in China during 1901–2005. J Geophys Res Biogeosci 121:1372–1393
Shibu ME, Leffelaar PA, Van Keulen H, Aggarwal PK (2006) Quantitative description of soil organic matter dynamics - a review of approaches with reference to rice-based cropping systems. Geoderma 137:1–18
Siebert S, Döll P (2010) Quantifying blue and green virtual water contents in global crop production as well as potential production losses without irrigation. J Hydrol 384:198–217
Six J, Frey SD, Thiet RK, Batten KM (2006) Bacterial and fungal contributions to carbon sequestration in agroecosystems. Soil Sci Soc Am J 70:555–569
Smith KA, Conen F (2004) Impacts of land management on fluxes of trace greenhouse gases. Soil Use Manage 20:255–263
Smith P, Martino D, Cai Z, Gwary D, Janzen H, Kumar P, McCarl B, Ogle S, O'Mara F, Rice C, Scholes B, Sirotenko O, Howden M, McAllister T, Pan G, Romanenkov V, Schneider U, Towprayoon S, Wattenbach M, Smith J (2008) Greenhouse gas mitigation in agriculture. Phil Trans R Soc B 363:789–813
Smith TM, Shugart HH, Woodward FI, Burton PJ (1993) Plant functional types. In: Soloon AM, Shugart HH (eds) Vegetation dynamics and global change. Chapman and Hall, New York, pp 272–292
Still CJ, Berry JA, Collatz GJ, DeFries RS (2003) Global distribution of C_3 and C_4 vegetation: carbon cycle implications. Global Biogeochem Cyc 17, 1006. https://doi.org/10.1029/2001gb001807

Subedi KD, Ma BL, Liang BC (2006) New method to estimate root biomass in soil through root-derived carbon. Soil Biol Biochem 38:2212–2218

Suyker AE, Verma SB (2010) Coupling of carbon dioxide and water vapor exchanges of irrigated and rainfed maize–soybean cropping systems and water productivity. Agric For Meteorol 150:553–563

Tollenaar M, Fridgen J, Tyagi P, Stackhouse PW Jr, Kumudini S (2017) The contribution of solar brightening to the US maize yield trend. Nat Clim Change 7:275–280

Tian H, Chen G, Liu M, Zhang C, Sun G, Lu C, Xu X, Ren W, Pan S, Chappelka A (2010) Model estimates of net primary productivity, evapotranspiration, and water use efficiency in the terrestrial ecosystems of the southern United States during 1895–2007. For Ecol Manage 259:1311–1327

Torn MS, Swanston CW, Castanha C, Trumbore SE (2009) Storage and turnover of natural organic matter in soil. In: Senesi N, Xing B, Huang PM (eds) Biophysico-chemical processes involving natural nonliving organic matter in environmental systems. Wiley, Hoboken, New Jersey, USA, pp 219–272

Treseder KK, Cross A (2006) Global distributions of arbuscular mycorrhizal fungi. Ecosystems 9:305–316

USGS United States Geological Survey (2017) Global 30 m Cropland Extent 2015. https://web.croplands.org/app/map?lat=0.17578&lng=0&zoom=2

Wagle P, Zhang Y, Jin C, Xiao X (2016) Comparison of solar-induced chlorophyll fluorescence, light-use efficiency, and process-based GPP models in maize. Ecol Appl 26:1211–1222

Wang X, Wang J, Xu M, Zhang W, Fan T, Zhang J (2015) Carbon accumulation in arid croplands of northwest China: pedogenic carbonate exceeding organic carbon. Sci Rep 5:11439. https://doi.org/10.1038/srep11439

Wang Z, Xiao X, Yan X (2010) Modeling gross primary production of maize cropland and degraded grassland in northeastern China. Agric For Meteorol 150:1160–1167

Whalen JK, Sampedro L (2009) Primary production. In: Whalen JK, Sampedro L (eds) Soil ecology and management. CAB International, Wallingford, UK, pp 109–133

Wickings K, Grandy AS (2011) The oribatid mite *Scheloribates moestus* (Acari: Oribatida) alters litter chemistry and nutrient cycling during decomposition. Soil Biol Biochem 43:351–358

Wilkinson MT, Richards PJ, Humphreys GS (2009) Breaking ground: pedological, geological, and ecological implications of soil bioturbation. Earth Sci Rev 97:254–269

Wolf J, West TO, Le Page Y et al (2015) Biogenic carbon fluxes from global agricultural pro-duction and consumption. Global Biogeochem Cycles 29:1617–1639. https://doi.org/10.1002/2015GB005119

Wright AL, Dou F, Hons FM (2007) Crop species and tillage effects on carbon sequestration in subsurface soil. Soil Sci 172:124–131

Wu LS, Wood Y, Jiang PP, Li LQ, Pan GX, Lu JH, Chang AC, Enloe HA (2008) Carbon sequestration and dynamics of two irrigated agricultural soils in California. Soil Sci Soc Am J 72:808–814

Yang F, Ichii K, White MA, Hashimoto H, Michaelis AR, Votava P, Zhu A, Huete A, Running SW, Nemani RR (2007) Developing a continental-scale measure of gross primary production by combining MODIS and AmeriFlux data through Support Vector Machine approach. Remote Sens Environ 110:109–122

Yu GR, Fu Y, Sun X, Wen X, Zhang L (2006) Recent progress and future directions of ChinaFLUX. Science in China Series D 49(Suppl 1):1–23

Zavattaro L, Bechini L, Grignani C et al (2017) Agronomic effects of bovine manure: a review of long-term European field experiments. Europ J Agron 90:127–138

Zhang F, Wang Z, Glidden S, et al. (2017) Changes in the soil organic carbon balance on China's cropland during the last two decades of the 20th century. Sci Rep. https://doi.org/10.1038/s41598-017-07237-1

Zhang Y, Song C, Zhang K, Cheng X, Band LE, Zhang Q (2014) Effects of land use/land cover and climate changes on terrestrial net primary productivity in the Yangtze River Basin, China, from 2001 to 2010. J Geophys Res Biogeosci 119:1092–1109

Zhang Y, Xu M, Chen H, Adams J (2009) Global pattern of NPP to GPP ratio derived from MODIS data: effects of ecosystem type, geographical location and climate. Global Ecol Biogeogr 18:280–290

Zhao M, Running S, Heinsch FA, Nemani R (2010) MODIS-derived terrestrial primary production. In: Ramachandran B, Justice CO, Abrams MJ (eds) Land remote sensing and global environmental change. Springer, Dordrecht, The Netherlands, pp 635–659

Chapter 4
Carbon Sequestration in Grassland Soils

Contents

Abstract Grasslands, including rangelands, shrublands, pastureland, and cropland sown with pasture and fodder crops, cover 35 million km^2 or 26% of the global ice-free land area. Grasslands support the livelihoods of 1 billion people with pastoralism (rising of livestock) being the most widespread human land-use system globally with 20 million km^2 of grassland used for livestock feed production. The global is under pasture has, particularly, strongly increased since prehistoric times. Many grasslands have suffered losses of soil organic carbon (SOC) because of soil disturbance, vegetation degradation, fire, erosion, nutrient shortage, and water deficit. The nature, frequency, and intensity of disturbance may play a key role in the SOC balance of grassland. Less well known are the effects of disturbance processes on the soil inorganic carbon (SIC) stock. However, bedrock, irrigation practices, soil acidification, liming, and grazing management potentially affect the SIC stock of grassland. In comparison, better studied are the SOC dynamics in grassland. Grasslands have a high inherent SOC stock with up to 343 Pg SOC stored to 1 m depth with a sequestration rate of 0.5 Pg C yr^{-1}. Grasslands sequester large amounts of SOC because of a high belowground C allocation, root turnover, and rhizodeposition. Grassland gross primary production (GPP) is the major natural soil C input and has been estimated at 31.3 Pg C yr^{-1} for tropical savannas and grasslands, and to 8.5 Pg C yr^{-1} for temperate grasslands and shrublands, respectively. The net primary production (NPP) of grassland denotes C assimilation by plants before losses caused by grazing, harvest, herbivory, mowing, and other processes. However, the numerous processes of NPP loss are among the reasons

K. Lorenz and R. Lal, *Carbon Sequestration in Agricultural Ecosystems*,
https://doi.org/10.1007/978-3-319-92318-5_4

why direct measurements of NPP of grassland are challenging because not all of the biomass produced remains within the ecosystem. Further, all components of NPP of grassland must be measured in a single study. Additional research is also needed about the quantitative contribution of the major sources of SOC, and roots may play a critical role in maintaining SOC stocks in the future. Possible inputs of below-ground C include: (i) incorporated surface plant residues, (ii) plant root litter and rhizodeposition, (iii) dung and urine of grazing animals, and (iv) black carbon (BC) in fire-affected grasslands. Grazing management must be targeted toward SOC sequestration due to the large global grazing land area and potential for considerable rates of increase in SOC stock.

Keywords Grazing lands · Methanogenesis · Animal-based diet
Grazing management · Improved pastures

4.1 Introduction

Grasslands are ecosystems in which the dominant vegetation component is comprised of herbaceous species (Jones and Donnelly 2004).

Rangelands include grasslands, pastures, savannas, and shrublands and are found in tropical, subtropical, temperate, and boreal climates (Table 4.1). Further, temperate and tropical grasslands can also be distinguished. Savannas are mixed tree–grass systems characterized by a discontinuous tree canopy in a continuous grass layer and are an important source of food and income for roughly a quarter of the world's population (Scholes and Archer 1997). Some grasslands consist of natural vegetation and some are of anthropogenic origin. Cultural grasslands are primarily planted and maintained for agricultural reasons (Dixon et al. 2014). Specifically, grasslands can be managed by grazing or mowing, and the intensity of management can be intensive or extensive (Chang et al. 2016a). Grasslands may be

Table 4.1 Types of rangelands

Type	Description
Prairie	Grassland in North America
Grasslands	Landscape or region dominated by herbaceous plant species
Steppe	Temperate grassland in Central Asia
Pampas	Temperate grassland in South America
Shrubland	Landscape or region dominated by shrubs supporting grasses, herbs, and geophytes
Woodland	Low-density forest supporting shrubs and herbaceous plants including grasses
Savanna	Mixed tree–grass system supporting primarily C_4 grasses
Desert	Landscape or region receiving <250 mm of precipitation annually
Tundra	Biome where tree growth is limited by low temperatures and short growing season supporting shrubs, sedges, grasses, mosses, and lichens

continuously managed as multi-purpose systems for forage production (meadows), animal grazing (pastures), or a combination of both (Ciais et al. 2010). Nearly 100% of uncultivated grasslands are grazed by large mammals (McSherry and Ritchie 2013).

Grasslands are the natural climax vegetation in regions where rainfall is low enough to prevent the growth of forests (e.g., steppes of Central Asia, prairies of North America). Where rainfall is too low or unreliable and soils too poor to support regular cropping, rangelands which are relatively undisturbed ecosystems containing savannas, woodlands, and shrublands may be present (Fig. 4.1; Allen et al. 2010). Otherwise, in regions such as Northwestern and Central Europe, New Zealand, parts of North and South America and Australia, rainfall is higher, and thus, the climax vegetation is forest. In those regions, the productivity of non-natural grasslands is higher than those of natural grasslands (Jones and Donnelly 2004).

In the year 2000, permanent grasslands including rangelands, shrublands, pastureland, and cropland sown with pasture and fodder crops, covered about 35 million km^2 or 26% of the global ice-free land area (Conant 2012; Ramankutty et al. 2008). Further, the global area of managed grassland was about 12 million km^2 (Chang et al. 2016a). In 2015, the area (million km^2) under pasture land use in Tropical Africa, Latin America, China, former USSR, Oceania and North Africa and Middle East were 8.09, 5.64, 3.93, 3.65, 3.51, and 3.45, respectively (Houghton and Nassikas 2017).

The global pasture area increased strongly from 10,000 BC to AD 2000 (Table 4.2). Large parts of African savannas, in particular, have been inhabited by humans throughout the evolution of our species and, thus, have been modified by activities such as grazing and logging. Savannas cover about 40% of Africa and 20% of the global land area (Scholes and Walker 1993). Modeling results indicate that grazed grasslands comprised most of the managed grasslands, whereas

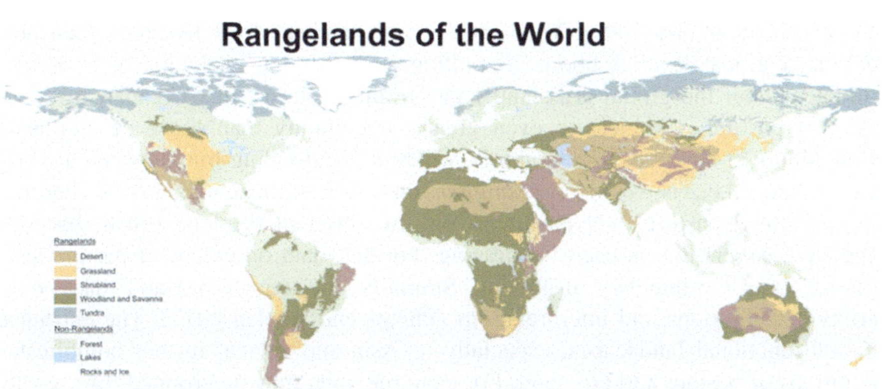

Fig. 4.1 Geographic distribution of global rangelands (By Dr. Karen Launchbaugh (University of Idaho—College of Natural Resources) [Public domain], via Wikimedia Commons)

Table 4.2 Range in global historical pasture area (million km^2) according to different scenarios (based on data reported in Klein Goldewijk et al. 2011)

Year	Range
10,000 BC	0.000–0.159
5000 BC	0.000–0.593
AD 1	0.53–3.32
AD 500	0.62–3.73
AD 1000	0.92–4.57
AD 1500	1.61–6.00
AD 1600	2.12–6.08
AD 1700	2.43–6.25
AD 1800	4.23–8.02
AD 1900	11.63–15.04
AD 2000	31.95–35.31

significant fractions of mown grasslands were only found in regions with high ruminant stocking density such as eastern China, India, Eastern and Northern Europe, and Eastern USA (Chang et al. 2016a). The largest fractions of managed grasslands were modeled in regions of high ruminant stocking density such as in Eastern Europe with a mean fraction of 90%, South Asia (59%), and Western Europe (55%). Further, the lowest managed grasslands fractions were modeled in the Russian Federation (17%; Chang et al. 2016a).

Grasslands support the livelihoods of ~1 billion people worldwide (Suttie et al. 2005). Specifically, pastoralism (rising of livestock) is the most widespread human land-use system on earth (Asner et al. 2004). Further, rangeland biomes characterized by livestock grazing were the most extensive among the anthropogenic biomes (Ellis and Ramankutty 2008). Depending on population density, residential, populated, and remote rangelands can be distinguished. Globally, rangelands covered nearly a third of global ice-free land and incorporating 73% of global pasture (28 million km^2), but these were found primarily in arid and other low-productivity regions with a high percentage of bare soil cover (around 50%; Ellis and Ramankutty 2008). Total area of grassland used for livestock feed production at global level is about 20 million km^2 (Mottet et al. 2017). However, estimates about the global grassland area are uncertain as, in particular, the definition of pasture is subject to even greater uncertainty than those of croplands (Ramankutty et al. 2008). For example, where is the dividing line between herding and grazing, i.e., is reindeer (*Rangifer tarandus* (Linnaeous, 1758)) herding reported under pastureland? Pasture is only a subset of the land (on herbaceous vegetative cover) that is used for grazing. Further, data on extent of pasture says nothing about the intensity of grazing. Similarly, the term 'rangeland' takes on a variety of definitions and interpretations (Phelps and Kaplan 2017). The existence of multifunctional landscapes, especially in Asia and Africa, further complicates definitions of pasture and cropland. For example, land may be cropped for a while, and then after harvest, is grazed for the remainder of the year. Thus, during the year the land is put to multiple uses, and it is not clear how to classify these lands, and how these lands are accounted for in census statistics (Ramankutty et al. 2008).

The global area of pasture increased by 59% from 1850 to 2015 (Houghton and Nassikas 2017). Agriculture and other land uses affect the potential natural grassland vegetation. Among terrestrial biomes, the biodiversity of grasslands is most affected by land-use pressures (Newbold et al. 2016). Specifically, pastures have mostly replaced grasslands, savannas, and shrubland (Ramankutty et al. 2008). Further, a substantial amount of savanna and grasslands have been converted to croplands in North America, Africa, and the Former Soviet Union. Also, a significant amount of forests in South America have been cleared for pastures. In total, up to 25% of the world's natural grasslands have been converted to croplands (Ramankutty et al. 2008). The potential natural grass vegetation has also been affected by urbanization. For example, >160,000 km^2 of land in the continental USA are cultivated with turf grasses in lawns (Milesi et al. 2016).

In 2000, 1.8 million km^2, or roughly 13% of the harvested area of all crops, were in perennial orchards, grasses, and plantations (Monfreda et al. 2008). However, data are uncertain as grasses for forage and fodder do not fall into the annual or perennial category, and the harvested area for grasses is evenly divided between both categories. The perennials are important for SOC sequestration as those are characterized by high soil carbon (C) input (Don et al. 2011). Perennial grasses are estimated to have covered 1.1 million km^2 in the year 2000. Roughly, 4% of the global harvested area is in perennial graminoids (i.e., grasses; Monfreda et al. 2008).

Grasses may be either belong to the C_3 or the C_4 photosynthetic pathway (Monfreda et al. 2008). C_4 photosynthesis in grass species causes a 19–88% daily growth enhancement compared to that in the C_3 photosynthesis (Atkinson et al. 2016). Leaves of C_4 grasses have less dense tissues, allowing more leaves to be produced for the same C cost compared to that of the C_3 grasses. Further, C_4 grasses invest more in roots than C_3 grasses (Atkinson et al. 2016). Faster growth and greater investment in roots, and water and nutrient acquisition may increase SOC stocks under C_4 compared to C_3 grasses growing at the same site. The global distribution of C_3 and C_4 grass species is strongly affected by temperature (Monfreda et al. 2008). Root biomass of C_3 and C_4 grass species may differently respond to the projected climate change (Nelson et al. 2017). The greater water use efficiency (WUE) of C_4 grasses gives them an advantage over C_3 grasses in hot, dry conditions. However, grasses are also cultivated outside of their optimal growing conditions. Nevertheless, broad patterns of C_3 and C_4 grasses are likely to follow the patterns derived from mean monthly temperature. C_4 fodder grass species produce animal feedstock in the Corn Belt of the USA and in northern Argentina and southern Brazil (Monfreda et al. 2008).

The total net C emission from global pasture land use was 16.3 Pg C for the period 1850–2015 (Houghton and Nassikas 2017). Thus, many grasslands have suffered losses of soil C due to intensive livestock grazing or agricultural uses and may have a high potential to store an appreciable fraction of atmospheric carbon dioxide (CO_2) as protected or stable C in the soil. The following sections discuss soil inorganic carbon (SIC) and soil organic carbon (SOC) stocks in grasslands.

Phytoliths (i.e., silicified features of plants formed as a result of biomineralization) and the importance of the phytolith C biosequestration flux are not discussed (see e.g., Santos and Alexandre 2017; Song et al. 2016, 2017).

4.2 Grassland Soil Inorganic Carbon

The SIC stock and its response to land use in grassland ecosystems are poorly understood (Monger et al. 2015). Grassland soils developed on limestone bedrock contain SIC as carbonates. As natural grasslands are primarily located in drier regions (i.e., arid and semiarid regions), they may contain, in particular, larger SIC stocks than those in more humid regions. Further, SIC may accumulate faster than previously thought, contrary to the common view that SIC is a mineral that takes centuries to millennia to accumulate in soil. For example, Wang et al. (2016) reported that converting desert shrubs to turfgrass may increase SIC in less than 10 years. This process may contribute to sequestration of atmospheric CO_2 in desert environments. However, it needs to be demonstrated that the Ca content in carbonates is derived directly from silicate minerals rather than from the preexisting carbonates, that the carbonate thus formed is durable and not subject to rapid dissolution, and that upstream CO_2 emissions are accounted for (Wang et al. 2016).

Reeder et al. (2004) emphasized the importance of including SIC in assessments of the amount and distribution of C in the soil–plant system in semiarid and arid ecosystems, and of the impacts of management on C sequestration. For example, for a long-term (56 years) heavy-grazing treatment compared to its adjacent non-grazed exclosure, average annual increase in SOC stock was 0.13 Mg C ha^{-1} yr^{-1}, compared with that of average annual increase in SIC stock of 0.29 Mg C ha^{-1} yr^{-1}. Thus, total soil C stock was 22.8 Mg C ha^{-1} higher in 0–90 cm depth with heavy grazing compared to that in the non-grazed exclosure, of which 16.3 Mg C ha^{-1} (68%) was due to increase in SIC, and 7.5 Mg C ha^{-1} (32%) in SOC. Further, SIC stock was higher (0.7 Mg C ha^{-1}) in the 0 to 60 cm depth of the light-grazing treatment compared to its exclosure, whereas SIC stock was 10.3 Mg C ha^{-1} higher in the 45 to 90 cm depth of the heavy-grazing treatment compared to its exclosure. Reeder et al. (2004) hypothesized that the large observed increase in SIC stock with heavy grazing was a combination of newly sequestered C and redistributed C from deeper in the soil profile.

During the 2000s, average SIC stocks to 10 cm depth of China's grassland were 3.1 Mg C ha^{-1}, ranging between 1.5 and 5.3 Mg C ha^{-1} for meadow and alpine steppe, respectively (Yang et al. 2012). The SIC stocks were more dynamic than previously assumed. Specifically, SIC stocks in the top 10 cm decreased significantly since the 1980s, with a mean rate of 26.8 g C m^{-2} yr^{-1}. Larger decreases were observed in those regions with stronger soil acidification and richer soil carbonates. The largest decrease (41.1 g C m^{-2} yr^{-1}) occurred in typical steppe on the Inner Mongolian Plateau, whereas the lowest decrease (5.3 g C m^{-2} yr^{-1}) was observed in alpine meadow on the Tibetan Plateau. The lost SIC may have been released to the atmosphere as CO_2, redistributed to the deeper soil layer, and/or transferred to the nearby regions. In particular, the fraction of soil carbonates entering the atmosphere may diminish the strength of terrestrial C sequestration and amplify the positive C-climate feedback (Yang et al. 2012).

Detailed analysis of native and cultivated grasslands in the Northern Great Plains of the USA and in the Kursk region of Russia demonstrated an increase in SIC stocks of Mollisols as a result of cultivation (Mikhailova et al. 2009). Before cultivation, the SIC stocks in the Northern Great Plains were 77–78 Mg C ha^{-1} to 1 m and 4 Mg C ha^{-1} to 0.75 m depth, respectively, depending on soil type and associated landscape position. In the Russian Chernozems, the SIC stocks were 9 Mg C ha^{-1} and 107 Mg C ha^{-1} to 1 m and 2 m depths, respectively. The Mollisols developed on calcareous substrate contained both lithogenic and pedogenic carbonates. Most of carbonate formation in Mollisols occurred below the root zones and, thus, below the zone of most biological activity. This trend is contrasted by carbonates in arid soils where they usually precipitate in the upper soil horizons (Mikhailova et al. 2009).

Non-cultivated native grasslands in two semiarid ecosystems in the USA had SIC stocks to 75 cm depth of 1.9 and 5.1 Mg C ha^{-1}, respectively (Denef et al. 2008). The SIC stocks did not change with depth. Cultivation generally increased soil CO_2, water infiltration and availability, cation concentration and decreased surface alkalinity—all factors which promote the formation of secondary carbonates. Thus, much larger amounts of SIC were found below 20 cm depth in the cultivated sites at both locations than in the grassland sites (Denef et al. 2008).

The carbonate C stock to 60 cm depth at an extensively grazed grassland site in Germany was more than twice as high as the SOC stock (195 vs. 86 Mg C ha^{-1}; Don et al. 2007). The soil developed on limestone bedrock and the grassland was converted from arable land 23 years ago. Carbonate content increased steadily below the former tillage horizon. However, carbonate was distributed irregularly in patches, and some patches were even free of carbonates up to 60 cm depth. Patches with high carbonate content were characterized by high clay content and higher bulk density. In contrast, patches free of carbonate originated from the loess cover which had been decalcified over time. They contained more SOC than that in profiles rich in clay and carbonate, especially in the subsoil (Don et al. 2007).

Similar to the SIC in cropland, irrigation practices, soil acidification, and liming may alter the SIC stocks in grasslands. In addition, grazing management may also affect the SIC stock in grassland, but the net effects of grassland management practices on SIC stocks are unclear (Sanderman 2012). Nevertheless, SIC stocks in deeper soil depths must be included in any assessments of grassland C. Additional research is needed to ascertain the importance of grassland management-induced changes in SIC stocks and on net greenhouse gas (GHG) emissions. Specifically, all reactants and products must be fully accounted for to elucidate whether or not specific processes in grasslands lead to a net sequestration of SIC.

4.3 Grassland Soil Organic Carbon

Global grasslands are important components of the terrestrial C cycle, storing 119–121 Pg C in grasslands and grazing lands (Erb et al. 2018), and 10–30% of SOC with a sequestration rate of 0.5 Pg C y^{-1} (Follett and Reed 2010). The SOC stock in grasslands is determined by the balance of C inputs and outputs. Carbon inputs are derived naturally from the annual photosynthetic C uptake of all leaves in a grassland or gross primary production (GPP), in the form of root exudates and litter. The net primary production (NPP) of grasslands is the net C stored as new plant material before harvest and other losses. Additional C inputs are derived from animal manure which may also be added. Carbon losses occur through natural processes of respiration, decomposition, erosion, leaching, fire and removal of biomass by grazing animals, and by human interventions through biomass harvesting. Management interventions such as mowing versus pasture and grazing intensity (i.e., the fraction of NPP consumed by grazing animals) may particularly influence the SOC stock. Grasslands exchange C as CO_2 between plants and soils, and the atmosphere by photosynthesis and respiration (Chang et al. 2015). In some regions, fire removes grassland C. For example, about 1.6 kg CO_2 is emitted per kg of dry matter burned in savannas and grasslands (Andreae 2004). For the period 2001–2010, fire in pastures accounted for about 43% (203 Mha) of global annual burned areas and about 22% (537 Tg C) of global C emissions from fire (Rabin et al. 2015). Global burned land area declined by 24.3% between 1998 and 2015, with large decreases in tropical savannas of South America and Africa, and grasslands across the Asian steppe (Andela et al. 2017). The declining savanna and grassland fires may have accounted for 7% of the contemporary global net land C flux.

For those grasslands that are regularly mowed to produce fodder, harvested biomass is later returned to the atmosphere, often within the same farm, in the form of CO_2 and methane (CH_4) emitted by animal digestion or by manure and slurry decomposition (Chang et al. 2015). When grasslands are grazed, biomass ingested by animals contains digestible and non-digestible organic compounds. The non-digestible C fraction (25–40%) of the intake is returned to the soil through excreta (feces and urine). The digestible part is respired as CO_2 shortly after intake (Chang et al. 2015). Only a small fraction serves to increase animal mass (e.g., meat) or to form animal products (e.g., milk and butter) which are exported from the grassland ecosystem (Soussana et al. 2010). Another small part of the digested C is emitted in the form of CH_4 by ruminant enteric fermentation (Sejian et al. 2012).

The C balance of grassland at local, regional, and continental scale is profoundly impacted by the nature, frequency, and intensity of disturbance (e.g., fire, mowing, grazing, and manure application, Soussana et al. 2007). Most C pool-related variables in grasslands could eventually recover to a stable state that is similar to the pre-disturbance state (Fu et al. 2017). However, some C-related variables, e.g., leaf area index (LAI) and NPP, are likely to exceed the pre-disturbance state, which may be due to regeneration of new leafs during recovering. In contrast, the soil C pool at the post-disturbance stable state showed nonsignificant changes compared to the pre-disturbance state (Fu et al. 2017). In Europe, the net biome productivity (NBP) of managed grassland ecosystems is significantly correlated with the total C removed by grazing and mowing (Soussana et al. 2007). Therefore, scientific knowledge of management intensity (intensive or extensive) is crucial to simulating the C balance (Chang et al. 2015). Aside the natural processes, organic C is also entering grassland soils by plant and soil management (i.e., addition of manure, and of non-harvested and non-burned residues). The fraction of organic C accumulating in grasslands which enter the soil (NBP_{soil}) primarily determines the amount of SOC sequestered (Schulze et al. 2010). For example, the NBP of European grassland is estimated to have been a net sink of 15 g C m^{-2} yr^{-1} during 1961–2010, equivalent to a 50-year continental cumulative SOC sequestration of 1.0 Pg C (Chang et al. 2015). However, a major change in grassland management intensity occurred across Europe during 1991–2010 from reduced livestock numbers. This change has 'inadvertently' enhanced SOC sequestration, over and above that caused through moderate increase in NBP by land-cover change, climate change, and rising CO_2 (Chang et al. 2016b). The total NBP for grasslands in China was estimated at 0.023 Pg C yr^{-1} during 1901–2005 with a high uncertainty of 65% (Shao et al. 2016).

4.3.1 Soil Carbon Input in Grassland

Grasslands are commonly low-productivity and water-limited ecosystems, sensitive, and vulnerable to climate perturbations and human intervention. Higher biomass and productivity characterizes natural grassland ecosystems with high species richness (Duffy et al. 2017). The amount of SOC sequestered in grasslands is primarily determined by the NBP fraction which enters the soil (Schulze et al. 2010). Grasslands sequester large amounts of SOC because of a high belowground C allocation and root turnover, i.e., by active rhizodeposition (Jones and Donnelly 2004). Further, activities of earthworms and other fauna in grasslands promote formation of macroaggregates in which microaggregates are formed (the reverse hierarchy) that protect SOC from decomposition for extended periods (Bossuyt et al. 2005). Rhizodeposition may favor SOC storage because direct incorporation into the soil matrix allows a high degree of physical protection of SOC. In particular, root litter transformation is an important determinant of the C cycle in grassland ecosystems, which is affected both by the root litter quality and by the rhizosphere activity (Personeni and Loiseau 2005). In addition, the vesicular–arbuscular mycorrhizae of grasses are specialized in mobilizing mineral ions, especially phosphorus, but are less efficient in breaking down organic matter (OM) than the ectomycorrhizae (Smith and Read 2008).

4.3.1.1 Natural Input Processes

The portion of NPP at non-erosional grasslands remaining after accounting for the losses by harvest, herbivory, and volatile organic compound (VOCs) is deposited as above- and belowground residues. Decomposition of aboveground residues but more important that of the belowground residues together with root exudates, and biotic and hydrologic belowground C transfer is processes of soil C input in natural grasslands (Lorenz and Lal 2005). For example, the annually produced litter of grassland species (i.e., grass, forb, sedge) is comprised of 26% leaf litter, 41% fine stems (e.g., culm, flowering stem), and 33% fine roots, respectively (Freschet et al. 2013). Fine stems and fine root litter decompose roughly twice as slow as compared to the leaf litter. Further, C_4 grass species may contribute relatively more to inputs of residues than the C_3 grass species. For example, more than 20% of terrestrial GPP is conducted by C_4 vegetation (Beer et al. 2010). However, this estimate is uncertain as C flux data for C_4-dominated ecosystems are scanty, especially those for tropical C_4 savannas.

Gross Primary Production
The grassland GPP is the sum of the net photosynthesis by all photosynthetic tissues in a grassland area measured at the ecosystem scale (Chapin and Eviner 2014). Grassland GPP is controlled by leaf area, nitrogen (N) supply, season, temperature, light, and atmospheric CO_2 concentration (Chapin et al. 2002). Further, grassland ecosystems harboring more plant species achieve higher GPP (Milcu et al. 2014). Differences in GPP between grassland ecosystems depend more strongly on differences in the quantity of light absorbed and length of photosynthetic season than on the efficiency of converting light to carbohydrates (Chapin and Eviner 2014). The light use efficiency (LUE) is low in low-resource ecosystems such as arid grasslands and declines with increase in temperature and is strongly reduced at extremely low temperatures (Chapin and Eviner 2014).

Globally, GPP has been estimated at 31.3 Pg C yr^{-1} for tropical savannas and grasslands, and to 8.5 Pg C yr^{-1} for temperate grasslands and shrublands, respectively (Beer et al. 2010). Savannas accounted for 26% of the global GPP and were the second most important biome in terms of global GPP following that of the tropical forests. The large area of savannas (about twice the surface area of tropical forests) explains their high contribution. Precipitation is the major climate control affecting GPP on 55 and 69% of the tropical savanna and grassland, and of the temperate grassland and shrubland areas, respectively (Beer et al. 2010). Similarly, globally spatial gradients (71%) and interannual changes (51%) in grassland GPP were mainly driven by precipitation (Liang et al. 2017). The arid/semiarid regions, in particular, strongly controlled the grassland ecosystems C cycle. The global grassland GPP was estimated at 11 Pg C yr^{-1}, and increasing at both annual and seasonal scales, with an annual increase of 0.023 Pg C (0.2%) from 1982 to 2011 (Liang et al. 2017).

Both GPP and ecosystem respiration of grasslands can be estimated by partitioning the net ecosystem CO_2 exchange (NEE) as assessed by the eddy covariance (EC) technique (He et al. 2014). However, the flux observations need to be upscaled from sites to regions in order to quantify the terrestrial C dynamics over large grassland areas. Modeling is the commonly used approach for estimating GPP at large spatial scales. Depending on the upscaling scheme, annual GPP per area for tropical savannas and grasslands ranged globally from 900.4 to 1,693.7 g C m^{-2} yr^{-1}, with a median of 1,134.9 g C m^{-2} yr^{-1} (Beer et al. 2010). For temperate grasslands and shrublands, the GPP also varied widely between 372.1 and 736.7 g C m^{-2} yr^{-1}, with a median of 478.2 g C m^{-2} yr^{-1}. However, an accurate estimation of regional GPP for grasslands is difficult (He et al. 2014). Also, data for environmental drivers of NEE of savanna ecosystems are scanty (Räsänen et al. 2017). Some examples of estimates for grassland GPP are given in the following section.

Based on tower CO_2 flux measurements from 20 European grasslands, the average annual GPP was estimated to range widely between 466.8 g C m^{-2} yr^{-1} for dry semi-natural pastures of South-central Europe and 1,872.8 g C m^{-2} yr^{-1} for intensively managed grasslands in Atlantic climate (Gilmanov et al. 2007). Possibly due to higher levels of N-fertilization in intensively managed European grasslands in Atlantic climate, grasslands achieved higher GPP levels than, e.g., unfertilized North American grasslands in comparable climate. Based on CO_2 flux measurements at CarboEurope eddy-covariance sites across geographic Europe as bordered in the east by the Ural mountains, the Caspian Sea, the Caucasus, and the Black Sea, Schulze et al. (2010) calculated a mean GPP of 1,120 g C m^{-2} yr^{-1}. When the land-based GHG balance derived mainly from ecosystem measurements was confronted with the atmospheric-based balance derived from measurements of GHG concentrations in the atmosphere and inversion models, mean grassland GPP of EU 25 between 2000 and 2005 and of the European continent was 1,343 g C m^{-2} yr^{-1} (Schulze et al. 2009). The available light and the length of the growing season were the limiting factors in Central and Northern Europe. Drought was an additional limitation mainly in Southern Europe (Schulze et al. 2010).

For an unfertilized warm-temperate Duke grassland in North Carolina (Table 4.3), the GPP was 273–555 g C m^{-2} yr^{-1} lower than in highly fertilized (though not as warm) European grasslands. In comparison, for a moderately fertilized grassland in Pennsylvania, GPP was considerably closer to that for the European grasslands. In accord with the previous studies, precipitation had a strong positive effect on GPP of grassland in Europe in the range between 300 and 800 mm yr^{-1}. At sites with higher precipitation, however, GPP was apparently controlled by other environmental factors (e.g., temperature), but data were scanty (Gilmanov et al. 2007).

Table 4.3 Annual gross primary production (GPP) of some grasslands in the USA

Location	GPP (g C m^{-2} yr^{-1})	References
North Carolina	1,208	Novick et al. (2004)
Pennsylvania	1,253–1,441	Skinner (2005); Skinner et al. (2008)
Northern Great Plains	402–431	Zhang et al. (2007)

The GPP for the grasslands in the Northern Great Plains (Table 4.3), was strongly impacted by soil properties, C_3/C_4 grasses, and land cover. The mean annual GPP for the Tibetan alpine grasslands between 2003 and 2008 was estimated at 312.3 g C m^{-2} yr^{-1}, ranging from 297.7 g C m^{-2} yr^{-1} in 2008 to 326.9 g C m^{-2} yr^{-1} in 2006 (He et al. 2014). The regional GPP exhibited a large spatial variability due to changes in temperature, precipitation, and vegetation distribution.

In contrast to grasslands without trees, GPP of savanna is comprised of overstorey and understorey components. Specifically, total GPP is a product of the interplay between trees and grass that co-dominate the savanna landscapes and are maintained through interactions with climate and disturbance (i.e., fire, land-use change, herbivory). For the first time, Moore et al. (2016) studied the temporal dynamics of savanna understorey and overstorey components of C flux for tropical savanna and estimated it at 2,267.1 g C m^{-2} yr^{-1}, of which the understorey contributed 32%. Further, contribution of the understorey was strongly seasonal, with most GPP occurring during the wet season (40% of total ecosystem GPP during the wet season and 18% in the dry; Moore et al. 2016).

Net Primary Production

The NPP of grassland is a measure of the C incorporated into the biomass of all autotrophs within the ecosystem. Specifically, it denotes C assimilation by plants before losses caused by grazing, harvest, herbivory, mowing, and other processes. Similar to cropland, the numerous processes of NPP loss are among the reasons why direct measurements of NPP of grassland are not possible because not all of the biomass produced remains within the ecosystem. Further, all components of NPP of grassland have not yet been measured in a single study. Also, no consensus exists for the preferred methods of calculation of NPP from field data, and assumed mathematical relationships between proxy measures and NPP are, therefore, required to estimate NPP of grassland (Lauenroth et al. 2006). Among several approaches consist of a collection of methods to approximate the amount of biomass or C that autotrophs assimilate over a specific time interval. Several methods of estimation of NPP are available for the conceptual, but not measurable attribute NPP of grassland ecosystems (Lauenroth et al. 2006). Sándor et al. (2016) obtained substantial differences in the outputs of 12 grassland models, indicating uncertainty in simulated grassland processes. However, uncertainties for biomass yield reduced after calibration with detailed production and phenology data.

During the 1990s, modeled production of unmanaged grassland comprised 63% of the total global production (Chang et al. 2016a). The production of grazed grasslands accounted for 34% of the total while that of the mown grasslands was only 3% of the total global production. The aboveground NPP (consumable NPP) was 5,778 Tg (million tonnes) of dry matter (DM), compared with an estimate of 147–654 Tg of DM for unmanaged grassland, thereby simulating the total annual consumption by wild herbivores. Addition of N fertilizer in the central USA, Europe, northeastern India, southern China, South Korea, Japan, and southern Brazil has increased productivity (Chang et al. 2016a). Some examples of NPP of grasslands for different regions are given in the following section.

For the period 1990–1999, the data-oriented NPP for grassland in EU-25 was estimated to range from 588 to 758 g C m^{-2} yr^{-1} compared with 750–797 g C m^{-2} yr^{-1} estimated by using a process-oriented model (Ciais et al. 2010). The modeled average NPP of pastures, in particular, was 1,237 g C m^{-2} yr^{-1} compared with 646 g C m^{-2} yr^{-1} for meadows. The higher NPP of pasture was explained by more intense N recycling in pasture compared to meadows (Ciais et al. 2010). However, the NPP of grassland in EU-25 was estimated from uncertain data as observations at representative sites were lacking and there were no direct measurements. Yet, the NPP of grassland in EU-25 was about 20% higher than that for croplands as grasslands can sustain photosynthesis all year round in suitable climates, whereas crops often have a shorter growing season (Gilmanov et al. 2010; Schulze et al. 2010).

The NPP of grassland in EU-28 plus Norway and Switzerland was estimated using the process-based biogeochemical model ORCHIDEE-GM for the period of 1961–2010. The average modeled NPP was 559 g C m^{-2} yr^{-1}, and 86% of it respired back into the atmosphere by heterotrophic processes in soil and 4% by the livestock grazing (Chang et al. 2015). A trend toward increasing NPP was modeled for this period overall European grasslands—except for the Mediterranean region. This trend was supported by multiple evidence from experimental studies and by the satellite vegetation indices. The increase could be induced by climate change and elevated CO_2 concentration, as well as addition of N and other changes in management (e.g., resowing with improved varieties of grass; Chang et al. 2016b). The increase in NPP of extensively managed grasslands (e.g., British Isles, Western Europe, Alpine), where N fertilizer is not applied, can be mainly attributed to climate change and to the CO_2 fertilization effect. Increase in NPP of the intensively managed grasslands is also induced by the intensified addition of N during the period of 1961–1990, and the simple parameterization of N effects on photosynthesis (Chang et al. 2015).

The NPP of grassland in China based on field surveys was estimated at 145–310 g C m^{-2} yr^{-1} (Zhang et al. 2014). Further, Zhang et al. (2016) used a modified Carnegie–Ames–Stanford Approach (CASA) model to estimate NPP of natural grassland in China for the period of 2004–2008. The mean estimated NPP of grassland of 504.78 g C m^{-2} yr^{-1} was consistent with that of the observed NPP data of 553.52 g C m^{-2} yr^{-1}. The estimated average annual NPP ranged from 443.23 to 554.40 g C m^{-2} yr^{-1}. There existed an increasing trend of NPP from the northwest to the southeast due to the zonal distribution of vegetation. The accumulation of NPP primarily occurred between April and October with an estimated average NPP of 482.19 g C m^{-2}, or about 88.78% of the annual total. The combination of the water and thermal regimes was the main determinants of the NPP of natural grasslands in China (Zhang et al. 2016). Otherwise, Shao et al. (2016) estimated a mean NPP of 214 g C m^{-2} yr^{-1} for grasslands in China during 1901–2005 based on published literature and the Multi-scale Synthesis and Terrestrial Model Intercomparison Project.

The NPP of grassland in Southern USA increased by about 15% from 1895 to 2007 (Tian et al. 2010). Specifically, the annual mean NPP for C_3 and C_4 grasslands based on a process-based model was 310 g C m^{-2} yr^{-1} for 1895–1950, and 355 g C m^{-2} yr^{-1} for 1951–2007, respectively. Similarly, the NPP of grassland of Canadian agroecosystems, based on C allocation coefficients, ranged from 270 g C m^{-2} yr^{-1} for legume species to 1,217 g C m^{-2} yr^{-1} for grassland/pasture (calculated from Bolinder et al. 2007).

Belowground Carbon Input
In comparison with croplands, grasslands are characterized by less disturbance such as by removing biomass, changing the vegetation cover, and altering soil functions (Conant 2012). This low disturbance, together with the permanent soil cover of grasslands, may result in high inputs of belowground C (Rumpel 2011). Otherwise, inputs of C into the soil may be decreased, in particular, by removing a large proportion of plant biomass to enhance yield of some useful material (e.g., for forage or fuel). However, little is generally known about the quantitative contribution of the major sources of SOC. Possible inputs of belowground C include: (i) incorporated surface plant residues, i.e., senescent plant leaves not consumed by animals and plant leaves which escaped harvesting, (ii) plant root litter and rhizodeposition, (iii) dung and urine of grazing animals, and (iv) black carbon (BC) in fire-affected grasslands (Rumpel 2011).

Fate of Surface Plant Residues
The aboveground NPP (ANPP) of grasslands is a source for the inputs of belowground C. The ANPP for 42 grassland sites spanning five continents, ranged widely between 64 and 1,503 g m^{-2} yr^{-1} (Fay et al. 2015). Before entering the soil, ANPP in non-erosional grasslands may be lost to fire, grazing, harvesting, decomposition, and photo- and thermal degradation.

Fire in pasture accounts for about 22% of global C emissions from fire (Rabin et al. 2015), with about 2.5 Pg C emitted annually by vegetation fires between 2001 and 2010 (Surawski et al. 2016). Further, on average 15% or more of the grazing land NPP is consumed by livestock (Fetzel et al. 2017; Irisarri et al. 2017). Under intensive grazing, up to 60% of ANPP is ingested by domestic herbivores (Lemaire and Chapman 1996). However, this percentage can be much lower under extensive grazing (Soussana and Lemaire 2014). The largest part of ingested C is digestible by herbivores and respired shortly after intake. A small part of digested ANPP may be emitted in the form of CH_4 through enteric fermentation by ruminants (Chang et al. 2015). For example, enteric CH_4 emissions compensated 34% of the on-site C sequestration in nine European grasslands (Soussana et al. 2010). The non-digestible C (25–40% of the intake according to the digestibility of the grazed herbage) is returned to the pasture in excreta, mainly as feces (Soussana and Lemaire 2014). More than 80% of ANPP is harvested and exported as hay or silage in a cutting regime. For example, 263 g C m^{-2} yr^{-1} is removed by cutting across the EU-25, and the grazed fraction of total NPP is only 0.18 in pasture, compared to a cut fraction of 0.4 in meadows (Ciais et al. 2010). Losses of ANPP occur by decomposition. By controlling lignin biosynthesis, silicon (Si) accumulation in grasses may also affect SOC cycling (Klotzbücher et al. 2018). Some of the remaining C is incorporated into the soil. For example, Cotrufo et al. (2015) estimated that 19% of the loss in the aboveground litter C in a tallgrass prairie is not mineralized into CO_2 but is transferred into the mineral soil. The importance of abiotic degradation of ANPP of grassland (e.g., by photo- and thermal degradation) especially in the arid region is widely recognized (van Asperen et al. 2015). However, to what extent the combined processes of photo- and thermal degradation of aboveground grass litter account for ANPP losses remains to be unclear (Lee et al. 2012).

Plant Roots and Soil Carbon Input

Roots of grass species translocate C deep into the soil but how deep they grow and why is uncertain (Fan et al. 2017). For example, mean rooting depths for better-studied grama grass (*Bouteloua gracilis*), feather grass (*Stipa* L.), bromegrass (*Bromus* Scop.), fescue (*Festuca* L.), wheatgrass (*Agropyron cristatum* Gaertn.), and beard grass (*Polypogon monspeliensis* (L.) Desf.) were 0.84, 1.01, 1.19, 1.23, 1.34, and 1.60 m, respectively (Fan et al. 2017). Through roots, pastures and grasslands allocate a large proportion of assimilates (30–50%) and NPP (50–67%) belowground (Kuzyakov and Domanski 2000; Saugier et al. 2001). However, the estimates by Saugier et al. (2001), based on biomass harvests, are uncertain as NPP that is not available to harvest was excluded due to consumption by herbivores, root exudation, transfer to mycorrhizae, and volatile emissions. Also unclear is how much of belowground NPP is used for root production as, for example, mean root production based on 11 methods at a single shortgrass steppe site ranged widely from 61 to 1,107 g OM m^{-2} yr^{-1} (Milchunas 2009). To overcome difficulties in root measurements in grasslands, root biomass can also be estimated based on root: shoot ratios, which range widely between 0.38 for tropical/subtropical grasslands and 26.027 for cool temperate grasslands (Mokany et al. 2006). Median root: shoot ratio increase from tropical (1.89) to temperate (4.22) to cool temperate grasslands (4.50). However, vegetation-specific root-to-shoot ratios must be used to predict root biomass (Mokany et al. 2006). Based on revised root-to-shoot ratios and biomass estimates, root biomass was estimated to be about 500 and 1,700 g m^{-2} in temperate grassland, and tropical savanna and grassland, respectively (Robinson 2007).

Rhizosphere processes, such as root turnover, rhizodeposition, root respiration, and microbial response, may be important for SOC dynamics in grassland soil (Rumpel 2011). These processes may lead to stabilization as well as destabilization of SOC, but are poorly understood (Cheng and Kuzyakov 2005). The amount of C entering grassland soils alone from root exudation is estimated to be in the range 0.1–5 g C kg $soil^{-1}$ $month^{-1}$ which is similar to that derived from root turnover (Jones et al. 2009). Further, there is a rapid flux of C through mycorrhizal hyphae in grassland. For example, at least 5.4–7.7% of the C lost by plant roots being respired from AM fungal mycelium and 3.9–6.2% being fixed in mycorrhizal mycelium within 21 h (Johnson et al. 2002).

Rhizodeposition

Rhizodeposition describes the release of organic C compounds by roots with particularly large amounts of C released by exudates with minor releases by mucilage and sloughed-off root border cells (Nguyen 2009). The C flow in the rhizosphere is extremely complex, being highly plant and environment dependent and also varying both spatially and temporally along the root. Thus, amount and type of rhizodeposits are highly context-specific (Jones et al. 2009). About a quarter of the

1.5–2.2 Mg C ha^{-1} translocated by pasture grasses belowground during one vegetation period is released into soil (Kuzyakov and Domanski 2000; Jones et al. 2009). However, many of the published data on C flow in both soil and roots are severely biased (Kuzyakov 2006). Specifically, it is almost impossible to separate the rhizosphere respiration from the mycorrhizal respiration. However, the root exudation of grassland species may stimulate mineralization of SOM the (rhizosphere priming effect; Shahzad et al. 2015). It is through their exudates that grass species provide energy to soil microorganisms and control the way the energy is used in order to maximize SOM mineralization and drive their own nutrient supply. However, co-evolution over hundreds of thousands of years of plant–microbe may be the reasons for lower SOC stocks compared to soils with less rhizodeposition (Sanderman et al. 2010).

Mycorrhizal Fungi

Mycorrhizal fungi live in symbiosis with the roots of plants. Almost all grasses are mycorrhizal plants with arbuscular mycorrhiza (AM) fungi most abundant in temperate grasslands and savannas among terrestrial biomes (Treseder and Cross 2006). However, the impacts of both AM and ectomycorrhiza (EM) on the SOC stock are poorly understood (Soudzilovskaia et al. 2015). Indirectly, AM fungi may benefit SOC sequestration by promoting NPP through facilitating the uptake of P, N, and K in certain grass species (Marschner and Dell 1994). Otherwise, whether the production rates of glomalin which are positively related to AM biomass (Treseder and Cross 2006), contribute to SOC sequestration in grassland soils is a matter of debate (Verbruggen et al. 2016).

4.3.1.2 Management-Derived Carbon Input

Grasslands are usually characterized by permanent vegetation cover, but management differences (e.g., mowing versus pasture and grazing intensity) may influence the soil C input and its dynamics (Parsons et al. 2011). Aside natural processes, grassland SOC stocks also depend on managerial interventions. Important among interventions in grazing lands include sowing of legumes and more productive grass species, fire management, irrigation, and fertilization (Henderson et al. 2015). All of these practices can increase forage production, increase returns of plant litter and dung (where more animals are introduced to make use of additional forage) to the soil, thereby increase the amount of organic C that is incorporated into soils and increase the SOC stocks (Henderson et al. 2015). Further, manure application may directly and indirectly (i.e., through increased NPP) add organic C to grassland soils (Maillard and Angers 2014). Improved grazing, irrigation, plant species management, and the use of organic or inorganic fertilizers of grazing lands can significantly increase SOC stocks (Conant et al. 2017).

Manure-carbon input

Application of manure directly adds C to grassland soils. For example, up to

1,450 g C m^{-2} yr^{-1} is added with manure and slurry to grasslands in Europe (Soussana et al. 2007). The amount of applied manure C is generally the dominant driver determining the extent of SOC increase following manure application (Maillard and Angers 2014). Further, animal species may also affect the response as cattle (*Bos taurus* Linnaeus, 1758) manure tends to increase SOC stocks more than that of pig (*Sus scrofa domesticus* Erxleben, 1777) and poultry manure (Maillard and Angers 2014). For example, annual application of liquid hog manure to grassland for twenty years did not increase the SOC stocks to 70 cm depth due to limited quantities and high decomposability of the C supplied by manure, and possibly the stimulation of the mineralization of both native SOC and fresh root-derived material (Angers et al. 2010). This priming effect contributed to lower SOC stocks (86 Mg C ha^{-1}) to 70 cm depth in the 50 m^3 ha^{-1} liquid hog manure treatment than in the 0 or 100 m^3 ha^{-1} treatments (100 Mg C ha^{-1}). Similarly, eighteen years of pig slurry addition to perennial ryegrass (*Lolium perenne* L.) had insignificant effects on SOC stock to 20 cm depth (Thomsen and Christensen 2004). Further, 5 years of broiler (*Gallus gallus domesticus*) litter addition had no significant effect on SOC stocks to 6 cm depth under 'Coastal' bermudagrass [*Cynodon dactylon* (L.) Pers.] (Franzluebbers et al. 2001). Although broiler litter added significant organic C over twelve years (2.4 Mg C ha^{-1} yr^{-1}), its decomposition was high, resulting in statistically undetectable changes in SOC stocks to 20 cm depth under pasture cultivated with 'Coastal' bermudagrass and tall fescue (*Lolium arundinaceum* [Schreb.] Darbysh.) (Franzluebbers and Stuedemann 2010). Application of cattle manure to switchgrass (*Panicum virgatum* L.) for four years increased SOC stocks by 8.59 and 15.01 Mg C ha^{-1} in 0–30 and 0–90 cm depth compared to the unfertilized control (Lee et al. 2007). The additional C input with manure application increased, in particular, C sequestration at the depth of 0–5 cm by 4.90 Mg C ha^{-1}. Up to 20% of applied dung C may be incorporated into surface soil horizons several weeks after application, with up to 8% remaining in the soil profile after one year (Dungait et al. 2010). Excreta deposition especially of cattle dung may transfer C into soil, whereas urine deposition may enhance loss of SOC by the priming effect, at least in the short term (Cai et al. 2017). However, the molecular mechanism through which dung and urine C contributes to SOC stocks needs additional research (Dungait et al. 2009).

Changes in soil carbon input by management-induced changes in net biome productivity

Changes in policy and socioeconomic influences may impact NBP of managed grasslands and, thus, SOC accumulation. The management intensity may be particularly important for increasing grassland SOC stocks (Chang et al. 2016b). For example, during the period 1991–2010, changes in grassland management intensity of European grasslands made the largest contribution to the increasing NBP trend, i.e., about 36–43% (0.65–0.88 g C m^{-2} yr^{-1}) of the trend among the other drivers including land-cover change, climate change, rising atmospheric CO_2 and N addition. Management intensity changes may have enhanced SOC stocks by 164–214 Tg C in 20 years over 1.3×10^6 km^2 grassland in Europe. In comparison, newly established grassland (i.e., increase in grassland area) caused a comparable but lower NBP positive trend of 0.48–0.51 g C m^{-2} yr^{-1} (Chang et al. 2016b). SOC accumulation in grassland converted from cropland in Europe depends on enhanced available soil N, resulting in high productivity (Gosling et al. 2017). The increase in NBP of European grasslands attributed to increasing CO_2 and climate change was 0.54–55, and 0.54–0.56 g C m^{-2} yr^{-1}, respectively (Chang et al. 2016b). However, the net effect of changes in N addition (including fertilization and atmospheric deposition) on NBP trends was only marginal. Among the reasons for this trend were (i) the rates of change in N addition were small over most European grasslands (within ±6% per decade) or (ii) N addition rate was already high throughout the period 1991–2010 (e.g., over 100 kg N ha^{-1} yr^{-1} in the south of Ireland, France, the Netherlands, and eastern Germany; Chang et al. 2016b). In comparison, animal manure N and mineral N fertilizers on global grasslands were 12,386 and 4,331 thousand tonnes in 1995, respectively (Bouwman et al. 2002). However, fully accounting for the N cycle would be required to produce better estimates of the grassland C balance and its trend (Chang et al. 2016b).

Soil Organic Carbon Loss

The grassland SOC dynamic is likely affected by climate change but magnitude and direction of change are not fully understood. For example, Dintwe and Okin (2018) projected that an increase in soil temperature may result in a loss of SOC from Kalahari savannas, whereas doubling atmospheric CO_2 concentration would cause an increase in SOC. The projected increase in air temperature may cause soil respiration to increase, while it may cause NPP to decrease. The rate of SOC loss from Kalahari savannas due to anthropogenic climate change was estimated at 1.1–2.0 Tg C yr^{-1} until the end of the twenty-first century. The net SOC loss from the global extent of savannas was estimated at 28.4–64.1 Tg C y^{-1} (Dintwe and Okin 2018).

The grassland SOC can also be rapidly lost through a number of processes such as soil disturbance, vegetation degradation, fire, erosion, nutrient shortage, and water deficit (Soussana and Lemaire 2014). The management of livestock grazing can also contribute to SOC losses (McSherry and Ritchie 2013). Specifically, the nature, frequency, and intensity of disturbance may contribute to the SOC loss and, thus, play a key role in the SOC balance of grassland (Soussana et al. 2004). For example, in a cutting and mowing regime, a large part of the NPP is exported from the soil as hay or silage, but part of these exports is compensated by farm manure and slurry application. Disturbance of the continuous grassland vegetation cover, and hence its continuous protection of SOC, contributes to increased losses of SOC (Soussana et al. 2004). Otherwise, moderate mowing may facilitate plant species richness, plant productivity, root biomass, and root exudates, and, thus, increase SOC stocks (Cong et al. 2014). However, it is not fully understood how the SOC stocks of grassland respond to land and soil degradation (Dlamini et al. 2016). For example, multiple drivers of SOC changes were addressed in a detailed analysis of C and N fluxes from a grazed, cut, and fertilized temperate grassland ecosystem (Jones et al. 2017). Over 9 years, all relevant imports and exports to the field were measured or modeled as well as changes in SOC and soil N stocks. The largest component of the C budget was NEE (218 g C m^{-2} y^{-1}) and was offset by C export from the field mainly as grass removed for silage (48.9 g C m^{-2} y^{-1}) and by leaching (16.4 g C m^{-2} y^{-1}). Other export terms (e.g., soil CH_4 emissions, manure applications, and enteric fermentation) were negligible and only contributed 0.02–4.2% of the total C losses. However, the grassland soil lost SOC to 60 cm depth at a rate of 29.08 g C m^{-2} y^{-1}. The GHG balance was -366 g CO_2 eq. m^{-2} y^{-1} and was strongly affected by CH_4 and nitrous oxide (N_2O) emissions. Specifically, the GHG sink strength of the NEE was reduced by 54% by CH_4 and N_2O emissions. Typically, areas with large N_2O emissions within a small area (hotspots) are those with high stocking density, high excretal inputs (resulting in high soil N), and situations when soil water-filled pore space is elevated (Luo et al. 2017). The enteric fermentation from ruminating sheep (*Ovis aries* Linnaeus, 1758) can also be an important CH_4 source, exceeding the contribution of N_2O to the GHG budget in some years (Jones et al. 2017). High plant productivity, stimulated by fertilization, resulted in high plant CO_2 fixation. However, increased N losses through N_2O emissions counteracted the benefits of C sequestration in terms of GHG emissions. Jones et al. (2017) concluded that the CO_2 exchange alone was not sufficient for the estimation of the global warming potential of this managed temperate grassland ecosystem.

Some examples of the effects of specific degradation processes on SOC are given in the following section.

Effects of grazing management on soil organic carbon stocks

In the hypothetical absence of land use, potential vegetation of grassland and grazing land would store 374–379 Pg C under current climate conditions, but actually stores only 119–121 Pg C (Erb et al. 2018). Grazing contributes one-third to the management-induced difference in global biomass C stocks. However, net effects of grazing management on SOC stocks are discussed controversially (Garnett et al. 2017). Grazing profoundly affects grassland SOC by removing NPP. For example, livestock grazed an estimated 1.27 Pg C yr^{-1} in 2011 (Wolf et al. 2015), and >15% of the grazing land NPP is consumed by livestock (Irisarri et al. 2017). The intensity and timing of grazing influence the removal of vegetation by herbivory and C allocation to roots as well as the grassland flora (Jones 2010). Ecological limits to grazing intensity are also known; i.e., seasonal limitations were responsible for grazing intensity below its potential on 39% of the total global natural grassland area in the year 2000 (Fetzel et al. 2017). Grazing also compacts soils, increases erosion, and alters soil hydrological processes (Lunt et al. 2007). The effects of grazing are largely driven by (i) the type of herbivore, (ii) the intensity of grazing pressure, (iii) the level of plant productivity, and (iv) the evolutionary history of grazing (Eldridge et al. 2016). All these influence the amount and distribution of SOC accumulating in the soil. For example, declines in SOC at low grazing contrasts may be due to the breakdown and loss of volatile and labile C forms (Eldridge and Delgado-Baquerizo 2017). Otherwise, increasing levels of grazing may increase SOC by increasing the proportion of NPP allocated to the roots. Thus, with increasing grazing contrast declines in labile C may be offset by increases in the contribution from dung and urine, as well as by increases in root-derived C. Together, these effects may partially compensate for the loss of more labile C. For example, dung deposition can ultimately enhance SOC sequestration by promoting C inputs from plants, whereas the effects of urine patches on SOC stocks and how grazing can be managed to reduce SOC losses from dung and urine patches are less well known (Cai et al. 2017). However, declines in ecosystem structure associated with the removal of vegetation cover, soil destabilization, and erosion are likely to lead to substantial reductions in total C and SOC at high levels of grazing (Eldridge and Delgado-Baquerizo 2017). Further, grazing intensity may alter not only the SOC stocks but also the soil respiration rate (Yu et al. 2017).

Grazing of relatively unmanaged rangelands may be a stronger driver of SOC loss than previously acknowledged (Sanderman et al. 2017). Management determines the grazing intensity, and the ratio between biomass grazed and biomass produced. Further, increasing livestock numbers and poor management causes widespread overgrazing and degradation of rangelands (Petz et al. 2014). Rangelands with the highest grazing intensity are located in the Sahel, Pakistan, West India, Middle East, North Africa, and parts of Brazil. However, because of the many types of grazing practices and the diversity of plant species, soils and climates, the effects of grazing on SOC stocks are inconsistent. Thus, decreases, increases, and no effects of grazing on grassland SOC stocks have been reported in the literature. For example, SOC losses in 0–90 cm depth can be as high as 56% of the SOC stock after 12 years of grazing (Franzluebbers and Stuedemann 2009).

McSherry and Ritchie (2013) performed a meta-analysis of the grazing effects on SOC by considering only interactions among factors with a wide range of values across experimental sites. The grazing effects on SOC, whether positive or negative, were potentially large, with cumulative changes of up to 16 Mg C ha^{-1} over 10–30 years, and annual changes as high as ± 1.5 Mg C ha^{-1} y^{-1}. However, the equal distribution of positive and negative effects suggested that the sign of the effect of grazers on SOC is highly context-dependent. Specifically, grazer effects on SOC may likely depend on precipitation, C3/C4 balance and soil texture. However, considerable knowledge gaps remained as, for example, there were no studies from tropical grasslands and savannas. In addition, potentially important factors, such as fire frequency, soil pH, temperature, type of grazing management (rotational vs. continuous), and wild versus domestic grazers were not considered (McSherry and Ritchie 2013).

On average, grassland degradation by grazing induced a 9% decline in SOC stocks to 30 cm depth based on a meta-analysis comparing non-degraded and degraded grasslands under temperate, humid, subhumid, tropical, and semiarid conditions (Dlamini et al. 2016). The intensity of degradation by grazing was important with a reduction in SOC stocks by 13% when the grassland was heavily degraded (<25% plant basal cover) and by 7% when moderately degraded (50–75% plant basal cover). Furthermore, there were no differences in SOC stock to 30 cm depth between degraded and non-degraded grasslands when the grasslands were lightly degraded (75–100% plant basal cover). There was also a general trend for grassland degradation to have a more pronounced impact on the reduction of SOC stocks under wet climates with acidic soils compared to that on basic soils in dry climates. Thus, proper management of soil pH in grasslands may be an important strategy to mitigate the degradation-induced losses of SOC caused by grazing (Dlamini et al. 2016).

Zhou et al. (2017) performed a comprehensive meta-analysis to examine the responses of a range of variables associated with belowground C cycling to livestock grazing in global grasslands. Grazing decreased the SOC stock to 100 cm depth which may be explained by decreased aboveground biomass leading to reduced litterfall and litter mass. However, grazing-induced decreases in the SOC stock were linearly correlated with that in root C stock but not with that in aboveground inputs. This supports the suggestion by Rasse et al. (2005) that C inputs from root production and biomass are more important to SOC storage compared to that from aboveground biomass. Further, grazing intensity can affect the belowground C cycles with different magnitudes and directions (Zhou et al. 2017). While light intensity grazing enhanced the SOC stock by 0.78%, moderate and heavy grazing decreased it by 3.45 and 9.92%, respectively. Specifically, light grazing resulted in increased above- and belowground biomass (including livestock foraging), and stimulated more photosynthetically fixed C inputs to belowground roots, leading to increased root exudates and root biomass. The increased root exudates may have enhanced SOC accumulation. In contrast, under moderate and heavy grazing decreased C inputs through plant removal by livestock and decreased litter mass as well as decreased root C stock and microbial biomass may have contributed to reduction in SOC stock. With regard to climatic factors, mean annual temperature (MAT) is negatively correlated with the response of the SOC pool to grazing (Zhou et al. 2017). Grasslands in tropical and temperate regions have high MAT and greater soil microbial activity which accelerates decay of SOM and increase turnover rate, resulting in lower SOC stock. Further, the SOC stock can be positively correlated with soil pH indicating that fungal biomass was suppressed at high pH conditions resulting in reduced decomposition and SOC accumulation in grazed ecosystems. However, the meta-analysis had several limitations (Zhou et al. 2017). For example, the majority of studies were performed in temperate climate, especially in Eastern Asia and Northern America. Thus, more long-term manipulative experiments in cold and hot regions, especially in Africa and Australia, need to be conducted to examine the effects of grazing on SOC cycling at temporal and spatial scales. Further, experimental duration of livestock grazing in most studies was less than 10 yr. Thus, there is a need to undertake studies over more than one decade to better understand the effects of grazing on SOC cycling (Zhou et al. 2017).

The previous meta-analyses by McSherry and Ritchie (2013) and Zhou et al. (2017) pooled the data of different studies together without considering the differences in soil depth at which the SOC was measured, thus producing highly uncertain/contradictory results (Abdalla et al. 2018). When the SOC data were normalized to 30 cm depth, meta-analysis studies indicated that taken across all climatic zones and grazing intensities, grazing (below the carrying capacity of the systems) resulted in a decrease in SOC storage, although its impact on SOC was climate-dependent. All grazing intensity levels increased SOC stocks under moist warm climate (+7.6%) while there were reductions under moist cool climate

(−19%). Under dry warm and dry cool climates, only the low (+5.8%) and low to medium (+16.1%) grazing intensities, respectively, were associated with increased SOC stocks. High grazing intensity increased SOC for C4-dominated grassland compared to C3-dominated grassland and C3–C4 mixed grasslands. Thus, to protect grassland soils from degradation, grazing intensity and management practices should be optimized according to climate region and grassland type (C3, C4, or C3–C4 mixed; Abdalla et al. 2018).

In many temperate regions, artificial drainage is a common practice to increase the grazing capacity of soils and extend the grazing season. This may cause increased mineralization and the loss of SOC (Meersmans et al. 2011). However, studies on changes in subsoil SOC stocks after artificial drainage in grassland are scanty. Torres-Sallan et al. (2017) suggested that specifically in soils subject to clay illuviation, installation of artificial drainage systems has a significant influence on the long-term stability of sequestered SOC at depth. Thus, drainage for increased productivity may reduce the capacity of some temperate grassland soils for longer-term storage of SOC. The potential trade-off between enhanced productivity and reduced SOC sequestration by artificial drainage must, therefore, be addressed (O'Sullivana et al. 2015).

In contrast to those under grazing management, the SOC dynamics in grasslands ecosystem with grazing exclusion have not been well characterized. For example, grazing exclusion for >27 years as part of China's 'Returning Grazing Land to Grassland' project had a positive impact on SOC stocks at most sites (Deng et al. 2017). Specifically, the mean rate of SOC stock change was 0.27, 0.23, 0.18, 0.09 Mg C ha^{-1} yr^{-1} in 0–10, 10–20, 20–30, and 30–100 cm, respectively. However, it was unclear whether uncertainties have been be introduced by dataset standardization. The rates of SOC stock changes reached a steady state followed by those for aboveground and belowground biomass C stocks (Deng et al. 2017).

Effects of fire on grassland soil organic carbon stocks

Fire is the most common anthropogenic grassland management practice because of easy application both in difficult terrains and on large areas (Abdalla et al. 2016). Burning results in increased biomass growth period and biomass production (Ojima et al. 1994), while at the same time improving grass cover and biodiversity (Boakye et al. 2013). Through combustion of OM in the upper soil layer, burning may cause a decline in grassland SOC stocks but long-term effects are debatable. For example, SOC stocks tended to be higher in 0–20 cm depth in annually burned plots on a 62-year-old field trial compared to unburned plots associated with tree encroachment, and to annual mowing plots (Abdalla et al. 2016). Fire can increase SOC content in areas where gentle topography or rapid vegetation cover limit post-fire erosion risks (Santín and Doerr 2016).

Fire-induced changes in the chemical composition of SOM may also occur, especially, formation of pyrogenic OM which is a more recalcitrant soil C input than fresh litter. For example, formation of N-heterocyclic compounds and/or more recalcitrant SOM in 0–4 cm soil depth was found at two long-term experimental grassland sites burned annually for 33 years (Kiersch et al. 2012). As total C contents were similar for the burned grassland sites to those of successional and mulched sites, the higher proportion of more recalcitrant SOM indicated that C was sequestered at the burned grassland sites. Similarly, while litter-derived C was incorporated into soil microbes, nematodes, and roots, pyrogenic OM contributed minimally at annually burned long-term experimental tallgrass prairie sites (Soong et al. 2017). However, the overall effects of long-term pyrogenic OM inputs on SOC stocks depend on how much C is lost by subsequent combustion and also by erosion (Soong and Cotrufo 2015). To sum up, frequent litter removal and replacement with pyrogenic OM can alter the belowground biological activity and, thus, impact C cycling in frequently burned grassland ecosystems.

4.3.2 Grassland Soil Organic Carbon Stocks

Globally, grassland soil C stocks are estimated at 343 Pg C in the top 1 m depth (Conant et al. 2017). In 0–30 cm depth, grassland SOC stocks may be as low as 2.5 kg C m^{-2} for herbaceous grasslands in deserts, and as high as 16.0 kg C m^{-2} for evergreen shrub grasslands in boreal regions (Petri et al. 2000). The highest SOC stocks in grasslands are observed under colder and wetter conditions (boreal, temperate) and the lowest under desert conditions. Total SOC stocks to 30 cm depth of natural grasslands ranged between 122 Tg C under evergreen shrub cover and 344 Tg C under sparse shrub/herbaceous cover. In degraded grasslands, total SOC stocks were as low as 37 Tg C under evergreen shrub cover, and as high as 76 and 77 Tg C under deciduous shrub and herbaceous cover, respectively. Improved grasslands stored in total between 49 Tg C under evergreen shrub cover and 334 Tg C under herbaceous cover (Petri et al. 2000).

Based on meta-analysis by Dlamini et al. (2016), SOC stocks to 30 cm depth were estimated to range between 0.1 and 38.8 kg C m^{-2}, with an average of 5.0 kg C m^{-2}. However, spatial variation in this dataset was also high. The greatest SOC stocks were found in temperate regions which were explained by lower average temperatures compared to humid, subhumid, tropical, and semiarid regions resulting in lower decomposition rates, hence in the accumulation of SOC. Greater SOC stocks under wet climates could also be attributable to the high productivity of grasslands in wet environments. In comparison, the lowest SOC stocks were observed in arid to semiarid grassland soils which were explained by the low rainfall amounts resulting in low biomass production and OM decomposition, thus reducing C inputs into the soil. In conclusion, MAP and MAT explained much of the variability of SOC stock in grassland soils to 30 cm depth as both climatic factors are positively correlated with SOC stocks. SOC stocks in grassland soils did also decrease with an increase in altitude which may be explained by lower MAT conditions at higher altitudes (Dlamini et al. 2016).

Among regions, the SOC stock to 100 cm depth in northern China's grasslands was estimated at 24.4 Pg C based on measured data from spatial stratified sampling (Ma et al. 2017). The SOC density decreased from the southeast to the northwest, with the lowest in Desert grassland (4 kg C m^{-2}) and the highest in southeastern Alpine grassland ($\sim$ 11 kg C m^{-2}). In Alpine grassland, SOC density decreased the most with increase in soil depth, and that in desert grassland did the least (Ma et al. 2017). The total SOC stock in the Kalahari savannas was estimated at 0.9 Pg C in the top meter (Dintwe and Okin 2018).

4.4 Grassland Soil Carbon Sequestration

Grazing management can be particularly targeted toward SOC sequestration due to the large global grazing land area and potential for considerable rates of increase in soil C stock (Conant et al. 2017). While decreased grazing intensity in drier areas (MAP < 333 mm) may result in SOC losses, substantial SOC sequestration is predicted for wetter areas by decreasing the grazing intensity (Conant and Paustian 2002). Most SOC sequestration potential may occur in areas with MAP < 1800 mm. Simply cessation of overgrazing and implementation of moderate grazing intensity is promising approaches for rehabilitating SOC in overgrazed areas (Conant and Paustian 2002). A case study from northwestern China indicated that it is possible to considerably reduce animal numbers by using an energy-balance/market-based approach while improving household incomes, providing conditions within which grassland recovery is possible (Kemp et al. 2013). Additional gains in grassland rehabilitation may be fostered through targeted environmental payment schemes.

Conant et al. (2017) reported the data based on meta-analysis of 50 studies on effects of improved grazing management on SOC including 137 data points. The average sequestration rate by improvement in grazing management was 0.28 Mg C ha^{-1} yr^{-1}. However, this did not apply uniformly to all grazing lands and extrapolating the results regionally or globally required information about where there is scope for improvement of grassland management. Specifically, improved grazing management may not always lead to an increase in soil C stocks. In general, rotational grazing can lead to increased productivity and potentially to increased soil C stocks. However, as was the case for productivity, there are few studies available for inclusion in the synthesis (Conant et al. 2017). Similarly, for a global dataset of managed temperate grasslands, Rutledge et al. (2015) noted a general trend for increased C sequestration with increase in NEP, suggesting that it may be possible to meet the dual goal of increased pasture production (i.e., milk, meat, and fiber) and increasing SOC storage. However, more standardized reporting is needed between NECB studies, and experiments involving side-by-side comparison of treatment and control plots (Rutledge et al. 2015).

Different management practices may be combined to enhance SOC storage. For example, avoiding fires and overgrazing, using a grazing rotation plan and a mixture of C_3 and C_4 species may ensure a continuous SOC storage in tropical pastures establishes after deforestation (Stahl et al. 2017). For example, pastures in French Amazonia stored between 1.27 and 5.31 Mg C ha^{-1} y^{-1}. This C was mainly sequestered in the humus of deep soil layers (20–100 cm), whereas no SOC storage was observed in the 0 to 20 cm layer (Stahl et al. 2017).

Aside animal grazing, common grassland management practices affecting SOC include the repeated addition of lime and nutrient fertilizers to soils. For example, using data spanning 129 years of the Park Grass Experiment at Rothamsted, UK, net SOC sequestration in 0–23 cm depth was 2–20 times greater in limed than in unlimed soils (Fornara et al. 2011). This was mainly caused by greater biological activity in limed soils, which despite increasing soil respiration rates, led to plant C inputs being processed and incorporated into resistant soil organo-mineral pools. In mesotrophic grasslands, 19 years of N-only additions strongly increased SOC storage to 20 cm depth (Fornara et al. 2013). This SOC accrual was explained by long-term accumulation of partially undecomposed plant-derived detritus. However, the addition of P and K either in isolation or in combination with N and without N was associated with low SOC sequestration. Further, liming increased SOC content under the PK treatment, and this was possibly due to positive liming- and PK-induced effects on the abundance of N-fixing legume species which are known to have positive effects on SOC accumulation (Fornara et al. 2013).

Egan et al. (2018) reported that liming increased SOC stocks of small soil aggregate fractions in the 23rd year of a temperate grassland experiment. Heyburn et al. (2017) suggested that regular liming applications may provide management solutions for increasing the long-term sustainability of permanent grasslands. Grazing also contributed to enhance SOC content of control plots (and K-only treated plots), and this was possibly due to grazing-induced increases in soil N availability (Fornara et al. 2013). Using 43 years of data from a permanent grassland experiment, Fornara et al. (2016) showed that soils have not yet reached SOC saturation. Even unfertilized control soils showed SOC sequestration rates of 0.35 Mg C ha^{-1} yr^{-1} in 0–15 cm depth. High application rates of cattle slurry further increased SOC sequestration to 0.86 Mg C ha^{-1} yr^{-1}, and a key cause of this SOC accrual was greater C inputs from cattle slurry. However, 85% of C added yearly through liquid manure was lost possibly via CO_2 fluxes and organic C leaching. Soils fertilized with NPK had the lowest C-gain efficiency (i.e., unit of C gained per unit of N added), and lowest SOC sequestration (similar to control soils). Further, soils receiving cattle slurry showed higher C gain compared to soils receiving NPK or pig slurry (Fornara et al. 2016).

The net effects of grazing management on the GHG balance must also be considered. For example, the SOC sequestration potential through grassland management potential offsets only 20–60% of C emissions from grazing systems (Garnett et al. 2017). This includes 4–11% of total livestock C emissions, and between 0.6 and 1.6% of total annual GHG emissions. In conclusion, improved grassland management can lead to increases in soil C stocks but economic and policy challenges exist toward its contribution as a GHG mitigation measure (Conant et al. 2017).

Sequestration of C in soils of grassland and agroecosystem contributes to limiting climate change and reducing GHG emissions by the private sector, i.e., the food and agriculture industry. Persson and Rockström (2011) observed a greater willingness, as well as readiness, by business to make bold investment decisions for a low carbon future. For example, General Mills, Inc., aims to cut GHGs by 28% by 2025 across the entire value chain (WES 2016). By 2050, General Mills proposes to slash emissions by up to 72% at a pace that may keep them in line with scientific consensus. Further, Kellogg Company's 2050 target is to cut its own and supplier emissions by 65 and 50%, respectively (WES 2016). Through release of a sustainability toolkit, Wal-Mart Stores, Inc. (Walmart), recently asked suppliers to reduce GHG emissions by 1 Pg (https://news.walmart.com/2017/04/19/walmart-launches-project-gigaton-to-reduce-emissions-in-companys-supply-chain). Several suppliers have already committed to the Project Gigaton with new or existing emissions-reduction projects. Companies from every segment of the food, farming, and agriculture industries across the USA have joined force with the Field to Market Alliance to advance the sustainability of food, fiber, and fuel production (https://fieldtomarket.org/). SUSTAIN is a partnership between farmers, companies, and Land O'Lakes that aligns and harmonizes on-farm conservation and company-led sustainability targets to maximize positive impacts on the environment (https://www.landolakessustain.com/). With the signature 'One Planet, One Health,' Danone together with consumers, retailers, farmers, suppliers, and non-governmental organizations (NGOs) is aiming at designing, producing, marketing, and consuming food in a new way—a way to protect and nourish health of the people and health of the planet (http://www.danone.com/en/for-all/our-mission-in-action/danone-one-planet-one-health/). Unilever has climate policies that extend across the entire value chain of their products which include food and beverages. Unilever has a clear target to halve the life-cycle emissions of their products by or before 2030, and to have a strategy for net-zero emissions from their operations by the same date (http://bit.ly/2i5uW1r). To sum up, investors will play a major role, whether active or passive, in climate change mitigation (Millar et al. 2018).

4.5 Conclusions

Grasslands cover more than a quarter of the global ice-free land area with managed grazing, in particular, having a larger geographic extent than any other form of land use. Most importantly, the livelihoods of 1 billion people depend on grassland ecosystems. Soils of grassland developed on limestone bedrock contain SIC which may also be altered by irrigation practices, soil acidification, liming, and grazing management. However, scientific knowledge regarding the response of the grassland SIC stock to land use must be improved. Up to 30% of the global SOC stock to 1 m depth may be stored in grasslands with a sequestration rate of 0.5 Pg C yr^{-1}. However, additional research is needed about the quantitative contribution of above- and belowground plant residues, rhizodeposition, dung, and urine of grazing animals, and BC to the SOC stock. Many grassland soils have a high potential to store additional amounts of SOC as intensive livestock grazing and degradation by agricultural uses have caused C losses in the past. Specifically, soil disturbance, vegetation degradation, fire, erosion, nutrient shortage, and water deficit may contribute to SOC losses. However, the positive response of grassland SOC to land and soil degradation should be improved. Further, additional studies are needed for assessing the effects of improved grazing management on SOC stocks, and implementing the available information to improve the management of grassland. In this regard, the importance of the private sector (i.e., Danone) to implement improved management practices cannot be overemphasized.

4.6 Review Questions

1. Contrast and compare the different grassland ecosystems.
2. How does grazing affect grasslands?
3. Describe the SIC dynamic in grassland developed on limestone bedrock and effects of anthropogenic disturbance.
4. Why do natural grasslands contain relatively large SOC stocks?
5. How can land use and management of grasslands be improved to strengthen SOC sequestration?
6. Contrast and compare effects of C_3 grasses, C_4 grasses, perennial grasses, legumes, trees, and grazing animals on SOC stocks.
7. How can grassland SOC stocks be adapted to a changing climate?
8. How can the large population directly depending on grassland ecosystems be incentivized to manage SOC for multiple benefits?

References

Abdalla K, Chivenge P, Everson C, Mathieu O, Thevenot M, Chaplot V (2016) Long-term annual burning of grassland increases CO_2 emissions from soils. Geoderma 282:80–86

Abdalla M, Hastings A, Chadwick DR et al (2018) Critical review of the impacts of grazing intensity on soil organic carbon storage and other soil quality indicators in extensively managed grasslands. Agric Ecosyt Environ 253:62–81

Allen DE, Pringle MJ, Page KL, Dalal RC (2010) A review of sampling designs for the measurement of soil organic carbon in Australian grazing lands. Rangeland J 32:227–246

Andela N, Morton DC, Giglio L et al (2017) A human-driven decline in global burned area. Science 356:1356–1362

Andreae MO (2004) Assessment of global emissions from vegetation fires. International Forest Fire News 31:112–121

Angers DA, Chantigny MH, MacDonald JD, Rochette P, Côté D (2010) Differential retention of carbon, nitrogen and phosphorus in grassland soil profiles with long-term manure application. Nutr Cycl Agroecosyst 86:225–229

Asner GP, Elmore AJ, Olander LP, Martin RE, Harris AT (2004) Grazing systems, ecosystem responses, and global change. Annu Rev Environ Resour 29:261–299

Atkinson RRL, Mockford EJ, Bennett C, Christin PA, Spriggs EL, Freckleton RP, Thompson K, Rees M, Osborne CP (2016) C4 photosynthesis boosts growth by altering physiology, allocation and size. Nature Plants 16038. https://doi.org/10.1038/nplants.2016.38

Beer C, Reichstein M, Tomelleri E, Ciais P, Jung M, Carvalhais N, Rödenbeck C, Altaf Arain M, Baldocchi D, Bonan GB, Bondeau A, Cescatti A, Lasslop G, Lindroth A, Lomas M, Luyssaert S, Margolis H, Oleson KW, Roupsard O, Veenendal E, Viovy N, Williams C, Ian Woodward F, Papale D (2010) Terrestrial gross carbon dioxide uptake: global distribution and covariation with climate. Science 329:834–838

Boakye MK, Little IT, Panagosl MD, Jansen R (2013) Effects of burning and grazing on plant species percentage cover and habitat condition in the highland grassland of Mpumalanga Province, South Africa. J Anim Plant Sci 23:603–610

Bolinder MA, Janzen HH, Gregorich EG, Angers DA, VandenBygaart AJ (2007) An approach for estimating net primary productivity and annual carbon inputs to soil for common agricultural crops in Canada. Agric Ecosyst Environ 118:29–42

Bossuyt H, Six J, Hendrix PF (2005) Protection of soil carbon by microaggregates within earthworm casts. Soil Biol Biochem 37:251–258

Bouwman A, Boumans L, Batjes N (2002) Estimation of global NH_3 volatilization loss from synthetic fertilizers and animal manure applied to arable lands and grasslands. Global Biogeochem Cy 16:8-1–8-14

Cai Y, Chang SX, Cheng Y (2017) Greenhouse gas emissions from excreta patches of grazing animals and their mitigation strategies. Earth Sci Rev 171:44–57

Chang J, Ciais P, Herrero M, Havlik P, Campioli M, Zhang X, Bai Y, Viovy N, Joiner J, Wang X, Peng S, Yue C, Piao S, Wang T, Hauglustaine DA, Soussana JF, Peregon A, Kosykh N, Mironycheva-Tokareva N (2016a) Combining livestock production information in a process-based vegetation model to reconstruct the history of grassland management. Biogeosciences 13:3757–3776

Chang J, Ciais P, Viovy N, Vuichard N, Herrero M, Havlík P, Wang X, Sultan B, Soussana JF (2016b) Effect of climate change, CO_2 trends, nitrogen addition, and land-cover and management intensity changes on the carbon balance of European grasslands. Glob Change Biol 22:338–350

Chang J, Ciais P, Viovy N, Vuichard N, Sultan B, Soussana JF (2015) The greenhouse gas balance of European grasslands. Glob Change Biol 21:3748–3761

Chapin FS III, Eviner VT (2014) Biogeochemical interactions governing terrestrial net primary production. In: Karl D, Schlesinger W (eds) Treatise on geochemistry (Second Edition)—Volume 10—Biogeochemistry. Elsevier, Amsterdam, Netherlands, pp 189–216

Chapin FS III, Matson PA, Mooney HA (2002) Principles of terrestrial ecosystem ecology. Springer, New York

Cheng W, Kuzyakov Y (2005) Root effects on soil organic matter decomposition. In: Wright S, Zobel R (eds) Roots and Soil management: interactions between roots and the soil, agronomy monograph no. 48, American Society of Agronomy, Crop Science Society of America, Soil Science Society of America, Madison, Wisconsin, USA, pp 119–143

Ciais P, Soussana J-F, Vuichard N, Luyssaert S, Don A, Janssens IA, Piao SL, Dechow R, Lathière J, Maignan F, Wattenbach M, Smith P, Ammann C, Freibauer A, Schulze E-D, Synthesis Team CARBOEUROPE (2010) The greenhouse gas balance of European grass-lands. Biogeosciences Discuss 7:5997–6050

Conant RT (2012) Grassland soil organic carbon stocks: status, opportunities, vulnerability. In: Lal R, Lorenz K, Hüttl RFJ, Schneider BU, von Braun J (eds) Recarbonization of the biosphere—ecosystems and the global carbon cycle. Springer, pp 275–302

Conant RT, Cerri CEP, Osborne BB, Paustian K (2017) Grassland management impacts on soil carbon stocks: a new synthesis. Ecol Appl 27:662–668

Conant RT, Paustian K (2002) Potential soil carbon sequestration in overgrazed grassland ecosystems. Glob Biogeochem Cycle 16:1143

Cong W, van Ruijven J, Mommer L, De Deyn GB, Berendse F, Hoffland E (2014) Plant species richness promotes soil carbon and nitrogen stocks in grasslands without legumes. J Ecol 102:1163–1170. https://doi.org/10.1111/1365-2745.12280

Cotrufo MF, Soong JL, Horton AJ, Campbell EE, Haddix ML, Wall DH, Parton WJ (2015) Formation of soil organic matter via biochemical and physical pathways of litter mass loss. Nat Geosci 8:776–781

Denef K, Stewart CE, Brenner J, Paustian K (2008) Does long-term center-pivot irrigation in-crease soil carbon stocks in semi-arid agro-ecosystems? Geoderma 145:121–129

Deng L, Shangguan ZP, Wu GL, Chang XF (2017) Effects of grazing exclusion on carbon sequestration in China's grassland. Earth Sci Rev 173:84–95

Dintwe K, Okin GS (2018) Soil organic carbon in savannas decreases with anthropogenic climate change. Geoderma 309:7–16

Dixon AP, Faber-Langendoen D, Josse C, Morrison J, Loucks CJ (2014) Distribution mapping of world grassland types. J Biogeogr 41:2003–2019

Dlamini P, Chivenge P, Chaplot V (2016) Overgrazing decreases soil organic carbon stocks the most under dry climates and low soil pH: a meta-analysis shows. Agric Ecosyst Environ 221:258–269

Don A, Schumacher J, Freibauer A (2011) Impact of tropical land-use change on soil organic carbon stocks—a meta-analysis. Glob Change Biol 17:1658–1670

Don A, Schumacher J, Scherer-Lorenzen M, Scholten T, Schulze ED (2007) Spatial and vertical variation of soil carbon at two grassland sites—implications for measuring soil carbon stocks. Geoderma 141:272–282

Duffy JE, Godwin CM, Cardinale BJ (2017) Biodiversity effects in the wild are common and as strong as key drivers of productivity. Nature 549:261–265. https://doi.org/10.1038/nature23886

Dungait JAJ, Bol R, Bull ID, Evershed RP (2009) Tracking the fate of dung-derived carbohydrates in a temperate grassland soil using compound-specific stable isotope analysis. Org Geochem 40:1210–1218

Dungait JAJ, Bol R, Lopez-Capel E et al (2010) Applications of stable isotope ratio mass spectrometry in cattle dung carbon cycling studies. Rapid Commun Mass Spectrom 24:495–500

Egan G, Crawley MJ, Fornara DA (2018) Effects of long-term grassland management on the carbon and nitrogen pools of different soil aggregate fractions. Sci Tot Environ 613–614:810–819

Eldridge DJ, Delgado-Baquerizo M (2017) Continental-scale impacts of livestock grazing on ecosystem supporting and regulating services. Land Degrad Develop 28:1473–1481

Eldridge DJ, Poore AGB, Ruiz-Colmenero M, Letnic M, Soliveres S (2016) Ecosystem structure, function, and composition in rangelands are negatively affected by livestock grazing. Ecol Appl 26:1273–1283

Ellis EC, Ramankutty N (2008) Putting people in the map: anthropogenic biomes of the world. Front Ecol Environ 6:439–447

Erb KH, Kastner T, Plutzar C et al (2018) Unexpectedly large impact of forest management and grazing on global vegetation biomass. Nature 553:73–76

Fan Y, Miguez-Macho G, Jobbágy EG, Jackson RB, Otero-Casal C (2017) Hydrologic regulation of plant rooting depth. Proc Natl Acad Sci USA 114:10572–10577

Fay PA, Prober SM, Harpole WS, et al (2015) Grassland productivity limited by multiple nutrients. Nature Plants 1:UNSP15080

Fetzel T, Havlik P, Herrero M, Erb KH (2017) Seasonality constraints to livestock grazing intensity. Glob Change Biol 23:1636–1647. https://doi.org/10.1111/gcb.13591

Follett RF, Reed DA (2010) Soil carbon sequestration in grazing lands: societal benefits and policy implications. Rangeland Ecol Manag 63:4–15

Fornara DA, Banin L, Crawley MJ (2013) Multi-nutrient vs. nitrogen-only effects on carbon sequestration in grassland soils. Glob Change Biol 19:3848–3857

Fornara DA, Steinbeiss S, McNamara NP et al (2011) Increases in soil organic carbon sequestration can reduce the global warming potential of long-term liming to permanent grassland. Glob Chang Biol 17:1925–1934

Fornara DA, Wasson EA, Christie P, Watson CJ (2016) Long-term nutrient fertilization and the carbon balance of permanent grassland: any evidence for sustainable intensification? Biogeosciences 13:4975–4984. https://doi.org/10.5194/bg-13-4975-2016

Franzluebbers AJ, Stuedemann JA (2009) Soil-profile organic carbon and total nitrogen during 12 years of pasture management in the Southern Piedmont USA. Agric Ecosyst Environ 129:28–36

Franzluebbers AJ, Stuedemann JA (2010) Surface soil changes during twelve years of pasture management in the Southern Piedmont USA. Soil Sci Soc Am J 74:2131–2141

Franzluebbers AJ, Stuedemann JA, Wilkinson SR (2001) Bermudagrass management in the Southern Piedmont USA: I. Soil and surface residue carbon and sulfur. Soil Sci Soc Am J 65:834–841

Freschet GT, Cornwell WK, Wardle DA et al (2013) Linking litter decomposition of above and belowground organs to plant-soil feedbacks worldwide. J Ecol 101:943–952

Fu Z, Li D, Hararuk O et al (2017) Recovery time and state change of terrestrial carbon cycle after disturbance. Environ Res Lett 12:104004

Garnett T, Godde C, Muller A et al (2017) Grazed and Confused? Ruminating on cattle, grazing systems, methane, nitrous oxide, the soil carbon sequestration question – and what it all means for greenhouse gas emissions. University of Oxford, FCRN

Gilmanov TG, Aires L, Barcza Z, Baron VS, Belelli L, Beringer J, Billesbach D, Bonal D, Brad-ford J, Ceschia E, Cook D, Corradi C, Frank A, Gianelle D, Gimeno C, Gruenwald T, Guo H, Hanan N, Haszpra L, Heilman J, Jacobs A, Jones MB, Johnson DA, Kiely G, Li S, Mag-liulo V, Moors E, Nagy Z, Nasyrov M, Owensby C, Pinter K, Pio C, Reichstein M, Sanz MJ, Scott R, Soussana J-F, Stoy PC, Svejcar T, Tuba Z, Zhou G (2010) Productivity, respi-ration, and light-response parameters of world grassland and agroecosystems derived from flux-tower measurements. Rangeland Ecol Manage 63:16–39

Gilmanov TG, Soussana JF, Aires L, Allard V, Ammann C, Balzarolo M, Barcza Z, Bernhofer C, Campbell CL, Cernusca A, Cescatti A, Clifton-Brown J, Dirks BOM, Dore S, Eugster W, Fuhrer J, Gimeno C, Gruenwald T, Haszpra L, Hensen A, Ibrom A, Jacobs AFG, Jones MB, Lanigan G, Laurila T, Lohila A, Manca G, Marcolla B, Nagy Z, Pilegaard K, Pinter K, Pio C, Raschi A, Rogiers N, Sanz MJ, Stefani P, Sutton M, Tuba Z, Valentini R, Williams ML, Wohlfahrt G (2007) Partitioning European grassland net ecosystem CO_2 exchange into gross primary productivity and ecosystem respiration using light response function analysis. Agr Ecosyst Environ 121:93–120

Gosling P, van der Gast C, Bending GD (2017) Converting highly productive arable cropland in Europe to grassland: a poor candidate for carbon sequestration. Sci Rep 7:10493. https://doi.org/10.1038/s41598-017-11083-6

He H, Liu M, Xiao X, Ren X, Zhang L, Sun X, Yang Y, Li Y, Zhao L, Shi P, Du M, Ma Y, Ma M, Zhang Y, Yu G (2014) Large-scale estimation and uncertainty analysis of gross primary production in Tibetan alpine grasslands. J Geophys Res Biogeosci 119:466–486

Henderson BB, Gerber PJ, Hilinski TE et al (2015) Greenhouse gas mitigation potential of the world's grazing lands: Modeling soil carbon and nitrogen fluxes of mitigation practices. Agric Ecosyst Environ 207:91–100

Heyburn J, McKenzie P, Crawley MJ, Fornara DA (2017) Long-term belowground effects of grassland management: the key role of liming. Ecol Appl 27:2001–2012

Houghton RA, Nassikas AA (2017) Global and regional fluxes of carbon from land use and land cover change 1850–2015. Glob Biogeochem Cycles 31:456–472

Irisarri JGN, Aguiar S, Oesterheld M, Derner JD, Golluscio RA (2017) A narrower gap of grazing intensity. Reply to Fetzel et al., 2017. Seasonality constrains to livestock grazing intensity. Glob Change Biol 23:3965–3966. https://doi.org/10.1111/gcb.13800

Johnson D, Leake JR, Ostle N, Ineson P, Read DJ (2002) In situ $^{13}CO_2$ pulse-labelling of upland grassland demonstrates that a rapid pathway of carbon flux from arbuscular mycorrhizal mycelia to the soil. New Phytol 153:327–334

Jones DL, Nguyen C, Finlay RD (2009) Carbon flow in the rhizosphere: carbon trading at the soil-root interface. Plant Soil 321:5–33

Jones MB (2010) Potential for carbon sequestration in temperate grassland soils. In: Abberton M, Conant R, Batello C (eds) Grassland carbon sequestration: management, policy and economics —proceedings of the Workshop on the role of grassland carbon sequestration in the mitigation of climate change. Integrated crop management, vol 11. FAO, Rome, pp 1–18

Jones MB, Donnelly A (2004) Carbon sequestration in temperate grassland ecosystems and the influence of management, climate and elevated CO_2. New Phytol 164:423–439

Jones SK, Helfter C, Anderson M et al (2017) The nitrogen, carbon and greenhouse gas budget of a grazed, cut and fertilised temperate grassland. Biogeosciences 14:2069–2088

Kemp DR, Guodong H, Xiangyang H et al (2013) Innovative grassland management systems for environmental and livelihood benefits. Proc Natl Acad Sci USA 110:8369–8374

Kiersch K, Kruse J, Eckhardt KU et al (2012) Impact of grassland burning on soil organic matter as revealed by a synchrotron- and pyrolysis–mass spectrometry-based multi-methodological approach. Org Geochem 44:8–20

Klein Goldewijk K, Beusen A, van Drecht G, de Vos M (2011) The HYDE 3.1 spatially explicit database of human-induced global land-use change over the past 12,000 years. Global Ecol Biogeogr 20:73–86

Klotzbücher T, Klotzbücher A, Kaiser K et al (2018) Variable silicon accumulation in plants affects terrestrial carbon cycling by controlling lignin synthesis. Glob Change Biol 24:e183–e189. https://doi.org/10.1111/gcb.13845

Kuzyakov Y (2006) Sources of CO_2 efflux from soil and review of partitioning methods. Soil Biol Biochem 38:425–448

Kuzyakov Y, Domanski G (2000) Carbon input by plants into the soil. Review. J Plant Nutr Soil Sci 163:421–431

Lauenroth WK, Wade AA, Williamson MA, Ross BE, Kumar S, Cariveau DP (2006) Uncertainty in calculations of net primary production for grasslands. Ecosystems 9:843–851

Lee DK, Owens VN, Doolittle JJ (2007) Switchgrass and soil carbon sequestration response to ammonium nitrate, manure, and harvest frequency on conservation reserve program land. Agron J 99:462–468

Lee H, Rahn T, Throop H (2012) An accounting of C-based trace gas release during abiotic plant litter degradation. Glob Change Biol 18:1185–1195

Lemaire G, Chapman D (1996) Tissue flows in grazed plant communities. In: Hodgson J, Illius AW (eds) The ecology and management of grazing systems. CAB International, Wallingford, UK, pp 3–35

Liang W, Lü Y, Zhang W et al (2017) Grassland gross carbon dioxide uptake based on an improved model tree ensemble approach considering human interventions: global estimation and covariation with climate. Glob Change Biol 23:2720–2742. https://doi.org/10.1111/gcb.13592

Lorenz K, Lal R (2005) The depth distribution of soil organic carbon in relation to land use and management and the potential of carbon sequestration in subsoil horizons. Adv Agron 88:35–66

Lunt ID, Eldridge DJ, Morgan JW, Witt GB (2007) A framework to predict the effects of livestock grazing and grazing exclusion on conservation values in natural ecosystems in Australia. Austr J Bot 55:401–415

Luo J, Wyatt J, van der Weerden TJ et al (2017) Potential hotspot areas of nitrous oxide emissions from grazed pastoral dairy farm systems. Adv Agron 145:205–268

Ma Q, Kuang W, Liu Z et al (2017) Spatial pattern of different component carbon in varied grasslands of northern China. Geoderma 303:27–36

Maillard É, Angers DA (2014) Animal manure application and soil organic carbon stocks: a meta-analysis. Glob Change Biol 20:666–679

Marschner H, Dell B (1994) Nutrient uptake in mycorrhizal symbiosis. Plant Soil 159:89–102

McSherry ME, Ritchie ME (2013) Effects of grazing on grassland soil carbon: a global review. Glob Change Biol 19:1347–1357

Meersmans J, van Wesemael B, Goidts E et al (2011) Spatial analysis of soil organic carbon evolution in Belgian croplands and grasslands, 1960–2006. Glob Change Biol 17:466–479

Mikhailova E, Post C, Cihacek L, Ulmer M (2009) Soil inorganic carbon sequestration as a result of cultivation in the Mollisols. In: McPherson BJ, Sundquist ET (eds) Carbon sequestration and

its role in the global carbon cycle. Geophysical Monograph Series 183, American Geophysical Union, Washington, D.C., pp 129–133
Milchunas DG (2009) Estimating root production: comparison of 11 methods in shortgrass steppe and review of biases. Ecosystems 12:1381–1402
Milcu A, Roscher C, Gessler A, Bachmann D, Gockele A, Guderle M, Landais D, Piel C, Escape C, Devidal S, Ravel O, Buchmann N, Gleixner G, Hildebrandt A, Roy R (2014) Functional diversity of leaf nitrogen concentrations drives grassland carbon fluxes. Ecol Lett 17:435–444
Milesi C, Elvidge CD, Dietz JB, Tuttle BD, Nemani RR, Running SW (2016) A strategy for mapping and modeling the ecological effects of US lawns. Ecology (submitted)
Millar RJ, Hepburn C, Beddington J, Allen MR (2018) Principles to guide investment towards a stable climate. Nat Clim Change 8:2–12
Mokany K, Raison RJ, Prokushkin AS (2006) Critical analysis of root: shoot ratios in terrestrial biomes. Glob Change Biol 12:84–96
Monfreda C, Ramankutty N, Foley JA (2008) Farming the planet: 2. Geographic distribution of crop areas, yields, physiological types, and net primary production in the year 2000. Global Biogeochem Cyc 22, GB1022, https://doi.org/10.1029/2007gb002947
Monger HC, Kraimer RA, Khresat S, Cole DR, Wang X, Wang J (2015) Sequestration of inorganic carbon in soil and groundwater. Geology 43:375–378
Moore CE, Beringer J, Evans B et al (2016) The contribution of trees and grasses to productivity of an Australian tropical savanna. Biogeosciences 13:2387–2403
Mottet A, de Haan C, Falcucci A et al (2017) Livestock: on our plates or eating at our table? A new analysis of the feed/food debate. Glob Food Secur 14:1–8
Nelson L, Blumenthal DM, Williams DG, Pendall E (2017) Digging into the roots of belowground carbon cycling following seven years of Prairie Heating and CO_2 Enrichment (PHACE), Wyoming USA. Soil Biol Biochem 115:169–177
Newbold T, Hudson LN, Arnell AP et al (2016) Has land use pushed terrestrial biodiversity beyond the planetary boundary? A global assessment. Science 353:288–291
Nguyen C (2009) Rhizodeposition of organic C by plant: mechanisms and controls. In: Lichtfouse E, Navarrete M, Debaeke P, Véronique S, Alberola C (eds) Sustainable agriculture. Springer, Netherlands, pp 97–123
Novick KA, Siqueira MBS, Juang J, Oren R, Stoy PC, Katul GG, Ellsworth DS (2004) Carbon dioxide and water vapor exchange in a warm temperate grassland. Oecologia 138:259–274
Ojima D, Schimel DS, Parton WJ, Owensby CE (1994) Long and short-term effects of fire on nitrogen cycling in tallgrass prairie. Biogeochemistry 24:67–84
O'Sullivana L, Creamer RE, Fealy R et al (2015) Functional land management for managing soil functions: a case-study of the trade-off between primary productivity and carbon storage in response to the intervention of drainage systems in Ireland. Land Use Pol 47:42–54
Parsons A, Rowarth J, Thornley J, Newton P (2011) Primary production of grasslands, herbage accumulation and use, and impacts of climate change. In: Lemaire G, Hodgson J, Chabbi A (eds) Grassland productivity and ecosystem services. CAB International, Wallingford, U.K., pp 3–18
Personeni E, Loiseau P (2005) Species strategy and N fluxes in grassland soil—a question of root litter quality or rhizosphere activity? Europ. J. Agron. 22:217–229
Persson Å, Rockström J (2011) Business leaders. Nat Clim Change 1:426–427
Petri M, Batello C, Villani R, Nachtergaele F (2000) Carbon status and carbon sequestration potential in the world's grasslands. In: Abberton M, Conant R, Batello C (eds) Grassland carbon sequestration: management, policy and economics—Proceedings of the Workshop on the role of grassland carbon sequestration in the mitigation of climate change. Integrated Crop Management, vol 11. fao, rome, pp 19–31
Petz K, Alkemade R, Bakkenes M, et al (2014) Mapping and modelling trade-offs and synergies between grazing intensity and ecosystem services in rangelands using global-scale datasets and models. Glob Environ Change 29:223–234

Phelps LN, Kaplan JO (2017) Land use for animal production in global change studies: defining and characterizing a framework. Glob Change Biol 23:4457–4471. https://doi.org/10.1111/gcb.13732

Rabin SS, Magi BI, Shevliakova E, Pacala SW (2015) Quantifying regional, time-varying effects of cropland and pasture on vegetation fire. Biogeosciences 12:6591–6604

Räsänen M, Aurela M, Vakkari V et al (2017) Carbon balance of a grazed savanna grassland ecosystem in South Africa. Biogeosciences 14:1039–1054

Ramankutty N, Evan AT, Monfreda C, Foley JA (2008) Farming the planet: 1. Geographic distribution of global agricultural lands in the year 2000. Global Biogeochem Cyc 22, GB1003. https://doi.org/10.1029/2007gb002952

Rasse DP, Rumpel C, Dignac MF (2005) Is soil carbon mostly root carbon? Mechanisms for a specific stabilization. Plant Soil 269:341–356

Reeder JD, Schuman GE, Morgan JA, Lecain DR (2004) Response of organic and inorganic carbon and nitrogen to long-term grazing of the shortgrass steppe. Environ Manage 33:485–495

Robinson D (2007) Implications of a large global root biomass for carbon sink estimates and for soil carbon dynamics. Proc R Soc B 274:2753–2759

Rumpel C (2011) Carbon storage and organic matter dynamics in grassland soils. In: Lemaire G, Hodgson J, Chabbi A (eds) Grassland productivity and ecosystem services. CAB International, Wallingford, U.K., pp 65–72

Rutledge S, Mudge PL, Campbell DI et al (2015) Carbon balance of an intensively grazed temperate dairy pasture over four years. Agric Ecosyst Environ 206:10–20

Sanderman J (2012) Can management induced changes in the carbonate system drive soil carbon sequestration? A review with particular focus on Australia. Agric Ecosyst Environ 155:70–77

Sanderman J, Farquharson R, Baldock J (2010) Soil carbon sequestration potential: a review for Australian agriculture—a report prepared for Department of Climate Change and Energy Efficiency, CSIRO Land and Water. http://www.csiro.au/resources/Soil-Carbon-Sequestration-Potential-Report.html

Sanderman J, Hengl T, Fiske GJ (2017) Soil carbon debt of 12,000 years of human land use. Proc Natl Acad Sci USA Early Edition

Sándor R, Ehrhardt F, Basso B et al (2016) C and N models intercomparison—benchmark and ensemble model estimates for grassland production. Adv Animal Biosci 7:245–247. https://doi.org/10.1017/S2040470016000297

Santín C, Doerr SH (2016) Fire effects on soils: the human dimension. Phil Trans R Soc B 371:20150171

Santos GM, Alexandre A (2017) The phytolith carbon sequestration concept: fact or fiction? A comment on "Occurrence, turnover and carbon sequestration potential of phytoliths in terrestrial ecosystems by Song et al. https://doi.org/10.1016/j.earscirev.2016.04.007". Earth Sci Rev 164:251–255

Saugier B, Roy J, Mooney HA (2001) Estimations of global terrestrial productivity: converging toward a single number? In: Roy J, Saugier B, Mooney HA (eds) Terrestrial global productivity. Academic Press, San Diego, pp 543–557

Scholes RJ, Archer SR (1997) Tree–grass interactions in savannas. Ann Rev Ecol System 28:517–544

Scholes R, Walker H (1993) An African Savanna: synthesis of the Nylsvley study. Cambridge University Press, Cambridge, UK

Schulze E-D, Ciais P, Luyssaert S, Schrumpf M, Janssens IA, Thiruchittampalam B, Theloke J, Saurat M, Bringezu S, Lelieveld J, Lohila A, Rebmann C, Jung M, Bastviken D, Abril G, Grassi G, Leip A, Freibauer A, Kutsch W, Don A, Nieschulze J, Börner A, Gash JH, Dolman AJ (2010) The European carbon balance. Part 4: integration of carbon and other trace-gas fluxes. Global Change Biol 16:1451–1469

Schulze E-D, Luyssaert S, Ciais P, Freibauer A, Janssens IA, et al (2009) Importance of methane

Sejian C, Naqvi SMK, Ezeji T, Lakritz J, Lal R (eds) (2012) Environmental stress and amelioration in livestock production. Springer-Verlag, Berlin

Shao J, Zhou X, Luo Y et al (2016) Uncertainty analysis of terrestrial net primary productivity and net biome productivity in China during 1901–2005. J Geophys Res Biogeosci 121:1372–1393

Shahzad T, Chenu C, Genet P, Barot S, Perveen N, Mougin C, Fontaine S (2015) Contribution of exudates, arbuscular mycorrhizal fungi and litter depositions to the rhizosphere priming effect induced by grassland species. Soil Biol Biochem 80:146–155

Skinner RH (2005) Management effects on carbon dioxide exchange over northeastern pastures (abstract). In: Greenhouse Gas Emissions and Carbon Sequestration Symposium, March 21–24, 2005, Wyndham Baltimore, Inner Harbor, Baltimore, Maryland. http://soilcarboncenter.k-state.edu/conference/

Skinner RH, Corson MS, Gilmanov TG (2008) Simulating gross primary productivity of humid-temperate pastures. Agron J 100:801–807

Smith SE, Read DJ (2008) Mycorrhizal symbiosis. Academic Press, London

Song Z, Liu H, Strömberg CAE, Yang X, Zhang X (2017) Phytolith carbon sequestration in global terrestrial biomes. Sci Tot Environ 603–604:502–509

Song Z, McGrouther K, Wang H (2016) Occurrence, turnover and carbon sequestration potential of phytoliths in terrestrial ecosystems. Earth Sci Rev 158:19–30

Soong JL, Cotrufo MF (2015) Annual burning of a tallgrass prairie inhibits C and N cycling in soil, increasing recalcitrant pyrogenic organic matter storage while reducing N availability. Glob Change Biol 21:2321–2333

Soong JL, Dam M, Wall DH, Cotrufo MF (2017) Below-ground biological responses to pyrogenic organic matter and litter inputs in grasslands. Funct Ecol 31:260–269

Soudzilovskaia NA, van der Heijden MGA, Cornelissen KHC et al (2015) Quantitative assessment of the differential impacts of arbuscular and ectomycorrhiza on soil carbon cycling. New Phytol 208:280–293

Soussana JF, Allard V, Pilegaard K et al (2007) Full accounting of the greenhouse gas (CO_2, N_2O, CH_4) budget of nine European grassland sites. Agr Ecosyst Environ 121:121–134

Soussana JF, Lemaire G (2014) Coupling carbon and nitrogen cycles for environmentally sustainable intensification of grasslands and crop-livestock systems. Agr Ecosyst Environ 120:9–17

Soussana JF, Loiseau P, Vuichard N et al (2004) Carbon cycling and sequestration opportunities in temperate grasslands. Soil Use Manage 20:219–230

Soussana JF, Tallec T, Blanfort V (2010) Mitigating the greenhouse gas balance of ruminant production systems through carbon sequestration in grasslands. Animal 4:334–350

Stahl C, Fontaine S, Klumpp K et al (2017) Continuous soil carbon storage of old permanent pastures in Amazonia. Glob Change Biol 23:3382–3392. https://doi.org/10.1111/gcb.13573

Surawski NC, Sullivan AL, Roxburgh SH, Meyer CPM, Polglase PJ (2016) Incorrect interpretation of carbon mass balance biases global vegetation fire emission estimates. Nature Commun 7:11536. https://doi.org/10.1038/ncomms11536

Suttie JM, Reynolds SG, Batello C (2005) Grasslands of the World. Food and Agriculture Organization of the United Nations, Plant Production and Protection Series (FAO, Rome), No. 34

Thomsen IK, Christensen BT (2004) Yields of wheat and soil carbon and nitrogen contents following long-term incorporation of barley straw and ryegrass catch crops. Soil Use Manage 20:432–438

Tian H, Chen G, Liu M, Zhang C, Sun G, Lu C, Xu X, Ren W, Pan S, Chappelka A (2010) Model estimates of net primary productivity, evapotranspiration, and water use efficiency in the terrestrial ecosystems of the southern United States during 1895–2007. For Ecol Manage 259:1311–1327

Torres-Sallan G, Schulte RPO, Lanigan GJ et al (2017) Clay illuviation provides a long-term sink for C sequestration in subsoils. Sci Rep 7:45635

Treseder KK, Cross A (2006) Global distributions of arbuscular mycorrhizal fungi. Ecosystems 9:305–316 van Asperen H, Warneke T, Sabbatini S, Nicolini G, Papale D, Notholt J (2015) The role of photo- and thermal degradation for CO_2 and CO fluxes in an arid ecosystem. Biogeosciences 12:4161–4174

Verbruggen E, Jansa J, Hammer EC, Rillig MC (2016) Do arbuscular mycorrhizal fungi stabilize litter-derived carbon in soil? J Ecol 104:261–269

Wang J, Monger C, Wang X, Serena M, Leinauer B (2016) Carbon sequestration in response to grassland–shrubland–turfgrass conversions and a test for carbonate biomineralization in desert soils, New Mexico, USA. Soil Sci Soc Am J 80:1591–1603. https://doi.org/10.2136/sssaj2016.03.0061

WES Winston Eco-Strategies (2016) Evaluation of General Mills' and Kellogg's GHG Emissions Targets and Plans. Winston Eco-Strategies, LLC. http://policy-practice.oxfam.org.uk/publications/evaluation-of-general-mills-and-kelloggs-ghg-emissions-targets-and-plans-indepe-610586

Wolf J, West TO, Le Page Y et al (2015) Biogenic carbon fluxes from global agricultural pro-duction and consumption. Global Biogeochem Cycles 29:1617–1639. https://doi.org/10.1002/2015GB005119

Yang Y, Fang J, Ji C, Ma W, Mohammat A, Wang S, Wang S, Datta A, Robinson D, Smith P (2012) Widespread decreases in topsoil inorganic carbon stocks across China's grasslands during 1980s–2000s. Global Change Biol 18:3672–3680

Yu SY, He H, Cheng P, Hou Z (2017) Depth heterogeneity of soil organic carbon dynamics in a heavily grazed alpine meadow on the Northeastern Tibetan Plateau: a radiocarbon-based approach. J Geophys Res Biogeosci 122:1775–1788. https://doi.org/10.1002/2016JG003567

Zhang L, Wylie B, Loveland T, Fosnight E, Tieszen LL, Ji L, Gilmanov T (2007) Evaluation and comparison of gross primary production estimates for the Northern Great Plains grasslands. Remote Sens Environ 106:173–189

Zhang M, Lal R, Zhao Y, Jiang W, Chen Q (2016) Estimating net primary production of natural grassland and its spatio-temporal distribution in China. Sci Tot Environ 553:184–195

Zhang Y, Song C, Zhang K, Cheng X, Band LE, Zhang Q (2014) Effects of land use/land cover and climate changes on terrestrial net primary productivity in the Yangtze River Basin, China, from 2001 to 2010. J Geophys Res Biogeosci 119:1092–1109

Zhou G, Zhou X, He Y et al (2017) Grazing intensity significantly affects belowground carbon and nitrogen cycling in grassland ecosystems: a meta-analysis. Global Change Biol 23:1167–1179

Chapter 5
Carbon Sequestration in Wetland Soils

Contents

Abstract Crops adapted to wetland conditions such as rice (*Oryza sativa* L.) have been cultivated on waterlogged anoxic soils for millennia. Grazing of livestock is another important agricultural activity in wetlands. Wetlands including peatlands may cover up to 26.9 million km^2 globally, and wetlands may contain up to 158 Pg soil organic carbon (SOC) to 1 m depth, but knowledge on wetland distribution, extent, and volume needs to be strengthened. Peatlands are organic-rich wetlands and cover ~4 million km^2 with large areas in the Northern Hemisphere. Over centuries to millennia, organic soils of peatlands have accumulated globally >750 Pg of carbon (C) as peat, sometimes to several meters depth as decomposition rates are greatly reduced under wet and acidic soil conditions. Thus, peatlands while covering only 3% of the global ice-free land area store more than one-fourth of the global SOC stock. Stocks in northern peatlands alone may store >600 Pg C, and their utilization for agriculture may release large amounts of carbon dioxide (CO_2) by peat oxidation. For example, ~1 Pg CO_2 is emitted annually from drained peatlands (including emissions from fire), with high emissions especially from drained organic soils in tropical regions often for cultivation of oil palms (*Elaeis guineensis* Jacq.). Further, cumulative net emissions from global peatland use have been estimated at 6 Pg C for the period 1850–2015. Wetlands are also among the major biogenic methane (CH_4) sources contributing to about 30% of the total CH_4 emissions and will increasingly contribute to the projected climate change. Further,

K. Lorenz and R. Lal, *Carbon Sequestration in Agricultural Ecosystems*,
https://doi.org/10.1007/978-3-319-92318-5_5

C losses from wetlands may also increase in the future because of the projected climate change. Thus, sustainable intensification (SI) should be applied to reduce CH_4 and C losses from wetlands. Options include, for example, restoring drained agricultural land-use types to flooded conditions, improved fertilizer and water management of paddy fields, and breeding of new crop cultivars better adapted to anoxic wetland soil conditions. In this respect, paludiculture of wetlands is promising as a suitable agricultural practice with the cobenefit of C sequestration. This chapter begins with a general overview on wetlands and peatlands. Then, the peatland C balance is discussed in more detail. Agricultural use and management of wetland soils are presented in the following section. The final section discusses options for more 'climate-friendly' agriculture in wetlands.

Keywords Organic soils · Paludiculture · Tropical peatlands · Oil palm cultivation · Rice cultivation · Methanogenesis · Sustainable intensification Fire · Gaseous emissions

5.1 Introduction

Wetlands are far more important in the biosphere than their extent suggests. Unmanaged wetlands are hot spots of biological diversity (Junk et al. 2006), ecosystem productivity (Rocha and Goulden 2009), and economic activity (aquaculture, tourism, timber; Mitsch and Gosselink 2015). Wetlands are key regulators of biogeochemical cycles, including water flows and associated nutrients (C, N, P), pollutants and sediments, coastal erosion, and land stabilization (Junk et al. 2013). Wetlands also play fundamental roles in climate change regulation and mitigation (Gumbricht et al. 2017). Wetlands act as long-term soil organic carbon (SOC) reservoirs dating back to the Holocene. However, humans have practiced crop cultivation and livestock grazing in wetlands since thousands of years. For example, crops were cultivated 6,000 years ago in the floodplains of Mesopotamia (Verhoeven and Setter 2010). The cultivation has affected the large amounts of organic matter (OM) which have accumulated in wetland or hydromorphic soils (Jungkunst et al. 2012), and, especially, organic-rich peatland soils as oxygen deficiency results in far less efficient microbial decomposition (Schlesinger and Bernhardt 2013). OM enters wetlands through net primary production (NPP) and transport in both solid and dissolved forms, while decomposition and transport out of wetlands result in OM loss (Keller and Medvedeff 2016). NPP is the ultimate source for soil organic carbon (SOC) of wetlands, and aboveground NPP for wetlands varies widely from 10 to 4,600 g C m^{-2} y^{-1} with <1,000 g C m^{-2} y^{-1} for northern peatlands (Keller and Medvedeff 2016; Gopal and Masing 1990). However, it is generally recognized that belowground NPP can be significant but is rarely measured similar to that in the non-wetland soils.

While extensive agricultural practices have less severe effects on wetland ecosystems, complete conversion of wetlands for intensive agriculture by drainage, nutrient and water management, and vegetation clearance, and peat extraction may severely affect carbon (C) sequestration and stocks. For example, world resources of peat are estimated at 1,900 Pg of which 0.026 Pg peat was produced in 2012 (Mitsch and Gosselink 2015). However, a major issue is that wetlands and peatlands are defined differently across countries and scientific disciplines (Jackson et al. 2017). In the following section, wetlands and the main subclass peatlands will be characterized with a focus on peatland soils as those dominate wetland soil C stocks.

5.1.1 Wetlands

Wetlands are distinct ecosystems with intermittent to permanent waterlogged soil conditions resulting in anoxia because oxygen consumption exceeds the rates of its delivery (Schlesinger and Bernhardt 2013). However, there is no universal definition of wetlands as the characteristics distinguishing wetlands from other ecosystems make them also less easy to define (Mitsch and Gosselink 2015). Specifically, depth and duration of flooding vary considerably from wetland to wetland and from year to year. Wetlands may be located at the margins between deep water and uplands and be influenced by both. Wetland species include obligate species adapted to wet environments and facultative species adapted to either wet or dry conditions. Wetlands vary widely in size, and their location can also vary greatly. Also, wetland condition or the degree of modification varies greatly from region to region and from wetland to wetland (Mitsch and Gosselink 2015). The Ramsar Convention considers wetlands to be 'areas of marsh, fen, peatland or water, whether natural or artificial, permanent or temporary, with water that is static or flowing, fresh, brackish or salt, including areas of marine water the depth of which at low tide does not exceed six meters.' (Davis 1994). Under this definition, wetlands include lakes and rivers, swamps and marshes, wet grasslands and peatlands, oases, estuaries, deltas and tidal flats, nearshore marine areas, mangroves and coral reefs, and human-made sites such as fish ponds, rice (*Oryza sativa* L.) paddies, reservoirs, and salt pans. Ramsar (2013) differentiates the five major wetland types estuarine, lacustrine, marine, riverine, and palustrine wetlands. However, a wetland definition satisfactory to all users has not been developed (Mitsch and Gosselink 2015).

Estimates for the global wetland area vary fourfold based on modeled area simulations (7.1–26.9 million km^2) and threefold (4.3–12.9 million km^2) in observational mapping (Melton et al. 2013). Variability in areas and volumes are related to definition issues discussed previously and to the temporality of the inundation patterns which complicate comparisons among estimates (Page et al. 2011). Also, wetland ecosystem boundaries can be difficult to assess, and detailed inventories for the major regions South America, Africa, and Russia are missing (Schlesinger and Bernhardt 2013). Major wetland regions are located in the Neotropics (4.15 million km^2), Europe (2.58 million km^2), North America (2.42 million km^2), and Asia (2.04 million km^2) (Finlayson et al. 1999). Further, about

Table 5.1 Estimated global area (million km^2) of wetland ecosystem types (modified from Mitsch et al. 2009)

Type of wetland	Area
Tidal salt marshes	0.1
Tidal freshwater marshes	0.02
Mangrove wetlands	0.24
Freshwater marshes	0.95
Freshwater swamps and riparian forests	1.09
Peatlands	3.5

1.63 million km^2 under paddy rice cultivation on flooded mineral soils was harvested in the year 2014 (http://www.fao.org/faostat/en/#home). The largest wetland types are peatlands followed by freshwater swamps/riparian forests, and freshwater marshes (Mitsch et al. 2009; Table 5.1).

Recently, improved estimates for tropical and subtropical wetland areas and volumes were reported by Gumbricht et al. (2017). Estimates were based on an expert system approach to estimate wetland and peatland areas, depths, and volumes, which relies on three biophysical indices related to wetland and peat formation (i) long-term water supply exceeding atmospheric water demand; (ii) annually or seasonally waterlogged soils; and (iii) a geomorphological position where water is supplied and retained. Tropical and subtropical wetland area covers an estimated 4.7 million km^2. The American continent is the major contributor (45%), and Brazil, with its Amazonian interfluvial region, contains the largest tropical wetland area (800,720 km^2) (Gumbricht et al. 2017). Historically, the global wetland area was much larger as humans have been draining, in-filling, and converting both coastal and inland wetlands for many centuries. Wetland area conversion and loss were in excess of 50% of the original area and are continuing in all parts of the world, particularly rapidly in Asia (Davidson 2014).

Storage of large amounts of organic C is among the major benefits of wetlands. Decomposition rates in wetland soils are low due to anoxic conditions resulting in accumulation of dead plant material. However, data on SOC stocks are uncertain as the volume of wetland soils (Gumbricht et al. 2017), and its C mass is generally not known as there is no single definition of wetland. Globally, SOC stocks to 1 m depth were estimated at 82 Pg C for permanent wetlands and 158 Pg C for all wetlands except open water (Köchy et al. 2015). Tropical wetlands alone may contain 38.3–39.9 Pg SOC in the top 1 m. Köchy et al. (2015) suggested that the areal extent of organic soils, their depth, and the bulk density at different depths should receive the greatest focus of future soil mapping activities to improve those estimates. In contrast to peatlands, wetlands such as salt marshes can accumulate SOC quite rapidly (Kirwan and Mudd 2012). Specifically, blue carbon, i.e., C sequestered by vegetated coastal ecosystems for long-term storage (Howard et al. 2014), is a potential sequestration component for atmospheric CO_2. Tidally influenced wetlands are attributed with 36% of the total sequestration by all wetlands and 18% of the total C sequestration of all ecosystems in the conterminous USA (Bridgham et al. 2007). Across the continental USA, there are 1,153–1,359 Tg of

SOC in the upper 100 cm of soils across a total of 24,945.9 km^2 of tidal wetland area (Hinson et al. 2017).

Wetland soils not only sequester C but simultaneously emit both carbon dioxide (CO_2) and methane (CH_4) as respiratory by-products under waterlogged conditions (Bridgham et al. 2013). Globally, CO_2 emissions from wetlands are estimated at 2.1 Pg C yr^{-1} (Aufdenkampe et al. 2011). The complete mineralization of OM in wetlands requires a complex community of microbes. Specifically, CH_4 is produced by fermentation processes of methanogenic microbes. Among the controls of CH_4 emissions from wetland are soil temperature, water table, and vegetation, but these relationships are modified depending on wetland type (bog, fen, or swamp), region (subarctic to temperate), and disturbance (Turetsky et al. 2014). Microbial controls also regulate wetland CH_4 cycling (Bridgham et al. 2013). Globally, soils of wetlands are among the major biogenic CH_4 sources contributing to about 30% of the total CH_4 emissions with natural wetlands emitting 177–284 Tg CH_4 y^{-1} for the decade of 2000–2009 (Ciais et al. 2013). Unmanaged wetlands are the largest and most uncertain natural sources of CH_4 and also the presumed drivers of the interannual variations in CH_4 atmospheric growth rates (Petrescu et al. 2015).

Increasing temperature, which increases wetland area by causing high-latitude permafrost thawing (Tchebakova et al. 2011), drives a long-term trend in wetland CH_4 emissions (McNorton et al. 2016). Over several centuries, sustained CH_4 emissions in natural northern latitude wetland ecosystems are typically offset by CO_2 uptake (Petrescu et al. 2015). In contrast, high rates of NPP and litter decomposition in tropical wetlands result in both high CO_2 and CH_4 emissions. For example, Sjögersten et al. (2015) estimated that 4,540 and 90 Tg CH_4 y^{-1} are released from tropical wetlands globally. Under the 'worst-case' RCP8.5 scenario with no climate mitigation, boreal CH_4 emissions are enhanced by 18.1–41.7 Tg, due to thawing of inundated areas during the cold season and rising temperature (Zhang et al. 2017). By 2099, tropical CH_4 emissions are projected to accelerate with a total increment of 48.4 Tg to 87.4 Tg. Thus, climate mitigation policies must consider mitigation of wetland CH_4 feedbacks to maintain average global warming below 2 °C (Zhang et al. 2017). Wetlands may also emit nitrous oxide (N_2O) due to their high potential for denitrification. For example, under favorable hydrological conditions, undisturbed wetlands are reported to act as moderate N_2O sources (Frolking et al. 2011. However, there are no reliable estimates of the contribution of wetlands to global N_2O emissions (Schlesinger and Bernhardt 2013).

Aside from the utilization for agriculture, the wetland SOC stock is also under threat from the climate change. Specifically, increases in temperature changes in the total amount and distribution pattern of precipitation, and, in the case of coastal wetlands, sea level rise may increasingly alter wetland soil C stock and sequestration (Settele et al. 2014). However, there is currently no consensus on the impact of climate change on greenhouse gas (GHG) emissions and the SOC stock of global wetlands (Meng et al. 2016). For example, less frequent inundation is likely to foster oxidation of OM, but the loss of OM may be coupled with enhanced NPP and reduced CH_4 emissions (Schlesinger and Bernhardt 2013). Some freshwater wetlands may become brackish or saline as a result of saltwater intrusion into coastal wetlands from sea level rise or through drought-induced evaporative concentration of salts resulting in enhanced OM decomposition. Predicting the effect of rising temperatures on wetland biogeochemistry is, particularly, difficult due to complex positive and negative feedbacks. For example, permafrost melting may lead to wetland drainage, and OM decomposition in many boreal wetlands may be enhanced. Another threat to wetland C stocks from the climate change is the higher susceptibility to wildfire. Further, elevated CO_2 concentrations may stimulate enhanced rates of OM oxidation and CH_4 production in wetlands (Schlesinger and Bernhardt 2013).

5.1.2 Peatlands

Forested and non-forested peatlands are among the wetland types differentiated by the Ramsar Convention. Peatlands may broadly be defined as organic-rich wetlands and include Histosols, bogs, and fens (Keller and Medvedeff 2016). Peatland refers to a peat covered landscape. All peats are formed in wetlands, but not all wetlands are associated with peat because not all wetlands have the conditions to form peat (Rydin and Jeglum 2006). However, classification of peatland types is challenging because of gradients in alkalinity, pH, nutrient availability, and vegetation among peatland types. Thus, similar to wetland, there is no agreed definition of peatland. Most definitions require a minimum peat thickness of 30 cm for an ecosystem to be considered a peatland (Loisel et al. 2017). Boreal, temperate, tropical, high-altitude, and low-altitude peatlands can also be differentiated, and common characteristic is that they contain a large SOC stock (Biancalani and Avagyan 2014). Peat is the OM formed in situ which has been accumulated in soil over long periods of time as a result of incomplete breakdown of dead plant remains due to oxygen deficiency usually caused by waterlogging (Osvald 1937). For example, tropical peatlands may have been involved in the global C cycle (GCC) prior to the last glacial maximum (26,500–19,000 years ago) and, thus, are older than subarctic and temperate peatlands (Page et al. 2004).

Similar to peatland, there is also no universal definition for peat. For example, peat can be defined as sedentarily accumulated material consisting of at least 30% (dry mass) of dead organic material (Joosten 2010). If the plant remains are still recognizable in wetlands, organic materials are called peat (Schlesinger and Bernhardt 2013). While peat in tropical regions is largely composed of trees, mosses and sedges are major components of peat in boreal and temperate regions (Biancalani and Avagyan 2014). To differentiate organic from mineral soils, an OM content of >20–30%, together with a depth criterion, is often used to characterize organic soils. Peatland soils with a thick layer of strongly decomposed acidic organic material, 70 cm thick, with continuous rock at 80 cm and in an environment with a large excess of precipitation qualify as Histosols (FAO 2015). The total extent of Histosols in the world is estimated to be 3.25–3.75 million km^2. In comparison, the global peatland area has been estimated at 4.23 million km^2 based on the improved global peatland map PEATMAP (Xu et al. 2018). Major peatland complexes are located in the circum-arctic zone, particularly the Western Siberian Lowlands in Russia, and the Hudson and James Bay Lowlands in Canada. Other important concentrations at lower latitudes include extensive peat-dominated wetland or swamp forest landscapes such as the Congo and Amazon Basins, and those of Southeast Asia (Xu et al. 2018). However, estimates are uncertain as there is a basic lack of consistent peatland mapping at the global scale (Biancalani and Avagyan 2014). Further, global wetland and soil datasets are poorly suited to estimating peatland distribution (Xu et al. 2018). Thus, previous global peatland inventories likely underestimated peat extent in the tropics and overestimated it in parts of mid- and high-latitudes of the Northern Hemisphere.

Large peatland areas are found with estimated 1.38 million km^2 in Russia, 1.13 million km^2 in Canada, 0.27 million km^2 in Indonesia, and 0.22 million km^2 in the USA (Joosten 2010). Based on PEATMAP, different numbers were reported with 0.15 million km^2 for Indonesia and 0.20 million km^2 for the USA (Xu et al. 2018). Recently, Tanneberger et al. (2017) provided the most accurate representation of current peatland distribution across the whole of Europe that was possible on the basis of available national data, using a consistent definition of peatland. The overall area of peatland in Europe was estimated at 593,727 km^2 (528,337 km^2 based on PEATMAP; Xu et al. 2018). Mires, i.e., peatlands where peat is being formed, covered >320,000 km^2 (54% of the peatland area). If shallow peatlands (<30 cm peat) in European Russia were also taken into account, the total peatland area in Europe was >1,000,000 km^2, which was almost 10% of the total surface area (Tanneberger et al. 2017).

Aside from the northern peatlands, other peatland-rich areas include the Amazon River Basin, Congo, Indonesia, the Tibetan Plateau, and southern Patagonia with peat often occurring under swamp forests (Dargie et al. 2017; Loisel et al. 2017). For example, lowland peatlands in Southeast Asia may cover 0.248 million km^2 (Biancalani and Avagyan 2014). However, unprecedented extents and volumes of peatland in the tropics (1.7 million km^2 and 7,268 (6,076–7,368) km^3) were reported by Gumbricht et al. (2017) based on an expert system model. South America and not Asia may contribute the most to tropical peatland area and volume (44% for both), partly related to some yet unaccounted extended deep deposits but mainly to extended but shallow peat in the Amazon Basin. About 38% of both tropical peat area and volume may be located in Asia with Indonesia having the deepest and most extended peat areas in the tropics. Further, Africa may have more peat than previously reported but climatic and topographic contexts leave it as the least peat-forming continent. The results suggest large biases in current understanding of the distribution, area, and volumes of tropical peat and their continental contributions (Gumbricht et al. 2017).

Soil information systems are the most complete source of georeferenced information on the occurrence of organic soils for large areas at the continental or global scale. For example, the majority of Histosols are located in the boreal, subarctic, and low arctic regions of the Northern Hemisphere. Only one-tenth of all Histosols are found in the tropics. Extensive areas of Histosols occur in the US and Canada,

Western Europe and northern Scandinavia and in the West Siberian Plain (FAO 2015). The peatland C balance will be discussed in more detail in the following section as peatland soils dominate wetland soil C storage.

5.2 Peatland Carbon Balance

Globally, most soils are unsaturated and oxic, so CO_2 is the main respiration flux (Gougoulias et al. 2014). However, in waterlogged anoxic or hydromorphic soils of peatlands, CO_2 is reduced to CH_4 by hydrogenotrophic archaea in methanogenesis. The net flux of CH_4 produced depends on the relative activity of methanogens versus the activity of aerobic CH_4-oxidising bacteria (methanotrophs) residing in the surface, oxic layers of peatland soils. Further, the microbial anaerobic oxidation of CH_4 in anoxic layers of peatlands may also contribute to the CH_4 balance (Gougoulias et al. 2014). However, the current paradigm that microbial methanogenesis can only occur in anoxic habitats has been challenged. For example, Angle et al. (2017) reported CH_4 production in well-oxygenated soils of a freshwater wetland. In this wetland, up to 80% of CH_4 fluxes could be attributed to methanogenesis in oxygenated soils.

The drivers of the C balance of peatlands differ across different scales (Fig. 5.1; Limpens et al. 2008). At the local scale, the depth of the unsaturated zone together with vegetation composition is predictors for soil respiration through their effects on the availability of electron donors (i.e., OM) and electron acceptors. At the ecosystem scale, the vegetation composition is controlled by water level, nutrient availability, and pH. At the landscape and regional scales, the percentage of peatlands (land cover) and their connections to other ecosystems through subsurface and surface hydrology and topography affect C export to water and atmosphere. However, full integration of C flux across spatial scales is difficult. Yet, a large part of the drivers is related to hydrology, i.e., surface wetness, depth of the unsaturated layer, water table depth, and water movement. Integrating these aspects into a single parameter (e.g., by calculating water residence times) and relating it to biogeochemical processes and vegetation shifts may be promising. To sum up, the main drivers controlling C fluxes in peatlands are largely scale dependent and most are related to hydrology. Despite high spatial and annual variability in net ecosystem exchange (NEE), the differences in annual NEE are more a function of broad-scale geographic location and physical setting than internal factors. Thus, there may be strong feedbacks. In contrast, CO_2 and CH_4 emissions may be mainly controlled by local factors (Limpens et al. 2008).

The waterlogged anoxic, cold, and acidic conditions limiting microbial decomposer activity together with plant and peat recalcitrance may be main reasons for accumulation of dead plant material in boreal and temperate peatlands (Rydin and Jeglum 2006). The same factors, except for low temperatures, may contribute to peat accumulation in tropical peatlands. There, trees are the main peat-forming plants, whereas those are less important in northern latitude peatlands. Here, peat

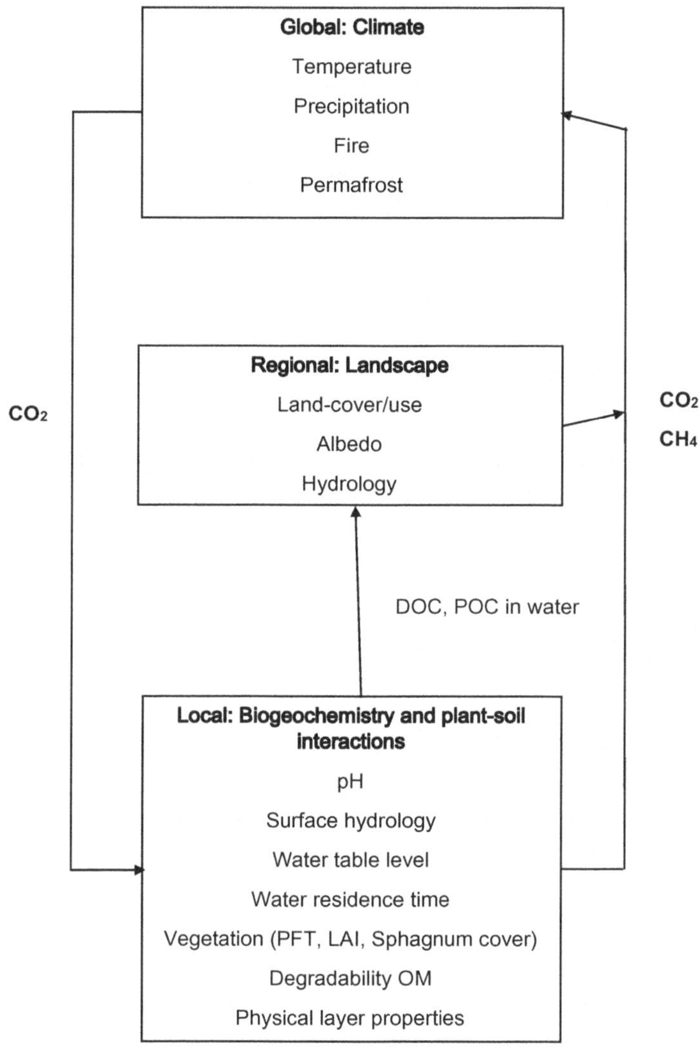

Fig. 5.1 Biogeochemical and biophysical drivers of the peatland carbon balance at different spatial scales (PAR = Photosynthetically active radiation, DOC = Dissolved organic carbon, POC = Particulate organic carbon, PFT = Plant functional type, LAI = Leaf area index; redrawn from Limpens et al. 2008)

mosses (*Sphagnum*), 'brown' mosses (Amblystegiaceae), and sedges (*Carex*) are the major peat-forming plants (Biancalani and Avagyan 2014). However, plant NPP and peat accumulation rates vary widely among peatlands. For example, the 'efficiency' of peatlands (i.e., the ratio between peat accumulation and NPP) varies between 1 and 20% depending on geographical location, age, and type of the peatland (Biancalani and Avagyan 2014). Gaseous C losses occur by CO_2 and CH_4

emissions from tropical peatlands, whereas boreal and temperate peatlands are net emitters of CH_4 only.

Peatlands dominate the wetland soil C storage and are very efficient at sequestering C relatively slowly over millennia (Loisel et al. 2017). However, the C sequestration rates have been highly variable during the Holocene (the past 12,000 years after the last glaciation) with peak C accumulation rates occurring during warmer climate intervals (Yu 2012). Nevertheless, C sequestration in peatlands during the Holocene is on the same order as the total land C sink with the majority occurring in northern peatlands (Stocker et al. 2017). Current total peat C stocks are estimated at 547, 50, and 15 Pg C for northern, tropical, and southern peatlands, respectively (Yu 2011). Peatland C stocks are much higher than those for wetlands as peatland soils can be much deeper than wetland soils. The range of Indonesia's total peat C stock alone is within 13.6 and 40.5 Pg C, with a best estimate of 28.1 Pg C (Warren et al. 2017). This is more than the estimated C stock in above- and belowground biomass of all Indonesian forests. When field measurements from one of the world's most extensive regions of swamp forest, the Cuvette Centrale depression in the central Congo Basin, are included, global tropical peatland C stocks increase substantially to ~105 Pg C (Dargie et al. 2017). Peat layers in tropical regions are typically thicker than those in temperate and boreal peatlands explaining the relatively large SOC stocks in tropical peatlands despite relatively low areal extent (Page et al. 2011). However, similar to the spatial extent and peat volume, there is high uncertainty of C stocks for tropical peatlands (Draper et al. 2014).

Joosten (2010) computed slightly different numbers and estimated that the largest peat stocks are located in Canada (155 Pg C) followed by those in Russia (138 Pg C), Indonesia (54 Pg C), and the USA (29 Pg C). However, the greatest source of uncertainty is the lack or insufficient representation of data, including peat depth, bulk density, and C accumulation data, especially from the world's large peatlands (Yu 2012). For example, using the peat volume approach, the C density approach and the time history approach, estimates of C stock in northern peatland are 180–455, 125–466, and 473–621 Pg C, respectively (Yu 2012). Further, major uncertainty is linked to the area covered by high-latitude peatlands (published estimates vary between 1.2 and 2.7 million km^2), which alone results in a range of 94–215 Pg SOC (Köchy et al. 2015). Scarcity of information about SOC at depths >1 m and variation in definitions of 'peatland' need to be addressed, in particular, to improve the estimates of peat C stock. Yu (2012) proposed to include the estimates of C stocks by regions and further utilization of widely available basal peat ages to reduce uncertainty. Improved extent and volume mapping of peat are required to reduce data uncertainty (Robinson et al. 2017). Nevertheless, peatlands cover only ~3% of the global ice-free land area of 130 million km^2 (Hooke et al. 2012), but store more than one-fourth of the global SOC stock of 3,000 Pg C (Köchy et al. 2015). Thus, drainage and utilization of peatlands for agriculture may release large amounts of C.

The properties of organic soils of peatlands have major effects on the SOC balance. The major difference in C flow between organic soils and those of drier soils outside of peatlands is that lateral transfer of C with water plays a more prominent role. Peatland C inputs occur by (i) plant photosynthesis; (ii) dissolved inorganic C (DIC) and dissolved organic C (DOC) deposition with rainwater; and (iii) intake of inorganic C from the weathering of underlying strata and lateral inflows of organic and inorganic C from other sites (Biancalani and Avagyan 2014). Peatland C release follows also lateral pathways as C is transported to streams in the form of particulate organic C (POC), DOC, or DIC (Biancalani and Avagyan 2014). Generally, DOC is the main component of the lateral peatland C flux (Roulet et al. 2007). The export of DOC is a key part of the peatland C cycle and has also important implications for downstream water chemistry (Biancalani and Avagyan 2014).

Similar to wetlands, the future of the peatland C stock is uncertain even without agricultural use. During the Holocene, northern peatlands have probably been a persistent atmospheric CO_2 sink and a CH_4 source, with an overall negative radiative impact or cooling effect on the global climate (Loisel et al. 2017). For example, CH_4 emissions from northern peatland are estimated to range from 31 to 106 Tg CH_4 yr^{-1} (Zhuang et al. 2004). The future of the large C stock of northern peatland is particularly uncertain under the projected climate change (Loisel et al. 2017). Specifically, there is no consensus on the direction and magnitude of the impact of climate change on the peatland C sink capacity. On the one hand, enhanced peat decomposition and subsequent C emission to the atmosphere may occur in a warmer and drier world. Otherwise, warmer temperatures would prolong the growing season for peatland vegetation and, thus, increase the NPP and C accumulation in peatlands that are not water-limited. However, not well understood are the impact of drought on fire severity, intensity, and recurrence on CO_2 emissions and the impact of warming temperature on permafrost thaw and subsequent peatland collapse on CH_4 emissions (Loisel et al. 2017). While Ise et al. (2008) simulate quick losses of labile SOC from northern peatlands during dry periods to the expected warming in the twenty-first century, Wang et al. (2015) propose that the projected ‘positive feedback loop’ between C emission and drought in peatlands may not occur in the long term. Rather, historically accumulated SOC stock in northern peatlands will potentially be protected by a vegetation shift from low-phenolic *Sphagnum*/herbs to high-phenolic shrubs. However, the functional identity and functional redundancy of European peat bog communities as a whole remain unchanged (Robroek et al. 2017). This strongly suggested that species turnover across environmental gradients is restricted to functionally similar species. Thus, plant taxonomic and functional turnover are decoupled, which may allow these peat bogs to maintain ecosystem functioning including C sequestration when subject to future climate change (Robroek et al. 2017).

The effects of permafrost thawing on peatland C accumulation are uncertain (Limpens et al. 2008). The SOC in permafrost may be protected by microscale structures and higher aggregate stability (Mueller et al. 2017; Oztas and Fayetorbay 2003). Thawing may accelerate the decomposition of this previously frozen and protected SOC pool (Walz et al. 2017). However, the amount of C released and whether C will be released as CO_2 or as CH_4 will largely depend on in situ thaw and hydrological conditions. For example, anaerobic environments created by melting ice may replace freezing temperatures as a mechanism for SOC stabilization, keeping CO_2 and CH_4 emissions lower than they would otherwise be (Zona et al. 2012). Otherwise, the water released by melting ice may increase water residence time, promote peat formation and, thus, local C accumulation but increase CH_4 emissions (Limpens et al. 2008). For example, the C sequestration rate of collapse scars is among the highest reported for peatlands (Meyers-Smith et al. 2008). Otherwise, if the permafrost layer acts as a water impermeable layer, thawing will lead to drainage and stimulate decomposition processes. Further, in peatlands with discontinuous permafrost, severe fire events may contribute to permafrost thawing leading to more permanent vegetation changes, potentially increasing C accumulation in the long term (Limpens et al. 2008). Chaudhary et al. (2017) applied a model accounting for feedbacks between hydrology, peat properties, permafrost, and dynamics of vegetation across a heterogeneous peatland landscape. This model was able to reproduce broad, observed patterns of peatland C and permafrost dynamics across the pan-Arctic region. Under a business-as-usual future climate scenario, Chaudhary et al. (2017) showed that non-permafrost peatlands may become a C source due to soil moisture limitations, while permafrost peatlands gain C due to an initial increase in soil moisture, which suppresses decomposition while enhancing plant production.

Tropical peatlands may increasingly lose C by fire following drought periods. As fire frequency is expected to increase with climate change, understanding the interactive effects of altered hydrology and fire on SOC and GHG emissions from peatlands is crucial (Keller and Medvedeff 2016). In the following section, the consequences of utilization of wetland soils for agriculture will be discussed.

5.3 Management and Use of Wetland Soils for Agriculture

Soils of wetlands and, in particular, peatlands contain large amounts of C. Thus, their conversion for agriculture and intensive agricultural use such as high degree of soil amelioration, drainage, and fertilizer use may result in a significant increase of the atmospheric radiative forcing (Petrescu et al. 2015). For example, total net emissions by peatland use have been estimated at 6 Pg C during the period 1850–2015 (Houghton and Nassikas 2017). Any management plans for these complex ecosystems should, therefore, carefully account for the potential biogeochemical effects on climate.

Aside from utilizing wet and peatlands as croplands and pastures, wetland SOC and peat have been used historically by local communities for many purposes, including as growing media and soil improvers, building material, livestock bedding, and fuel for generating heat and energy (Biancalani and Avagyan 2014). Naturally forested peatlands are also a valuable source of timber. However, when disturbed by agricultural practice, the large peat C stock accumulated over millennia may rapidly subside. Peat C is lost, in particular, because agricultural land use alters the balance between C inputs of dead plant OM to the soil, and the mineralization and export of SOC (Biancalani and Avagyan 2014). Today, ~0.26

million km^2 of organic soils is utilized for agriculture, i.e., 0.18 million km^2 for cropland and 0.08 million km^2 for pastures (FAO 2014). Between 2000 and 2006, the area of organic soils being cropped increased across EU-25, but no further increase occurred after 2006 (Robinson et al. 2017).

Large areas of flooded mineral soils, especially in Asia, are under rice cultivation. Globally, the wetland category 'Rice Paddy and Field' covers 2.14 and 1.06 million km^2 in tropical regions (Köchy et al. 2015). SOC stock estimates to 1 m depth are 17.1 Pg C globally with tropical wetland soils under flooded rice storing 8.4 Pg C. Rice paddies are a substantial CH_4 source and have emitted 33–40 Tg CH_4 y^{-1} for the decade of 2000–2009 (Ciais et al. 2013). Bridgham et al. (2013) reported a median value of 53 Tg CH_4 y^{-1} based on both bottom-up and top-down methods. Rice paddies share the same fundamental set of controls over CH_4 emissions as natural wetlands. However, the dominant groups of methanogens in rice paddies differ from those observed in natural sites. Specifically, biogenic CH_4 is produced in rice fields by anaerobic methanogenic archaea (Alpana et al. 2017). Variations in the ecophysiology of methanogens are likely factors accounting for differences in CH_4 production (Bridgham et al. 2013). Rice paddies emit also N_2O as plant residue-N and N fertilizer application enhance N_2O emissions. In the year 2000, 42–122 Gg N_2O was emitted from flooded rice soils (Gerber et al. 2016).

5.3.1 Wetlands

Wetlands do not necessarily be drained for agricultural use (Mitsch and Gosselink 2015). For example, some use is made of more or less undisturbed wetlands such as harvesting salt marsh hay (*Spartina patens*) as bedding and fodder for cattle. Further, there is a renewed interest in the ancient Mexican practice of marceno (Mitsch and Gosselink 2015). Specifically, during the dry season, small areas in freshwater wetlands are planted in corn (*Zea mays* L.) tolerant to flooding. After harvest, the marshes are naturally flooded, and native grasses reestablished until the next dry season (Mitsch and Gosselink 2015). Major examples of agricultural uses of wetlands are crop fields on river floodplain soils and rice fields (Verhoeven and Setter 2010). Some of the agricultural wetlands may be converted to some degree but maintain a modified range of ecosystem services (Ramsar Convention on Wetlands; FAO; International Water Management Institute 2014). Examples are small seasonal wetlands in Africa, rice paddies, and coastal grazing marshes. However, the use of salt marshes for livestock production affects multiple ecosystem properties, creating trade-offs and synergies with other ecosystem services (Davidson et al. 2017).

Floodplains in many parts of the world are managed by flood recession agriculture. The higher parts of the floodplains are highly suitable for growing crops, while the lower parts are wetter but are often suitable for grazing (Verhoeven and Setter 2010). Floodplain soils are nutrient-rich and are naturally 'fertilized' as a result of flooding events. Other wetlands are dependent on continued agricultural activities to maintain their ecological character, such as mowing and grazing in wet grasslands (Ramsar Convention on Wetlands; FAO; International Water Management Institute 2014). Some wetlands are maintained in a natural state for production and harvesting of specific products such as wild rice (*Zizania* spp.) in the USA. Other wetland systems are constructed or managed solely for agricultural purposes such as cranberry (*Vaccinium macrocarpon*) bogs in North America (Ramsar Convention on Wetlands; FAO; International Water Management Institute 2014). In the long term, however, intensification of land use with high levels of fertilizer and pesticide use may result in non-sustainable wetland agriculture (subsidence) and be devastating for wetland biodiversity such as observed in the Dutch wetlands (Verhoeven and Setter 2010).

Some more or less undisturbed wetlands are also used for aquaculture, the farming of fish and shellfish (Mitsch and Gosselink 2015). Aquaculture is the fastest growing food sector and continues to expand with an annual rate of 7.8% worldwide between 1990 and 2010 (Troell et al. 2014). Currently, aquaculture provides half of the fish consumed worldwide. However, aquaculture is increasingly dependent on terrestrial crops and wild fish for feeds, draws on freshwater and land resources for a large portion of its aggregate production, and can be damaging to aquatic ecosystems and fisheries. Thus, it is unclear whether aquaculture adds resilience to the global food system (Troell et al. 2014).

A total of 0.255 million km^2 of organic soils has been drained for agriculture, with about 60% of total drained organic soils in boreal and temperate cool, 34% in tropical, and 5% in warm temperate regions (Tubiello et al. 2016). Further, more than 90% of drained organic soils are under cropland. Total GHG emissions are 0.91 Pg CO_2 eq. y^{-1} with 0.86 Pg CO_2 eq. y^{-1} from drained cropland and 0.05 Pg CO_2 eq. y^{-1} from drained grassland. Emissions of CO_2 alone represent more than 85% of the total of CO_2 and N_2O emissions (Tubiello et al. 2016). Globally, CO_2 emissions from drained organic soils represented one-third of net CO_2 emissions from the ‘agriculture, forestry, and other land-use’ (AFOLU) sector, on average over the period 1990–2010. Further, the CO_2 component alone represented more than one-fourth of the net CO_2 emissions from the AFOLU sector. A few key countries, largely in Southeast Asia, make the largest contribution to the global trends as rates of C loss are much higher under tropical climates because biological decomposition is a temperature-dependent process (Tubiello et al. 2016).

5.3.2 Peatlands

In their natural state, peatlands capture C as CO_2 but may also act as CO_2 sources in some years (Strack 2008). In contrast, agricultural peatlands commonly act as sources for C and GHGs (Schrier-Uijl et al. 2014). Contemporary agriculture techniques, in particular, heavily impact peatlands through land clearance, drainage, and fertilization, a process that too often involves fire (Wijedasa et al. 2017). When only slightly disturbed, anthropogenic modification of peatlands may result in reduced biomass production, resulting in a decreased input of OM to the peat C store (Biancalani and Avagyan 2014). For example, grazing of temperate peatlands by livestock and the production of hay may lead to changes in the composition of plant species and affect C inputs. Overgrazing, in particular, contributes to degradation of the Qinghai–Tibetan Plateau peatlands in China (Yang et al. 2017). Here, soil drainage facilitated earthworm invasion and subsequent SOC loss (Wu et al. 2017).

Compared to grazing, crop cultivation is a more intensive form of peatland use that disturbs the ecosystem to much greater extent (Biancalani and Avagyan 2014). For example, 80% of all Indonesian peatlands are less than 300 cm deep and are, thus, allowable for conversion under current regulations (Warren et al. 2017). The past, ongoing, and eventual conversion of these shallower peatlands may release 10.6 Pg C to the atmosphere assuming total peat loss. Drastic changes occur also by conversion of tropical peatlands for oil palm (*Elaeis guineensis* Jacq.) cultivation in Southeast Asia. Palm oil is used for bioenergy among other things. Converting peatlands to produce food crop-based biofuels create a 'biofuel carbon debt' (Fargione et al. 2008). For example, converting tropical peatland rainforest in Indonesia and Malaysia to palm biodiesel would result in a biofuel carbon debt of 3,450 Mg of CO_2 ha^{-1} that would take 423 years to repay. The required drainage of peatland causes a sustained emission of 55 Mg of CO_2 ha^{-1} yr^{-1} from oxidative peat decomposition. Until the carbon debt is repaid, producing and using palm biodiesel from this land would cause greater GHG release than would refining and using an energy-equivalent amount of petroleum diesel (Fargione et al. 2008).

Drainage changes peatlands from long-term C reservoirs to net sources of GHG emissions (Wijedasa et al. 2017). Drainage results in a rapid increase in decomposition rates, leading to increased emissions of CO_2, and some N_2O from microbial breakdown of peat, and vulnerability to further emissions of CO_2, CO, CH_4 and black carbon (BC) emissions by fire (Smith et al. 2014). For example, in 1997, about 24,000 km^2 of peatland was burned in Southeast Asia, which released an estimated 0.81–0.95 Pg C (Page et al. 2002). Further, during the summer of 2010, an extreme period of high temperatures and low rainfall resulted in widespread and prolonged fires on abandoned drained peatlands in the Russian Federation (Biancalani and Avagyan 2014).

The type of land use, including the management type, and not SOC content may be a major control of CO_2 emission from drained peatland soils. For example, lower NEE values (−6 to 1707 g CO_2–C m^{-2} yr^{-1}) were found at arable sites and higher values (1354–1823 g CO_2–C m^{-2} yr^{-1}) at grassland sites on drained organic soils independent of differences in SOC contents (Eickenscheidt et al. 2015). Further, cultivated peatlands are not necessarily net C sources to the atmosphere as were shown in a study from Sweden (Hadden and Grelle 2017). Young and old SOC may respond differently to peatland use. For example, in strongly degraded organic soils, drained and managed since decades, fresh OM additions suppressed the decomposition of old SOC, i.e., peat (Bader et al. 2018). Fresh OM addition induced positive SOC decomposition in grassland soils, but no priming effects were observed for cropland soils (Bader et al. 2018).

There is a clear relation between CO_2 emissions and the water table depth below the peat surface. Emissions increase proportionally as water tables are lowered below 0.5 m (Biancalani and Avagyan 2014). The CO_2 emission factors are the highest for drained tropical peatlands and as high as 14 Mg C ha^{-1} y^{-1} for croplands and 9.6 Mg C ha^{-1} y^{-1} for grassland (IPCC 2014). Hooijer et al. (2012) estimated CO_2 emissions of >73 Mg C ha^{-1} yr^{-1} from tropical peatlands under plantation agriculture. In total, 355–855 Mg CO_2 may be released annually more than 5 years after drainage of tropical peatlands in Southeast Asia (Hooijer et al. 2010). Emissions are predicted to rapidly increase until at least 2020, and emissions from fire and decomposition in the first years after drainage may be of similar magnitude (Biancalani and Avagyan 2014).

In contrast to other GHGs, CH_4 emissions from drained peat soils may be absent (Biancalani and Avagyan 2014). However, drainage ditches and canals typically represent hot spots for CH_4 emissions. Hot spots may also occur at interfaces between different microtopographic units and may be responsible for the bulk of a site's CH_4 emissions. Drainage also causes an increase in N_2O emissions from peat decomposition. The application of fertilizers contributes to increasing the overall negative balance of GHG emissions from drained peatlands. N_2O fluxes from agricultural peatlands may be dominated by emission pulses which are often observed after fertilization or thaw events. Generally, there is no thorough understanding of the magnitudes and drivers of N_2O emissions (Butterbach-Bahl and Wolf 2017). However, in terms of CO_2 equivalents, peatland N_2O emissions are much lower than CO_2 or CH_4 emissions (IPCC 2014).

About a quarter of the global peatland area drained for agriculture is located in the former Soviet Union (Biancalani and Avagyan 2014). Drained fen peatlands have provided some of the most fertile soils for crops and support high levels of production. In undrained fens, precipitation exceeds evapotranspiration. In contrast, precipitation equals evapotranspiration in natural bogs. Drained bogs have problems of high acidity and waterlogging producing low yields of grasses and crops. Thus, the agricultural use of peatlands has changed over time with many areas having been abandoned. For example, while peatland drainage and conversion to agriculture have virtually ceased in boreal and temperate zone countries, it is increasing in the tropics. The palm oil and paper industries, in particular, are fueling rapid agricultural development of tropical peatlands. In contrast, paludiculture is now receiving increasing attention (Joosten et al. 2015). Paludiculture uses biomass from wet and rewetted peatlands under conditions that maintain the peat mass, promotes peat accumulation, and provides some of the ecosystem services (ESs) associated with natural peatlands (Biancalani and Avagyan 2014).

To sum up, C stocks in tropical peatland are particularly affected by drainage as tropical areas account for 34% of all drained organic soils, which is about three times greater than their share of total organic soils (Biancalani and Avagyan 2014). Besides the effects on GHG emissions, drained peatlands also release more DOC than undrained peatlands (Moore et al. 2013). As older C from deeper in the peat profile is also mobilized, drainage severely diminishes C sequestration and climate change mitigation in peatlands as a C that has accumulated over millennia is returned to the atmosphere. Any management practices that lower the water table lead to losses of C and N from peatlands (Biancalani and Avagyan 2014).

The subsidence components and their impacts on oxidation, compaction, shrinkage, and consolidation need to be addressed separately to assess drainage effects on the peatland C balance (Hooijer et al. 2012). As build-up of peat generally requires that water levels be near the soil surface, drainage is required to allow biomass production by non-adapted vegetation such as major upland crops. The drop in the water table ranges from 0.4 m for grasslands to 1.2 m for crop production. Following lowering the water table, the soil surface subsides by consolidation as the soil moisture content declines and the peat volume contracts. While the peat surface may fall up to 1 m in the first year, the rate of surface lowering varies afterward from 1–2 cm per year for temperate peatlands to 3–5 cm per year for tropical peatlands drained for agricultural production (Biancalani and Avagyan 2014). For example, apparent subsidence rates for the Willard muck agricultural area in Ohio ranged from 33 to 216 cm over a 40-year period (Hidlebaugh 1982). Elder and Lal (2008) estimated subsidence rates (cm yr^{-1}) of 1.34 under moldboard plow, 1.07 under no-till and 1.25 under bare management in the same region. Thus, within 214–268 years, the deepest organic soil deposits will be completely depleted assuming current subsidence rates. Further, all tillage treatments were persistent sources of CO_2 of 18.9–22.5 Mg CO_2–C ha^{-1} yr^{-1}, but emissions did not differ between treatments (Elder and Lal 2008). Effects of tillage practices on emissions of GHGs from organic soils are generally less well studied.

When drainage exposes underlying mineral deposits unsuitable for agriculture, agricultural production in some peatlands may end (Biancalani and Avagyan 2014). Some examples of improved management of agricultural wetlands are discussed in the following section.

5.4 Improved Management of Wetland Soils

Wetlands have been reclaimed for agriculture in many parts of the world with ever more effective drainage and land amelioration measures (Verhoeven and Setter 2010). Much of their original character may have been lost leading to reduced biodiversity and reduced performance of functions other than crop productivity. In contrast, extensive agriculture, i.e., activities in a more or less intact wetland ecosystem without the use of machinery and chemicals, may have less severe effects than intensive agricultural use of wetlands (Verhoeven and Setter 2010). Emissions of CH_4 from paddy fields, for example, can be reduced by improved management. Specifically, increased availability of mineral fertilizer contributed to a decrease in use of organic amendments in paddy rice fields, and this resulted in a decrease in CH_4 emissions (van der Gon 1999). Reductions in water use contributed also to a reduction in CH_4 emissions (Frolking et al. 2004). Managed peatlands can also be turned back into GHG and C sinks within 15 years of abandonment and rewetting if appropriate measures (i.e., reductions in management intensity) are taken (Schrier-Uijl et al. 2014).

Drainage of organic soils results in soil subsidence and contributes to increased atmospheric CO_2 concentrations as OM decomposition is accelerated. Rewetting drained peatland soils is considered an important climate change mitigation tool to reduce GHG emissions and create suitable conditions for SOC sequestration. However, knowledge about the exchange of CO_2 and CH_4 following rewetting during restoration of disturbed peatlands is limited. Restoring drained agricultural land-use types to flooded conditions is among the options to reduce C losses from drained peatlands. For example, converting a drained agricultural area in the Sacramento-San Joaquin Delta, California, back to flooded conditions for rice agriculture created an atmospheric sink for CO_2 (Knox et al. 2015). Simultaneously, CH_4 emissions from the rice paddy increased. Although the rice paddy was a net C source, it lost 230–454 g C m^{-2} y^{-1} less than the drained agricultural peatlands under corn and pasture. However, wetland restoration provided the most GHG benefit, with the potential of converting drained peatlands from GHG sources to GHG sinks (Knox et al. 2015). Emissions for a rewetted industrial cutaway peatland in Ireland were −104 g CO_2–C m^{-2} yr^{-1} (i.e., CO_2 sink) and 9 g CH_4–C m^{-2} yr^{-1} (i.e., CH_4 source), while N_2O emissions were not detected nearly a decade after rewetting (Wilson et al. 2016). However, although the GHG balance was reduced noticeably (i.e., less warming) in comparison with a drained site, it was still higher than comparative intact peatland sites. In contrast, during the eighth year following rewetting, a restored disturbed peatland ecosystem in British Columbia, Canada, was almost CO_2 eq. neutral (CO_2 eq. (g) = −103.1 g CO_2 eq. m^{-2} yr^{-1}) over a 100-year time horizon (Lee et al. 2017). To sum up, disturbed peatlands may become net annual C sink after restoration by rewetting. However, long-term monitoring of new rewetted ecosystems is needed to provide critical information for land managers, policymakers, and other stakeholders (Wilson et al. 2016).

The major cereal, legume, and fiber crops are domesticated dryland plants (Biancalani and Avagyan 2014). Cultivating them in wetlands is only possible after drainage which severely affects C sequestration and other ecosystem services. In contrast to conventional drainage-based agriculture, low-intensity agricultural activities in wetlands may be associated with a high biodiversity and other ecosystem services (Verhoeven and Setter 2010). Many traditional agricultural systems have resulted in a very diverse landscape with high species densities in a human-created setting. Examples are the moist grasslands and herbaceous fens in Europe, which have been used for grazing and hay-making. Salt marsh grazing management can be based on local context and desired ecosystem services (Davidson et al. 2017). For example, grazing leads to reductions in blue C, i.e., long-term C storage in coastal vegetated habitats in the Americas but not in Europe. Further, grazing may compromise coastal protection and the provision of a nursery habitat for fish while creating provisioning and cultural benefits through increased wildfowl abundance (Davidson et al. 2017). Rice has traditionally been cultivated in strongly modified landscapes with rice paddies on hill slopes or lowlands with a human-controlled water regime (Verhoeven and Setter 2010).

Another strategy is to make major crop species suitable for growth in wetland environments (Verhoeven and Setter 2010). A range of crop varieties is known that have better waterlogging tolerance than the regular cultivars. However, waterlogging tolerance needs to be further enhanced in new cultivars by crop breeding. In the future, more flood-tolerant and salt-tolerant crops may be cultivated in wetlands (Verhoeven and Setter 2010).

Paludiculture has the cobenefit of C sequestration (Joosten et al. 2015). Spontaneous vegetation on natural sites is used for paludiculture, or crops are artificially established on rewetted sites. Peat is generally formed by roots and rhizomes in the temperate, subtropical, and tropical zones, and peatlands hold vegetation of which aboveground parts can be harvested without substantially harming peat conservation and formation (Joosten et al. 2015). Traditional agricultural products such as food, feed, fiber, and fuel are derived from paludiculture. For example, common reed (*Phragmites australis* (Cav.) Trin. ex Steud.) may be used as construction material, for paper production and as bioenergy feedstock. Cattails (*Typha* spp.) and reed canary grass (*Phalaris arundinacea* L.) may also be feedstock for bioenergy production. In addition, reed canary grass provides fodder and can be used for pasture and the production of silage and hay. Further, peat moss (*Sphagnum* spp.) can replace fossil peat in horticultural growing media. Aside temperate peatlands, opportunities may also exist for paludiculture in degraded Indonesian peatlands. However, more research is needed on the suitability of paludiculture to replace drainage-based agriculture. Similar, the potential to establish sustainable grazing practices on peatlands is less well known (Biancalani and Avagyan 2014). Paludiculture can often compete effectively with drainage-based peatland agriculture. However, technical and political constraints hamper large-scale implementation of this type of peatland use (Joosten et al. 2015).

5.5 Conclusions

Wetlands are characterized by high C density similar to permafrost soils. Wetlands have been used by humans for growing crops and grazing animals for millennia. Due to the anoxic conditions, large amounts of C have accumulated particularly in peatland soils as decomposition rates are greatly reduced. Thus, draining wetlands and converting them for growing crops under non-flooded conditions releases large amounts of C and enhance climate change, and this may be exacerbated by increasing incidence of fire. Wetland CH_4 emissions will increasingly contribute to the projected climate change. Otherwise, climate change may further enhance wetland C losses. Thus, less intensive agricultural practices such as paludiculture should be promoted to reduce C losses while benefiting from other ecosystem services of wetlands.

5.6 Review Questions

1. Describe traditional agricultural practices in wetlands.
2. What are the principal differences between wetlands and peatlands?
3. How can wetlands and their use be adapted to climate change?
4. Biochar has been proposed to replace peat in horticultural media—describe the differences between both and the challenges to realize this potential.
5. Describe 'wetland-friendly' agricultural practices.
6. What are the major challenges in plant breeding for adapting crops to wetland conditions?
7. Rewetting of wetlands has been proposed as a climate change mitigation strategy—discuss its potential in relation to 'negative emission technologies'. What are the implications for the three dimensions of sustainability?

References

Alpana S, Vishwakarma P, Adhya TK, Inubushi K, Dubey SK (2017) Molecular ecological perspective of methanogenic archaeal community in rice agroecosystem. Sci Tot Environ 596–597:136–146

Angle JC, Morin TH, Solden LM et al (2017) Methanogenesis in oxygenated soils is a substantial fraction of wetland methane emissions. Nat Commun 8:1567. https://doi.org/10.1038/s41467-017-01753-4

Aufdenkampe AK, Mayorga E, Raymond PA et al (2011) Riverine coupling of biogeochemical cycles between land, oceans and atmosphere. Front Ecol Environ 9:53–60

Bader C, Müller M, Szidat S, Schulin R, Leifel J (2018) Response of peat decomposition to corn straw addition in managed organic soils. Geoderma 309:75–83

Biancalani R, Avagyan A (eds) (2014) Towards climate-responsible peatlands management. Mitigation of climate change in agriculture Series 9. FAO, Rome

Bridgham SD, Cadillo-Quiroz H, Keller JK, Zhuang Q (2013) Methane emissions from wetlands: biogeochemical, microbial, and modeling perspectives from local to global scales. Glob Change Biol 19:1325–1346

Bridgham SD, Megonigal JP, Keller JK, Bliss NB, Trettin C (2007) Wetlands. Chapter 13 and Appendix F. In: King AW, Dilling L, Zimmerman G, Fairman D, Houghton R, Marland G, Rose A, Wilbanks T (eds) The first state of the carbon cycle report (SOCCR): The North American carbon budget and implications for the global carbon cycle. Washington, DC: U.S. Climate Change Science Program, 177–192, pp 139–148

Butterbach-Bahl K, Wolf B (2017) Warming from freezing soils. Nat Geosci 10:248–249

Chaudhary N, Miller PA, Smith B (2017) Modelling past, present and future peatland carbon accumulation across the pan-Arctic region. Biogeosciences 14:4023–4044

Ciais P, Sabine C, Bala G, et al (2013) Carbon and Other Biogeochemical Cycles. In: Stocker TF, Qin D, Plattner GK, et al (eds) Climate change 2013: the physical science basis. contribution of working group I to the fifth assessment report of the intergovernmental panel on climate change. Cambridge University Press, Cambridge, United Kingdom and New York, NY, USA

Dargie GC, Lewis SL, Lawson IT, et al. (2017) Age, extent and carbon storage of the central Congo Basin peatland complex. Nature 542:86+

Davidson KE, Fowler MS, Skov MW et al (2017) Livestock grazing alters multiple ecosystem properties and services in salt marshes: a meta-analysis. J Appl Ecol 54:1395–1405. https://doi.org/10.1111/1365-2664.12892

Davidson NC (2014) How much wetland has the world lost? Long-term and recent trends in global wetland area. Mar Freshw Res 65:934–941

Davis TJ (1994) The Ramsar Convention manual: a guide to the convention on wetlands of international importance, especially as waterfowl habitat. Ramsar Convention Bureau, Gland, Switzerland

Draper FC, Roucoux KH, Lawson IT et al (2014) The distribution and amount of carbon in the largest peatland complex in Amazonia. Environ Res Lett 9:124017

Eickenscheidt T, Heinichen J, Drösler M (2015) The greenhouse gas balance of a drained fen peatland is mainly controlled by land-use rather than soil organic carbon content. Biogeosciences 12:5161–5184

Elder JW, Lal R (2008) Tillage effects on gaseous emissions from an intensively farmed organic soil in North Central Ohio. Soil Till Res 98:45–55

Fargione J, Hill J, Tilman D, Polasky S, Hawthorne P (2008) Land clearing and the biofuel carbon debt. Science 319:1235–1238

Finlayson CM, Davidson NC, Spiers AG, Stevenson NJ (1999) Global wetland inventory—current status and future priorities. Mar Freshwater Res 50:717–727

Food and Agriculture Organization of the United Nations FAO (2014) Agriculture, forestry and other land use emissions by sources and removals by sinks – 1990–2011 Analysis. Tubiello FN, Salvatore M, Cóndor Golec RD, et al. ESS Working Paper 2, Rome, 87pp

Food and Agriculture Organization of the United Nations FAO (2015) World reference base for soil resources 2014. World Soil Resources Report 106, Rome

Frolking S, Li CS, Braswell R, Fuglestvedt J (2004) Short- and long-term greenhouse gas and radiative forcing impacts of changing water management in Asian rice paddies. Glob Change Biol 10:1180–1196

Frolking S, Talbot J, Jones M et al (2011) Peatlands in the Earth's 21st century climate system. Environ Res 19:371–396

Gerber JS, Carlson KM, Makowski D et al (2016) Spatially explicit estimates of N_2O emissions from croplands suggest climate mitigation opportunities from improved fertilizer management. Glob Change Biol 22:3383–3394

Gopal B, Masing V (1990) Biology and ecology. In: Pattan BC (ed) Wetlands and shallow continental water bodies, vol 1. SPB Academic Publishing, The Netherlands, pp 91–239

Gougoulias C, Clark JM, Shaw LJ (2014) The role of soil microbes in the global carbon cycle: tracking the below-ground microbial processing of plant-derived carbon for manipulating carbon dynamics in agricultural systems. J Sci Food Agric 94:2362–2371

Gumbricht T, Roman-Cuesta RM, Verchot L et al (2017) An expert system model for mapping tropical wetlands and peatlands reveals South America as the largest contributor. Glob Change Biol 23:3581–3599. https://doi.org/10.1111/gcb.13689

Hadden D, Grelle A (2017) The impact of cultivation on CO_2 and CH_4 fluxes over organic soils in Sweden. Agric For Meteorol 243:1–8

Hidlebaugh TA (1982) Subsidence of the cultivated organic soils of the Celeryville-Willard marsh of North-Central Ohio. Master of Arts Thesis. Bowling Green State University

Hinson AL, Feagin RA, Eriksson M et al (2017) The spatial distribution of soil organic carbon in tidal wetland soils of the continental United States. Glob Change Biol 23:5468–5480. https://doi.org/10.1111/gcb.13811

Hooijer A, Page S, Canadell JG et al (2010) Current and future CO_2 emissions from drained peatlands in Southeast Asia. Biogeosciences 7:1505–1514

Hooijer A, Page S, Jauhiainen J et al (2012) Subsidence and carbon loss in drained tropical peatlands. Biogeosciences 9:1053–1071

LeB Hooke R, Martín-Duque JF, Pedraza J (2012) Land transformation by humans: a review. GSA Today 22:4–10

Houghton RA, Nassikas AA (2017) Global and regional fluxes of carbon from land use and land cover change 1850–2015. Glob Biogeochem Cycles 31:456–472

Howard J, Hoyt S, Isensee K, Telszewski M, Pidgeon E (2014) Coastal blue carbon: methods for assessing carbon stocks and emissions factors in mangroves, tidal salt marshes, and seagrasses.

Conservation International, Intergovernmental Oceanographic Commission of UNESCO, International Union for Conservation of Nature Arlington, Virginia, USA, 184
IPCC (2014) 2013 Supplement to the 2006 IPCC guidelines for national greenhouse gas inventories: wetlands. In: Hiraishi T, Krug T, Tanabe K, et al (eds) IPCC. Switzerland
Ise T, Dunn AL, Wosfy SC, Moorcroft PR (2008) High sensitivity of peat decomposition to climate change through water-table feedback. Nat Geosci 1:763–766
Jackson RB, Lajtha K, Crow SE et al (2017) The ecology of soil carbon: pools, vulnerabili-ties, and biotic and abiotic controls. Annu Rev Ecol Evol Syst 48:419–445
Joosten H (2010) The global peatland CO_2 picture. Wetlands International. https://www.wetlands.org/publications/the-global-peatland-co2-picture/
Joosten H, Gaudig G, Krawczynski R, et al. (2015) Managing soil carbon in Europe: paludicultures as a new perspective for peatlands. In: Banwart SA, Noellemeyer E, Milne E (eds) Soil carbon: science, management and policy for multiple benefits. CAB International, pp 297–306
Jungkunst HF, Krüger JP, Heitkamp F et al (2012) Accounting more precisely for peat and other soil carbon resources. In: Lal R, Lorenz K, Hüttl RF, Schneider BU, von Braun J (eds) Recarbonization of the biosphere—ecosystems and the global carbon cycle. Springer, Dordrecht, The Netherlands, pp 127–157
Junk W, An S, Finlayson C et al (2013) Current state of knowledge regarding the world's wetlands and their future under global climate change: a synthesis. Aquat Sci 75:151–167
Junk W, Brown M, Campbell C et al (2006) The comparative biodiversity of seven globally important wetlands: a synthesis. Aquat Sci 68:400–414
Keller JK, Medvedeff CA (2016) Soil organic matter. In: Vepraskas MJ, Craft CB (eds) Wetland soils: genesis, hydrology, landscapes, and classification. CRC Press, Boca Raton, FL, pp 165–188
Kirwan ML, Mudd SM (2012) Response of salt-marsh carbon accumulation to climate change. Nature 489:550–553
Knox SH, Sturtevant C, Matthes JH et al (2015) Agricultural peatland restoration: effects of land-use change on greenhouse gas (CO_2 and CH_4) fluxes in the Sacramento-San Joaquin Delta. Global Change Biol 21:750–765
Köchy M, Hiederer R, Freibauer A (2015) Global distribution of soil organic carbon—Part 1: masses and frequency distributions of SOC stocks for the tropics, permafrost regions, wetlands, and the world. SOIL 1:351–365
Lee SC, Christen A, Black AT et al (2017) Annual greenhouse gas budget for a bog ecosystem undergoing restoration by rewetting. Biogeosciences 14:2799–2814
Limpens J, Berendse F, Blodau C et al (2008) Peatlands and the carbon cycle: from local processes to global implications—a synthesis. Biogeosciences 5:1475–1491
Loisel J, van Bellen S, Pelletier L et al (2017) Insights and issues with estimating northern peatland carbon stocks and fluxes since the last glacial maximum. Earth Sci Rev 165:59–80
McNorton J, Gloor E, Wilson C, et al. (2016) Role of regional wetland emissions in atmospheric methane variability. Geophys Res Lett 43:11,433–11,444
Melton J, Wania R, Hodson E et al (2013) Present state of global wetland extent and wetland methane modelling: conclusions from a model inter-comparison project. (WETCHIMP). Biogeosciences 10:753–782
Meng L, Roulet N, Zhuang Q, Christensen TR, Frolking S (2016) Focus on the impact of climate change on wetland ecosystems and carbon dynamics. Environ Res Lett 11:100201
Mitsch WJ, Gosselink JG (2015) Wetlands, 5th edn. Wiley, Hoboken, NJ
Mitsch WJ, Gosselink JG, Anderson CJ, Zhang L (2009) Wetland Ecosystems. Wiley, Hoboken, NJ
Myers-Smith IH, Harden JW, Wilmking M et al (2008) Wetland succession in a permafrost collapse: interactions between fire and thermokarst. Biogeosciences 5:1273–1286
Moore S, Evans CD, Page SE et al (2013) Deep instability of deforested tropical peatlands revealed by fluvial organic carbon fluxes. Nature 493:660–663
Mueller CW, Hoeschen C, Steffens M et al (2017) Microscale soil structures foster organic matter stabilization in permafrost soils. Geoderma 293:44–53

Osvald H (1937) Myrar och myrodling. Kooperativa förbundets bokförlag, Stockholm (In Swedish)

Oztas T, Fayetorbay F (2003) Effect of freezing and thawing processes on soil aggregate stability. CATENA 52:1–8

Page SE, Rieley JO, Banks C (2011) Global and regional importance of the tropical peatland carbon pool. Glob Change Biol 17:798–818

Page SE, Siegert F, Rieley JO et al (2002) The amount of carbon released from peat and forest fires in Indonesia during 1997. Nature 420:61–65

Page SE, Wüst RAJ, Weiss D et al (2004) A record of Late Pleistocene and Holocene carbon accumulation and climate change from an equatorial peat bog (Kalimantan, Indonesia): implications for past, present and future carbon dynamics. J Quaternary Sci 19:625–635

Petrescu AMR, Lohila A, Tuovinen JP et al (2015) The uncertain climate footprint of wetlands under human pressure. Proc Natl Acad Sci USA 112:4594–4599

Ramsar (2013) The Ramsar convention manual: a guide to the convention on wetlands, 6th ed. Retrieved from http://www.ramsar.org/sites/default/files/documents/library/manual6-2013-e.pdf

Ramsar Convention on Wetlands; FAO; International Water Management Institute (IWMI) (2014) Wetlands & agriculture: Partners for growth. Ramsar Convention on Wetlands; Gland, Switzerland; FAO, Rome, Italy; International Water Management Institute (IWMI), Colombo, Sri Lanka. http://www.ramsar.org/sites/default/files/wwd14_leaflet_en.pdf

Robinson DA, Panagos P, Borrelli P et al (2017) Soil natural capital in Europe; a framework for state and change assessment. Sci Rep. https://doi.org/10.1038/s41598-017-06819-3

Robroek BJM, Jassey VEJ, Payne RJ et al (2017) Taxonomic and functional turnover are decoupled in European peat bogs. Nat Commun 8:1161. https://doi.org/10.1038/s41467-017-01350-5

Rocha A, Goulden M (2009) Why is marsh productivity so high? New insights from eddy covariance and biomass measurements in a Typha marsh. Agric For Meteorol 49:159–168

Roulet NT, Lafleur PM, Richard PJH et al (2007) Contemporary carbon balance and late Holocene carbon accumulation in a northern peatland. Glob Change Biol 13:397–411

Rydin H, Jeglum J (2006) The Biology of Peatlands. Oxford University Press, Oxford

Schlesinger WH, Bernhardt ES (2013) Wetland ecosystems. Schlesinger WH, Bernhardt ES, Biogeochemistry: an analysis of global change. Academic Press, Amsterdam, Netherlands, pp 233–273

Schrier-Uijl AP, Kroon PS, Hendriks DMD et al (2014) Agricultural peatlands: towards a greenhouse gas sink—a synthesis of a Dutch landscape study. Biogeosciences 11:4559–4576. https://doi.org/10.5194/bg-11-4559-2014

Settele J, Scholes R, Betts R, et al (2014) In Field CB, Barros VR, Dokken DJ, et al (eds) Climate change 2014: impacts, adaptation, and vulnerability. Part A: global and sectoral aspects. Contribution of Working Group II to the Fifth Assessment Report of the Intergovernmental Panel on Climate Change. Cambridge University Press, Cambridge, United Kingdom and New York, NY, USA, pp 271–359

Sjögersten S, Black CR, Evers S et al (2015) Tropical wetlands: a missing link in the global carbon cycle? Global Biogeochem Cycles 28:1371–1386

Smith P, Bustamante M, Ahammad H, et al (2014) Agriculture, forestry and other land use (AFOLU). In: Edenhofer OR, Pichs-Madruga Y, Sokona E, et al (eds) Climate change 2014: mitigation of climate change. Contribution of working group III to the fifth assessment report of the intergovernmental panel on climate change. Cambridge University Press, Cambridge, United Kingdom and New York, NY, USA

Stocker BD, Yu Z, Massa C, Joos F (2017) Holocene peatland and ice-core data constraints on the timing and magnitude of CO_2 emissions from past land use. Proc Natl Acad Sci USA 114:1492–1497

Strack M (ed) (2008) Peatlands and climate change. International Peat Society, Jyväskylä, Finland

Tanneberger F, Tegetmeyer C, Busse S et al (2017) The peatland map of Europe. Mires Peat 19:1–17. https://doi.org/10.19189/MaP.2016.OMB.264

Tchebakova NM, Parfenova EI, Lysanova GI, Soja AJ (2011) Agroclimatic potential across central Siberia in an altered twenty-first century. Environ Res Lett 6:045207. https://doi.org/10.1088/1748-9326/6/4/045207

Troell M, Naylor RL, Metian M et al (2014) Does aquaculture add resilience to the global food system? Proc Natl Acad Sci USA 111:13257–13263

Tubiello FN, Biancalani R, Salvatore M, Rossi S, Conchedda G (2016) A worldwide assessment of greenhouse gas emissions from drained organic soils. Sustainability 8:371. https://doi.org/10.3390/su8040371

Turetsky MR, Kotowska A, Bubier J et al (2014) A synthesis of methane emissions from 71 northern, temperate, and subtropical wetlands. Glob Change Biol 20:2183–2197

van der Gon HG (1999) Changes in CH_4 emission from rice fields from 1960 to 1990s. 2. The declining use of organic inputs in rice farming. Glob Biogeochem Cycles 13:1053–1062

Verhoeven JTA, Setter TL (2010) Agricultural use of wetlands: opportunities and limitations. Ann Bot 105:155–163

Walz J, Knoblauch C, Böhme L, Pfeiffer EM (2017) Regulation of soil organic matter decomposition in permafrost affected Siberian tundra soils—impact of oxygen availability, freezing and thawing, temperature, and labile organic matter. Soil Biol Biochem 110:34–43

Wang H, Richardson CJ, Ho M (2015) Dual controls on carbon loss during drought in peatlands. Nat Clim Change 5:584–588

Warren M, Hergoualc'h K, Kauffman JB, Murdiyarso D, Kolka R (2017) An appraisal of Indonesia's immense peat carbon stock using national peatland maps: uncertainties and potential losses from conversion. Carbon Balance Manage 12:12. https://doi.org/10.1186/s13021-017-0080-2

Wijedasa LS, Jauhiainen J, Kononen M et al (2017) Denial of long-term issues with agriculture on tropical peatlands will have devastating consequences. Glob Change Biol 3:977–982. https://doi.org/10.1111/gcb.13516

Wilson D, Farrell CA, Fallon D et al (2016) Multiyear greenhouse gas balances at a rewetted temperate peatland. Glob Change Biol 22:4080–4095. https://doi.org/10.1111/gcb.13325

Wu X, Cao R, Wei X et al (2017) Soil drainage facilitates earthworm invasion and subsequent carbon loss from peatland soil. J Appl Ecol 54:1291–1300. https://doi.org/10.1111/1365-2664.12894

Xu J, Morris PJ, Liu J, Holden J (2018) PEATMAP: Refining estimates of global peatland distribution based on a meta-analysis. CATENA 160:134–140

Yang G, Peng C, Chen H et al (2017) Qinghai-Tibetan Plateau peatland sustainable utilization under anthropogenic disturbances and climate change. Ecosyst Health Sustain 3:e01263

Yu Z (2011) Holocene carbon flux histories of the world's peatlands: global carbon-cycle implications. Holocene 21:761–774

Yu ZC (2012) Northern peatland carbon stocks and dynamics: a review. Biogeosciences 9:4071–4085

Zhang Z, Zimmermann NE, Stenke A et al (2017) Emerging role of wetland methane emissions in driving 21st century climate change. Proc Natl Acad Sci USA 114:9647–9652

Zhuang Q, Melillo J, Kicklighter D, et al (2004) Methane fluxes between terrestrial ecosystems and the atmosphere at northern high latitudes during the past century: a retrospective analysis with a process-based biogeochemistry model. Global Biogeochem Cy 18, GB3010. https://doi.org/10.1029/2004gb002239

Zona D, Lipson DA, Paw KT, et al (2012) Increased CO_2 loss from vegetated drained lake tundra ecosystems due to flooding. Global Biogeochem Cycles 26:GB2004. https://doi.org/10.1029/2011gb004037

Chapter 6
Agroforestry Systems

Contents

Abstract Agroforestry implies the inclusion of trees or other woody perennials within farming systems to capture the interactive benefits of perennials and seasonals, and/or animals for sustainable agricultural production. Among the benefits is the greater ability of agroforestry systems to capture and utilize growth resources (i.e., light, nutrients, water) compared to single-species systems. Agroforestry systems are estimated to cover about 10 million km^2 of agricultural land globally. Estimates for carbon (C) sequestration potential above- and belowground over 50 years range between 1.1 and 2.2 Pg (1 Pg = 1Gt = 10^{15}g) C yr^{-1}, but these numbers are highly uncertain. Land-use practices for agroforestry systems are very diverse. In the tropics, this includes alley cropping, homegardens, improved fallows, multi-purpose trees on farms and rangelands, silvopastoral grazing systems, shaded perennial-crop systems, shelterbelts, windbreaks, and taungya (i.e., growing agricultural crops during early stages of establishment of forestry plantations). Furthermore, alley cropping, forest farming, riparian buffer strips, silvopasture, and windbreaks are agroforestry practices in temperate regions. Thus, agroforestry systems are structurally and functionally more complex than either croplands or pastures or tree monocultures. Also, the greater efficiency of growth resource capture and utilization enhances net carbon (C) sequestration in soils under agroforestry compared to those under crops and pastures. Trees capture large amounts of atmospheric carbon dioxide (CO_2) during photosynthesis, and transfer a fraction to the soil as surface and subsurface input which may eventually be sequestered. However, data on soil organic carbon (SOC) stocks and sequestration of

K. Lorenz and R. Lal, *Carbon Sequestration in Agricultural Ecosystems*,
https://doi.org/10.1007/978-3-319-92318-5_6

agroforestry systems are scanty. The few published data vary greatly depending on the agroforestry system, species composition and age, geographical location, environmental factors, and management practices. For example, 1.25 and 302 Mg C ha^{-1} may be stored to 40 cm depth in a Canadian alley cropping system and to 100 cm depth in a cacao (*Theobroma cacao* L.) agroforestry system in Brazil, respectively. Furthermore, sequestration rates of up to 7.4 Mg C ha^{-1} yr^{-1} in the top 0–10 or 0–20 cm soil depth have been reported. Land-use conversion from less complex systems such as agricultural system to agroforestry systems may, particularly, increase SOC stocks. However, lack of standard methods and procedures does not allow any firm conclusions about the C sequestration potential of agroforestry soils. Furthermore, any generalizations about the SOC sequestration potential of agroforestry systems are hampered by interrelated and site-specific factors such as agroecological conditions and management practices. This chapter begins with a comparison of different tropical and temperate agroforestry practices. This is followed by a discussion about afforestation of denuded lands and associated changes in SOC stocks and its importance for the role of trees for SOC sequestration in agroforests. Then, an overview is given about SOC stocks and sequestration in existing agroforestry systems, and also for those after conversion from other land uses. This chapter concludes with a list of recommendations to enhance SOC sequestration in agroforests.

Keywords Agroforestry systems · Trees · Temperate regions · Tropical regions Afforestation · Soil organic carbon sequestration · Root-derived carbon

6.1 Agroforestry Practices

Agroforestry refers to the practice of purposeful growing of trees and crops, and/or animals, in interacting combinations, for a variety of benefits and services (Nair et al. 2008). Both agriculture and forestry are combined into an integrated system to achieve maximum benefits by a greater efficiency in resource (i.e., nutrients, light, and water) capture and utilization (Kohli et al. 2008). Agroforestry is recognized as an integrated approach for sustainable land use because of its benefits for production and the environment (Jose 2009; Nair et al. 2009). Furthermore, it is also recognized as a mitigation strategy for greenhouse gases (GHGs) under the Kyoto Protocol as a greater net C sequestration is achieved compared to single-species systems. However, comparing agroforestry and adjacent agricultural lands, Kim et al. (2016) found only minor differences in net methane (CH_4) and nitrous oxide (N_2O) emissions, with no clear overall direction of change. Overall, agroforestry was estimated to contribute to mitigating 27 Mg CO_2 eq. ha^{-1} at least for the first 14 years after establishment (Kim et al. 2016). Climate change mitigation has been proposed as a low-hanging fruit of agroforestry (Nair 2012). However, enabling policies and rigorous long-term research are needed.

Globally, ~10 million km^2 of land may be under agroforestry practices (Nair et al. 2009). Agroforestry systems may be grouped into agrosilvicultural (crops + trees), silvopastoral (pasture/animals + trees), and agrosilvopastoral (crops + pasture/animals + trees) (Nair 1987). Numerous and diverse agroforestry systems are practiced in the tropics because of favorable climatic conditions and various socioeconomic factors (Fig. 6.1; Nair et al. 2008).

Tropical agroforestry practices, for example, include alley cropping (hedgerow cropping; Fig. 6.2) in which fast-growing, preferably leguminous, woody species are grown in crop fields (Kang et al. 1985; Lal 1989; Nair et al. 2009). The woody species are periodically pruned and the prunings either applied as mulch into the alleys or used as animal fodder. Homegardens are another practice which combines a large number of various trees and crops, and sometimes livestock in homesteads (Kumar and Nair 2004). In improved fallow systems, fast-growing, preferably leguminous, woody species are left to grow for site improvement during fallow phases between cropping years (Okigbo 1985; Vagen et al. 2004). Another tropical agroforestry practice is multi-purpose trees on farms and rangelands (Kumar and Nair 2004). Fruit trees among others are grown and provide fruits, fuelwood, fodder, timber, and other goods.

Silvopastural systems include grazing systems (Fig. 6.3), and cut and carry systems (protein banks) (Kumar and Nair 2004). Grazing systems allow cattle to graze under trees, whereas animals are fed in stables with fodder from trees grown on the farm in a cut and carry system.

Shade-tolerant species such as cacao (*Theobroma cacao* L.) and coffee (*Coffea arabica*, *Coffea canephora* and other species within the genus *Coffea*) can be grown under shade of overstory trees in a practice called shaded perennial-crop systems (Fig. 6.4; Muschler and Bonnemann 1997). For shelterbelts and windbreaks, trees are used to protect from wind damage, sea encroachment, floods, and other deleterious agents (Long and Nair 1999). Taungya refers to a tropical agroforestry practice where agricultural crops are grown during early stages of establishment of forestry plantations (Nair 1991). Numerous and diverse agroforestry systems can also be distinguished in temperate regions (Nair et al. 2009). For example, alley cropping in temperate agroforestry refers to a practice where trees are planted between the rows of agricultural or horticultural crops (Jose et al. 2004). Another practice is forest farming which utilizes forested areas for producing specialty crops (Long and Nair 1999).

In riparian buffer strips, perennial vegetation such as trees, shrubs, or grasses is planted between cropland or pasture and streams, lakes, wetlands, ponds, or other waterbodies (Fig. 6.5). Silvopasture and windbreaks are temperate agroforestry practices managed similar to tropical practices (Long and Nair 1999). Old temperate agroforestry practices are tree-based agriculture involving a large number of multi-purpose trees such as chestnuts (*Castanea* spp.), oaks (*Quercus* spp.), carob (*Ceratona siliqua* L.), olive (*Olea europa* L.), and figs (*Ficus* spp.) (Rigueiro-Rodriguez et al. 2008).

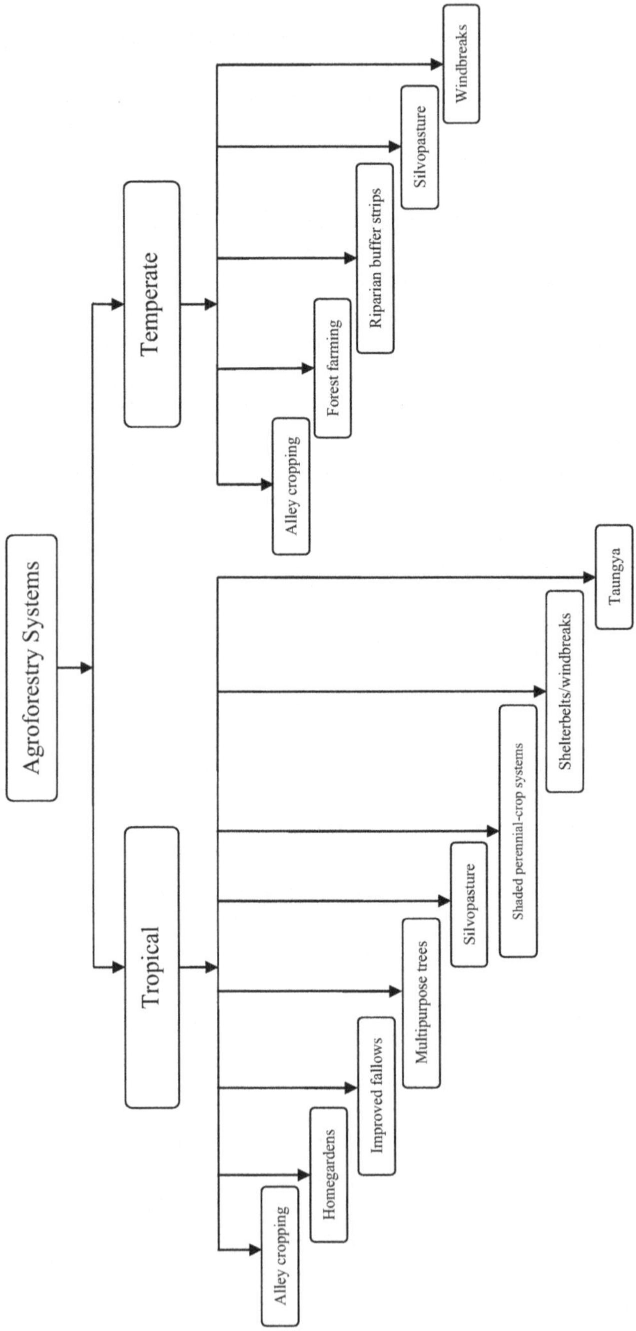

Fig. 6.1 Major agroforestry systems in tropical and temperate regions (modified from Nair et al. 2008)

Fig. 6.2 Alley cropping soybean (*Glycine max* (L.) Merr.) between walnut trees (*Juglans regia* L.) in Missouri, USA (Wikimedia Commons)

Fig. 6.3 Dehesa Boyal, en Bollullos Par del Condado (Huelva, Spain) (licensed under the Creative Commons Attribution-Share Alike 2.5 Generic license)

6.2 Importance of Agroforestry Trees for Soil Organic Carbon Sequestration

Terrestrial C sequestration efforts have largely focused on adaptive management of existing forests and conservation tillage of croplands (Perry et al. 2009). However, tree-based farm practices are a viable C sequestering option. The tree C stocks are profoundly lower than the soil organic carbon (SOC) stocks. For example, Zomer et al. (2017) estimated that tree C stocks on agricultural land are 36 Pg (1 Pg = 1 Gt = 10^{15}g) C while up to 715 Pg SOC may be stored to 1 m depth in agricultural biomes (Stockmann et al. 2013). Agroforestry systems have a higher potential to sequester atmospheric carbon dioxide (CO_2) as SOC than the croplands, pastures, or natural grasslands they replace (Nair et al. 2009). Among agricultural practices, the introduction of agroforestry has the highest technical potential to increase C sequestration in EU-27 estimated at 1.4 Pg CO_2 eq. y^{-1} (Aertsens et al. 2013). The agroforest SOC stock contributes relatively more to C sequestration for climate change mitigation than the biomass C stock as some of the agroforestry C has to be sequestered for millennia (Mackey et al. 2013). The incorporation of trees has many benefits, including improvement in soil properties, greater net C sequestration but also corollary benefits beyond C sequestration and tree-related income (Young

Fig. 6.4 Coffee (*Coffea* spp.) grown in association with *Canavalia* Adans. spp. cover crop under trees in San Jeronimo Community, Jesús de Otoro municipality, Intibucá, Honduras (photograph credit Maren Barbee; Creative Coomons Attribution 2.0 Generic (CC BY 2.0))

Fig. 6.5 Riparian buffer at Bear Creek, Story county, Iowa, USA (public domain)

1997; Jandl et al. 2007; Soto-Pinto et al. 2010; Pardon et al. 2017; Zomer et al. 2017). Negative effects of tree-crop associations from interspecific competition have also been reported but species interactions in intercropping systems are less well studied (Zhang et al. 2013).

Among the reasons for enhanced C sequestration is that trees have extensive root systems which can grow deep into the mineral soil and, thus, root-derived C inputs may be the major source for the SOC stock in deeper soil horizons (Canadell et al. 1996; Lorenz and Lal 2005; Rasse et al. 2005). Aside higher soil C inputs, the priming effect from tree roots may drive deep SOC dynamics in agroforests, and possibly limit SOC storage (Cardinael et al. 2018). Root-derived C is four times more likely to be stabilized as SOC by association with aggregates or clay minerals than an equivalent mass of aboveground litter C (Jackson et al. 2017; Prescott 2010). For example, the relative root contribution of European beech (*Fagus sylvatica* L.) to SOC was 1.55 times that of shoot (Scheu and Schauermann 1994). Similarly, in croplands, total root-derived C contributes 1.5 times to >3 times more C to SOC than shoot-derived C (summarized in Johnson et al. 2006). Thus, agroforests store more SOC in deeper soil layers near trees than away from trees (Nair et al. 2010). However, studies on tree root systems face many challenges. For example, roots break easily during the excavation process or the soil is not sufficiently excavated to expose the entire root system (Sinacore et al. 2017). Fine roots are often excluded from the calculations of total root biomass due to their fragility. Tree species within the same guild and grown in similar conditions were associated with a high diversity of rooting patterns and architectures. Thus, a better understanding of root architecture is necessary to understanding root function in terms of resource acquisition, and ecosystem function (Sinacore et al. 2017).

Quantitative information about belowground C inputs in agroforests is scanty (Albrecht et al. 2004; Mendez-Millan et al. 2010). Another reason for the promotion of SOC sequestration in agroforests is that tree roots have the potential to recover nutrients from below the crop rooting zone and enhance tree and crop plant growth by subsequent increase in N nutrition may result in increase of SOC sequestration (van Noordwijk et al. 1996; Lorenz and Lal 2010). Similarly, mixed plantings with N-fixing trees may cause higher biomass production and, thus, SOC sequestration and pools particularly in deeper soil horizons as N promotes humification rather than decay (Haggar et al. 1993; Nair et al. 2009; Prescott 2010). Also, changes in microbial decomposer community composition under N-fixing trees may result in greater retention of relatively stable SOC (Resh et al. 2002). N-fixing trees in mixtures with non-N-fixing trees may develop deeper root profiles due to niche-partitioning (da Silva et al. 2009). Mixed tree plantings in agroforestry systems may generally enhance SOC sequestration as increases in tree species diversity increases fine-root productivity (Meinen et al. 2009). Further, higher species richness and tree density result in higher SOC stocks in agroforests (Nair et al. 2010). Particularly preferred in agroforests are trees which are leafless during the active growth phase of understory crops to retain the positive effects of tree species while limiting the negative effects of competition with crops. Specifically, trees with reverse phenology such as the African tree *Faidherbia albida* (Delile) A. Chev which sheds the leaves during the rainy season and retains it during the dry months have recently received increased attention (Fig. 6.6; Garrity et al. 2010).

Among the reasons for the positive effects of trees on SOC sequestration are that trees modify the quality and quantity of belowground litter C inputs, and modify

Fig. 6.6 Agroforestry system in Burkina Faso, West Africa: sorghum (*Sorghum bicolor* (L.) Moench) grown under Faidherbia albida and *Borassus akeassii* (photograph credit Marco Schmidt; licensed under the Creative Commons Attribution-Share Alike 2.5 Generic license)

microclimatic conditions such as soil moisture and temperature regimes (Laganiere et al. 2010). Root litter usually decomposes more slowly than leaf litter of the same species (Cusack et al. 2009). Further, hydraulic lift of soil water by roots of a single tree may enhance soil water uptake by neighboring trees and other plants in the agroforestry system which may lead to increased SOC sequestration (Kizito et al. 2006; Liste and White 2008). Also, trees have a higher potential for SOC sequestration than crop and pasture plant species as trees may be associated with higher proportions of stabilized SOC in deeper mineral soil horizons (Jobbágy and Jackson 2000). Trees contribute to more C in the silt- + clay-sized (<53 μm diameter) or relatively stable fractions in deeper soil profiles than other agroforest species (Nair et al. 2010). Further, in surface soil horizons of intensively managed agricultural landscapes trees potentially reduce SOC losses by reducing soil erosion (Lal 2005). Other factors that contribute to the enrichment of SOC under trees in agroforests are less soil disturbance due to reduction or cessation of tillage operations which reduces the decomposition and mineralization of soil organic matter (SOM). Furthermore, changes in soil microbial communities and activities, and biodiversity under trees may enhance SOC sequestration. For example, addition of a single tree species to moorland resulted in changes in belowground soil microbial communities and in nutrient cycling (Mitchell et al. 2010). However, field studies

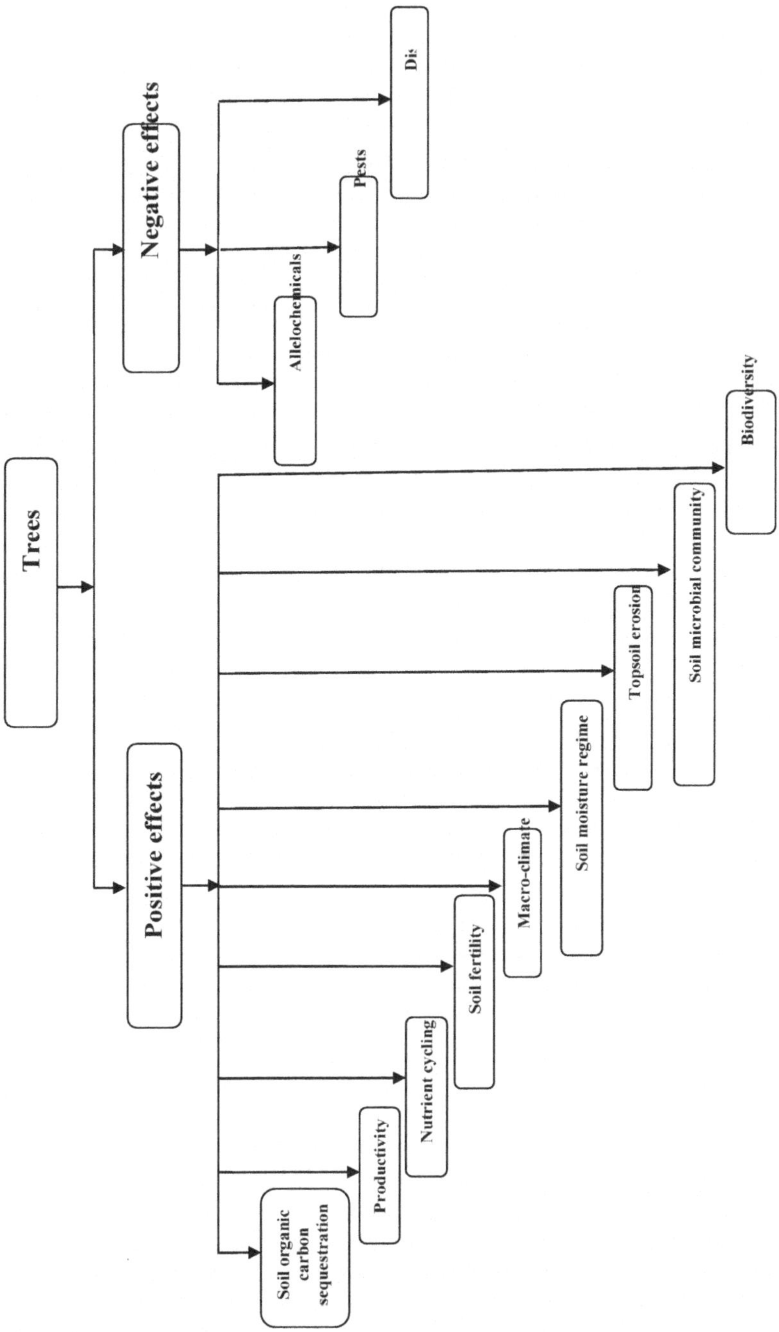

Fig. 6.7 Schematic summarizing potential positive and negative effects of including trees in agroecosystems

on the mechanisms and processes associated with SOC dynamics and storage in tree-based systems such as agroforests are scanty (Nair et al. 2010).

The integration of trees into agricultural production systems may create positive interactions such as enhanced productivity, cycling of nutrients, soil fertility, and macroclimate (Fig. 6.7; Nair et al. 2010). For example, nitrogen (N) transfer from dinitrogen (N_2)-fixing trees to perennial crops within low-input tropical agroforestry systems contributes to soil N replacement (Munroe and Isaac 2014). The increased soil N availability may improve soil fertility and crop yields. However, increasing N availability may also catalyze the release of N into the surrounding environment, in particular in form of the potent GHG N_2O (Rosenstock et al. 2014). Other negative tree–agroecosystem interactions have also been reported such as the creation of allelopathic, pest, and disease vectors.

Allelochemicals are present in many types of plants and are released into the soil by a variety of mechanisms, including leaching from tree canopy, decomposition of residues, volatilization, and root exudation (Jose et al. 2004). Mulching with plant residues, in particular, may result in the liberation of allelochemicals into the soil (John et al. 2006). Allelochemicals affect germination, growth, development, distribution, and reproduction of a number of plant species (Inderjit and Malik 2002). Most of the tropical agroforestry species compared by Rizvi et al. (1999) have negative allelopathic effects on food and fodder crops. For example, litter extracts of *Azadirachta indica* (commonly known as neem) inhibit seed germination of a range of crop species and root growth. Further, eucalypts release volatile and non-volatile allelochemicals that affect growth of the associated vegetation. *Leucaena leucocephala* (Lam.) de Wit improves the overall productivity of land but a range of crop species is negatively sensitive to its allelochemicals. Allelochemicals may also contribute to pest management as trees live long and produce a large amount of leaves and litter. Thus, species mixtures with no or positive allelopathic effects on the companion crops must be created in agroforestry systems (Rizvi et al. 1999). In comparison, less well studied are allelopathic effects of temperate agroforestry species (Jose et al. 2004). The most widely used examples are those of black walnut (*Juglans nigra* L.) and pecan (*Carya illinoinensis* [Wangenh.] K. Koch). Both species produce the phenolic compound juglone (5-hydroxy-1,4-naphthoquinone) with inhibitory effects on survival and growth of some crop species (Jose et al. 2004). However, allelopathic investigations in agroforestry systems are often lacking conclusive field verification. Specifically, separating allelopathic effects of trees from root competition is challenging (John et al. 2006).

Studies about afforestation (i.e., the introduction of trees on previously treeless cropland, pasture, or natural grassland) may provide some insight on the potential effects of trees on SOC sequestration in agroforestry systems. Observations about effects of afforestation on SOC have been synthesized and reviewed by Post and Kwon (2000), Guo and Gifford (2002), and Paul et al. (2002). However, Laganiere et al. (2010) showed that conclusions based on these observations are limited by inappropriate experimental design, sampling methods, and/or soil analysis techniques. Thus, for a more accurate evaluation of tree effects on SOC sequestration studies designed specifically to test afforestation must be compared. Probably, the most powerful and least biased study design is the retrospective design, i.e., re-sampling the same soils over a given period of time (repeated measurements) (Laganiere et al. 2010). In contrast, paired sites (afforested *vs.* treeless control) are uncertain with respect to uniformity of certain variables assumed to be fixed among sites and uncertain as sampling constitutes a single measurement in time. Extending the paired sites design to a chronosequence partially resolves this problem (Laganiere et al. 2010). However, the validity of chronosequence studies has been questioned because the basic assumptions that the sequence only differs in age and that each chronosequence has the same biotic and abiotic history are seldom tested (Johnson and Miyanishi 2008). Thus, paired site design probably overestimates the change in SOC stocks relative to the retrospective design (Laganiere et al. 2010).

Laganiere et al. (2010) compared afforestation effects on SOC stocks by meta-analysis of observations from studies designed specifically to test afforestation. For first-rotation plantations, the mean age was 23.3 years and mean sampling depth was 34.2 cm. Afforestation resulted in an increase in SOC stocks by 26% for croplands, 3% for pastures, and <10% for natural grasslands, but changes for pastures and natural grasslands were not significantly different from zero. Soil sampling was probably done too early as it may take much longer until a new SOC equilibrium in the soil profile is reached after plantation establishment. It takes >100 years after plantation establishment to create a significant increase in SOC stock in the boreal zone (Ritter 2007). Thus, the relatively young age of the plantations is probably the reason why SOC losses of 1.5% were observed for afforestation projects in the boreal zone, whereas SOC gains between 7 and 17% were observed in other climate zones, with the temperate maritime zone accruing the highest gains. Otherwise, tree root systems are generally deeper than the sampling depths for the studies compared by Laganiere et al. (2010). Thus, whole profile studies after long periods of time are needed to accurately determine SOC stock changes following afforestation of croplands, pastures, and natural grasslands.

Some effects of tree species on SOC stocks are also reported by Laganiere et al. (2010). Specifically, SOC stocks increased by 12% when *Eucalyptus* spp. and *Pinus* spp. were planted, but more than 25% when afforestation was done with other broadleaf trees. However, effects of planting conifer trees other than *Pinus* spp. on SOC stocks are negligible. The planting of N-fixing trees for afforestation can increase the SOC stocks as is indicated by the >30% increase in SOC stocks when N-fixers are present in forest stands (Johnson and Curtis 2001; Resh et al. 2002). However, conclusions based on the meta-analysis by Johnson and Curtis (2001) are also hampered as mean sampling depth was not specified and some data assumptions were made. In the long-term, tree productivity in plantations and, thus, C inputs to soils may be reduced as trees take up considerable amounts of nutrients from the soil which may be partially removed by repeatedly harvesting tree biomass (Berthrong et al. 2009). Otherwise, little evidence is available for the effect of tree species in forest stands on the SOC stock other than that observed in the forest floor C (Jandl et al. 2007).

In summary, major factors contributing to restoring SOC stocks after afforestation on agricultural soils are previous land use, tree species, soil clay content, pre-planting disturbance and, to a lesser extent, climate zone (Laganiere et al. 2010). In particular, the positive impact of afforestation on SOC stocks is more pronounced in croplands relative to pastures or natural grasslands. Broadleaf tree species have a greater capacity to enhance SOC stocks, most probably due to their higher root biomass-to-aboveground biomass ratio than conifer trees. Also, clay-rich soils (>33%) have a greater capacity to enhance SOC than clay-poor soils (<33%). Yet, it may take several decades after afforestation until effects on SOC stocks can be observed in deeper soil horizons (Laganiere et al. 2010). Recommendations for agroforestry systems are that soil disturbance must be minimized during tree establishment and tree species with a high root biomass-to-aboveground biomass ratio and/or N-fixing trees should be planted when SOC sequestration is among the objectives for establishing the agroforest. Furthermore, sustainable practices for harvesting tree biomass must be used to maintain long-term soil fertility and productivity.

6.3 Carbon Sequestration in Agroforestry Systems

Carbon sequestration in agroecosystems and agroforestry systems occurs in aboveground biomass (i.e., stem, branch, and foliage), in belowground biomass (i.e., roots), and in soil. The potential of agroforestry systems to accumulate C may range between 12 and 228 Mg C ha^{-1}, with a median of 95 Mg C ha^{-1} (Albrecht and Kandji 2003). Further, estimates for the C sequestration potential of agroforests over a 50-year period ranged between 1.1 and 2.2 Pg C yr^{-1} but, in particular, estimates of land area are highly uncertain (Dixon 1995; Nair et al. 2009). Also, the above- and belowground vegetation C sequestration potential is highly variable (Nair et al. 2009). Aboveground sequestration rates, for example, ranged from 0.29 Mg C ha^{-1} yr^{-1} for a 7.5-year-old fodder bank in the West African Sahel to 15.21 Mg C ha^{-1} yr^{-1} for a 4-year-old mixed species stand in Puerto Rico (Parrotta 1999; Takimoto et al. 2008). Age in these studies, however, was not clearly defined and supposedly the number of years since the establishment of trees (Nair et al. 2009). Ajit et al. (2017) estimated that over 30-year-agroforestry systems under various districts in India may sequester 0.05–1.03 Mg C ha^{-1} yr^{-1} (with average value of 0.21 Mg C ha^{-1} yr^{-1}). In comparison, the estimated rate of soil C sequestration at district level may range from 0.003 to 0.51 Mg C ha^{-1} yr^{-1} over the same period. To which soil depth was not specified. The C sequestration potential of agroforestry systems may offset one-third of the total GHG emissions from India's agricultural sector (Ajit et al. 2017).

In general, agroforests on fertile humid sites have higher vegetation C sequestration rates than those on arid, semiarid and degraded sites, and tropical agroforests have higher vegetation C sequestration rates than temperate agroforestry systems. Climate change threatens C sequestration in agroforestry worldwide through increased heat, drought, surface evaporation, and associated soil drying (Borland et al. 2015). Among adaptation strategies discussed is the introduction of bioengineered crassulacean acid metabolism into short-rotation forestry bioenergy trees as a potentially viable path to sustaining agroforestry production systems. Agroforestry systems improve, in particular, the resilience of smallholder farmers through more efficient water utilization, improved microclimate, enhanced soil productivity and nutrient cycling, control of pests and diseases, improved farm productivity, and diversified and increased farm income while at the same time sequestering C (Lasco et al. 2014). However, robust models on climate change impacts for agroforestry systems are not available (Luedeling et al. 2014).

Higher SOC stocks in agroforests can be achieved by increasing the amount of biomass C returned to the soil and by strengthening SOM stabilization and/or by decreasing the rate of biomass decomposition and SOM destabilization (Lal 2005; Sollins et al. 2007). However, insufficient information is available on effects of agroforestry practices on the chemical composition of SOM. Dhillon et al. (2017) reported varying effects of tree species on SOM in shelterbelts. Some species caused an increase in labile C forms, such as carbohydrates, while others resulted in higher abundance of recalcitrant aliphatic C forms. The SOM composition is an important factor in regulating SOC dynamics, especially during the initial phase of decomposition, and in surface soil layers (Dhillon et al. 2017).

Compared to monocultures, agroforestry systems are more efficient in capturing the resources available at the site for biomass growth and the increased biomass may result in higher C inputs to the soil. Higher inputs of C and energy positively influence the soil microbial community, and their functions including SOM decomposition, N fixation, uptake of phosphorus (P) by mycorrhiza, and the promotion of plant growth (Araujo et al. 2012). Direct C inputs to the soil can potentially be increased by some of the agroforestry practices described in Sect. 6.1. These include: (i) returning prunings of woody species to the soil as mulch and allowing abundant tree litter to decompose on site, (ii) allowing livestock to graze and add dung to the soil, (iii) allowing woody species to grow and add surface and belowground litter during crop fallow phases, (iv) integrating trees and their litter input in animal production systems, (v) allowing litter inputs to the soil from shade-tolerant species growing under trees, and (vi) benefiting from the soil C inputs of agricultural crops grown during early stages of the establishment of forestry plantations. Whether mechanisms of SOM stabilization and destabilization can potentially be affected by agroforestry practices is less well known. For example, Baah-Acheamfour et al. (2014) found most SOC to 10 cm depth in hedgerows, shelterbelts, and silvopastures in Alberta, Canada, in the fine fraction which is more stable than that in the coarse fraction.

Agroforestry practices that promote the depth transfer of SOM may result in higher profile SOC stocks as decomposition is slower in deeper soil layers and the proportion of stabilized SOC is higher (Lorenz and Lal 2005). Otherwise, the rate of biomass decomposition can potentially be directly reduced by manipulating litter chemical and physical properties through selection of species mixtures in agroforests. For example, lower decomposition rates are observed when litter is more 'recalcitrant,' i.e., when it contains larger proportions of biopolymers of higher molecular weight and irregular structure that are less accessible to enzymes and also more hydrophobic (Preston et al. 2009a). Higher molecular weight structures include condensed tannins, cutin, lignin, or modified lignin (Preston et al. 2009b). However, lignin is generally not preserved with decomposition or lost only slightly more slowly than other components. Only wood decomposition by brown-rot fungi causes large relative increases in lignin concentrations (Preston et al. 2009b). Aside from the amount of 'recalcitrant' material, decomposability of soil C inputs may also be related to their C:N ratio and N content (Horwarth 2007).

Agroforesty systems can affect SOC stocks also by avoiding burning and conserving soil by minimizing soil disturbance and erosion control (Soto-Pinto et al. 2010). The inclusion of trees in perennial crops for alley cropping or hedgerow intercropping can serve as erosion control measures (Albrecht and Kandji 2003). Among the practices listed in Sect. 6.1, erosional SOC losses can directly be reduced by practices which never leave the soil un-vegetated such as improved fallows and Taungya. Shelterbets, windbreaks, and riparian buffer strips are other agroforestry practices with a potential to reduce SOC losses caused by wind and water erosion. Occlusion of C in soil macroaggregates is proposed as a major mechanism of C protection in cacao agroforest soils in Brazil (Gama-Rodrigues et al. 2010). The low level of soil disturbance in these agroforests may, thus, promote SOC stabilization.

Similar to that of agroforest vegetation, SOC sequestration also varies widely depending on the agroforestry system but the number of published studies is small (Nair et al. 2009). For example, SOC stocks ranged from 1.25 Mg C ha^{-1} in the top 40-cm of a 13-year-old alley cropping system in Southern Canada to 173 Mg C ha^{-1} in the top 100 cm of 10–16-year-old silvopastoral systems at the Atlantic Coast of Costa Rica (Amézquita et al. 2005; Oelbermann et al. 2006). Very high SOC stocks of 302 Mg C ha^{-1} to 100 cm depth have been reported for 30-year-old cacao agroforestry systems in Brazil (Gama-Rodrigues et al. 2010). These high values have been explained by comparatively high amounts of above- and belowground plant litter inputs, and low removal of C with the harvested cacao beans and low SOC losses by erosion. Compared to other land-use practices with the exception of forests, agroforests have higher SOC contents and can be ranked in the order forests > agroforests > tree plantations > arable crops (Nair et al. 2009). However, agroforestry systems may not be superior to traditional systems in avoiding SOC loss during initial phases after converting forest for agricultural land use. For example, Fonte et al. (2010) reported similar SOC losses from 0 to 15 cm depth in a traditional slash-and-burn compared to a slash-and-mulch agroforestry system in Honduras. However, agroforestry systems offer a greater potential to improve soil fertility, and biological and human health (Pinho et al. 2012; Schroth et al. 2000; Vinceti et al. 2013).

Ideally, SOC sequestration in agroforestry systems should be reported as rates (i.e., mass SOC per units of area and time), similar to vegetation C sequestration rates but SOC sequestration data are often reported as stocks (Nair et al. 2009). Rates of SOC sequestration including soil depths for some agroforestry systems are presented in Table 6.1. Kim et al. (2016) reported also a high variability in net SOC sequestration rates depending on the type of agroforestry, with reported increments

Table 6.1 Soil organic carbon sequestration rates (Mg C ha^{-1} y^{-1}) in some agroforestry systems

Agroforestry system/species	Location	Age (years)	Soil depth (cm)	Sequestration rate (Mg C ha^{-1} y^{-1})	References
Alley cropping system: hybrid poplar (*Populus deltoides x nigra* DN-177) + wheat (*Triticum aestivum* L.), soybean (*Glycine max* L.), and maize (*Zea mays* L.) rotation	Southern Canada	13	0–20	0.30	Oelbermann et al. (2006)
			0–40	0.39	
Intercropping system: Norway spruce (*Picea abies* L.) + barley (*Hordeum vulgare* L. cv. OAC Kippen)	Southern Canada	13	0–20	0	Calculated from Peichl et al. (2006)
Intercropping system: hybrid poplar + barley	Southern Canada	13	0–20	1.04	Calculated from Peichl et al. (2006)
Alley cropping system: *Erythrina poeppigiana* (Walp.) O.F Cook + maize and bean (*Phaseolus vulgaris* L.)	Costa Rica	19	0–20	1.79	Oelbermann et al. (2006)
			0–40	2.34	Oelbermann et al. (2006)
Multi-strata agroforest: cacao (*Theobroma cacao* L.) + *Erythrina poeppigiana* (Walp.) O.F Cook	Costa Rica	10	0–45	4.16	Calculated from Beer et al. (1990)
Multi-strata agroforest: cacao + *Cordia alliodora* (Ruiz & Pav.) Oken	Costa Rica	10	0–45	1.55	Calculated from Beer et al. (1990)
Multi-strata agroforest: cacao + canopy trees	Ghana, West Africa	15	0–15	−0.39*	Calculated from Isaac et al. (2005)
		25	0–15	0.06	Calculated from Isaac et al. (2005)

*Indicates soil organic carbon loss

ranging from 1.0 to 7.4 Mg C ha^{-1} y^{-1} to 0–10 or 0–20 cm soil depth. The SOC sequestration rates were about 2 Mg C ha^{-1} y^{-1} in youngest stands that gradually diminished with time since stand establishment. Kim et al. (2016) primarily compared tree-crop coexistence types where trees and agricultural crops are grown together. More SOC sequestration to 30 cm depth occurs in agroforestry systems classified as silvopastoral (4.38 Mg C ha^{-1} yr^{-1}) compared to other systems (Feliciano et al. 2018). With sequestration rates of 2.23 Mg SOC C ha^{-1} yr^{-1}, SOC benefits are greater for agroforestry systems under tropical climates when compared to those located in other climates. In terms of land-use change, the greatest SOC sequestration to 30 cm depth (4.38 Mg C ha^{-1} yr^{-1}) results from the transition of a grassland system to a silvopastoral system (Feliciano et al. 2018).

Some SOC losses may occur initially after converting natural forests to agroforests (Beer et al. 1990). Even 13 years after establishing, no SOC sequestration was observed in 0 to 20 cm depth in an intercropping system in southern Canada (calculated from Peichl et al. 2006). However, a 10-year-old multi-strata agroforest in Costa Rica sequestered SOC at a mean rate of 4.16 Mg C ha^{-1} yr^{-1} in 0–45 cm depth (calculated from Beer et al. 1990). Although tropical agroforests may have higher SOC sequestration rates, temperate systems may be more effective in stabilization of the residue C inputs from tree prunings, litterfall, and crop residues in the soil (Oelbermann et al. 2006). SOC sequestration in agroforestry systems may be strengthened when the proportion of the stabilized SOC fraction in deeper soil horizons increases. This trend may be the result of major C inputs from the decomposition of dead tree roots, root exudates, and associated microorganisms (Lorenz and Lal 2005; Haile et al. 2008). For example, the introduction of slash pine (*Pinus elliottii* Englem) in bahiagrass (*Paspalum notatum* Flueggé) pasture results in SOC increases deeper in the soil profile to 125 cm depth and in increases in relatively stable SOC, i.e., C associated with silt + clay in deeper soil horizons (Haile et al. 2010). Similarly, the SOC content associated with silt + clay to 1 m depth follows a trend of increasing amount with increasing tree density in tropical homegardens in Kerala, India (Saha et al. 2010). In comparison, less attention has been given to how integrating trees in the agricultural landscape in the form of tree-based intercropping systems affect the growth of trees. For example, Rivest et al. (2009) observed that leafless aboveground biomass of hybrid poplars (*Populus* spp.) in intercropping treatments is up to 40% higher than for no intercrop management. This growth stimulation may result from the uptake of inorganic fertilizer nutrients supplied for the intercrop needs. However, competition between trees and crops for water may diminish this response (Burgess et al. 2004).

The changes in SOC stocks after land conversion to agroforestry were recently meta-analyzed by De Stefano and Jacobson (2017). Overall, SOC stocks increased when land use changed to agroforestry from less complex systems, such as agricultural systems. For example, conversion from agriculture to agroforestry increased SOC stocks by 26, 40, and 34% in 0–15, 0–30, and 0–100 cm depth, respectively. Further, conversion from pasture/grassland to agroforestry resulted in SOC stock increases of 9% in 0–30 cm and 10% in 0–60 cm. In contrast, land-use change from forest to agroforestry decreased SOC stocks by 26 and 24% in 0–15 cm and 0–30 cm depth, respectively. However, the meta-analysis may have been influenced by heterogeneity, inconsistencies in study design, lack of standardized sampling procedures, failure to report variance estimators, and lack of important explanatory variables (De Stefano and Jacobson 2017).

In summary, available results indicate that agroforests store higher amounts of C above- and belowground than single-species cropping and grazing systems (Nair et al. 2010). The SOC storage potential in soil is high below 50 cm soil depth. Conversion of agricultural into agroforestry systems may result in increases of SOC stocks.

6.4 Enhancing Carbon Sequestration in Agroforest Soils

The management of integrated tree and crop production systems may alter rate and magnitude of C sequestration but rigorous datasets are required to identify the underlying mechanisms for improved agroforestry practices aimed at SOC sequestration (Nair et al. 2010). For example, net increases in the SOC stock may be managed through selection of agroforestry systems and soil management practices that affect the amount and quality of C inputs by tree and non-tree components (Nair et al. 2009). Litter fall and in turn SOC sequestration may be affected by stand density management as, for example, higher stocking levels of trees enhance the vegetation C stock (Nair et al. 2010). In contrast, thinning and pruning of trees may reduce SOC sequestration by reducing litter fall and accelerating decomposition due to changes in understory light, air/soil temperature, and soil moisture regimes. Soil properties such as soil structure and aggregation also affect SOC sequestration. The biomass C input to the soil in croplands, pastures, and natural grasslands can be increased and its decomposition rate decreased by adoptions of conservation-effective measures that reduce losses of nutrients and water, increase biomass production, and protect SOC against losses through enhancing biological, chemical, and physical stabilization mechanisms (Lal and Follett 2009). Thus, a decrease in cultivation intensity may result in an increase in SOC stocks in agroforests (Nair et al. 2010). Similar to agricultural systems, reducing soil disturbance such as tillage, adding amendments such as fertilizers and composts, supplying water through irrigation, and incorporating OM into soil may enhance SOC sequestration in agroforestry systems. However, the extent of tillage operations in agroforests is considerable less than in agricultural systems. Further, fertilization studies on most tropical tree species are scanty (Nair et al. 2010). Nevertheless, manure additions may influence formation and stability of soil aggregates in agroforests. Also, while herbicide applications show mixed effects on SOC stocks and aggregation, there is little information about effects of pesticides on SOC sequestration (Nair et al. 2010). In summary, the effects of improved management practices on SOC in agroforestry systems are site-specific.

Multi-species agroforestry systems have the potential to be more productive than the best-performing monocultures and, thus, may sequester more SOC due to enhanced belowground interactions, but the experimental evidence is not conclusive (Ong et al. 2004; Rao et al. 2004). The management of agroforestry systems for C sequestration includes the selection of tree species and their silvicultural management such as stand density and rotation length (Nair et al. 2009). Functionally, important tree species, i.e., those having deep and extensive root systems to enhance C input into the soil, may have a high potential to enhance SOC sequestration in agroforestry systems (Lorenz and Lal 2010). Broadleaf trees, in particular, have a larger and more deeply anchored root system (i.e., higher root biomass-to-aboveground biomass ratio) than coniferous tree species and may therefore generate higher C inputs from roots at soil depth (Laganiere et al. 2010). Data on belowground inputs form agroforestry tree species are, however, scanty

(Albrecht et al. 2004). Thus, similar to non-agroforest trees it is not known whether agroforest tree species differ in their ability to sequester SOC in deeper mineral soils (Jandl et al. 2007). Otherwise, mixed plantings including N-fixing trees produce more biomass and this may result in increased SOC sequestration (Nair et al. 2009). However, positive, negative, and neutral effects of N-fixing trees on SOC accretion are reported. Thus, appropriate tree species must be selected to enhance SOC sequestration (Oelbermann et al. 2006).

Whether mixed tree plantings in agroforests generally enhance SOC sequestration is also unknown as manipulative biodiversity experiments with multiple tree species are scanty (Scherer-Lorenzen et al. 2005). A decline in total species richness was observed in cacao and coffee agroforests compared to natural forests (De Beenhouwer et al. 2013). Further, biodiversity responses to intensification differed between Asia and Latin America, and between different species groups. Management intensification decreased provision of ecosystem services with 37% when comparing forest with agroforest, and with 27% when comparing agroforest with plantation (De Beenhouwer et al. 2013). In some situations, agroforests can be designed to optimize both biodiversity and crop production benefits (Clough et al. 2011). For example, agroforestry can enhance biodiversity and ecosystem service provision relative to conventional agriculture and forestry in Europe (Torralba et al. 2016).

In summary, it remains to be studied whether agroforestry systems and their biodiversity can be specifically designed and managed to maximize the belowground C sequestration in the soil by more fully exploring the SOC storage potential in the entire mineral soil through the inclusion of trees and their associated root-derived C inputs.

6.5 Conclusions

Agroforestry practices can be sustainable soil and land-use management practices if managed to maximize site productivity by increasing the efficiency in nutrient, light, and water capture and utilization. This is achieved by purposeful growing trees or other woody perennials in interacting combinations with crops and/or animals. A large number of traditional and improved agroforestry systems have been identified, in particular, in the tropics because of favorable climatic conditions and favorable socioeconomic factors. Existing uncertainties can be addressed by strengthening research data on the global areal extent of agroforestry practices with a wide variation in published statistics. Thus, the global potential for C sequestration by agroforestry can also be improved. Simple estimations obtained by multiplying the estimates of area under agroforests with an average of C sequestration potential can be improved by using the modern modeling techniques and strengthening the statistical database. By adding trees, reducing soil degradation and erosion, agroforests sequester C in above- and belowground biomass and as SOC. SOC sequestration rates of up to 7.4 Mg C ha^{-1} yr^{-1} in the top 20 cm soil depth have been reported. There is also a strong need for adoption of standard methods and procedures to determine SOC sequestration of the global area under agroforests. It is important to realize the low-cost environmental benefits of agroforestry, which are at present underappreciated and underexploited. More science and the adoption of consistent standards are needed to fully realize the C mitigation potential of agroforestry systems.

6.6 Review Questions

1. Describe merits and limitations of major agroforestry practices.
2. How do soil and land-use management practices of agroforests affect the biomass addition to the soil, and C losses by decomposition and erosion?
3. What are the processes underlying major advantages of SOC sequestration in agroforests?
4. What may be major constraints for establishing agroforestry practices?
5. Tabulate ecosystem C budgets for agricultural *vs.* agroforest systems.
6. How would you determine the SOC sequestration potential of agroforestry practices?
7. What is land equivalent ratio (LER)?
8. Some trees have allelopathic effects on crops, why?
9. Sketch root system of trees and crops in agroforestry systems.
10. Tabulate merits and limitations of agroforestry systems.

References

Aertsens J, De Nocker L, Gobin A (2013) Valuing the carbon sequestration potential for European agriculture. Land Use Pol 31:584–594

Ajit Dhyani SK, Handa AK et al (2017) Estimating carbon sequestration potential of existing agroforestry systems in India. Agroforest Syst 91:1101–1118. https://doi.org/10.1007/s10457-016-9986-z

Albrecht A, Cadisch G, Blanchart E, Sitompul SM, Vanlauwe B (2004) Below-ground inputs: relationships with soil quality, soil C storage and soil structure. In: van Noordwijk M, Cadisch G, Ong CK (eds) Below-ground interactions in tropical agroecosystems-concepts and models with multiple plant components. CABI, Wallingford, UK, pp 193–207

Albrecht A, Kandji ST (2003) Carbon sequestration in tropical agroforestry systems. Agric Ecosyst Environ 99:15–27

Amézquita MC, Ibrahim M, Llanderal T, Buurman P, Amézquita E (2005) Carbon sequestration in pastures, silvo-pastoral systems and forests in four regions of the Latin American tropics. J Sust For 21:31–49

Araujo ASF, Carvalho LF, Iwata LBF et al (2012) Microbiological process in agroforestry systems. A review. Agron Sustain Dev 32:215–226. https://doi.org/10.1007/s13593-011-0026-0

Baah-Acheamfour M, Carlyle NC, Bork EW, Chang SX (2014) Trees increase soil carbon and its stability in three agroforestry systems in central Alberta, Canada. For Ecol Manage 328:131–139

Beer J, Bonnemann A, Chavez W, Fassbender HW, Imbach AC, Martel I (1990) Modelling agroforestry systems of cacao *(Theobroma cacao)* with laurel *(Cordia alliodora)* or poro *(Erythrina poeppigiana)* in Costa Rica. Agroforest Syst 12:229–249

Berthrong ST, Jobbágy EG, Jackson RB (2009) A global meta-analysis of soil exchangeable cations, pH, carbon, and nitrogen with afforestation. Ecol Appl 19:2228–2241

Borland AM, Wullschleger SD, Weston DJ et al (2015) Climate-resilient agroforestry: physiological responses to climate change and engineering of crassulacean acid metabolism (CAM) as a mitigation strategy. Plant Cell Environ 38:1833–1849. https://doi.org/10.1111/pce.12479

Burgess PJ, Incoll LD, Corry DT, Beaton A, Hart BJ (2004) Poplar (*Populus* spp) growth and crop yields in a silvoarable experiment at three lowland sites in England. Agroforest Syst 63:157–169

Canadell J, Jackson RB, Ehleringer JR, Mooney HA, Sala OE, Schulze E-D (1996) Maximum rooting depth of vegetation types at the global scale. Oecologia 108:583–595

Cardinael R, Guenet B, Chevallier T et al (2018) High organic inputs explain shallow and deep SOC storage in a long-term agroforestry system—combining experimental and modeling approaches. Biogeosciences 15:297–317

Clough Y, Barkmann J, Juhrbandt J et al (2011) Combining high biodiversity with high yields in tropical agroforests. Proc Natl Acad Sci USA 108:8311–8316

Cusack DF, Chou WW, Yang WH, Harmon ME, Silver WL, Team The LIDET (2009) Controls on long-term root and leaf litter decomposition in neotropical forests. Glob Change Biol 15:1339–1355

da Silva EV, de Moraes Gonçalves JL, de Freitas Coelho SR, e Moreira RM, de Miranda Mello SL, Bouillet J-P, Jourdan C, Laclau J-P (2009) Dynamics of fine root distribution after establishment of monospecific and mixed-species plantations of *Eucalyptus grandis* and *Acacia mangium*. Plant Soil 325:305–318

De Beenhouwer M, Aerts R, Honnay O (2013) A global meta-analysis of the biodiversity and ecosystem service benefits of coffee and cacao agroforestry. Agric Ecosyst Environ 175:1–7

De Stefano A, Jacobson MG (2017) Soil carbon sequestration in agroforestry systems: a meta-analysis. Agroforest Syst. https://doi.org/10.1007/s10457-017-0147-9

Dhillon GS, Gillespie A, Peak D, Van Rees KCJ (2017) Spectroscopic investigation of soil organic matter composition for shelterbelt agroforestry systems. Geoderma 298:1–13

Dixon RK (1995) Agroforestry systems: sources or sinks of greenhouse gases? Agroforest Syst 31:99–116

Feliciano D, Ledo A, Hillier J, Nayak DR (2018) Which agroforestry options give the greatest soil and above ground carbon benefits in different world regions? Agric Ecosyst Environ 254:117–129

Fonte SJ, Barrios E, Six J (2010) Earthworms, soil fertility and aggregate-associated soil organic matter dynamics in the quesungual agroforestry system. Geoderma 155:320–328

Gama-Rodrigues EF, Nair PKR, Nair VD, Gama-Rodrigues AC, Baligar VC, Machado RCR (2010) Carbon storage in soil size fractions under two cacao agroforestry systems in Bahia, Brazil. Environ Manage 45:274–283

Garrity DP, Akinnifesi FK, Ajayi OC, Weldesemayat SG, Mowo JG, Kalinganire A, Larwanou M, Bayala J (2010) Evergreen agriculture: a robust approach to sustainable food security in Africa. Food Sec 2:197–214

Guo LB, Gifford RM (2002) Soil carbon stocks and land use change: a meta analysis. Glob Change Biol 8:345–360

Haggar JP, Tanner EVJ, Beer JW, Kass DCL (1993) Nitrogen dynamics of tropical agroforestry and annual cropping systems. Soil Biol Biochem 25:1363–1378

Haile SG, Nair PKR, Nair VD (2008) Carbon storage of different soil-size fractions in Florida silvopastoral systems. J Environ Qual 37:1789–1797

Haile SG, Nair VD, Nair PKR (2010) Contribution of trees to carbon storage in soils of silvopastoral systems in Florida, USA. Glob Change Biol 16:427–438

Horwarth W (2007) Carbon cycling and formation of soil organic matter. In: Paul EA (ed) Soil microbiology, ecology, and biochemistry. Academic Press, Burlington, MA, pp 303–339

Inderjit Malik AU (2002) Chemical ecology of plants: allelopathy in aquatic and terrestrial ecosystems. Birkhäuser-Verlag, Berlin

Isaac ME, Gordon AM, Thevathasan NV, Oppong SK, Quashi-Sam J (2005) Temporal changes in soil carbon and nitrogen in west African multistrata agroforestry systems: a chronosequence of pools and fluxes. Agroforest Syst 65:23–31

Jackson RB, Lajtha K, Crow SE et al (2017) The ecology of soil carbon: pools, vulnerabilities, and biotic and abiotic controls. Annu Rev Ecol Evol Syst 48:419–445

Jandl R, Lindner M, Vesterdahl L, Bauwens B, Baritz R, Hagedorn F, Johnson DW, Minkkinen K, Byrne KA (2007) How strongly can forest management influence soil carbon sequestration? Geoderma 137:253–268

Jobbágy EG, Jackson RB (2000) The vertical distribution of soil organic carbon and its relation to climate and vegetation. Ecol Appl 10:423–436

John J, Patil RH, Joy M, Nair AM (2006) Methodology of allelopathy research: 1. agroforestry systems. Allelopathy J 18:173–214

Johnson DW, Curtis PS (2001) Effects of forest management on soil C and N storage: meta analysis. For Ecol Manag 140:227–238

Johnson EA, Miyanishi K (2008) Testing the assumptions of chronosequences in succession. Ecol Lett 11:419–431

Johnson JMF, Allmaras RR, Reicosky DC (2006) Estimating source carbon from crop residues, roots and rhizodeposits using the national grain-yield database. Agron J 98:622–636

Jose S (2009) Agroforestry for ecosystem services and environmental benefits: an overview. Agroforest Syst 76:1–10. https://doi.org/10.1007/s10457-009-9229-7

Jose S, Gillespie AR, Pallardy SG (2004) Interspecific interactions in temperate agroforestry. Agroforest Syst 61:237–255

Kang BT, Grimme H, Lawson TL (1985) Alley cropping sequentially cropped maize and cowpea with Leucaena on a sandy soil in southern Nigeria. Plant Soil 85:267–276

Kim DG, Kirschbaum MUF, Beedy TL (2016) Carbon sequestration and net emissions of CH_4 and N_2O under agroforestry: Synthesizing available data and suggestions for future studies. Agric Ecosyst Environ 226:65–78

Kizito F, Dragila M, Sène M, Lufafa A, Diedhiou I, Dick RP, Selker JS, Dossa E, Khouma M, Badiane A, Ndiaye S (2006) Seasonal soil water variation and root patterns between two semi-arid shrubs co-existing with Pearl millet in Senegal, West Africa. J Arid Environ 67:436–455

Kohli RK, Singh HP, Batish DR, Jose S (2008) Ecological interactions in agroforestry: an overview. In: Batish DR, Kohli RK, Jose S, Singh HP (eds) Ecological basis of agroforestry. CRC Press, Boca Raton, FL, pp 3–14

Kumar BM, Nair PKR (2004) The enigma of tropical homegardens. Agroforest Syst 61:135–152

Laganière J, Angers D, Paré D (2010) Carbon accumulation in agricultural soils after afforestation: a meta-analysis. Glob Change Biol 16:439–453

Lal (1989) Agroforestry systems and soil surface management of a tropical alfisol. Agroforestry Syst 8:1–6

Lal R (2005) Soil carbon sequestration in natural and managed tropical forest ecosystems. J Sust For 21:1–30

Lal R, Follett RF (2009) Soils and climate change. In: Lal R, Follett RF (eds) Soil carbon sequestration and the greenhouse effect. SSSA Special Publication 57, 2nd edn. Madison, WI, xxi–xxviii

Lasco RD, Delfino RJP, Espaldon MLO (2014) Agroforestry systems: helping smallholders adapt to climate risks while mitigating climate change. WIREs Clim Change 5:825–833. https://doi.org/10.1002/wcc.301

Liste H-H, White JC (2008) Plant hydraulic lift of soil water—implications for crop production and land restoration. Plant Soil 313:1–17

Long AJ, Nair PKR (1999) Trees outside forests: agro-, community-, and urban forestry. New For 17:145–174

Lorenz K, Lal R (2005) The depth distribution of soil organic carbon in relation to land use and management and the potential of carbon sequestration in subsoil horizons. Adv Agron 88:35–66

Lorenz K, Lal R (2010) Carbon sequestration in forest ecosystems. Springer, Dordrecht, The Netherlands

Luedeling E, Kindt R, Huth NI, Koenig K (2014) Agroforestry systems in a changing climate—challenges in projecting future performance. Curr Op Environ Sustain 6:1–7

Mackey B, Prentice IC, Steffen W et al (2013) Untangling the confusion around land carbon science and climate change mitigation policy. Nat Clim Change 3:552–557. https://doi.org/10.1038/NCLIMATE1804

Meinen C, Hertel D, Leuschner C (2009) Root growth and recovery in temperate broad-leaved forest stands differing in tree species diversity. Ecosystems 12:1103–1116

Mendez-Millan M, Dignac M-F, Rumpel C, Rasse DP, Derenne S (2010) Molecular dynamics of shoot vs. root biomarkers in an agricultural soil estimated by natural abundance ^{13}C labeling. Soil Biol Biochem 42:169–177

Mitchell RJ, Campbell CD, Chapman SJ, Cameron CM (2010) The ecological engineering impact of a single tree species on the soil microbial community. J Ecol 98:50–61

Munroe JW, Isaac ME (2014) N_2-fixing trees and the transfer of fixed-N for sustainable agroforestry: a review. Agron Sustain Dev 34:417–427. https://doi.org/10.1007/s13593-013-0190-5

Muschler RG, Bonnemann A (1997) Potentials and limitations of agroforestry for changing land-use in the tropics: experiences from Central America. For Ecol Manage 91:61–73

Nair PKR (1987) Agroforestry systems inventory. Agroforest Syst 5:301–317. https://doi.org/10.1007/BF00119128

Nair PKR (1991) State-of-the-art of agroforestry systems. For Ecol Manage 45:5–29

Nair PKR (2012) Climate change mitigation: A low-hanging fruit of agroforestry. In: Nair PKR, Garrity D (eds) Agroforestry—the future of global land use, advances in agroforestry 9. Springer, Dordrecht, Netherlands, pp 31–66. https://doi.org/10.1007/978-94-007-4676-3_7

Nair PKR, Gordon AM, Mosquera-Losada M-R (2008) Agroforestry. In: Jorgensen SE, Fath BD (eds) Ecological engineering, encyclopedia of ecology, vol 1. Elsevier. Oxford, UK, pp 101–110

Nair PKR, Kumar BM, Nair VD (2009) Agroforestry as a strategy for carbon sequestration. J Plant Nutr Soil Sci 172:10–23

Nair PKR, Nair VD, Kumar BM, Showalter JM (2010) Carbon sequestration in agroforestry systems. Adv Agron 108:237–307

Oelbermann M, Voroney RP, Gordon AM, Kass DCL, Schlönvoigt AM, Thevathasan NV (2006) Soil carbon dynamics and residue stabilization in a Costa Rican and southern Canadian alley cropping system. Agroforest Syst 68:27–36

Okigbo BN (1985) Improved permanent production systems as an alternative to shifting intermittent cultivation. In: Improved production systems as an alternative to shifting cultivation, FAO Soils Bulletin 53, FAO, Rome, Italy, pp 1–100

Ong CK, Kho RM, Radersma S (2004) Ecological interactions in multispecies agroecosystems: concepts and rules. In: Ong CK, Huxely P (eds) Tree-crop interactions, a physiological approach. CAB International, Wallingford, UK, pp 1–15

Pardon P, Reubens B, Reheul D et al (2017) Trees increase soil organic carbon and nutrient availability in temperate agroforestry systems. Agric Ecosyst Environ 247:98–111

Parrotta JA (1999) Productivity, nutrient cycling and succession in single- and mixed-species stands of *Casuarina equisetifolia*, *Eucalyptus robusta* and *Leucaena leucocephala* in Puerto Rico. For Ecol Manag 124:45–77

Paul KI, Polglase PJ, Nyakuengama JG, Khanna PK (2002) Change in soil carbon following afforestation. For Ecol Manag 168:241–257

Peichl M, Thevathasan NV, Gordon AM, Huss J, Abohassan RA (2006) Carbon sequestration potentials in temperate tree-based intercropping systems, southern Ontario, Canada. Agroforest Syst 66:243–257

Perry CH, Woodall CW, Liknes GC, Schoeneberger MM (2009) Filling the gap: improving estimates of working tree resources in agricultural landscapes. Agroforest Syst 75:91–101

Pinho RC, Miller RP, Alfaia SS (2012) Agroforestry and the improvement of soil fertility: a view from Amazonia. Appl Environ Sci Article 2012:616383. https://doi.org/10.1155/2012/616383

Post WM, Kwon KC (2000) Soil carbon sequestration and land-use change: processes and potential. Glob Change Biol 6:317–327

Prescott CE (2010) Litter decomposition: what controls it and how can we alter it to sequester more carbon in forest soils? Biogeochemistry 101:133–149

Preston CM, Nault JR, Trofymow JA (2009a) Chemical changes during 6 years of decomposition of 11 litters in some Canadian forest sites. Part 2. 13C abundance, solid-state 13C NMR spectroscopy and the meaning of "lignin". Ecosystems 12:1078–1102

Preston CM, Nault JR, Trofymow JA, Smyth C, CIDET Working Group (2009b) Chemical changes during 6 years of decomposition of 11 litters in some Canadian forest sites. Part 1. Elemental composition, tannins, phenolics, and proximate fractions. Ecosystems 12:1053–1077

Rao MR, Schroth G, Williams SE, Namirembe S, Schaller M, Wilson J (2004) Managing bewlo-ground interactions in agroecosystems. In: Ong CK, Huxely P (eds) Tree-crop interactions, a physiological approach. CAB International, Wallingford, UK, pp 309–328

Rasse DP, Rumpel C, Dignac M-F (2005) Is soil carbon mostly root carbon? Mechanisms for a specific stabilization. Plant Soil 269:341–356

Resh SC, Binkley D, Parrotta JA (2002) Greater soil carbon sequestration under nitrogen-fixing trees compared with *Eucalyptus* species. Ecosystems 5:217–231

Rigueiro-Rodriguez A, McAdam JH, Mosquera-Losada MR (2008) Agroforestry in Europe. Springer, Dordrecht, The Netherlands

Ritter E (2007) Carbon, nitrogen and phosphorus in volcanic soils following afforestation with native birch (*Betula pubescens*) and introduced larch (*Larix sibirica*) in Iceland. Plant Soil 295:239–251

Rivest D, Cogliastro A, Olivier A (2009) Tree-based intercropping systems increase growth and nutrient status of hybrid poplar: a case study from two Northeastern American experiments. J Environ Manage 91:432–440

Rizvi SJH, Tahir M, Rizvi V, Kohli RK, Ansari A (1999) Allelopathic interactions in agroforestry systems. Crit Rev Plant Sci 19:773–796

Rosenstock TS, Tully KL, Arias-Navarro C et al (2014) Agroforestry with N_2-fixing trees: sustainable development's friend or foe? Curr Op Environ Sustain 6:15–21

Saha SK, Nair PKR, Nair VD, Kumar BM (2010) Carbon storage in relation to soil size-fractions under tropical tree-based land-use systems. Plant Soil 328:433–446

Scherer-Lorenzen M, Potvin C, Koricheva J, Schmid B, Hector A, Bornik Z, Reynolds G, Schulze E-D (2005) The design of experimental tree plantations for functional biodiversity research. In: Scherer-Lorenzen M, Körner C, Schulze E-D (eds) Forest diversity and function. Ecological Studies, vol 176. Springer, Berlin, pp 347–376

Scheu S, Schauermann J (1994) Decomposition of roots and twigs: effects of wood type (beech and ash), diameter, site of exposure and macrofauna exclusion. Plant Soil 241:155–176

Schroth G, Krauss U, Gasparotto L, Aguilar JAD, Vohland K (2000) Pests and diseases in agroforestry systems of the humid tropics. Agroforest Syst 50:199–241

Sinacore K, Hall JS, Potvin C et al (2017) Unearthing the hidden world of roots: root biomass and architecture differ among species within the same guild. PLoS ONE 12:e0185934. https://doi.org/10.1371/journal.pone.0185934

Stockmann U, Adams MA, Crawford JW et al (2013) The knowns, known unknowns and unknowns of sequestration of soil organic carbon. Agric Ecosyst Environ 164:80–99

Sollins P, Swanston C, Kramer M (2007) Stabilization and destabilization of soil organic matter-a new focus. Biogeochemistry 85:1–7

Soto-Pinto L, Anzueto M, Mendoza J, Ferrer GJ, de Jong B (2010) Carbon sequestration through agroforestry in indigenous communities of Chiapas, Mexico. Agroforest Syst 78:39–51

Takimoto A, Nair PKR, Nair VD (2008) Carbon stock and sequestration potential of traditional and improved agroforestry systems in the West African Sahel. Agri Ecosyst Environ 125: 159–166

Torralba M, Fagerholm N, Burgess PJ, Moreno G, Plieninger T (2016) Do European agroforestry systems enhance biodiversity and ecosystem services? A meta-analysis. Agric Ecosyst Environ 230:150–161

Vagen TG, Lal R, Singh BR (2004) Soil carbon sequestration in sub-Saharan Africa: a review. Land Degrad Develop 16:53–71

Van Noordwijk M, Lawson G, Soumare A, Groot JJR, Hairiah K (1996) Root distribution of trees and crops: competition and/or complementarity. In: Ong CK, Huxely P (eds) Tree-crop interactions, a physiological approach. CAB International, Wallingford, UK, pp 319–364

Vinceti B, Termote C, Ickowitz A et al (2013) The contribution of forests and trees to sustainable diets. Sustainability 5:4797–4824. https://doi.org/10.3390/su5114797

Young A (1997) Agroforestry for soil management. C.A.B International and ICRAF, Wallingford, U.K.

Zhang W, Ahanbieke P, Wang BJ et al (2013) Root distribution and interactions in jujube tree/ wheat agroforestry system. Agroforest Syst 87:929–939. https://doi.org/10.1007/s10457-013-9609-x

Zomer RJ, Neufeldt H, Xu J, et al (2017) Global tree cover and biomass carbon on agricultural land: The contribution of agroforestry to global and national carbon budgets. Sci Rep 6:29987. https://doi.org/10.1038/srep29987

Chapter 7
Biomass and Bioenergy

Contents

Abstract Bioenergy from biomass can replace fossil fuels in the production of heat, electricity, and liquid fuels for transport but the potential contribution is in need of further research and objective discussions. Biomass can also provide feedstock for the chemical industry to replace petroleum. Feedstock for this purpose are major plant oil crops such as oil palm (*Elaeis guineensis* or *E. oleifera*), soybean [*Glycine max* (L.) Merr.], rapeseed (*Brassica napus* L.), and sunflower (*Helianthus annuus* L.). In 2015, traditional biomass accounted for 9.1%, and biofuels for transportation accounted for 0.8% of the global final energy consumption. In principle, biomass could meet up to one-third of the projected global energy demand in 2050 by bringing new land under cultivation and/or increasing productivity. However, aside physically possible, socially acceptable biomass potential scenarios must be assessed. The main feedstocks for generating heat, electricity, or gaseous, liquid, and solid fuels are forestry, agricultural and livestock residues, short-rotation forest plantations, energy crops, and the organic component of municipal residues and wastes. Traditional biomass such as fuelwood, charcoal, and animal dung is source for about 99% of all bioenergy. Of minor importance is 'modern' biomass such as sugar, grain, and vegetable oil crops for the production of liquid biofuels. However, in the future the bulk of liquid biofuels may be produced from lignocellulosic crops cultivated on marginal, degraded, and surplus

K. Lorenz and R. Lal, *Carbon Sequestration in Agricultural Ecosystems*,
https://doi.org/10.1007/978-3-319-92318-5_7

agricultural land. Dedicated lignocellulosic energy crops include perennial plants such as switchgrass (*Panicum virgatum* L.), *Miscanthus x giganteus*, sugarcane (*Saccharum* spp.), *Agave* spp., and short-rotation woody crops such as hybrid poplar (*Populus* spp.) and willow (*Salix* spp.). Compared to conventional crops such as corn (*Zea mays* L.), energy crops are less depending on favorable climatic and soil conditions and require fewer inputs of agrochemicals. Thus, using energy crops would reduce the direct competition for land with food production and ecosystem services, and potentially have lower net energy and greenhouse gas (GHG) effects. However, the carbon costs of dedicating land to bioenergy will exceed the benefits. For example, conversion of native ecosystems for bioenergy often results in soil organic carbon (SOC) loss. The long-term potential of energy crops depends largely on land availability, choice of crop species, improvements by biotechnology, water availability, and effects of climate change. Aside from the dedicated bioenergy plantations, other potential feedstocks are the large volumes of unused organic residues and wastes but it is unclear whether their share can be increased. However, agricultural residues are also required on site to maintain SOC stocks, soil health, and agricultural productivity, and to reduce soil erosion. The SOC sequestration may be the key component in determining the GHG reduction potential of biofuels compared to fossil fuels. Life cycle assessment (LCA) is a widely used approach to assess the GHG balance of biomass production. Removing 25 and 100% of corn residues, for example, jeopardizes agroecosystem services and causes losses of up to 3 and up to 8 Mg SOC ha^{-1} in 0–30 cm soil depth after 10 years, respectively. In comparison, SOC accumulates in the top 30 cm under perennial grasses at rates of up to 1 Mg SOC ha^{-1} yr^{-1}. Thus, more intense harvest for bioenergy adversely affects the SOC stock. Also, producing energy crop feedstock by converting previously uncultivated land will cause a reduction in the SOC stock. Otherwise, adding residues from forest harvest, processing, and after end use may be beneficial to the SOC stock compared to establishing woody crop plantations. Sugarcane, perennial grasses, and trees can be cultivated sustainably for bioenergy but estimates for the potential of global bioenergy plantations when environmental and agricultural constraints are taken into account vary widely. Specifically, long-term, large-scale biomass cultivation plots, in particular, of switchgrass and *Miscanthus x giganteus* are scanty. While biofuels and, in particular, liquid biofuels will offset only a modest share in fossil energy use over the next decade, the impacts on agriculture and food security may be drastic. This chapter begins with a section about biomass as feedstock alternative to petroleum. Then, agroecosystem land use and management types for producing traditional and energy crop feedstocks are discussed with a focus on non-woody plants. The chapter concludes with a section about the effects of agricultural biomass production systems for bioenergy and biofuel on SOC sequestration. Additional information about the potential of woody biomass from agroforestry and plantations as feedstock for bioenergy can be found elsewhere (e.g., Buchholz et al. 2016; Lorenz and Lal 2010).

Keywords Plant oil crops · Traditional biomass · Biofuels · Dedicated energy crops · Crop residues · Carbon debt · Life cycle assessment

7.1 Biomass Feedstock for the Chemical Industry

About 10% of petroleum consumed annually is used by the (petro)chemical industry as raw material in the manufacturing of products such as chemical feed stocks, solvents, lubricants, plastics, synthetic rubbers and fibers, detergents (Carlsson 2009). However, petroleum needs to be replaced by alternative and sustainable sources of industrial feedstocks as conventional oil production may go into decline and the era of plentiful, low-cost petroleum is coming to an end (Owen et al. 2010). Advanced bio-based products are discussed to substitute their fossil-based counterparts (Schipfer et al. 2017). Thus, biomass for chemical production may increasingly compete with biomass as feedstock for various energy options (Daioglou et al. 2015). The sustainability of biomass production is evaluated by life cycle assessment (LCA). LCA must include information about where and how land-use change and related impacts will occur in supply chain and innovation decisions (Chaplin-Kramer et al. 2017).

7.1.1 Biomass Production Systems

Plant oils and oleochemicals can replace petroleum as feedstock for the chemical industry (Carlsson 2009). Further, algal residues retrieved after the abstraction of oil for bioenergy are potential new carbon (C) resource to produce important chemicals (Yamaguchi et al. 2017). Major agricultural commodities for plant oils are oil crops which comprised 336 Tg (1 Tg = 10^{12} g) in 2014, with about 15% used as industrial feedstocks (FAO 2017). About 90% of the plant oil used annually is produced from oil palm (*Elaeis guineensis* or *E. oleifera*; Fig. 7.1), soybean [*Glycine max* (L.) Merr.], rapeseed (*Brassica napus* L.; Figure 7.2), sunflower (*Helianthus annuus* L.), peanut (*Arachis hypogaea* L.; Figure 7.3), and cotton (*Gossypium hirsutum* L.) (FAO 2017). Plastics can also be produced from sugarcane (*Saccharum* spp.; Figure 7.4). For example, Dow Chemical is building a bioplastics plant in Brazil that will produce plastic from sugarcane in volumes competitive with plastics generated from petroleum (http://www.technologyreview.com/energy/38114/). Specifically, the plant will produce 240 million liters of ethanol that will be converted into hundreds of thousands Mg of polyethylene. Many plant-based monomers are already in use or under development to replace petrochemical-based source materials for commodity plastics (Hillmyer 2017). Among the important monomers are ethylene, ethylene glycol, isoprene, propylene, terephthalic acid, and styrene. However, polymers from renewable plant resources

Fig. 7.1 Oil palms (*Elaeis guineensis* or *E. oleifera*) in Malaysia (public domain)

Fig. 7.2 Rapeseed (*Brassica napus* L.) field in Germany (licensed under the Creative Commons Attribution 3.0 Unported license)

Fig. 7.3 Peanut (*Arachis hypogaea* L.) field, Pelham, Mitchell County, Georgia, USA (licensed under the Creative Commons Attribution-Share Alike 3.0 Unported license)

do not yet outcompete traditional petrochemical-based materials in both price and performance (Hillmyer 2017).

Oil palm, soybean, rapeseed, and sunflower are feedstocks for multi-purpose oils (Carlsson 2009). Palm oil production was 57 Tg in 2014 at an average yield of about 4 Mg ha^{-1} with an increasing share for biodiesel production (FAO 2017). Oil palm is the most productive world oil crop (Bourgis et al. 2011). Most palm oil is derived from the fruit mesocarp that contains by far the highest oil content reported for any plant tissue. However, expansion of oil palm plantations by deforestation in the major producer countries like Indonesia and Malaysia releases large quantities of C from biomass and peat, soil organic carbon (SOC), and greenhouse gases (GHGs; Beer et al. 2007; Koh et al. 2011). Soybean is second among the world plant oil producers with about 46 Tg of oil produced 2014, and increasingly used as feedstock for biodiesel. Rapeseed is the third largest oil crop with about 26 Tg of oil produced in 2014, and also an increasing share of it used as a source for biodiesel. Sunflower is the fourth largest oil crop with about 16 Tg sunflower oil produced in 2013 (FAO 2017).

In contrast to multi-purpose oils, technical oils are not suited for food purposes. Examples of technical oil crops include crambe (*Crambe abyssinica* L.), flax (*Linum usitatissimum* L.), cotton, Ethiopian mustard (*Brassica carinata* A. Braun), and camelina (*Camelina sativa* L.) (Carlsson 2009). Plant oils as feedstock

Fig. 7.4 Field of sugarcane (*Saccharum* spp.) over 2 m tall in northern Queensland, Australia (licensed under the Creative Commons Attribution 2.0 Generic license)

alternatives to petroleum should be produced from non-food oil crops and not compete for land resources with food crops. For example, Ethiopian mustard has a better agronomic performance in dry climates, and a high pest and disease resistance compared to rapeseed (Carlsson 2009). Also, camelina may have a large potential as it is adapted to a wide range of climatic conditions in temperate regions. For tropical and subtropical countries, the oil-bearing shrub Jatropha (*Jatropha curcas* L.) may have potential as non-food oil producing plant (FAO 2008). Jatropha is highly adapted to arid and semiarid conditions with yields ranging from 680 to 1700 Mg oil ha^{-1} (Carlsson 2009). However, the seed and oil productivity of Jatropha is hugely variable (Kant and Wu 2011). Thus, the results of massive Jatropha planting programs of unprecedented scale are anything but encouraging.

7.1.2 Biomass Production and Soil Organic Carbon

Biomass is a valuable feedstock for the chemical industry but is also needed on-site to maintain soil fertility and health (Lal 2009, 2016). The conversion of native land to feedstock production may result in SOC losses similar to those following conversion of native land to cropland or pasture (Houghton 2010). For example, clearing carbon (C)-rich habitats such as tropical primary forests for biomass production may result in high carbon dioxide (CO_2) emissions and SOC losses with no new SOC equilibrium reached within decades (Don et al. 2011; Fargione et al. 2008). GHG emissions may increase temporarily by 800% if rainforest and by 2000% if peat forest is converted to oil palm due to mobilization of C stocks in vegetation and soil (Beer et al. 2007). Annually, an estimated 4.6 Tg of belowground C may be emitted by tropical peatland conversion to oil palm (Koh et al. 2011). In temperate regions, SOC stock changes associated with land-use changes may occur for centuries (Poeplau et al. 2011). However, most land-use change studies only consider the role of soil in CO_2 release but exclude its potential role in SOC storage (Dale et al. 2011). Further, changes in SOC stocks may also occur during the cultivation of plants as biomass feedstock for the chemical industry. There is, however, no clear evidence for consistent changes in SOC stocks over the crop cycle of oil palms (Chase and Henson 2010). For example, in the majority of cases, conversion of grassland to oil palm plantations in Papua New Guinea resulted in net sequestration of SOC to 1.5 m depth (Goodrick et al. 2015). Further, SOC stocks to 30 cm depth increased after conversion of Atlantic forest into oil palm production in Bahia State, Brazil (Frazão et al. 2014). In contrast, the SOC loss for rapeseed cultivation without straw removal in France was about 0.24 Mg C ha^{-1} yr^{-1} (Arrouays et al. 2002).

7.2 Bioenergy Feedstock Production in Agroecosystems

During the last decades, issues of energy security, increasing prices of fossil fuels, and climate change increased efforts to use biomass for the production of heat, electricity, and transport fuels (Bauen et al. 2009; Edenhofer et al. 2011). However, biofuels are not a new technology as liquid fuels made from sources such as food crops have been researched for more than a century (Nuffield Council on Bioethics 2011). Biomass, the source of bioenergy, consists of any organic matter (OM) of plant or animal origin. It is available in many forms and from many different sources such as forestry products (biomass from logging and silvicultural treatments, process residues such as sawdust and black liquor), agricultural products (crops, harvest residues, food processing waste, animal dung), and municipal and other waste (waste wood, sewage sludge, organic components of municipal solid waste). Renewable energy sources such as energy crops or modern biomass (i.e., all biomass except traditionally used wood, charcoal, agricultural residues, and animal dung for cooking and heating) from dedicated plantations may play a central role in moving the world onto a more secure, reliable, and sustainable energy path (IEA 2016). Bioenergy carbon capture and storage (BECCS) systems that convert biomass to heat, electricity, or liquid or gas fuels, coupled with CO_2 capture and sequestration (CCS), storing the CO_2 terrestrially or in the ocean are among the most frequently discussed carbon dioxide removal (CDR) options (IPCC 2014). However, BECCS received no direct reference throughout the Paris agreement (Anderson 2015). Further, how bioenergy production contributes to sustainable development has a limited scientific basis for policymaking (Robledo-Abad et al. 2017). Otherwise, Souza et al. (2017) suggested that sustainable biomass production can make a significant contribution to reaching the 2 °C target of the Paris agreement. Many bioenergy cropping systems may bring multiple benefits and off-set environmental problems associated with fossil fuels use as well as intensive food production and urbanization. Souza et al. (2017) provided evidence that there are many approaches to land use for bioenergy expansion that do not lead to competition for food or other needs.

7.2.1 Bioenergy Production Potential

Bioenergy is the energy available from biological sources such as biomass. Biofuels are combustible material derived from biomass, commonly produced from plant, animals and microorganisms, and organic wastes (UNEP 2009). Biofuels may be solid, liquid, or gaseous (Fig. 7.5). Biomass is the only current renewable source of liquid transportation fuel (US DOE 2005). If sustainable biomass is used and the fossil energy used in processing the biomass is not excessive, biofuels can have significant benefits compared to fossil fuels (Bauen et al. 2009). Technically, the limits of bioenergy are fundamentally dependent on land availability,

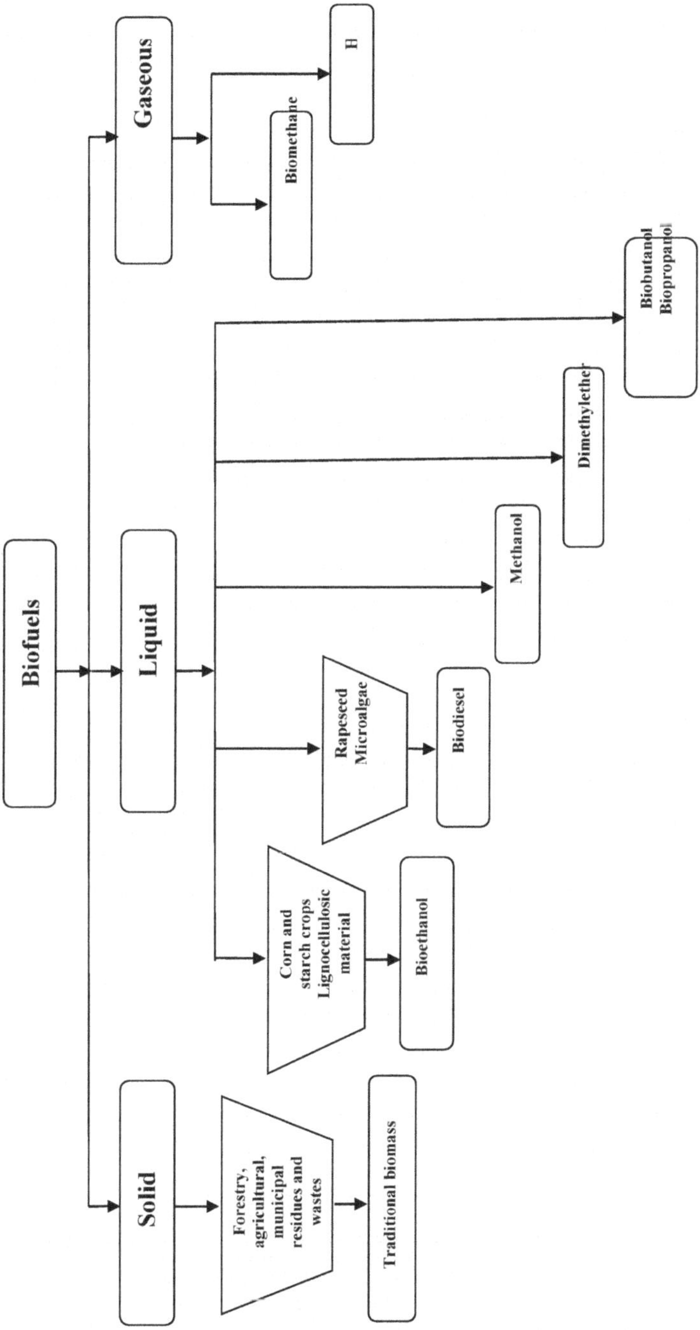

Fig. 7.5 Examples for different types of biofuels and feedstocks (modified from Bauen et al. 2009)

photosynthetic constraints, the sustainable management of nutrients and water resources, and the nature and rate of investments (Strapasson et al. 2017). Further, environmental differences across years and across sites can have a large impact on biomass production as shown for the production of herbaceous energy crops in the USA (Lee et al. 2018).

To understand the sustainability of biofuels in view of their social, economic, and ecological dimensions, and in consideration of the sustainability of upscaling, Goetz et al. (2017) identified several issues that need to be addressed. This included the need for (i) accounting for issues of national and global environmental and social justice; (ii) the distribution of benefits (and costs); (iii) critical perspective regarding the assumptions of modeled futures; (iv) re-evaluation the prospects of states and enterprises to govern for sustainability; and (v) consideration of short-term as well as long-term opportunity costs of the decision to maintain and/or expand bioenergy production and consumption and the related dangers of creating new energy infrastructures that lock in a different set of unsustainable energy production and consumption patterns; without necessarily delivering on associated policy goals, such as improving energy security (Goetz et al. 2017). For example, there is a disjunct between the evidence that global bioenergy studies can provide and policymakers' desire for estimates that can straightforwardly guide policy targets (Slade et al. 2014). Thus, there is a need for bottom-up assessments informed by empirical studies, experimentation, and cross-disciplinary learning to better inform the policy debate. Many technical projections and aspirational goals for future bioenergy use could be difficult or impossible to achieve sustainably (Searle and Malins 2015).

Solar power can provide at least 100 times more useable energy per hectare on three quarters of the world's land compared to bioenergy (Searchinger et al. 2017). Thus, bioenergy is unlikely to compete successfully in the long term with inanimately harvested solar energy (Pickard 2010). For example, harvesting of terrestrial solar radiation to perform useful work is at least an order of magnitude more efficient when carried out by solar-thermal or solar-photovoltaic processes that when carried out by way of biomass conversion. Among the reasons is that plant photosynthesis is inherently inefficient. For example, maximum conversion rates of solar energy into OM are 1.6% for Napier grass (*Pennisetum purpureum* Schumach. 1827) and sugarcane grown under the most favorable conditions, but more usual rates for temperate crops are 0.2–0.4%. Thus, per square meter of land and averaged over a cycle of the seasons, inanimate harvesting of solar radiation can collect and deliver to the grid more kilowatt-hours than biomass harvesting (Pickard 2010). For example, in terms of delivering electrical or mechanical power, ethanol from one hectare (10,000 m^2) of switchgrass (*Panicum virgatum* L.) in the USA is equivalent to 30 m^2 of low-efficiency photovoltaic cells (Patzek 2010). Thus, in the long term, biomass may only be successful in replacing petroleum as feedstock in niche applications for producing biodiesel, aviation fuel or petrochemicals (Pickard 2010).

In principle, global biomass potential could meet between 15% and up to one-third of the projected global energy demand in 2050 (Dornburg et al. 2010;

Beringer et al. 2011). However, biomass potential scenarios must be assessed with respect to what is socially acceptable in addition to what is physically possible (Goetz et al. 2017; Slade et al. 2011). Policies related to diet, agriculture, and land use may be at least as important as those focused on biomass for bioenergy per se. Further, global biomass potentials vary widely with the highest potentials assigned to Asia, Africa, and South America while Europe, North America, and the Pacific region contribute a minor share (Offermann et al. 2011). The greatest potential may have land that was not previously used for agriculture or pasture but has been abandoned and not converted to forest or urban areas (Field et al. 2008). Globally, potential aboveground plant growth on these lands may have an energy content of about 5% of world energy consumption in 2006. For example, producing 1 Pg (1 Pg = 10^{15} g) of *Miscanthus x giganteus* biomass annually on 1 million km^2 of marginal and degraded land in China may account for 45% of the year 2007 electricity output (Sang and Zhu 2011). Yet, growing *Miscanthus x giganteus* on marginal agricultural lands in China has tremendous potential to meet future biofuel needs (Qin et al. 2011). Globally, the potential for biomass energy may also be increased through agricultural intensification (Johnston et al. 2011). Low-input high-diversity (LIHD) mixtures of native grassland perennials may also provide feedstock for biofuel (Tilman et al. 2006). LIHD biofuels may be produced on agriculturally degraded lands and, thus, need to neither displace food production nor cause loss of biodiversity via habitat destruction. Further, pastures converted to LIHD grasslands, especially in tropical developing countries, may have a large potential for biofuel production (Machovina and Feeley 2017).

Advanced biofuels for transportation are on a path to becoming cost competitive in the next few years (Balagopal et al. 2010). In 2015, traditional biomass accounted for 9.1% of the global final energy consumption while biofuels for transport contributed 0.8% (REN21 2017). About 27% of global transport fuel may be provided by biofuel in 2050 (IEA 2011). Traditional biomass (i.e., forestry, agricultural and municipal residues, and wastes) are the main feedstocks for generating electricity and heat, constituting about 99% of all bioenergy (Bauen et al. 2009). The global primary energy supply from traditional biomass parallels the world's industrial wood production but is of low efficiency with average efficiencies of about 10–20% (Arvizu et al. 2011). All other biomass can be defined as modern biomass but with higher average efficiencies of about 58%.

Liquid biofuels from modern biomass contribute only a small share with transport biofuels representing only 2% of total bioenergy. While ethanol accounts for 82% of the global biofuel production, biodiesel may be a locally important biofuel. For example, in Germany, biodiesel provides 75% of the renewable liquid fuel production as diesel automobiles comprise a large percentage of the market (Hambrick et al. 2010). Further, the EU produced 54% of the global biodiesel supply in 2008. Biofuels may also be distinguished according to their origin and production technology as first-, second-, or third-generation biofuels (Fig. 7.6; IEA 2011). Emerging at the basic research level are the fourth-generation biofuels including photobiological solar fuels and electrofuels (Aro 2016). Their development depends on synthetic biology of algae and cyanobacteria. Fourth-generation

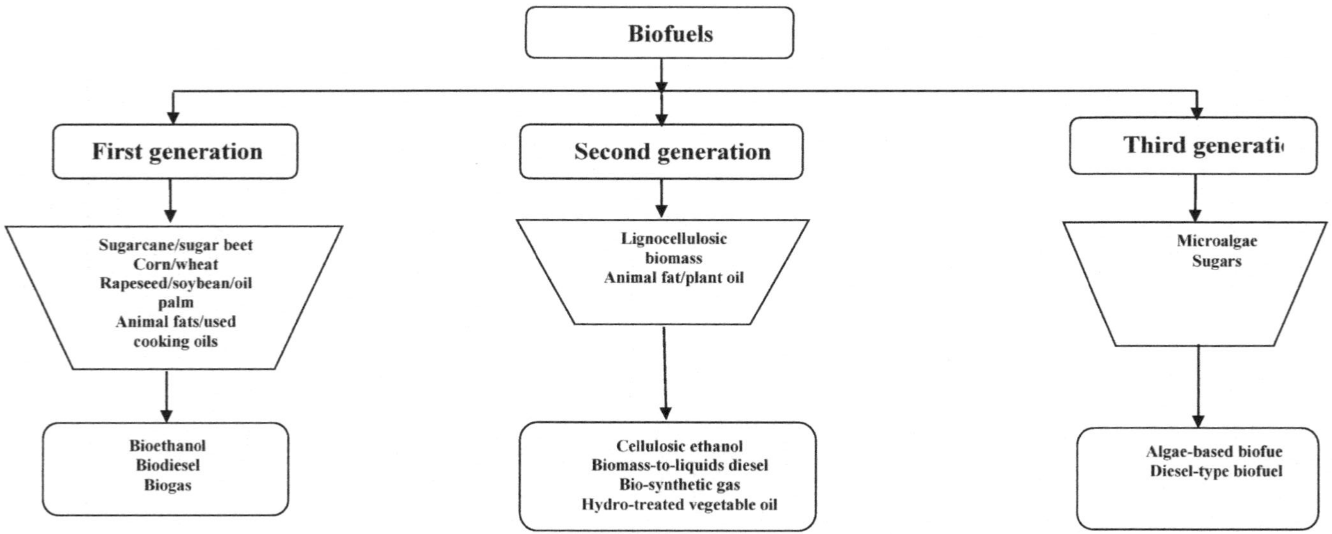

Fig. 7.6 Examples for first-, second-, and third-generation biofuels and feedstocks (based on IEA 2011)

biofuels are expected to be produced by direct conversion of solar energy to a fuel from sunlight, water, and CO_2. In the future, such solar biofuels may be produced in engineered photosynthetic microorganisms or in synthetic living factories (Aro 2016).

First-generation biofuels are produced commercially using well-established conversion technologies. Feedstocks are seeds, grains, or whole plants from crops such as corn (*Zea mays* L.), sugarcane, rapeseed, wheat (*Triticum* spp.), sunflower seeds, or oil palm (UNEP 2009). As most of these plants are used as fodder or to feed people, feedstock must rather be produced from non-fodder/food biomass sources to maintain global food security in face of the projected population growth (Godfray et al. 2010). Bioethanol is the most common first-generation biofuel, followed by biodiesel, vegetable oil and biogas (UNEP 2009). By closing the yield gap (i.e., difference between current agricultural yield and future potential based on climate and biophysical characteristics of the growing region), additional bioethanol and biodiesel may also be produced (Johnston et al. 2011).

Because of concerns regarding the sustainability of many first-generation biofuels, second-generation biofuels from non-food, lignocellulosic or modern biomass receive increased attention (Bartle and Abadi 2010; IEA 2010). These materials are available as primary (in the field) or secondary (processing) residues from agriculture and forestry, tertiary waste (from urban/industrial activity). Sources include the stalks of wheat, corn stover, wood, and dedicated energy or biomass crops (UNEP 2009). Large potentials to dedicate agricultural land to second-generation feedstock production in the future exist in many developing regions such as Latin America and sub-Saharan Africa (IEA 2010). However, it is unclear whether and how those potentials can be realized. For example, progress of the US cellulosic biofuel industry has been very disappointing due to overly optimistic estimates for the readiness of the technology and for the cost of production (Schnoor 2011). Otherwise, cellulosic biofuel concerns related to climate mitigation, biodiversity, reactive N loss, and crop water use can be addressed in the USA with appropriate crop, placement, and management choices (Robertson et al. 2017). In particular, growing native perennial species on marginal lands may provide substantial potential for climate mitigation and other benefits.

The herbaceous and woody species being targeted for development as second-generation energy crops are robust perennial species that can resprout from rootstocks after harvest. In contrast to first-generation feedstocks, processing of second-generation feedstocks has a more favorable ratio of energy output to input (Bartle and Abadi 2010). Thus, reduced CO_2 emissions are a potential benefit of second-generation biofuels if sustainable biomass is used and the fossil energy used in processing the biomass is not too excessive (IEA 2010). However, many second-generation biofuels are under development and assumed to enter the energy market not before 2020. Even later relevant may be third-generation biofuels such as biodiesel from algae and, in particular, biopropanol and biobutanol (IEA 2011). Second- and third-generation biofuels are sometimes also called advanced biofuels (UNEP 2009). However, it needs to be technically proven whether algae and cyanobacteria are suitable as a source of bioenergy, especially with respect to demands on water and land resources (Larkum 2010; Wigmosta et al. 2011). Further, microalgal biofuel production is associated with high costs and energy requirements for agitation, harvesting, and drying of biomass (Singh and Dhar 2011). The greatest challenge for oil from algae may be political and, in particular, continuing support (Hall and Benemann 2011). Advanced 'drop-in replacement' biofuels that are more compatible than bioethanol with the existing petroleum infrastructure are also developed using metabolic engineering and synthetic biology (Simmons 2011).

In 2050, biofuels may displace enough petroleum to avoid the equivalent of 2.1 Pg C emissions each year if produced sustainably (IEA 2011). Another important potential effect of biomass removal for bioenergy and growing biofuel crops is the alteration of the SOC balance (Anderson-Teixeira et al. 2009). SOC sequestration, in particular, may be the key in determining the GHG reduction potential of biofuels relative to fossil fuels. The changes in SOC stocks have a dominant effect on total GHG emissions of bioenergy cropping systems (Brandão et al. 2011). However, SOC sequestration and soil GHG emissions are not adequately addressed in the GHG balance of biofuel production using LCA (Whitaker et al. 2010). Further, the projected increase in energy crop cultivation requires additional nitrogen (N) fertilization and may cause a strong increase in global soil nitrous oxide (N_2O) emissions. N_2O emissions may be an important term in the GHG balance of biofuels produced from energy crops (Bouwman et al. 2010; Crutzen et al. 2008). Optimization of N-use efficiency is among the options to reduce the disturbance of the N cycle caused by increased N fertilization for producing biomass-based biofuels (Erisman et al. 2010).

Bioenergy production by the agricultural sector may potentially compromise the overall goal of enhancing agroecosystem C for mitigating climate change (Morgan et al. 2010). If high yields were to be achieved, massive fertilizer application would be needed with likely harmful consequences for global and local ecosystems (Creutzig 2016). The future deployment of bioenergy as a mitigation option is, thus, uncertain as it depends on many factors (Creutzig et al. 2015). For example, biophysical effects of altered surface energy and water balance may affect the GHG mitigation potential of transitioning to bioenergy crops at regional scales (Zhu et al. 2017). Globally, the extent of cropland on which conversion to energy crops or forest would result in a net C loss was estimated to be $\sim$4.2 million km^2 or 35.6% of the total cropland in Africa, 40.3% in Asia and Russia Federation, 30.8% in EU25, 48.4% in North America, 13.7% in South America, and 58.5% in Oceania (Albanito et al. 2016). *Miscanthus x giganteus* and switchgrass cultivars are the bioenergy feedstock with the highest climate mitigation potential. Both provide the best climate mitigation option on $\sim$4.9 million km^2 of cropland worldwide with $\sim$42% of this land characterized by a terrain slope equal or above 20%. If this land-use change would occur, $\sim$58.1 Pg fossil fuel C equivalent would be replaced (Albanito et al. 2016).

7.2.2 Traditional Biomass for Heat and Power Production

In many developing countries, traditional biomass (i.e., fuel wood, dried manure) is used as feedstock for power and heat generation although with low efficiency (Arvizu et al. 2011; Bauen et al. 2009; UNEP 2009). Dried biomass (mainly wood) contributes 10% to the world primary energy mix, and the direct combustion of biomass for the production of heat is the leading bioenergy application throughout the world. Further, wood, grass cuttings, switchgrass, charcoal, domestic refuse, and dried manure are densified by torrefaction or carbonization for processing into first-generation solid biofuels (UNEP 2009). In developed countries, residuals and waste are also used for generating electricity and heat in combined heat-and-power (CHP) plants.

7.2.2.1 Harvest Residues as Bioenergy Feedstock

Heat production from agricultural residues and waste is common in developing countries with an increasing share of rice (*Oryza sativa* L.) residues and coconut (*Cocos nucifera* L.) husks. Further, in countries with a large sugarcane industry, bagasse (i.e., sugarcane after juice extraction) is an important feedstock. In contrast to dedicated energy crops such as switchgrass and hybrid poplar, agricultural residues are a more immediately available bioenergy feedstock as they are already produced as a coproduct of food and fiber production (Gregg and Izaurralde 2010). Considering the global magnitude of agriculture and forestry production, residue biomass is potentially a large and under-utilized resource. Thus, agricultural and forest residues could be the feedstock of choice in the initial stage of second-generation biofuel production (IEA 2010). However, little research and development is being conducted in utilizing residue biomass compared to energy crops and plantations (Gregg and Smith 2010). Most importantly, harvest residues provide essential environmental services such as feed and bedding, and residues left on the soil contribute nutrients and soil C inputs while SOC replacement contributes to GHG emissions (Smil 1999). Thus, further research is particularly necessary to determine the proper balance of residues that should remain on the field or in the forest to maintain soil fertility and SOC stock, and the amount that can be removed for energy, as well as with regard to nutrient recycling after energy recovery (UNEP 2009). As commercial-scale cellulosic ethanol plants are expected to operate for three decades or more, steady access to harvest residues and other feedstock is essential (Service 2010). However, the supply chain to provide cellulosic material to meet the increasing share of cellulosic ethanol mandated by the US energy legislation, for example, has not been established. Asia and North America account for two-thirds of the available residue potential due to the production of crops with high residue yields and socioeconomic conditions which limit alternative uses (Daioglou et al. 2016).

Forestry residues (tree tops, branches, slash) are among the traditional biomass used as feedstock for heat and power production (Nabburs et al. 2007). Much of the forest biomass used in developing countries is from unsustainable collection and deforestation as removal of residues may influence forest biodiversity and deplete the soil mineral nutrient and SOC stocks (Gregg and Smith 2010; Riffell et al. 2011). Specifically, the SOC stock may decrease due to reduced C inputs, net primary productivity (NPP) may be reduced due to increase in nutrient export, and forest C losses may increase through accelerated soil erosion (Powers et al. 2005). Further, forest ecosystem functions may be altered with possible negative effects on the C balance (Marland and Obersteiner 2008). However, the major uncertainty for estimating sustainable removal rates is the amount of forestry residue that needs to be left behind to reduce soil erosion (Gregg and Smith 2010). Also, there is no clear consensus on the minimum amount of residues required to remain on forest sites to maintain ecosystem services (Daioglou et al. 2016).

7.2.2.2 Importance of Harvest Residues for Soil Organic Carbon

Similar to forestry residues, agriculture residues such as stalks, stover, and chaff and those from animal husbandry are traditionally used for heat and power production (Gregg and Smith 2010; UNEP 2009). However, there is a linear relationship between residue retention on a site and SOC sequestration (Campbell et al. 2002). Thus, removing agricultural residues from the field depletes the SOC stock (Beniston et al. 2015; Bessou et al. 2011). For example, soil C emissions are about 0.40 Mg C ha^{-1} yr^{-1} when rapeseed straw is removed at sites in France compared to 0.24 Mg C ha^{-1} yr^{-1} when straw is not removed (Arrouays et al. 2002). To maintain the SOC stock, it is assumed that for corn, wheat, barley (*Hordeum vulgare* L.), oats (*Avena sativa* L.), and rapeseed, about 30% of the harvest residue can be removed (Graham et al. 2007; Wilhelm et al. 2007). For example, <40% of the total aboveground residues other than grain can be removed from wheat fields in Saskatoon, Canada, without adversely affecting the SOC stock (Lafond et al. 2009). It is also assumed that for potatoes (*Solanum tuberosum* L.) and sugar beets (*Beta vulgaris* var. *saccharifera*) where the entire plant is harvested, 95% of the residue is recoverable (i.e., 5% retained) (Gregg and Smith 2010). For rice and all other miscellaneous crops (fruits, vegetables), it is assumed that 75% of the residue is recoverable (25% retained).

Studies on sustainable residue harvest rate typically focus on a specific crop and often a single field site (Gregg and Izzauralde 2010). Much of the current research on crop residue has focused on corn residue (stover) owing to the large of amount of biomass corn produces. Using the most representative methodology for assessing long-term corn stover harvesting impacts, Jones et al. (2017) estimated that only 5.3 billion liters per year of ethanol, or 8.7% of the near-term US cellulosic biofuel demand, could be met under common no-till farming practices in the US Midwest. Adjustments to management practices, such as adding cover crops to no-till management, will be required to produce sufficient quantities of residue meeting the GHG emission reduction standard for cellulosic biofuels.

The retention of crop residues on agricultural soils has many direct and indirect benefits with strong impacts on soil health and quality (Lal 2009, 2016). For example, crop residue retention improves the diversity of soil bacterial communities and, in particular, favors beneficial bacterial groups (Ceja-Navarro et al. 2010). Directly, crop residues retained on the soil surface moderate water and energy balance, buffer against erosive forces of raindrops and wind, recycle plant nutrients, and serve as food and habitat for soil organisms (Lal 2009). Indirectly, crop residues affect soil processes through alterations of microclimate, soil moisture and temperature regimes, water and solute transport, and erosional processes. Returning crop residues to soil as amendments is particularly essential to (i) sequestering SOC at rates of 100–1000 kg C ha^{-1} y^{-1} depending on soil type and climate with a total potential globally of 0.6–1.2 Pg C $year^{-1}$ and (ii) recycling plant nutrients (20–60 kg of N, P, K, Ca per Mg of crop residues) amounting to 118 million Mg of N, P, K in residues produced annually in the world (83.5% of world's fertilizer consumption; Lal 2009). Thus, crop residues must be used as soil amendments for meeting world's food demands whereas biofuel plantations must be established on agricultural surplus/marginal soils and, in particular, on eroded/desertified/salinized/disturbed/polluted soils (Lal 2009).

The question of a sustainable residue harvest rate is a question of trade-offs in terms of yield, erosion, and nutrient balance which must be answered for different crops, crop rotations, soil types, climate and management practices. For example, for no-till soils in the USA, over eight times more corn stover is needed to protect against SOC loss than to protect against soil erosion (Wilhelm et al. 2007). Thus, removing residue from agricultural soils in the USA will, in most cases, reduce SOC stocks, reduce soil N, reduce yields, and increase erosion (Gregg and Izaurralde 2010). To enhance the sustainability of stover harvest, the additional nutrients removed with stover must be replaced, and practices should include annual or perennial cover crops, the use of no-tillage and crop rotations (Karlen et al. 2011a). Crop residue management decisions should balance energy, conservation, and soil health requirements for plant biomass (Karlen et al. 2015). Further, some fraction of biomass removal from the field would help lower both agronomic production cost and environmental impact (Shah et al. 2017).

The amount of crop residue needed to maintain SOC stocks can be assessed by the indicator effective organic matter (EOM; De Vries et al. 2010). It is defined as the quantity of OM that is still present in the soil one year after application of the crop residue. Based on environmental conditions similar to arable lands in the Netherlands, only oil palm, winter wheat, and corn produce enough residues to prevent a decline in SOC stocks. Further, SOC stocks are also not maintained when sugarcane or sweet sorghum [*Sorghum bicolor* (L.) Moench] residues are burned. However, in tropical regions, higher amounts of EOM may be required while residue yields are often smaller (De Vries et al. 2010). Thus, residue retention values vary depending on local climate, soil, topography, and management practices (Gregg and Izaurralde 2010). Using a single value of crop residues that could be removed without significant impacts on soil fertility and erosion, and, specifically SOC stocks, to assess biofuel production potentials for large regions such as EU27 is, therefore, highly questionable (Fischer et al. 2010).

Models can be used to simulate long-term changes of residue removal on SOC stocks (Anderson-Teixeira et al. 2009). For example, SOC loss from corn with residue harvest increased by 0.2% for every 1% increase in residue removal, or by about 0.06–0.09 Mg C ha^{-1} per 1% increase in residue removal. Soils with higher clay content had reduced rates of SOC loss. Thus, 10 years of corn biomass removal may result in losses of 3 Mg C ha^{-1} in the top 30 cm soil depth at 25% residue removal and of 8 Mg C ha^{-1} in the top 30 cm at 100% residue removal. Not only did SOC loss increase with percentage residue harvest, losses were also proportionally larger at shallower soil depths. Thus, reduced surface residue inputs were mainly responsible for SOC losses and not reduced belowground inputs (Anderson-Teixeira et al. 2009). For the USA, simulated corn residue removal increases soil loss at rates that vary with topography, crop rotation, and management (Gregg and Izaurralde 2010). For example, corn residue removal decreases 100-year mean yields by 0.07–0.08% for every percent of residue mass removed, decreases SOC by 40–90 kg C ha^{-1} yr^{-1} per Mg of residue harvested, and decreases soil N by 3 kg N ha^{-1} yr^{-1} per Mg residue harvested (Gregg and Izaurralde 2010).

SOC is a soil health and quality indicator that needs to be closely monitored in long-term studies to quantify crop residue removal effects (Karlen et al. 2011b). Changes in SOC by residue harvest depend on initial SOC stock, crop type, crop yield, crop rotation, soil type, climate, management practices (i.e., tillage), and physical characteristics such as slope, erodibility, and topography (Gollany et al. 2010). Removing crop residues from the field led to average SOC contents that were 12 and 18% lower than in soils in which crop residues were retained, in temperate and tropical climates, respectively (Raffa et al. 2015). The large dataset showed a wide variability as a result of the wide range of biophysical and management factors affecting net changes in SOC. For example, continuous corn was the only system that provided sufficient crop residue to maintain SOC stocks in the Central Great Plains of the USA compared to other crop rotations (Benjamin et al. 2010). Further, average SOC sequestration rate when corn crop residues were returned to a soil in China was 0.4 Mg C ha^{-1} yr^{-1} and those of no-tillage was 0.274 kg C ha^{-1} yr^{-1} (Lu and Zhang 2010). After 29 years of complete corn stover removal, corn-derived SOC stocks decreased by up to 39% at no-till (NT) plots in Minnesota, USA, but no changes in SOC stocks were observed at a plot in Connecticut, USA (Hooker et al. 2005; Wilts et al. 2004).

To mitigate loss of crop residue C and its impact on SOC stocks, manure or a cover crop/intensified crop rotation under NT may be options (Gollany et al. 2011). In comparison to disk tillage, conservation tillage somewhat mitigated the SOC loss by residue removal observed in winter wheat–corn–cotton and winter wheat–corn–soybean rotations in the mid-coastal region of South Carolina (Gollany et al. 2010). However, residue harvest from the studied loamy sand soils may have adverse effects on SOC stocks. The effects of residue removal on SOC are often restricted to surface layers with little changes in soil profile SOC stocks. For example, corn residue removal in a winter wheat–corn double cropping system resulted in lower SOC and total N to 30 cm depth but no differences among treatments were detected below this depth (Du et al. 2010). In contrast, the retention of sugarcane harvest residues results in SOC accumulation. Less clear are the effects of residue harvest on SOC accumulation under temperate perennial grasses (Anderson-Teixeira et al. 2009). Further, *Agave* species may have a large potential as bioenergy feedstock on rain-fed semiarid lands as certain species display annual aboveground productivities that are comparable to those of the most water-use-efficient C_3 or C_4 crops but with only 20% of the water required for cultivation (Borland et al. 2009). However, the effects of removing *Agave* leaves on SOC stocks are less well known (Davis et al. 2011).

In summary, the impact of crop residue management on SOC stocks differs among studies because of differences in site-specific conditions (Persson et al. 2010). For achieving sustainable residue removal rates, the intensity and frequency of tillage must be reduced, and cover crops and/or perennials added to the rotation (Johnson et al. 2010). Additional research is needed to determine the proper balance of crop residues that should remain in the field to maintain the SOC stock and the amount that can be removed for bioenergy depending on site-specific conditions (UNEP 2009).

7.2.3 Dedicated Energy Crops

Aside residue and waste biomass, energy crops are grown as feedstock for bioenergy. First-generation liquid biofuels are produced from starch, sugar, and oilseed food commodities to replace petroleum-based transportation fuels (UNEP 2011). Bioethanol and biodiesel are predominantly based on food commodity feedstocks such as corn, sugarcane, soy, canola (*Brassica napus* L. or *B. campestris* L. or *B. Rapa* var.), sunflower, and oil palm. Depending on location, plot size, and management, yields of dedicated bioenergy crops vary widely, i.e., 5–44 Mg dry mass ha^{-1} yr^{-1} for *Miscanthus x giganteus*, 1–35 Mg dry mass ha^{-1} yr^{-1} for switchgrass, and 0–51 Mg dry mass ha^{-1} yr^{-1} for woody species (Searle and Malins 2014).

In 2016, 98.6 billion liters of bioethanol and 30.8 billion liters of biodiesel were produced globally (REN21 2017). The USA and Brazil together alone accounted for 90% of global bioethanol production. However, both land use and land-use change for growing dedicated energy crops may increase GHG emissions and result in SOC losses ('carbon debt'; Fargione et al. 2008; Searchinger et al. 2008). For example, among 14 bioenergy feedstocks, sugarcane had the lowest land-use requirements while those were highest for grain sorghum, soybeans, and rapeseed (Miller 2010). Further, among sugarcane, sugar beet, corn, rice, and cassava (*Manihot esculenta* Crantz) growing under the same conditions of soil and climate in Brazil, sugarcane and corn crops had currently and over the next 30 years the lowest CO_2 emissions (Machado et al. 2017). Brazilian sugarcane ethanol may displace 3.8–13.7% of crude oil consumption by 2045 under projected climate change while protecting forests under conservation, and accounting for future land demand for food and animal feed production (Jaiswal et al. 2017). However, in most LCAs, the impact of land-use and land-cover change on biofuel GHG emissions is not properly accounted for (Gasparatos et al. 2011). Life cycle emissions from bioenergy are substantial (100 g CO_2 eq kWh^{-1}), but highly uncertain (Pehl et al. 2017). In comparison, projected life cycle emission from fossil fuel carbon capture and sequestration plants is 78–110 g CO_2 eq kWh^{-1}. Bioenergy's indirect GHG emissions may exceed even those of fossil-fuel-based electricity, due to high land-use and land-cover change (LULCC) emissions. Nevertheless, bioenergy can play an important role for climate protection by providing substantial net-negative emissions if combined with CCS, and if LULCC emissions are subject to comparably stringent regulations to those from the energy and industry sectors (Pehl et al. 2017).

The use of LCA to assess the C balance of US biofuel production and use is discussed controversially (DeCicco 2017; DeCicco et al. 2016; De Kleine et al. 2017). To understand the full life cycle impacts of producing biofuel, (i) biomass production, (ii) farm to biorefinery transportation, (iii) biorefining, and (iv) biorefinery to fueling station transportation must be considered (Scown et al. 2012). The SOC balance of biofuel crops depends, in particular, on (i) the conversion of uncultivated land to biofuel crops, (ii) the ability of biofuel crops to sequester SOC, and (iii) the effects of biomass harvest on SOC sequestration (Anderson-Teixeira et al. 2009). Net SOC sequestration or emissions during feedstock cultivation may be a major driver for overall GHG impact of biofuels but more researches on factors influencing the SOC balance are needed (Scown et al. 2012).

7.2.3.1 Indirect Land-Use Changes and Soil Organic Carbon

Indirect land-use changes (ILUCs) occur when grassland and forest are converted to cropland somewhere on the globe to meet the demand for commodities displaced by the production of biofuel feedstock. ILUC may be associated with cultivating bioenergy crops but its importance is a debatable issue (Searchinger et al. 2008; Mathews and Tan 2009). Specifically, there is a wide variation in GHG emissions attributable to ILUC (Zilberman 2017). Similarly, the ILUC carbon intensity, i.e., the amount of CO_2 emitted per unit of biofuel produced, differs widely depending on the feedstock, on the type of land used or displaced and on the amortization period (Djomo and Ceulemans 2012). Nevertheless, to deliver transport fuel targets under the Renewable Energy Directive in the European Union, ILUC will affect an area equivalent to just larger than Belgium to just under that of the Republic of Ireland (Bowyer 2010). As SOC and biomass C are released by ILUC, use of additional conventional (first-generation) biofuels would lead to more GHG emissions than meeting the same need through fossil fuel use. Thus, accounting for ILUC is a priority for European biofuels policy (van Renssen 2011).

ILUC in C-rich ecosystems such as those in the tropics may result in significant SOC losses. For example, conversion of primary tropical forest into cropland causes a 25% loss of the SOC stock whereas 30% of SOC stock is lost by conversion into perennial crops and 12% by conversion into grassland (Don et al. 2011). In Brazil, direct land-use changes for the planned expansion of biofuel plantations may have a small impact on C emissions from soil because most plantations would replace rangeland areas (Lapola et al. 2010). However, ILUC especially those pushing the rangeland frontier into the Amazonian forest may offset C savings from biofuels (sugarcane ethanol and soybean diesel). Otherwise, land-use changes and associated C debt for biodiesel may be reduced when oil palm is cultivated as feedstock (Lapola et al. 2010).

In temperate regions, grassland conversion to cropland causes a rapid SOC loss (−36%) with a new SOC equilibrium being reached after 17 years (Poeplau et al. 2011). In contrast, deforestation and grassland establishment in temperate regions cause a long-lasting C sink with a relative SOC stock change of 128% but no new equilibrium may be reached within 120 years. ILUC effects can probably be reduced by new technology and approaches. For example, land-efficient technologies for producing animal feed, including leaf protein concentrates, pretreated forages, and double crops appear to be technically feasible to increase the total amount of plant biomass available for biofuels in the USA (Dale et al. 2010). This approach may also reduce GHG emissions while increasing soil fertility and promoting biodiversity. In principle, GHG emissions associated with ILUC can be quantified and reductions implemented by appropriate policies (Fritsche et al. 2010). However, estimates of SOC losses and GHG emissions associated with ILUC are uncertain (Plevin et al. 2010). For example, the range of published ILUC values is enormous, i.e., ranging from about 200% below, up to 1,700% above the C footprint values of fossil fuels (Finkbeiner 2014). Thus, scientific robustness is not sufficient for political and corporate decision making with regard to ILUC.

7.2.3.2 Direct Land-Use Changes and Soil Organic Carbon

Rates of grassland conversion to cropland for ethanol feedstock production may increase linearly with proximity to a biorefinery location (Wright et al. 2017). Similar to the land-use conversion for feedstock production for the chemical industry, conversion to energy crop plantations may result in SOC losses (Houghton 2010). However, few empirical data exist about the consequences of land-use changes for biofuels on SOC sequestration or SOC release rates (McKone et al. 2011). Initially, large SOC losses (i.e., 22% or 20 Mg C ha^{-1} to a depth of 15 cm) were observed when forest or grassland was converted to sugarcane (Anderson-Teixeira et al. 2009). The changes in SOC stock upon conversion of native vegetation to sugarcane are probably less important compared to C losses from biomass but thorough analyses and paired site comparison are scanty (Lisboa et al. 2011). Small SOC losses are observed when grassland is converted to *Miscanthus x giganteus* but data are uncertain as, for example, among eight sites the

SOC change ranged from −3.3 to 3.3 Mg C ha^{-1} yr^{-1} in the top 30 cm (Anderson-Teixeira et al. 2009). Data on SOC changes are also limited when cropland, fallow land, or grassland is converted to switchgrass.

Model results indicate that converting existing cropland or prairie to switchgrass in the USA increases SOC stocks (Robertson et al. 2011). Further, when cropland is converted to one or more native prairie species to produce biofuel feedstock, SOC accumulates but relatively less at sites that are burned or grazed. Simulation studies indicate that in the USA, the conversion of forest to corn production for biofuels under current tillage practices could reduce SOC by 22% at 30 years after the conversion and up to by 29% at 100 years after the conversion (Kim et al. 2009). Further, converting grassland to cornfield under the current tillage practice may reduce SOC by 15% after 30 years and by 14% after 100 years. Measurements of the GHG consequences of converting former cropland enrolled in the US Conservation Reserve Program (CRP) to continuous corn, corn–soybean rotation, or perennial grass for biofuel production indicated that the foregone SOC sequestration makes one of the largest contributions to overall C debt (Gelfand et al. 2011). Specifically, the total C debt includes future debt created by additional changes in SOC stocks and the loss of substantial future SOC sequestration. Projected C debt repayment periods under NT management ranged from 29 to 40 years for corn–soybean and continuous corn, respectively. However, under conventional tillage, repayment periods were three times longer, from 89 to 123 years, respectively (Gelfand et al. 2011).

In general, greatest increases in SOC occur after conversion from annual species to perennial biofuel crops on poorer-quality sites (Dale et al. 2011). The conversion of annual crops to perennial energy crops potentially increases SOC sequestration, which may not be the case when perennial energy crops are introduced after grassland (Bessou et al. 2011). Also, there is no clear difference between perennial energy crops (i.e., *Miscanthus x giganteus*, switchgrass). In summary, converting native ecosystems to biofuel production frequently results in large C debts or large amounts of CO_2 released as result of burning or microbial decomposition of organic C stored in plant biomass and soil, i.e., when tropical and peatland rainforests are converted (Table 7.1; Fargione et al. 2008). However, experimental evidence for the extent of SOC loss after land conversion to energy crop cultivation is limited as reaching a new SOC stock equilibrium may take decades to centuries.

7.2.3.3 Energy Crop Cultivation and Soil Organic Carbon

Bioenergy crops are projected to cover large areas of lesser-studied landscapes unsuited for food crop production, and this will change the soil microbiome and also affect SOC stocks. Soil microorganisms can promote biofuel feedstock yields directly or indirectly by fixing atmospheric N, increasing P acquisition, recycling nutrients, improving soil aggregation, and suppressing plant disease (Tiedje and Donohue 2008). For example, after 6 years the cropping system at a site in Michigan, USA, significantly impacted the bacterial community composition

Table 7.1 Carbon debts (Mg C ha^{-1}) from direct land-use change for nine scenarios of biofuel production (modified from Fargione et al. 2008)

Biofuel	Former ecosystem	Carbon debt (Mg C ha^{-1})				
		Aboveground	Belowground	Root	Soil	Total
Palm biodiesel	Tropical rainforest	145	47	47	0	191
	Peat rainforest	145	797	47	750	941
Soybean biodiesel	Tropical rainforest	142	59	37	21	201
	Cerrado grassland	4	19	13	6	23
Sugarcane ethanol	Cerrado wooded	16	29	23	6	45
Corn ethanol	Grassland	2	35	7	28	37
	Abandoned cropland	2	17	7	10	19
Prairie biomass ethanol	Abandoned cropland	2	0	0	0	2
	Marginal cropland	0	0	0	0	0

within the top 50 cm depth, with corn and switchgrass communities the most different within the 0–25 cm and 25–50 cm depths, respectively (Zhang et al. 2017). Differences in rooting depth and biomass among crops may have contributed to the changes in soil bacterial community composition. Specifically, belowground plant C inputs such as roots, rhizomes, exudates, and associated soil microorganisms are the primary source for SOC under biofuel crops as most of the aboveground biomass is harvested (Lemus and Lal 2005; Lorenz and Lal 2005). Thus, belowground allocation of C to roots plays a critical role in driving changes in SOC under biofuel crops, and this process may differ among species (Anderson-Teixeira et al. 2009). Otherwise, aboveground inputs to SOC (to 1 m depth) ranged from 70 to 81% for sugarcane and corn vs. 40% for miscanthus (Carvalho et al. 2017). Modeled aboveground C inputs (to 30 cm depth) ranged from 54 to 82% for sugarcane, but were 67% for miscanthus. This study suggested that aboveground crop residues were the main C residue source to the soil in the current bioethanol sector (corn and sugarcane). The indiscriminate removal of crop residues to produce cellulosic biofuels can reduce SOC stocks and reduce the environmental benefits of bioenergy (Carvalho et al. 2017).

Perennial plant species may introduce more C to the soil than annual energy crops such as corn. However, SOC sequestration associated with transition from cropland, grassland, and forest to perennial bioenergy systems is uncertain (Qin et al. 2016). For example, amounts of crop residues found on the soil surface or within the soil were much higher for the perennial crops *Miscanthus x giganteus* and switchgrass than for semi-perennial and annual bioenergy crops at a site in northern France (Ferchaud et al. 2016). More 'recalcitrant' residues but, specifically, higher belowground biomass for the perennial species contributed to this difference. During 6 years, new SOC accumulation was higher for the semi-perennial species fescue (*Festuca arundinacea* Schreb.) and alfalfa (*Medicago* sativa L.) compared to the perennial species (1.50 vs. 0.58 Mg C ha^{-1} yr^{-1}, respectively). Thus, SOC stocks to 33 cm depth increased for semi-perennial species (0.98 Mg C ha^{-1} yr^{-1}) while those did not change for perennial and annual crops. However, SOC stocks will have to be monitored over a longer period to confirm the differences between bioenergy crops and test the occurrence of long-term effects (Ferchaud et al. 2016). Gauder et al. (2016) stressed the importance of including subsoil SOC stocks. At a site in southwestern Germany, perennial energy crops increased SOC stocks in 0–90 cm soil depth by 15–20 Mg C ha^{-1} over 11 years compared with annual energy crops.

The establishment of perennial, i.e., certain grassland species in a common garden experiment in Minnesota in the USA, declined SOC within five years but the response was species-specific and varied with soil depth (Laungani and Knops 2009). For example, *Schizachyrium scoparium* (Michx.) Nash lost 15 and 20% SOC from depths 0–10 and 10–25 cm, respectively. In contrast, no change in total SOC stock to 50 cm depth was observed for *Poa pratensis* L., and the forbs *Lespedeza capitata* Michx. and *Solidago canadensis* L. However, in 0–10 cm depth, SOC increased for *S. canadensis* whereas it decreased for *L. capitata* and *P. pratensis*. The increased corn production for bioethanol in the USA is estimated to decrease the SOC stock when conservation tillage practices are not more widely adopted (Larson et al. 2010). Biomass removal from perennial warm-season grasses may be a desirable alternative to corn residue removal for biofuel production and maintenance of soil ecosystem services (Blanco-Canqui et al. 2017).

The agricultural management of crop plantations including cropping practices, irrigation, use of fertilizers, and types of tillage may also affect the SOC stock. For example, decreases in SOC stocks were 0.28–0.40 Mg C ha^{-1} yr^{-1} for oil crops, 0.76–1.00 Mg C ha^{-1} yr^{-1} for potatoes, 0.76–1.30 Mg C ha^{-1} yr^{-1} for beets, and 0.29–0.40 Mg C ha^{-1} yr^{-1} for cereals (summarized in Scholz et al. 2010). In contrast, SOC increased by 0.6–0.8 Mg C ha^{-1} yr^{-1} under grass cultivation. On a global scale, however, agricultural management has not been included in analyses of changes in SOC density (Houghton 2010). For example, soil inventories are inconclusive whether European cropland soils are a C sink or a C source (Ciais et al. 2010). Ongoing SOC loss in spite of good practices (e.g., inputs of manure and crop residues) is mainly related to soils with high C concentrations which are not in equilibrium but SOC losses may also be a result of already ongoing climate change (Kutsch et al. 2010). Thus, a modification in the good practice guidelines to increase C inputs may be required. Possibly as a result of tillage, cropland soils in Europe are much less efficient in C sequestration than those of other biomes (Ciais et al. 2010).

The amount and rate of SOC change when cropland, grassland, or forest is converted to bioenergy systems are highly dependent on the specific land transition (Qin et al. 2016). For example, changes in SOC cropland conversions resulted in an overall SOC gain of 6–14% relative to initial SOC stocks based on data from worldwide field observations. Otherwise, conversion from grassland or forest to corn (without residue removal) or poplar caused significant SOC loss (9–35%). No significant SOC changes were observed in land converted from grasslands or forests to switchgrass, *Miscanthus x giganteus*, or willow. Further, SOC response ratios were similar in both 0–30 and 0–100 cm soil depths in most cases. Thus, changes in SOC stock by land transitions to bioenergy systems should incorporate SOC changes in deep soil which at present is not a common practice (Qin et al. 2016).

The land conversion SOC loss when native ecosystems are converted to sugarcane appears to recover within a century at rates of about 0.3 yr^{-1} to 0.5 Mg C ha^{-1} yr^{-1} in the top 30 cm (Anderson-Teixeira et al. 2009). This SOC accumulation is proportionally greater at depth and the rate of recovery increases with retention of harvest residues. Cultivation of *Miscanthus x giganteus*, switchgrass, or native mixes may increase SOC by an average of 0.1–1 Mg C ha^{-1} yr^{-1} in the top 30 cm (Anderson-Teixeira et al. 2009). In particular, *Miscanthus x giganteus* has a great potential to increase SOC as, for example, this perennial grass has 61% greater annual solar radiation interception efficiency by the plant canopy and can produce 59% more aboveground biomass than heavily fertilized, highly domesticated annual corn in the US Corn Belt (Dohleman and Long 2009). Further, *Miscanthus x giganteus* allocates large quantities of C to rhizomes and roots which are contributing to the significant amounts of *Miscanthus x giganteus*-derived C accumulating in the soil (Amougou et al. 2011). Also, *Miscanthus x giganteus* may have the potential to accumulate SOC at a faster rate compared to switchgrass, native mixes, or continuous corn (Davis et al. 2010). *Miscanthus x giganteus* plantations in Ireland, for example, have the potential to store between 2 and 3 Mg C ha^{-1} yr^{-1} depending on crop yield and the initial SOC stock (Dondini et al. 2009). Specifically in degraded soils, switchgrass increases the amount of SOC sequestered because of its extensive root system, root secretion of organic compounds that bind soil particles and stabilize SOC with inputs into soil of up to 2.2 Mg C ha^{-1} yr^{-1} (Lemus and Lal 2005). Further, clayey soils (clay content > 35%) under sugarcane in Brazil accumulated C at a rate of 2.04 Mg C ha^{-1} yr^{-1} and sandy soils at a rate of 0.73 Mg C ha^{-1} yr^{-1} (Galdos et al. 2010). Also, harvesting sugarcane mechanically without burning tended to increase SOC stocks providing a C sink of 1.5 Mg C ha^{-1} yr^{-1} to 30 cm depth. In general, SOC stocks under sugarcane were comparable to those under natural vegetation at least when mechanical harvest was applied (Lisboa et al. 2011).

In arid and semiarid regions of the world, cultivation of crassulacean acid metabolism (CAM) plants could play a significant role in sequestration of atmospheric CO_2 in the soil (Borland et al. 2009). Compared to C_3 and C_4 bioenergy crops, CAM plants have water-use efficiency (WUE) advantages due to nocturnal stomatal opening, rapid rectifier-like root hydraulic conductivity responses to fluctuating soil water potential, and the capacity to buffer against periods of drought (Owen et al. 2016). Thus, CAM bioenergy crops may be particularly suitable for biomass production on low-grade land which is needed to meet future energy demands and minimize resource conflicts. For example, the CAM bioenergy crops *Agave tequilana* F.A.C. Weber and *Opuntia ficus-indica* (L.) Mill. outperformed the C_4 bioenergy crop, *Panicum virgatum* L. (switchgrass) in arid zones in the latitudinal range 30°S–30°N (Owen et al. 2016).

In summary, the functional drivers of differences in the SOC balance among energy crops are not fully understood (Anderson-Teixeira et al. 2009), but the soil microbiome may play a role. Long-term common garden experiments with replicated monocultures under similar management practices (i.e., fertilization, no-till) and climate, and on similar soil types are required to study effects of energy crop species on SOC stocks (Laungani and Knops 2009).

7.2.3.4 Enhancing Soil Organic Carbon Under Energy Crops

The effects of energy crop cultivation on SOC stocks can be reduced by applying by-products of bioenergy production as amendments or fertilizers to the soil. The recycling of nutrients with wastes and ashes from the bioenergy sector is necessary to reduce increasing fertilizer nutrient demand for bioenergy production. For example, by 2030 nutrient fertilizer application rate in the USA may have to be increased by a factor of 5.5 relative to 1997 to attain the goal set by the US Biomass Roadmap (Han et al. 2011). Further, another 750,000 Mg of fertilizer would have to be applied by 2020 to cultivate 45% of China's required bioethanol feedstock on new marginal land (Qiu et al. 2011). Aside nutrients, by-products of bioenergy production generally contain large amounts of C and potentially the more 'recalcitrant' fractions (Cayuela et al. 2010). Thus, soil application of by-products may directly enhance the SOC stock or indirectly by improving physicochemical soil characteristics for plant growth. By-products may either be directly applied to soils or through animal manure after using these materials as fodder. During a laboratory incubation study, for example, first-generation biofuel by-products (i.e., rapeseed meal, distilled dried grains with solubles) emitted more CO_2 compared to second-generation biofuel by-products (i.e., non-fermentables from hydrolysis of different lignocellulosic materials). However, biochars showed the lowest CO_2 loss in this laboratory study and may, thus, have the highest SOC storage potential (Cayuela et al. 2010). While most studies focused either on biogas residues or biochar application to soils, long-term field studies with a range of bioenergy by-products are required to assess their effects on SOC stocks.

7.3 Sustainability of Biomass and Bioenergy Production

The complex interactions among bioenergy sustainability, food security, and resource management are not completely understood (Goetz et al. 2017; Kline et al. 2017). Bioenergy systems differ widely in terms of costs and CO_2 emissions (Azar 2011). In response to increasing C prices, certain bioenergy forms may become so attractive that food prices may increase significantly, that poor people may be evicted from their lands, and that rainforest and other sensitive ecosystems may be destroyed in order to pave the way for bioenergy plantations. Thus, expanding bioenergy production will require sophisticated land and water-use management (Arvizu et al. 2011). Specifically, avoiding negative environmental impacts of large-scale bioenergy production will require policies that balance associated water and land requirements (Bonsch et al. 2016). For example, characterization of the spatial distribution of the long-term production potential of biomass resources in the USA can support crop selection decisions (Daly et al. 2018). Increases in global feedstock productivity for food, fodder, fiber, forest products, and energy will be required as will be substantial conversion technology improvements. Needed is also a refined understanding of the complex social, energy, and environmental interactions associated with biomass and bioenergy production and use (Arvizu et al. 2011; Slade et al. 2011). To deal with those challenges, for example, bioenergy taxes are discussed as a policy instrument (Azar 2011).

Issues associated with the sustainability of first-generation biofuels are that they need to be (i) net energy providers, (ii) environmentally sustainable, (iii) economically competitive, (iv) not compete with food production, and (v) met additional social criteria (Hill et al. 2006; Rist et al. 2009). However, for the USA, the energy return on energy investment (EROI) ratio for corn-based ethanol is about 1:1 whereas it is 10:1 for oil and gas production (Hall and Benemann 2011). The liquid biofuels, in particular, have a low EROI. On one hand, biofuels provide a number of ecosystem services (ESs; e.g., fuel, climate regulation) but they also compromise other ESs (e.g., food, freshwater services) which are of paramount value to human well-being (Gasparatos et al. 2011). Specifically, the main impacts of biofuels on human well-being are (i) rural development, (ii) energy security and access to energy, (iii) food security and access to food, (iv) health, (v) land tenure, and (vi) gender issues. Thus, ESs may be suitable to assess biofuel sustainability (Gasparatos et al. 2011). However, when the Critical Zone (CZ), i.e., the zone that includes the biophysical processes occurring from the top of the vegetation canopy to the weathering zone below the groundwater table, is included, current corn feed production systems are more energy efficient and less environmentally costly than corn-based ethanol (Richardson and Kumar 2017). Thus, continued research, development and demonstration are needed in ways that are synergistic with food agriculture for producing biofuels that possess high productivity and sustainability profiles (Murphy et al. 2011). It may be possible to meet increasing demands for food using new technologies although this may not be easily or cheaply accomplished (Valentine et al. 2012). Appropriate policies are needed as the impact of biofuel policy on grain prices can be very large (de Gorter and Drabik 2012).

Similar to the sustainability of first-generation biofuels that of the second-generation biofuels will depend on energy security, food production, land productivity, SOC and erosion, life cycle GHG emissions, biodiversity, air quality, and water quantity and quality, how the complexity of feedstock logistics can be addressed, and whether jobs are created and regional development is promoted (Dale et al. 2011; IEA 2010). Aside biochemical consequences, bioenergy crops potentially impact the surface energy (albedo) and water balance (evapotranspiration) with consequences for the local, regional, and global climate. For example, growing high-albedo bioenergy crops such as switchgrass (*Panicum virgatum* L.) and *Miscanthus x giganteus* in the central USA may cool the regional climate (Georgescu et al. 2011). Also, expanding sugarcane into existing crop and pasture land in Brazil may have a direct local cooling effect (Loarie et al. 2011). The large-scale development of biomass energy technologies may produce or exacerbate water shortages globally, and be limited by the availability of freshwater (Mulder et al. 2010). For example, a net increase in evapotranspiration is projected for *Miscanthus x giganteus* and switchgrass expansion in the Midwestern USA leading to significant reduction in soil-moisture storage and specific surface runoff (Le et al. 2011). However, bioenergy development may also provide opportunities to improve water productivity and increase access to water by leveraging the introduction of efficient water management techniques and providing energy for water pumping and cleaning, and by providing a wider range of land-use options to optimize the use of land and water (Otto et al. 2011). Sustainable biofuel systems should, therefore, (1) have favorable economics, (2) conserve natural resources, (3) preserve ecology, and (4) promote social justice (Payne 2010). However, as large-scale commercial production was implemented only recently, the sustainable harvest rate of residues in relation to SOC sequestration, soil erosion, yield, and soil nutrients is unclear (Anderson-Teixeira et al. 2009; Gregg and Izaurralde 2010).

The sustainability of biofuel feedstock production can probably be achieved via appropriately designed crop residue and perennial lignocellulosic systems (Dale et al. 2011; Valentine et al. 2012). The bioenergy supply chains can be sustainably designed both efficiently and effectively, and also for fostering relationships with relevant external stakeholders (Gold 2011). However, the future agricultural expansion for large-scale cultivation of energy crops will be limited by land availability and irrigation area, and create increasing pressures on already fragmented and degraded lands, areas rich in biodiversity, and on habitats of endemic and endangered species (Beringer et al. 2011). Also, the projected increase in food demand will constrain energy crop cultivation on agricultural land (Dornburg et al. 2010). For example, China has little extra farmland or marginal land for producing biomass energy as it possesses about 9% of the world's arable land supporting 20% of the world's population (Qiu et al. 2011; Sang and Zhu 2011). Thus, farmland in other countries particularly those in Africa which has been traditionally used by local communities is increasingly leased or sold to outside investors from corporations and governments ('land acquisitions'; Deininger 2011).

For biomass and bioenergy production, the best possible standard of environmental sustainability must be developed, for example, by an international organization already working toward such a standard, such as the United Nations Environment Programme (Nuffield Council on Bioethics 2011). Specifically, different biofuel types must be certified on the basis of their life cycle GHG emissions according to attributional LCA, and based on a single international standard. Recently, global, government-level consensus was achieved on a set of 24 voluntary, science-based indicators for assessing the sustainable production and use of bioenergy (GBEP 2011). Among the indicators is soil quality, described as 'percentage of land for which soil quality, in particular in terms of SOC, is maintained or improved out of total land on which bioenergy feedstock is cultivated or harvested.' Thus, the responses of SOC to annual and perennial biofuel cropping systems must be determined. For example, total SOC stock is among a list of nineteen recently recommended measurable environmental indicators for bioenergy sustainability as it potentially affects climate change, N mineralization, humification, water holding capacity, infiltration, and cation exchange capacity (McBride et al. 2011). Other indicators were also related to soil quality, and to water quality and quantity, GHGs, biodiversity, air quality, and productivity. Biomass sustainability certification should be based on physically possible and socially acceptable scenarios (Slade et al. 2011). Sustainability certification of oil palm was associated with reduced deforestation in Indonesia from 2001 to 2015 (Carlson et al. 2018). However, certification had no causal impact on forest loss in peatlands or active fire detection rates.

With regard to biofuel-related scientific research, it has recently been suggested that GHGs, fuel production, and feedstock production are well represented in the literature (Ridley et al. 2012). However, this is not the case for trade, biodiversity, and human health. Gaps were particularly striking across the Southern Hemisphere where the greatest socioeconomic benefits but also significant environmental damages may occur. Thus, there is enough uncertainty to warrant caution toward the developing biofuel industry, especially in several understudied topics and regions (Ridley et al. 2012). Similarly, Searchinger et al. (2017) concluded that currently policy should not support bioenergy from energy crops and other dedicated uses of land. Franzluebbers (2015) emphasized that the design of sustainable landscapes to accommodate food, feed, fiber, and biofuel production has biophysical limitations, but also requires the input of a diversity of stakeholders, including farmers, communities, industries, and government policymakers. Attention to soil ecology and management to optimize multiple functions should be a part of the discussion.

7.4 Conclusions

Biomass has its own niches but in the long term may be surpassed by other sources of renewable energy such as inanimately harvested solar energy because bioenergy has the much lower solar radiation conversion efficiency. Yet, biomass has the potential to replace petroleum as feedstock for biodiesel, aviation fuel, and petrochemicals. Biomass production must not only be physically possible but also socially acceptable. The greatest potential for yielding biomass for bioenergy is probably by cultivation on abandoned land and land not converted to other uses. Further, agricultural intensification may contribute to an increase in global biofuel production. However, adequate research inputs are needed to assess the potential of species such as Jatropha on marginal lands. On contaminated land, short-rotation woody crops (SRWC) such as *Salix* spp. and *Populus* spp. may be grown. Among agricultural systems, issues of production costs, land demand, and effectiveness in reducing GHG emissions may be reasons why ethanol from sugarcane may be the only first-generation biofuel produced in the long term. However, sugarcane may have larger impacts on regional climate forcing and human health than previously thought due to not well characterized emissions of non-GHG gases and aerosols from burning sugarcane croplands. To effectively replace petroleum-based fuels, biomass may be increasingly turned more complex alcohols and hydrocarbons that closely resemble those in petroleum aside being converted into ethanol. However, logistics of biomass harvest, storage, processing, and transport for the rapid growth in demand for lignocellulosic bioenergy must be adapted to meet the challenges of scale. For example, while Africa and South America have more than half of the total land available for biofuel production, large fuel demands exist in the USA, Europe, China, and India.

7.5 Review Questions

1. Describe the principal differences in plant biomass feedstocks for bioenergy and the chemical industry
2. What are major issues facing the use of algal biomass as feedstock?
3. Discuss the pros and cons of the use of traditional biomass in developing countries
4. How much soil area of what productivity in a specific geographic region is needed on average to prepare a warm meal, supply an apartment with energy for electricity, warm water, and heating, drive 10 km by car, or travel 10,000 km by plane?
5. Describe the input data needed to evaluate the performance of bioenergy by life cycle analysis
6. How does bioenergy production affect the three dimensions of sustainability locally, regionally, and globally?
7. Which plant species are most promising feedstocks and why?
8. What are the implications of growing biomass for bioenergy for global soil resources?
9. What other low-GHG emitting alternatives exist for energy generation?

References

Albanito F, Beringer T, Corstanje R, Poulter B, Stephenson A, Zawadzka J, Smith P (2016) Carbon implications of converting cropland to bioenergy crops or forest for climate mitigation: a global assessment. GCB Bioenerg 8:81–95

Amougou N, Bertrand I, Machet JM, Recous S (2011) Quality and decomposition in soil of rhizome, root and senescent leaf from *Miscanthus x giganteus*, as affected by harvest date and N fertilization. Plant Soil 338:83–97

Anderson K (2015) Talks in the city of light generate more heat. Nature 528:437

Anderson-Teixeira KJ, Davis SC, Masters MD, DeLucia EH (2009) Changes in soil organic carbon under biofuel crops. GCB Bioenergy 1:75–96

Aro EM (2016) From first generation biofuels to advanced solar biofuels. Ambio 45(Suppl. 1): S24–S31. https://doi.org/10.1007/s13280-015-0730-0

Arrouays D, Balesdent J, Germon J, Jayet P, Soussana J, Stengel P (2002) Stocker du carbone dans les sols agricoles de France? Expertise scientifique collective. Rapport d'expertise re´ alise´ par INRA a` la demande du Ministe`re de l'Ecologie et du De´veloppement Durable. In: Contribution a` la lutte contre l'effet de serre. Paris, France: INRA; 2002

Arvizu D, Bruckner T, Edenhofer O, Estefen S, Faaij A, Fischedick M, Hiriart G, Hohmeyer O, Hollands KGT, Huckerby J, Kadner S, Killingtveit Å, Kumar A, Lewis A, Lucon O, Matschoss P, Maurice L, Mirza M, Mitchell C, Moomaw W, Moreira J, Nilsson LJ, Nyboer J, Pichs-Madruga R, Sathaye J, Sawin J, Schaeffer R, Schei T, Schlömer S, Seyboth K, Sims R, Sinden G, Sokona Y, von Stechow C, Steckel J, Verbruggen A, Wiser R, Yamba F, Zwickel T (2011) Technical summary. In: Edenhofer O, Pichs-Madruga R, Sokona Y, Seyboth K, Matschoss P, Kadner S, Zwickel T, Eickemeier P, Hansen G, Schlömer S, von Stechow C (eds) IPCC special report on renewable energy sources and climate change mitigation. Cambridge University Press, Cambridge, United Kingdom and New York, NY, USA

Azar C (2011) Biomass for energy: a dream come true. or a nightmare? WIREs Clim Change 2:309–323

Balagopal B, Paranikas P, Rose J (2010) What's next for alternative energy?. The Boston Consulting Group Inc., Boston, MA

Bartle JR, Abadi A (2010) Toward sustainable production of second generation bioenergy feedstocks. Energy Fuels 24:2–9

Bauen A, Berndes G, Junginger M, Londo M, Vuille F (2009) Bioenergy—a sustainable and reliable energy source. A review of status and prospects. IEA Bioenergy: ExCo:2009:06

Beer T, Grant T, Campbell PK (2007) The greenhouse and air quality emissions of biodiesel blends in Australia. CSIRO

Beniston JW, Shipitalo MJ, Lal R et al (2015) Carbon and macronutrient losses during accelerated erosion under different tillage and residue management. Eur J Soil Sci 66:218–225. https://doi.org/10.1111/ejss.12205

Benjamin JG, Halvorson AD, Nielsen DC, Mikha MM (2010) Crop management effects on crop residue production and changes in soil organic carbon in the central Great Plains. Agron J 102:990–997

Beringer T, Lucht W, Schaphoff S (2011) Bioenergy production potential of global biomass plantations under environmental and agricultural constraints. GCB Bioenerg 3:299–312

Bessou C, Ferchaud F, Gabrielle B, Mary B (2011) Biofuels, greenhouse gases and climate change. A review. Agron Sustain Dev 31:1–79

Blanco-Canqui H, Mitchell RB, Jin VL, Schmer MR, Eskridge KM (2017) Perennial warm-season grasses for producing biofuel and enhancing soil properties: an alternative to corn residue removal. GCB Bioenerg 9:1510–1521. https://doi.org/10.1111/gcbb.12436

Bonsch M, Humpenöder F, Popp A, Bodirsky B, Dietrich JP, Rolinski Biewald A, Lotze-Campen H, Weindl I, Gerten D, Stevanovic M (2016) Trade-offs between land and water requirements for large-scale bioenergy production. GCB Bioenerg 8:11–24

Borland AM, Griffiths H, Hartwell J, Smith JAC (2009) Exploiting the potential of plants with crassulacean acid metabolism for bioenergy production on marginal lands. J Exp Bot 60:2879–2896

Bourgis F, Kilaru A, Cao X, Ngando-Ebongue G-F, Drira N, Ohlrogge JB, Arondel V (2011) Comparative transcriptome and metabolite analysis of oil palm and date palm mesocarp that differ dramatically in carbon partitioning. Proc Natl Acad Sci USA 108:12527–12532

Bouwman AF, Van Grinsven JJM, Eickhout B (2010) Consequences of the cultivation of energy crops for the global nitrogen cycle. Ecol Appl 20:101–109

Bowyer C (2010) Anticipated indirect land-use change associated with expanded use of biofuels and bioliquids in the EU—an analysis of the national renewable energy action plans. www.ieep.eu

Brandão M, Milà i Canals L, Clift R (2011) Soil organic carbon changes in the cultivation of energy crops: implications for GHG balances and soil quality for use in LCA. Biomass Bioenerg 35:2323–2336

Buchholz T, Hurteau MD, Gunn J, Saah D (2016) A global meta-analysis of forest bioenergy greenhouse gas emission accounting studies. GCB Bioenerg 8:281–289

Campbell CA, McConkey BG, Gameda S, Izaurralde RC, Liang BC, Zentner RP, Sabourin D (2002) Efficiencies of conversion of residue C to soil C. In: Kimble JM, Lal R, Follett RF (eds) Agricultural practices and policies for carbon sequestration in soil. CRC Press, Boca Raton, FL, pp 305–314

Carlson KM, Heilmayr R, Gibbs HK et al (2018) Effect of oil palm sustainability certification on deforestation and fire in Indonesia. Proc Natl Acad Sci USA 115:121–126

Carlsson AS (2009) Plant oils as feedstock alternatives to petroleum—a short survey of potential oil crop platforms. Biochimie 91:665–670

Carvalho JLN, Hudiburg TW, Franco HCJ, Delucia EH (2017) Contribution of above- and belowground bioenergy crop residues to soil carbon. GCBBioenerg 9:1333–1343. https://doi.org/10.1111/gcbb.12411

Cayuela ML, Oenema O, Kuikman PJ, Bakker RR, van Groenigen JW (2010) Bioenergy by-products as soil amendments? Implications for carbon sequestration and greenhouse gas emissions. GCB Bioenerg 2:201–213

Ceja-Navarro JA, Rivera-Orduña FN, Patiño-Zúñiga L, Vila-Sanjurjo A, Crossa J, Govaerts B, Dendooven L (2010) Phylogenetic and multivariate analyses to determine the effects of different tillage and residue management practices on soil bacterial communities. Appl Environ Microbiol 76:3685–3691

Chaplin-Kramer R, Sim S, Hamel P et al (2017) Life cycle assessment needs predictive spatial modelling for biodiversity and ecosystem services. Nat Commun 8:15065. https://doi.org/10.1038/ncomms15065

Chase LDC, Henson IE (2010) A detailed greenhouse gas budget for palm oil production. Int J Agric Sustain 8:199–214

Ciais P, Wattenbach M, Vuichard N, Smith P, Piao SL, Don A, Luyssaert S, Janssens IA, Bondeau A, Dechow R, Leip A, Smith PC, Beer C, Van DerWerf GR, Gervois S, Van Oost K, Tomelleri E, Freibauer A, Schulze E-D, Synthesis Team CARBOEUROPE (2010) The European carbon balance. Part 2: croplands. Glob Change Biol 16:1409–1428

Creutzig F (2016) Economic and ecological views on climate change mitigation with bioenergy and negative emissions. GCB Bioenerg 8:4–10

Creutzig F, Ravindranath NH, Berndes G, Bolwig S, Bright R, Cherubini F, Chum H, Corbera E, Delucchi M, Faaij A, Fargione J, Haberl H, Heath G, Luconi O, Plevin R, Popp A, Robledo-Abad C, Rose S, Smith P, Stromman A, Suh S, Masera O (2015) Bioenergy and climate change mitigation: an assessment. GCB Bioenerg 7:916–944

Crutzen PJ, Mosier AR, Smith KA, Winiwarter W (2008) N_2O release from agro-biofuel production negates global warming reduction by replacing fossil fuels. Atmos Chem Phys 8:389–395

Daioglou V, Stehfest E, Wicke B, Faaij AC, Van Vuuren DP (2016) Projections of the availability and cost of residues from agriculture and forestry. GCB Bioenerg 8:456–470

Daioglou V, Wicke B, Faaij AC, Van Vuuren DP (2015) Competing uses of biomass for energy and chemicals: implications for long-term global CO_2 mitigation potential. GCB Bioenerg 7:1321–1334

Dale BE, Bals BD, Kim S, Eranki P (2010) Biofuels done right: land efficient animal feeds enable large environmental and energy benefits. Environ Sci Technol 44:8385–8389

Dale VH, Fl Kline, Wright LL, Perlack RD, Dowing M, Graham RL (2011) Interactions among bioenergy feedstock choices, landscape dynamics, and land use. Ecol Appl 21:1039–1054

Daly C, Halbleib MD, Hannaway DB, Eaton LM (2018) Environmental limitation mapping of potential biomass resources across the conterminous United States. GCB Bioenerg. https://doi.org/10.1111/gcbb.12496

Davis SC, Dohleman FG, Long SP (2011) The global potential for *Agave* as a biofuel feedstock. GCB Bioenergy 3:68–78

Davis SC, Parton WJ, Dohleman FG, Smith CM, Del Grosso S, Kent AD, DeLucia EH (2010) Comparative biogeochemical cycles of bioenergy crops reveal nitrogen-fixation and low greenhouse gas emissions in a *Miscanthus x giganteus* agro-ecosystem. Ecosystems 13:144–156

De Gorter H, Drabik G (2012) The effect of biofuel policies on food grain commodity prices. Biofuels 3:21–24

DeCicco (2017) Author's response to commentary on "Carbon balance effects of U.S. biofuel production and use". Clim Chang 144:123–129. https://doi.org/10.1007/s10584-017-2026-9

DeCicco JM, Liu DY, Heo J et al (2016) Carbon balance effects of U.S. biofuel production and use. Clim Chang 138:667–680

De Kleine R, Wallington TJ, Anderson JE, Kim HC (2017) Commentary on "Carbon balance effects of U.S. biofuel production and use" by DeCicco et al. (2016). Clim Chang 144:111–119. https://doi.org/10.1007/s10584-017-2032-y

Deininger K (2011) Challenges posed by the new wave of farmland investment. J Peasant Stud 38:217–247

De Vries SC, Van de Ven GWJ, Van Ittersum MK, Giller KE (2010) Resource use efficiency and environmental performance of nine major biofuel crops, processed by first-generation conversion techniques. Biomass Bioenerg 34:588–601

Djomo SN, Ceulemans R (2012) A comparative analysis of the carbon intensity of biofuels caused by land use changes. GCB Bioenerg 4:392–407. https://doi.org/10.1111/j.1757-1707.2012.01176.x

Dohleman FG, Long SP (2009) More productive than maize in the Midwest: how does Miscanthus do it? Plant Physiol 150:2104–2115

Don A, Schumacher J, Freibauer A (2011) Impact of tropical land-use change on soil organic carbon stocks—a meta-analysis. Glob Change Biol 17:1658–1670

Dondini M, Hastings A, Saiz G, Jones MB, Smith P (2009) The potential of *Miscanthus* to sequester carbon in soils: comparing field measurements in Carlow, Ireland to model predictions. GCB Bioenergy 1:413–425

Dornburg V, van Vuuren D, van de Ven G, Langeveld H, Meeusen M, Banse M, van Oorschot M, Ros J, van den Born GJ, Aiking H, Londo M, Mozaffarian H, Verweij P, Lyseng E, Faaij A (2010) Bioenergy revisited: key factors in global potentials of bioenergy. Energy Environ Sci 3:258–267

Du Z, Ren T, Hu C (2010) Tillage and residue removal effects on soil carbon and nitrogen storage in the North China Plain. Soil Sci Soc Am J 74:196–202

Edenhofer O, Pichs-Madruga R, Sokona Y, Seyboth K, Arvizu D, Bruckner T, Christensen J, Devernay J-M, Faaij A, Fischedick M, Goldstein B, Hansen G, Huckerby J, Jäger-Waldau A, Kadner S, Kammen D, Krey V, Kumar A, Lewis A, Lucon O, Matschoss P, Maurice L, Mitchell C, Moomaw W, Moreira J, Nadai A, Nilsson LJ, Nyboer J, Rahman A, Sathaye J, Sawin J, Schaeffer R, Schei T, Schlömer S, Sims R, Verbruggen A, von Stechow C, Urama K, Wiser R, Yamba F, Zwickel T (2011) Summary for policymakers. In: Edenhofer O, Pichs-Madruga R, Sokona Y, Seyboth K, Matschoss P, Kadner S, Zwickel T, Eickemeier P, Hansen G, Schlömer S, von Stechow C (eds) IPCC special report on renewable energy sources and climate change mitigation. Cambridge University Press, Cambridge, United Kingdom and New York, NY, USA

Erisman JW, van Grinsven H, Leip A, Mosier A, Bleeker A (2010) Nitrogen and biofuels; an overview of the current state of knowledge. Nutr Cycl Agroecosyst 86:211–223

FAO (2008) Forests and energy. FAO Forestry paper 154. FAO, Rome

FAO (2017) FAOSTAT. http://www.fao.org/faostat/en/#data

Fargione J, Hill J, Tilman D, Polasky S, Hawthorne P (2008) Land clearing and the biofuel carbon debt. Science 319:1235–1238

Ferchaud F, Vitte G, Mary B (2016) Changes in soil carbon stocks under perennial and annual bioenergy crops. GCB Bioenerg 8:290–306

Field CB, Campbell JE, Lobell DB (2008) Biomass energy: the scale of the potential resource. Trends Ecol Evol 23:65–72

Finkbeiner M (2014) Indirect land use change—help beyond the hype? Biomass Bioenerg 62:218–221

Fischer G, Prieler S, Velthuizen H, Berndes G, Faaij A, Londo M, de Wit M (2010) Biofuel production potentials in Europe: sustainable use of cultivated land and pastures, part II: land use scenarios. Biomass Bioenerg 34:173–187

Franzluebbers AJ (2015) Farming strategies to fuel bioenergy demands and facilitate essential soil services. Geoderma 259–260:251–258

Frazão LA, Paustian K, Cerri CEP, Cerri CC (2014) Soil carbon stocks under oil palm plantations in Bahia State, Brazil. Biomass Bioenerg 62:1–7

Fritsche UR, Sims REH, Monti A (2010) Direct and indirect land-use competition issues for energy crops and their sustainable production—an overview. Biofuels Bioprod Bioref 4:692–704

Galdos MV, Cerri CC, Lal R, Bernoux M, Feigl B, Cerri CEP (2010) Net greenhouse gas fluxes in Brazilian ethanol production systems. GCB Bioenerg 2:37–44

Gasparatos A, Stromberg P, Takeuchi K (2011) Biofuels, ecosystem services and human wellbeing: putting biofuels in the ecosystem services narrative. Agric Ecosyst Environ 142:111–128

Gauder M, Billen N, Zikeli S et al (2016) Soil carbon stocks in different bioenergy cropping systems including subsoil. Soil Till Res 155:308–317

GBEP (Global Bioenergy Partnership) (2011) GBEP Sustainability indicators, May 20, 2011. FAO/GBEP, Rome

Gelfand I, Zenone T, Jasrotia P, Chen J, Hamilton SK, Robertson GP (2011) Carbon debt of Conservation Reserve Program (CRP) grasslands converted to bioenergy production. Proc Natl Acad Sci USA 108:13864–13869

Georgescu M, Lobell DB, Field CB (2011) Direct climate effects of perennial bioenergy crops in the United States. Proc Natl Acad Sci USA 108:4307–4312

Godfray HCJ, Beddington JR, Crute IR, Haddad L, Lawrence D, Muir JF, Pretty J, Robinson S, Thomas SM, Toulmin C (2010) Food security: the challenge of feeding 9 billion people. Science 327:812–818

Goetz A, German L, Weigelt J (2017) Scaling up biofuels? A critical look at expectations, performance and governance. Energ Pol in press

Gold S (2011) Bio-energy supply chains and stakeholders. Mitig Adapt Strateg Glob Change 16:439–462

Gollany HT, Novak JM, Liang Y, Albrecht SL, Rickman RW, Follett RF, Wilhelm WW, Hunt PG (2010) Simulating soil organic carbon dynamics with residue removal using the CQESTR model. Soil Sci Soc Am J 74:372–383

Gollany HT, Rickman RW, Liang Y, Albrecht SL, Machado S, Kang S (2011) Predicting agricultural management influence on long-term soil organic carbon dynamics: implications for biofuel production. Agron J 103:234–246

Goodrick I, Nelson PN, Banabas M, Wurster CM, Bird MI (2015) Soil carbon balance following conversion of grassland to oil palm. GCB Bioenerg 7:263–272

Graham RL, Nelson R, Sheehan J, Perlack RD, Wright LL (2007) Current and potential U.S. corn stover supplies. Agron J 99:1–11

Gregg JS, Izaurralde RC (2010) Effect of crop residue harvest on long-term crop yield, soil erosion and nutrient balance: trade-offs for a sustainable bioenergy feedstock. Biofuels 1:69–83

Gregg JS, Smith SJ (2010) Global and regional potential for bioenergy from agricultural and forestry residue biomass. Mitig Adapt Strateg Glob Change 15:241–262

Hall CAS, Benemann JR (2011) Oil from algae? Bioscience 61:741–742

Hambrick W, Jungjohann A, Chiu A, Flynn H (2010) Beyond biofuels: renewable energy opportunities for US farmers. Heinrich Böll Stiftung. www.hbfus.org

Han FX, King RL, Lindner JS, Yu T-Y, Durbha SS, Younan NH, Monts DL, Su Y, Luthe JC, Plodine MJ (2011) Nutrient fertilizer requirements for sustainable biomass supply to meet U.S. bioenergy goal. Biomass Bioenerg 35:253–262

Hill J, Nelson E, Tilman D, Polasky S, Tiffany D (2006) Environmental, economic, and energetic costs and benefits of biodiesel and ethanol biofuels. Proc Natl Acad Sci USA 103:11206–11210

Hillmyer MA (2017) The promise of plastics from plants. Science 358:868-

Hooker BA, Morris TF, Peters R, Cardon ZG (2005) Long-term effects of tillage and corn stalk return on soil carbon dynamics. Soil Sci Soc Am J 69:188–196

Houghton RA (2010) How well do we know the flux of CO_2 from land-use change? Tellus 62B:337–351

IEA (International Energy Agency) (2010) Sustainable production of second-generation biofuels. Paris

IEA (2011) Technology roadmap—biofuels for transport. Paris

IEA (2016) World energy outlook 2016. OECD/IEA, Paris

Intergovernmental Panel on Climate Change (IPCC) (2014) Climate change 2014: mitigation of climate change, contribution of working group III to the fifth assessment report of the intergovernmental panel on climate change. Cambridge University Press, Cambridge

Jaiswal D, De Souza AP, Larsen S et al (2017) Brazilian sugarcane ethanol as an expandable green alternative to crude oil use. Nat Clim Change 7:788–794. https://doi.org/10.1038/NCLIMATE3410

Johnson JMF, Karlen DL, Andrews SS (2010) Conservation considerations for sustainable bioenergy feedstock production: if, what, where, and how much? J Soil Water Conserv 65:88A–91A

Johnston M, Licker R, Foley J, Holloway T, Mueller ND, Barford C, Kucharik C (2011) Closing the gap: global potential for increasing biofuel production through agricultural intensification. Environ Res Lett 6:034028

Jones CD, Zhang X, Reddy AD, Robertson GP, Izaurralde RC (2017) The greenhouse gas intensity and potential biofuel production capacity of maize stover harvest in the US Midwest. GCBBioenerg 9:1333–1343. https://doi.org/10.1111/gcbb.12411

Kant P, Wu S (2011) The extraordinary collapse of Jatropha as a global biofuel. Environ Sci Technol 45:7114–7115

Karlen DL, Beeler LW, Ong RG, Dale BE (2015) Balancing energy, conservation, and soil health requirements for plant biomass. J Soil Water Conserv 70:279–287

Karlen DL, Birell SJ, Hess JR (2011a) A five-year assessment of corn stover harvest in central Iowa, USA. Soil Till Res 115–116:47–55

Karlen DL, Varvel GE, Johnson JMF, Baker JM, Osborne SL, Novak JM, Adler PR, Roth GW, Birrell SJ (2011b) Monitoring soil quality to assess the sustainability of harvesting corn stover. Agron J 103:288–295

Kim H, Kim S, Dale BE (2009) Biofuels, land use change, and greenhouse gas emissions: some unexplored variables. Environ Sci Technol 43:961–967

Kline KL, Msangi S, Dale VH et al (2017) Reconciling food security and bioenergy: priorities for action. GCB Bioenerg 9:557–576

Koh LP, Miettinen J, Chin Liew SC, Ghazoul J (2011) Remotely sensed evidence of tropical peatland conversion to oil palm. Proc Natl Acad Sci USA 108:5127–5132

Kutsch WL, Aubinet M, Buchmann N, Smith P, Osborne B, Eugster W, Wattenbach M, Schrumpf M, Schulze ED, Tomelleri E, Ceschia E, Bernhofer C, Béziat P, Carrara A, Di Tommasi P, Grünwald T, Jones M, Magliulo V, Marloie O, Moureaux C, Olioso A, Sanz MJ, Saunders M, Sogaard H, Ziegler W (2010) The net biome production of full crop rotations in Europe. Agric Ecosyst Environ 139:336–345

Lafond GP, Stumborg M, Lemke R, May WE, Holzapfel CB, Campbell CA (2009) Quantifying straw removal through baling and measuring the long-term impact on soil quality and wheat production. Agron J 101:529–537

Lal R (2009) Soil quality impacts of residue removal for bioethanol production. Soil Till Res 102:233–241

Lal R (2016) Soil health and carbon management. Food Energ Sec 5:212–222

Lapola DM, Schaldach R, Alcamo J, Bondeau A, Koch J, Koelking C, Priess JA (2010) Indirect land-use changes can overcome carbon savings from biofuels in Brazil. Proc Natl Acad Sci USA 107:3388–3393

Larkum AWD (2010) Limitations and prospects of natural photosynthesis for bioenergy production. Curr Opin Biotechnol 21:271–276

Larson JA, English BC, De La Torre Ugarte DG, Menard RJ, Hellwinckel CM, West TO (2010) Economic and environmental impacts of the corn grain ethanol industry on the United States agricultural sector. J Soil Water Conserv 65:267–279

Laungani R, Knops JMH (2009) The impact of co-occurring tree and grassland species on carbon sequestration and potential biofuel production. GCB Bioenerg 1:392–403

Le PVV, Kumar P, Drewry DT (2011) Implications for the hydrologic cycle under climate change due to the expansion of bioenergy crops in the Midwestern United States. Proc Natl Acad Sci USA 108:15085–15090

Lee DK, Aberle E, Anderson EK et al (2018) Biomass production of herbaceous energy crops in the United States: field trial results and yield potential maps from the multiyear regional feedstock partnership. GCB Bioenerg. https://doi.org/10.1111/gcbb.12493

Lemus R, Lal R (2005) Bioenergy crops and carbon sequestration. Crit Rev Plant Sci 24:1–21

Lisboa CC, Butterbach-Bahl K, Mauder M, Kiese R (2011) Bioethanol production from sugarcane and emissions of greenhouse gases—known and unknowns. GCB Bioenerg 3:277–292

Lu W, Zhang T (2010) Life-cycle implications of using crop residues for various energy demands in China. Environ Sci Technol 44:4026–4033

Loarie SR, Lobell DB, Asner GP, Mu Q, Field CB (2011) Direct impacts on local climate of sugar-cane expansion in Brazil. Nature Clim Change 1:105–109

Lorenz K, Lal R (2005) The depth distribution of soil organic carbon in relation to land use and management and the potential of carbon sequestration in subsoil horizons. Adv Agron 88: 35–66

Lorenz K, Lal R (2010) Carbon sequestration in forest ecosystems. Springer, Dordrecht, The Netherlands

Machado KS, Seleme R, Maceno MM, Zattar IC (2017) Carbon footprint in the ethanol feedstocks cultivation—agricultural CO_2 emission assessment. Agr Syst 157:140–145

Machovina B, Feeley KJ (2017) Restoring low-input high-diversity grasslands as a potential global resource for biofuels. Sci Tot Environ 609:205–214

Marland G, Obersteiner M (2008) Large-scale biomass for energy, with considerations and cautions: an editorial comment. Clim Change 87:335–342

Mathews JA, Tan H (2009) Biofuels and indirect land use change effects: the debate continues. Biofuel Bioprod Bioref. https://doi.org/10.1002/bbb.147

McBride AC, Dale VH, Baskaran LM, Downing ME, Eaton LM, Efroymson RA, Garten CT, Kline KL, Jager HI, Mulholland PJ, Parish ES, Schweizer PE, Storey JM (2011) Indicators to support environmental sustainability of bioenergy systems. Ecol Indic 11:1277–1289

McKone TE, Nazaroff WW, Berck P, Auffhammer M, Lipman T, Torn MS, Masanet E, Lobscheid A, Santero N, Mishra U, Barrett A, Bomberg M, Fingerman K, Scown C, Strogen B, Horvath A (2011) Grand challenges for life-cycle assessment of biofuels. Environ Sci Technol 45:1751–1756

Miller SA (2010) Minimizing land use and nitrogen intensity of bioenergy. Environ Sci Technol 44:3932–3939

Morgan JA, Follett RF, Allen LH Jr, Del Grosso S, Derner JD, Dijkstra F, Franzluebbers A, Fry R, Paustian K, Schoeneberger MM (2010) Carbon sequestration in agricultural lands of the United States. J Soil Water Conserv 65:6A–13A

Mulder K, Hagens N, Fisher B (2010) Burning water: a comparative analysis of the energy return on water invested. Ambio 39:30–39

Murphy R, Woods J, Black M, McManus M (2011) Global developments in the competition for land from biofuels. Food Pol 36:S52–S61

Nabuurs GJ, Masera O, Andrasko K, Benitez-Ponce P, Boer R, Dutschke M, Elsiddig E, Ford-Robertson J, Frumhoff P, Karjalainen T, Krankina O, Kurz WA, Matsumoto M, Oyhantcabal W, Ravindranath NH, Sanz Sanchez MJ, Zhang X (2007) Forestry. In: Metz B, Davidson OR, Bosch PR, Dave R, Meyer LA (eds) Climate change 2007: Mitigation. Contribution of working group III to the fourth assessment report of the intergovernmental panel on climate change. Cambridge University Press, Cambridge, United Kingdom and New York, NY, USA, pp 541–584

Nuffield Council on Bioethics (2011) Biofuels: ethical issues. http://www.nuffieldbioethics.org

Offermann R, Seidenberger T, Thrän D, Kaltschmitt M, Zinoviev S, Miertus S (2011) Assessment of global bioenergy potentials. Mitig Adapt Strateg Glob Change 16:103–115

Otto M, Berndes G, Fritsche U (2011) The bioenergy and water nexus. Biofuels Bioprod Bioref 5:343–346

Owen NA, Fahy KF, Griffiths H (2016) Crassulacean acid metabolism (CAM) offers sustainable bioenergy production and resilience to climate change. GCB Bioenerg 8:737–749

Owen NA, Inderwildi OR, King DA (2010) The status of conventional world oil reserves—hype or cause for concern? Energy Pol 38:4743–4749

Patzek TW (2010) A probabilistic analysis of the switchgrass ethanol cycle. Sustainability 2:3158–3194

Payne WA (2010) Are biofuels antithetic to long-term sustainability of soil and water resources? Adv Agron 105:1–46

Pehl M, Arvesen A, Humpenöder F et al (2017) Understanding future emissions from low-carbon power systems by integration of life-cycle assessment and integrated energy modelling. Nat Energ 2:939–945

Persson T, Garcia Y Garcia A, Paz JO, Fraisse CW, Hoogenboom G (2010) Reduction in greenhouse gas emissions due to the use of bio-ethanol from wheat grain and straw produced in the south-eastern USA. J Agric Sci 148:511–527

Pickard WF (2010) The future of biomass energy: a Fermi-calculation perspective. Energy Pol 38:1672–1674

Plevin RJ, O'Hare M, Jones AD, Torn MS, Gibbs HK (2010) Greenhouse gas emissions from biofuels' indirect land use change are uncertain but may be much greater than previously estimated. Environ Sci Technol 44:8015–8021

Poeplau C, Don A, Vesterdal L, Leifeld J, Van Wesemael B, Schumacher J, Gensior A (2011) Temporal dynamics of soil organic carbon after land-use change in the temperate zone—carbon response functions as a model approach. Glob Change Biol 17:2415–2427

Powers RF, Scott DA, Sanchez FG, Voldseth RA, Page-Dumroese DS, Elioff JD, Stone DM (2005) The North American long-term soil productivity experiment: findings from the first decade of research. For Ecol Manage 220:31–50

Qin Z, Dunn JB, Kwon H, Mueller S, Wander MM (2016) Soil carbon sequestration and land use change associated with biofuel production: empirical evidence. GCB Bioenerg 8:66–80

Qin Z, Zhuang Q, Zhu X, Cai X, Zhang X (2011) Carbon consequences and agricultural implications of growing biofuel crops on marginal agricultural lands in China. Environ Sci Technol 45:10765–10772

Qiu H, Huang J, Keyzer M, van Veen W, Rozelle S, Fisher G, Ermolieva T (2011) Biofuel development, food security and the use of marginal land in China. J Environ Qual 40:1058–1067

Raffa DW, Bogdanski A, Tittonell P (2015) How does crop residue removal affect soil organic carbon and yield? A hierarchical analysis of management and environmental factors. Biomass Bioenerg 81:345–355

REN21 (2017) Renewables 2017 Global Status Report. REN21 Secretariat, Paris

Richardson M, Kumar P (2017) Critical Zone services as environmental assessment criteria in intensively managed landscapes. Earth's Future 4. https://doi.org/10.1002/2016ef000517

Ridley CE, Clark CM, LeDuc SD, Bierwagen BG, Lin BB, Mehl A, Tobias DA (2012) Biofuels: network analysis of the literature reveals key environmental and economic unknowns. Environ Sci Technol 46:1309–1315
Riffell S, Verschuyl J, Miller D, Wigley B (2011) Biofuel harvests, coarse woody debris, and biodiversity—a meta-analysis. For Ecol Manage 261:878–887
Rist J, Lee JSH, Koh LP (2009) Biofuels: social benefits. Science 326:1344
Robertson GP, Hamilton SK, Barham BL, et al (2017) Cellulosic biofuel contributions to a sustainable energy future: choices and outcomes. Science 356:eaal2324
Robertson GP, Hamilton SK, Del Grosso SJ, Parton WJ (2011) The biogeochemistry of bioenergy landscapes: carbon, nitrogen, and water considerations. Ecol Appl 21:1055–1067
Robledo-Abad C, Althaus HJ, Berndes G et al (2017) Bioenergy production and sustainable development: science base for policymaking remains limited. GCBBioenerg 9:541–556
Sang T, Zhu W (2011) China's bioenergy potential. GCB Bioenerg 3:79–90
Schipfer F, Kranzl L, Leclère D, Sylvain L, Forsell N, Valin H (2017) Advanced biomaterials scenarios for the EU28 up to 2050 and their respective biomass demand. Biomass Bioenerg 96:19–27
Schnoor JL (2011) Cellulosic biofuels disappoint. Environ Sci Technol 45:7099
Scholz V, Heiermann M, Kaulfuss P (2010) Sustainability of energy crop cultivation in Central Europe. In: Lichtfouse E (ed) Sociology, organic farming, climate change and soil science. Springer, Dordrecht, The Netherlands, pp 109–145
Scown CD, Nazaroff WW, Mishra U, Strogen B, Lobscheid AB, Masanet E, Santero NJ, Horvath A, McKone TE (2012) Lifecycle greenhouse gas implications of US national scenarios for cellulosic ethanol production. Environ Res Lett 7:014011. https://doi.org/10.1088/1748-9326/7/1/014011
Searchinger TD, Beringer T, Strong A (2017) Does the world have low-carbon bioenergy potential from the dedicated use of land? Energ Pol 110:434–446
Searchinger T, Heimlich R, Houghton RA, Dong F, Elobeid A, Fabiosa J, Tokgoz S, Hayes D, Yu T-H (2008) Use of U.S. croplands for biofuels increases greenhouse gases through emissions from land-use change. Science 319:1238–1240
Searle SY, Malins CJ (2014) Will energy crop yields meet expectations? Biomass Bioenerg 65:3–12. https://doi.org/10.1016/j.biombioe.2014.01.001
Searle S, Malins C (2015) A reassessment of global bioenergy potential in 2050. GCB Bioenerg 7:328–336. https://doi.org/10.1111/gcbb.12141
Service RF (2010) Is there a road ahead for cellulosic ethanol? Science 329:784–785
Shah A, Darr M, Khanal S, Lal R (2017) A techno-environmental overview of a corn stover biomass feedstock supply chain for cellulosic biorefineries. Biofuels 8:59–69
Simmons BA (2011) Opportunities and challenges in advanced biofuel production: the importance of synthetic biology and combustion science. Biofuels 2:5–7
Singh NK, Dhar DW (2011) Microalgae as second generation biofuel. A review. Agronomy Sust Developm 31:605–629
Slade R, Bauen A, Gross R (2014) Global bioenergy resources. Nat Clim Change 4:99–105
Slade R, Saunders R, Gross R, Bauen A (2011) Energy from biomass: the size of the global resource. Imperial College Centre for Energy Policy and Technology and UK Energy Research Centre, London
Smil V (1999) Crop residues: agriculture's largest harvest. Bioscience 49:299–308
Souza GM, Ballester MVR, de Brito Cruz CH et al (2017) The role of bioenergy in a climate-changing world. Environ Develop 23:57–64
Strapasson A, Woods J, Chum H et al (2017) On the global limits of bioenergy and land use for climate change mitigation. GCB Bioenerg 9:1721–1735. https://doi.org/10.1111/gcbb.12456
Tiedje J, Donohue T (2008) Microbes in the energy grid. Science 320:985
Tilman D, Hill J, Lehman C (2006) Carbon-negative biofuels from low-input high diversity grassland biomass. Science 314:1598–1600

UNEP (United Nations Environment Program) (2009) Towards sustainable production and use of resources: assessing biofuels. Working Group on biofuels of the International Panel on Sustainable Resource Management, UNEP, Paris

UNEP (2011) Towards a green economy: pathways to sustainable development and poverty eradication. www.unep.org/greeneconomy

U.S. DOE (United States Department of Energy) (2005) Biomass as a feedstock for a bioenergy and bioproducts industry: the technical feasibility of a billion-ton annual supply. http://www.osti.gov/bridge

Valentine J, Clifton-Brown J, Hastings A, Robson P, Allison G, Smith P (2012) Food vs. fuel: the use of land for lignocellulosic 'next generation' energy crops that minimize competition with primary food production. GCB Bioenerg 4.1–19

Renssen Van (2011) A biofuel conundrum. Nature Geosci 1:389–390

Whitaker J, Ludley KE, Rowe R, Taylor G, Howard DC (2010) Sources of variability in greenhouse gas and energy balances for biofuel production: a systematic review. GCB Bioenerg 2:99–112

Wigmosta MS, Coleman AM, Skaggs RJ, Huesemann MH, Lane LJ (2011) National microalgae biofuel production potential and resource demand. Water Resour Res 47, W00H04. https://doi.org/10.1029/2010wr009966

Wilhelm WW, Johnson JMF, Karlen DL, Lightle TD (2007) Corn stover to sustain soil organic carbon further constrains biomass supply. Agron J 99:1665–1667

Wilts R, Reicosky DC, Allmaras RR, Clapp CE (2004) Long-term corn residue effects: harvest alternatives, soil carbon turnover, and root derived carbon. Soil Sci Soc Am J 68:1342–1351

Wright CK Larson B, Lark TJ, Gibbs HK (2017) Recent grassland losses are concentrated around U.S. ethanol refineries. Environ Res Lett 12:044001

Yamaguchi S, Kawada Y, Yuge H, Tanaka K, Imamura S (2017) Development of new carbon resources: production of important chemicals from algal residue. Sci Rep 7:855

Zhang B, Penton R, Xue C et al (2017) Soil depth and crop determinants of bacterial communities under ten biofuel cropping systems. Soil Biol Biochem 112:140–152

Zhu P, Zhuang Q, Eva J, Bernacchi (2017) Importance of biophysical effects on climate warming mitigation potential of biofuel crops over the conterminous United States. GCBBioenerg 9:577–590

Zilberman D (2017) Indirect land use change: much ado about (almost) nothing. GCB Bioenerg 9:485–488

Chapter 8
Biochar

Contents

Abstract Biochar is a carbon (C) enriched by-product of bioenergy production obtained during pyrolysis (heating under limited supply of oxygen) of plant-derived feedstock including organic matter (OM). Biochar is defined as 'charcoal for application to soils.' It may be enriched in polycondensed aromatic C forms which enhance soil organic carbon (SOC) sequestration after soil application. For SOC sequestration to occur, all of the SOC sequestered must originate from the atmospheric CO_2 pool and be transferred into the soil humus through land unit plants, plant residues, and other organic solids. Biochar can also be described as an anthropogenically produced black carbon (BC) material. BC also known as pyrogenic organic carbon (PyOC), fire-derived organic matter or wildfire charcoal, constitutes certain ranges in the combustion continuum ranging from slightly charred plant material to highly condensed soot. Application of biochar to soil has recently received increased attention as a means to sequester C and to produce secondary agronomic benefits. In particular, it is thought that application of charcoal together with organic wastes, dung, and bones is the cause of high SOC concentrations and sustained soil fertility by ancient agricultural management practices that created *terra preta de Índio*, deep black soils in the Brazilian Amazon. However, natural charcoal and biochar are not well suited as proxies for each other. Also, whether new *terra preta* can be generated by biochar application

K. Lorenz and R. Lal, *Carbon Sequestration in Agricultural Ecosystems*,
https://doi.org/10.1007/978-3-319-92318-5_8

(together with organic wastes) to soils under other land uses, for different soil types and climates, is not known. In particular, the properties of organic materials (e.g., wood, manure, leaves) used as feedstock for biochar production and charring conditions (e.g., temperature, charring time) vary widely. Thus, the biological, chemical, and physical properties among biochars also vary widely. Biochar may be the most stable soil amendment with estimated mean residence times (MRTs) of several hundreds to thousands of years, and may increase nutrient availability beyond the fertilizer effect. Thus, applying biochar to the soil potentially improves soil productivity, SOC storage, and infiltration of percolating soil water in the long term through its porous structure and stability. However, interactions between soils and biochar are diverse and difficult to predict. For example, yield-stimulating effects of biochar are not universal, and may be restricted to the tropics as biochar increases yield through liming and fertilization, consistent with the low soil pH, low fertility, and low fertilizer inputs typical of arable tropical soils. Relatively few studies provide a quantitative assessment of biochar soil management scenarios. Needed are, in particular, large-scale long-term field studies on different crops at diverse locations designed to test effects of application of different biochar types on SOC sequestration and secondary agronomic benefits. Similarly, a routine standard method to quantify soil biochar C is essential but not yet available. This chapter begins with a comparison of biological, chemical, and physical properties of charred OM relevant to agricultural application. This is followed by a discussion about effects of application of charred OM on SOC sequestration. The chapter concludes with an overview of research gaps that need to be addressed to realize the full potential of biochar for SOC sequestration in agricultural soils. In this chapter, BC, char, charcoal, and PyOC are used interchangeably, whereas the term biochar will be used when anthropogenically charred biomass is purposefully applied to soil for agricultural and environmental benefits.

Keywords Charcoal · *Terra preta* · Black carbon · Pyrogenic organic carbon Pyrolysis · Crop yield · Decomposition · Erosion · Leaching · Greenhouse gas emissions

8.1 Properties of Black Carbon

Black carbon (BC) is an important component of soils, and engineered BC or biochar has received increased attention as a soil amendment to mitigate atmospheric increases in carbon dioxide (CO_2) levels and for secondary agronomic benefits. BC can be defined as a 'carbonaceous substance of pyrogenic origin resistant to thermal or chemical degradation by applying specific methods' (Hammes and Abiven 2013). BC is generally considered a very stable form of carbon (C), but its physicochemical properties are diverse, and less stable soil BC and biochar have also been found (Kuhlbusch 1998; Hammes et al. 2008; Lehmann 2007; Lian and Xing 2017).

Biochar is produced, in particular, when organic material such as wood chips and wood pellets, tree bark, crop residues, energy crop, organic wastes, chicken litter, dairy manure or sewage sludge is thermally decomposed at low-moderate temperatures and long heating time under limited supply of oxygen (O_2) (slow pyrolysis; Fig. 8.1: Sohi et al. 2010; Lehmann and Joseph 2009). This creates a C-enriched, fine-grained, and porous by-product of biomass pyrolysis (Sohi et al. 2009). The character of the biochar produced depends on the extent of pyrolysis (peak process temperature), and its completeness on particle size and kiln residence time (Sohi et al. 2010). As feedstock for biomass pyrolysis and process conditions vary widely, no consistent definition of biochar exists (Guo et al. 2016). To most researchers and users, biochar is the C-enriched black solids made from pyrolysis or gasification of biomass materials and is used exclusively as a soil amendment in agricultural and environmental applications. Biochar can be defined as 'charcoal for application to soils' (Verheijen et al. 2009). However, biochar and natural charcoal are not well suited as proxies for each other (Santín et al. 2017).

Biochar comprises part of a combustion continuum without distinct boundaries of materials described as BC which includes soot (Fig. 8.2), charcoal (Fig. 8.3), and char from vegetation-based fire (Schmidt and Noack 2000). The chemistry of BC varies with the degree of fire exposure and the production mechanism (Fig. 8.4; Table 8.1; Hammes et al. 2007). While char retains some plant chemistry and morphology, the geochemistry of soot reflects combustion conditions as it is

Fig. 8.1 Homemade biochar (licensed under the Creative Commons Attribution-Share Alike 3.0 Unported license)

produced during condensation of gas-phase intermediates. Char, charcoal, and soot may be formed simultaneously during combustion and coexist in soil. Methodological problems for BC quantification arise from the chemical variations of BC within the combustion continuum. For example, char and soot may have overlapping properties such as specific surface area and oxidative kinetics depending on formation conditions (Hammes et al. 2007). Sometimes, the generic term elemental carbon (EC) is used to represent results of methods designed to measure various manifestations of the metaphorical BC spectrum (Currie et al. 2002). Other than BC, the pyrogenic organic carbon (PyOC) continuum includes also graphite particles which are formed from soot and char precursors under heat and pressure (Preston and Schmidt 2006).

The characterization of biochar properties resulted in a growing number of peer-reviewed research papers. In comparison, a much greater amount of research has been dedicated to properties of char and charcoal, and their effect on soil properties and soil genesis (Ponomarenko and Anderson 2001; Glaser et al. 2002; Schmid et al. 2002; Kleber et al. 2003). Thus, the knowledge about biochar is changing due to the wide variety of biomass feedstocks and charring conditions used that affect biological, chemical, and physical properties (Keiluweit et al. 2010). Further, compared to controlled pyrolysis biochar, traditionally charred biomass, i.e., the 'biochar' most likely to be actually applied to soil in developing countries,

Fig. 8.2 Soot from a chimney (licensed under the Creative Commons Attribution-Share Alike 3.0 Unported, 2.5 Generic, 2.0 Generic, and 1.0 Generic license)

Fig. 8.3 Charcoal briquettes (photo credit Titus Tscharntke; public domain)

has a heterogeneous thermal and biochemical composition and unpredictable biological stability (Jegajeevagan et al. 2016). Also, while changes in agricultural soils after biochar application are well documented, the effects of biochar application on forest soils remain poorly understood (Li et al. 2017). The application of biochars to forest soils may result in the improvement of soil physical, chemical, and microbial properties while also mitigating soil greenhouse gas (GHG) emissions. Thus, application of biochar to forest soils and especially to plantation soils with low fertility may have considerable advantages (Li et al. 2017).

The first study on the direct comparability of corresponding natural and anthropogenic PyOC indicated that wildfire charcoals were formed under higher maximum temperatures and oxygen availabilities, but much shorter heating durations than slow-pyrolysis biochars, which is the most widespread biochar production technology (Santín et al. 2017). This resulted in different physicochemical properties which are particularly relevant regarding their respective roles as C sinks, as even the wildfire charcoals formed at the highest temperatures had lower C sequestration potentials than most slow-pyrolysis biochars (Santín et al. 2017). Thus, generalizations about the properties of biochar relevant to application on agricultural soils and SOC sequestration must be viewed cautiously (Lorenz and Lal 2014). The addition of biochar to soil may offer a large mitigation potential among agricultural pathways (Griscom et al. 2017). However, it has not been well demonstrated beyond research settings. Trade-offs, costs, and feasibility of

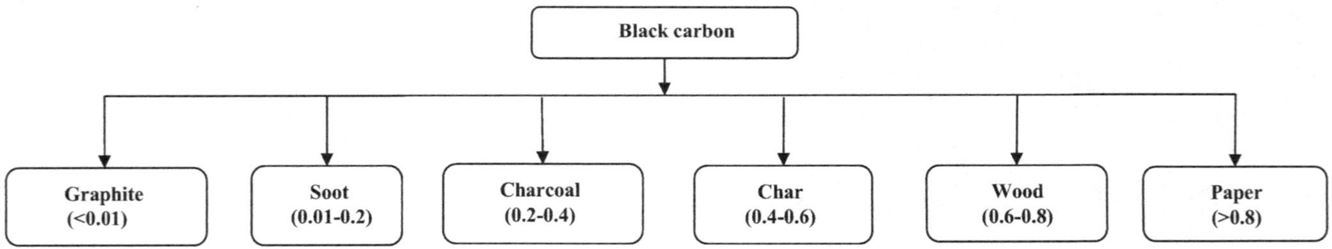

Fig. 8.4 Combustion continuum of black carbon. Increased heating yields increasingly carbon-rich materials indicated by decreases in atomic oxygen to carbon ratios shown in brackets (modified from Hedges et al. 2000)

Table 8.1 Characteristics of the combustion continuum of black carbon (modified from Masiello 2004; Preston and Schmidt 2006; Hammes et al. 2007)

Material	Fuel	Combustion residues		Combustion condensates
Terminology	Biomass, wood, litter	Char/charred biomass	Charcoal	Soot
Formation temperature		Low	Intermediate	High
Size	mm and larger	mm and larger	mm to submicron	Submicron
Plant structures	Abundant	Present	Few	None
Reactivity	High (degradable)	Intermediate	Intermediate	Low (inert, refractory)
Atomic H/C ratio	1.7–1.0	1.0–0.6	0.6–0.3	<0.3
Atomic O/C ratio	1.0–0.6	0.6–0.4	0.4–0.2	<0.2

large-scale implementation of biochar are poorly understood (Griscom et al. 2017). The first direct comparison between PyOC and SOC persistence in temperate arable soils indicated that soil PyOC contents decreased more rapidly than expected from current concepts; i.e., the mean residence time (MRT) of native PyOC was just 1.6 times longer than that of SOC (Lutfalla et al. 2017). Further, in the absence of a substantial increase in fire frequency, PyOC cycling might contribute to, rather than mitigate, the future increase in atmospheric CO_2 (Landry and Matthews 2017). Thus, claims that climate change could be mitigated through the incorporation of massive amounts of biochar into soils should be avoided (Baveye et al. 2018).

8.1.1 Physical Properties Relevant to Agricultural Application

Biochar potentially improves the soil physical environment (Blanco-Canqui 2017). Specifically, biochar may reduce soil bulk density by 3–31%, increase porosity by 14–64%, but may have only limited effects on penetration resistance. Biochar may increase wet aggregate stability by 3–226%, improve soil consistency, but may have mixed effects on dry soil aggregate stability. Further, biochar potentially increases available water by 4–130%. While saturated hydraulic conductivity may decrease in coarse-textured soils, it may increase in fine-textured soils following biochar application. Studies on other physical properties are scanty but suggest that biochar potentially reduces tensile strength and particle density, alters water infiltration, moderates soil thermal properties, but may have minimal effects on soil water repellency. Sandy soils may respond more to biochar than clayey soils. Further, biochar effectiveness potentially increases as the amount of biochar applied increases. A decrease in biochar particle size may increase water retention but may reduce saturated flow. However, only limited field data are available, and field-scale and long-term studies assessing all soil physical properties under different scenarios of biochar management are needed (Blanco-Canqui 2017; Mukherjee and Lal 2014).

Effects on the soil physical environment may be related to intrinsic physical characteristics of biochars which show a large variation in Downie et al. (2009). One of the most striking physical features of biochar produced at lower temperature is its highly porous structure clearly retaining the architecture of cellulosic plant material in the feedstock (Sohi et al. 2009). Surface area of wood and grass char strongly increases above 400 °C charring temperature (Keiluweit et al. 2010). Biochar specific surfaces are higher than those of sand and comparable to or higher than those of clay (Downie et al. 2009). Estimates range from 10 to several 100 $m^2 g^{-1}$ biochar. Thus, when added as a soil amendment to a coarse-textured soil, biochar may cause a net increase in total soil-specific surface area which may enhance water, air, and nutrient cycling, and accentuate microbial activity.

Scanning electron microscopy (SEM) can be used to describe the physical structure of biochar (Sohi et al. 2010). Most of the surface area of biochar is located in pores < 2 nm in diameter which may be responsible for high adsorptive capacities for small molecules such as gases. These nanopores are especially formed when biochar is produced at high charring intensity (Keiluweit et al. 2010). Pores of internal width between 2 and 50 nm are also present in biochar, and are important for many liquid–solid adsorption processes (Downie et al. 2009). Although total surface areas of pores < 2 nm in diameter are larger than those of pores of internal width > 50 nm, the volume of the larger pores can be broader in biochar. This trend may result in greater functionality than those of more narrow surface areas. The particle sizes of biochar are smaller than those of the feedstock material, but relatively increased particle sizes in biomass chars have also been observed (Cetin et al. 2004). While bulk densities are lower in biochar compared to the biomass feedstock, the opposite is the trend in mechanical strength (Downie et al. 2009).

8.1.2 Chemical Properties Relevant to Agricultural Application

Biochar contains most of the plant nutrients also contained in the biomass feedstock as much of the mineral content in the feedstock is carried over into the biochar where it is concentrated due to loss of C, H, and O during pyrolysis (Amonette and Joseph 2009). However, the high charcoal content in *terra preta de Índio* soils did not primarily contribute to high nutrient contents (Glaser 2007). Primary and secondary nutrient sources are required such as manure, organic by-products, ash, terrestrial and aquatic plant biomass. Important for SOC sequestration is the organo-chemical composition of biochar as pyrolysis converts organic compounds into relatively 'recalcitrant' aromatic structures (Baldock and Smernik 2002). The proportions of hemicellulose, cellulose, and lignin in the feedstock influence the proportion of C stabilized in biochar (Sohi et al. 2010). However, potentially toxic compounds such as acrolein and other toxic carbonyl compounds, cresols, dioxins, formaldehyde, polycyclic aromatic compounds (PAHs), xylenols, volatile organic compounds (VOCs), and carbon (aceous) nanoparticles may also be formed upon pyrolysis of any feedstock (Sohi et al. 2009; Thies and Rillig 2009). Such pollutants could have adverse effects on the plant growth as well as microbial community in soil (Dutta et al. 2017).

Inorganic chemistry

Woody chars may generally have low mineral ash contents (<1% by weight) compared to grass, straw, and grain husks chars (up to 24% by weight; Raveendran et al. 1995). Otherwise, switchgrass (*Panicum virgatum* L.) and corn (*Zea mays* L.) stover biochars may have high ash contents of 32 and 55% by weight, respectively, mainly because of silica (Brewer et al. 2009). Very high ash contents are observed in biochars from manures and rendering wastes. For example, chicken-litter biochars and bone biochars may have 45 and 84% mineral matter, respectively (Purevsuren et al. 2004; Koutcheiko et al. 2007). Even within a single biochar particle, however, there is a large variation in mineral content (Amonette and Joseph 2009). Only scant information is available about plant macro- and micronutrient contents in biochar, but contents appear to be highly variable (Chan and Xu 2009). For example, N contents may range from 1.7 to 78.2 g N kg^{-1}, P contents from 0.2 to 73 g P kg^{-1}, and K contents from 1.0 to 58 g K kg^{-1}, respectively. Thus, N and P contents in biochars cover wider ranges than those observed in other organic amendments used in agriculture for fertilization. Poultry (*Gallus domesticus*) litter biochars, on the other hand, are potential slow-release fertilizers and are relatively enriched in K, P, Ca, Na, and Mg but also in Cl and S compared to biochar derived from hardwood (Agblevor et al. 2010).

Alkalinity is, probably, one of the most influential biochar properties, because changes in pH have cascading impacts on many soil processes, including N mineralization, mineral precipitation, ion exchange, and GHG emissions (Fidel et al. 2017). In fact, many studies have shown that biochar amendments to soils increase and buffer soil pH. However, the nature of biochar alkalis, the influence of feedstock and biochar production conditions on biochar alkalinity, and the variability of alkali properties among biochars are poorly understood (Fidel et al. 2017). Similarly, the effects on SIC sequestration are less well known (Fig. 8.5).

Some inorganic compounds present in the feedstock for biomass chars are volatilized during pyrolysis, but the majority is being retained as discrete mineral phases or as part of the structure of the carbonaceous residue (Wornat et al. 1995). Functional groups on the surfaces of mineral phases, however, may be transformed into oxides which may have consequences for the surface charge (Amonette and Joseph 2009). Aside from the carboxylate groups on the biochar surface (Liang et al. 2006), transformation into oxides during pyrolysis contributes to the cation exchange capacity (CEC) of biochars which is not very high for fresh biochar but

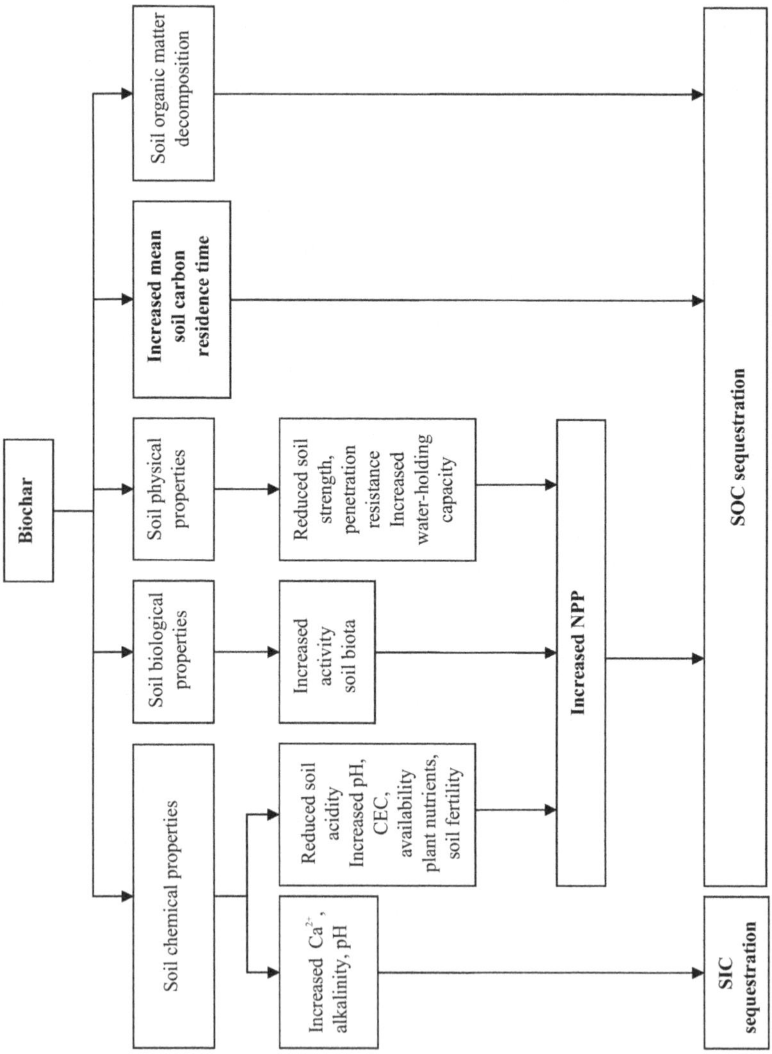

Fig. 8.5 Biochar effects on soil carbon sequestration

higher than that of whole soils, clay, and SOM after long periods of time in the soil (Sohi et al. 2009). For example, some old anthropogenic soils with larger proportions of charcoal have more CEC (Sombroek et al. 1993). Little is known about the distribution of mineral ash within different types of biochar, and on the distribution and the stability of heavy metals in biochar (Amonette and Joseph 2009). High contents of Cu, Zn, Cr, and Ni have been reported for char from sewage sludge, and very high Cr contents are observed in char from tannery waste (Muralidhara 1982; Bridle and Pritchard 2004). For example, poultry litter-based activated C is high in Cu and Zn which may pose an environmental risk if this material is applied to soil (Qiu and Guo 2010). Otherwise, broiler litter biochar contains only trace amounts of Cd and As (Agblevor et al. 2010). Arsenic in broiler litter originates from the chemical Roxarsone as an additive in bird feed (Hileman 2007). The bioavailability of biochar-derived metals, however, has not been widely studied (Chan and Xu 2009). Biochar and more generally pyrogenic OM can immobilize metal(loid)s by physical trapping, electrostatic interactions, ionic/proton exchange, surface complexation, precipitation, and redox reactions (Qi et al. 2017).

Organic chemistry

The defining chemical property of biochar is the high C content in the organic portion and the high aromatic (aryl) C content (Krull et al. 2009; Lehmann and Joseph 2009). Carbon contents may range from 172 to 905 g C kg^{-1} (Chan and Xu 2009). Thus, biochars cannot be commonly regarded as organic C-rich materials (i.e., $\gg$500 g organic C kg^{-1} dry weight) as, in particular, biochars from manures or grass species may contain much less than 500 g organic C kg^{-1} dry weight (Krull et al. 2009). Pyrolysis converts plant-derived OM into aromatic structures with furan-like compounds which increases the 'recalcitrance' of C (Baldock and Smernik 2002). However, relatively labile aliphatic compounds also occur in BC and biochar (Keiluweit et al. 2010; Cheng et al. 2006). The degree of aromaticity and maturation of BC and biochars can be assessed by the determination of atomic H/C and O/C ratios (Table 8.1). For example, unburned fuel materials have H/C ratios of about 1.5, and C enrichment during pyrolysis causes a decrease in this ratio although a H/C ratio of below 0.5 cannot be commonly applied to categorize biochars as by-products from burning of many residues have significantly higher H/C ratios (Krull et al. 2009).

A range of methods are used for characterizing the chemical structure and composition of biochar (Sohi et al. 2009). The surface chemistry of biochar has been studied by using diffuse reflectance infrared Fourier transform spectroscopy (FTIR), X-ray photoelectron spectroscopy (XPS), energy-dispersive X-ray spectroscopy (EDX), and near-edge X-ray absorption fine structure (NEXAFS) spectroscopy (Sohi et al. 2010). The composition of biochar has been studied by applying solid-state ^{13}C nuclear magnetic resonance (NMR) spectroscopy (with cross-polarization, CP, or Bloch decay, combined with magic angle spinning, MAS), mid-infrared (MIR) spectroscopy, pyrolysis–gas chromatography–mass spectrometry (PY–GC–MS), and matrix-assisted laser desorption ionization time-of-flight mass spectroscopy (MALDI-TOF MS). These methods can be used besides extractive techniques analogous to those applied to SOM (Sohi et al. 2009). Based on NMR spectroscopy, for example, biochar produced above 350 °C is dominated by aromatic (aryl) C, but only heating above 500 °C tends to remove plant-characteristic functional group C structures (Krull et al. 2009). In particular, alkyl and O-alkyl C are increasingly converted to aryl C at higher temperatures. Thus, biochar produced by temperatures < 500 °C varies substantially in chemical characteristics, and the stability most likely influenced by the degree of aromaticity. Similarly, charcoal produced from grass and wood biomass at 450 °C retains some plant-related chemical properties, in particular lignin units as detectable by PY-GC-MS (Song and Peng 2010). Both charcoal types are highly aromatic, with aromatic nuclei cross-linked by aliphatic chain-like bridges. Important differences are that grass charcoal is lower in 1-ring aromatic compounds, phenolic compounds, and furan derivatives derived from cellulose but higher in $\geq$3-ring aromatic compounds and N-containing compounds than wood charcoal.

Using a combination of FTIR and NEXAFS, Keiluweit et al. (2010) observed that increasing charring temperatures from 100 to 700 °C resulted in analogous but quantitatively different physical-chemical transitions in wood and grass char. Specifically, no FTIR-detectable changes are observed by heating up to 200 °C. Dehydration of cellulosic and ligneous compounds starts at 300 °C, lignin/cellulose-derived transformation products are visible at 400 °C, and charring > 500 °C results in increasing degree of condensation. NEXAFS spectra, on the other hand, reveal that aromatic and quinonic compounds become prevalent by charring at 400 °C, while O-containing and aliphatic groups are lost. The first evidence for condensation reactions is observed at 500 °C, and chars produced at 600 and 700 °C are poorly crystalline in nature. Keiluweit et al. (2010), thus, proposed that four distinct categories of char exist consisting of a unique mixture of chemical phases and physical states. At low charring intensity, transition chars are formed in which the crystalline character of the precursor material (i.e., cellulose) is preserved. With increasing charring intensity, amorphous chars are formed in which heat-altered molecules and aromatic polycondensates are randomly mixed. Intermediates such as pyranones, anhydrosugars, phenols, quinones, pyrroles, and furans as well as small (poly)aromatic units are formed, and stable aromatic lignin residues become relatively enriched (Knicker et al. 2008). Not only cyclic and aromatic volatiles but also aliphatic components such as cutans and lipids are retained in amorphous chars (Almendros et al. 2003). Further increases in charring intensity result in the formation of composite chars which consist of poorly ordered graphene stacks embedded in amorphous phases. Aromatic, aliphatic, and O-containing components are preserved. At highest charring intensity, turbostratic chars are formed which are dominated by disordered graphitic crystallites which clearly distinguish them from natural or synthetic graphite. Molecular variations among the four char categories affect the persistence of biochar in the soil and its sorptive properties (Keiluweit et al. 2010).

Aside from aromatics, BC is commonly considered to contain random stacks of graphitic layers (Schmidt and Noack 2000). However, assuming a graphite-like polyaromatic structure of char and biochar is probably an oversimplification as the structure may rather be seen as a heterogeneous mixture of thermally altered biomacromolecules (Knicker et al. 2008; Keiluweit et al. 2010). Thus, plant biochars may contain proportions of the lignin backbone, but also furans and anhydrosugars, and pyranones from cellulose (Song and Peng 2010). Further, pyrolyzing plant feedstock with relatively high N contents generates biochar containing N-heteroaromatic C derived from peptides. Environmentally recalcitrant N forms such as aromatic and heterocyclic N-containing structures are produced during pyrolysis by transformation of labile N compounds and influence biochar stability (Almendros et al. 2003).

The quantitative characterization of the C forms present in BC, char, and biochar is a major challenge. Thus, almost all methods for organo-chemical characterization of biochar are semiquantitative (Hammes et al. 2007). However, quantitative NMR techniques (i.e., direct polarization or Bloch decay MAS and a combination of both) have been applied recently to characterize switchgrass and corn stover char (Brewer et al. 2009). Their composition is dominated by aromatic C, the majority of which is not protonated, with aromaticities between 81 and 94%. Smaller signals of C=O and alkyl groups are also detected. Between seven and eight aromatic rings are fused to aromatic ring clusters (Brewer et al. 2009). In contrast, char produced during vegetation fires is characterized by small, linearly condensed structures with probably less than six aromatic rings per cluster (Knicker et al. 2008).

During the pyrolytic production of biochar, PAHs can form and be present on the surface of biochar (Wang et al. 2017). The complex mixture of organic compounds in biochar may contain potentially toxic compounds for humans and the environment, but these have been inadequately researched (Cui et al. 2009; Dutta et al. 2017; Sohi et al. 2009). For example, there is a lack of studies measuring dioxins and PAHs in pyrolysis products which is a major hurdle to evaluate the environmental impact of these compounds when biochar is applied to soil (Garcia-Perez 2008). Total PAH concentrations between 3 and 16 $\mu g\ g^{-1}$ for pyrolyzed *Pinus rigida* Mill. (pitch pine) wood compared to 28 $\mu g\ g^{-1}$ for *P. rigida* char from prescribed burning have been reported (Brown et al. 2006). Otherwise, biochar has also a very high affinity and capacity for sorbing organic compounds and, in particular, environmental contaminants such as PAHs, polychlorinated biphenyls (PCBs), and many herbicides and pesticides (Smernik 2009). Organic contaminants can be retained by pyrogenic OM and biochar through pore filling, electrostatic interactions, partitioning, hydrogen bonding, van der Waals interactions, and π–π electron–donor–acceptor interactions (Qi et al. 2017). The immobilization of soil contaminants by biochar is influenced by soil types and soil components, biochar feedstocks and production conditions, contaminant types and molecule properties as well as sorption competition and natural aging processes. Otherwise, the release of PAH from biochars may generally be expected to be negligible because of the strong PAH sorption potential of biochar (Hilber et al. 2017). Thus, any potential cytotoxicity of biochar may relate to its particulate nature and properties. For example, cytotoxic effects of biochar on fibroblast cells were reported for the first time by Sigmund et al. (2017). This cytotoxic effect was likely related to the particulate nature and size distribution of the biochar. Specifically, the biochar had particles similar in size to atmospheric particulate matter (PM2.5). Sigmund et al. (2017) recommended that practitioners should wear respiratory protective equipment during biochar production to minimize the risk of exposure. During field application, biochar should be applied in slurries and should always be mixed with a soil matrix to avoid secondary dust formation (Sigmund et al. 2017).

Pyrogenic OM including biochar may not only be passive sorbents but promote chemical reactions of sorbed contaminants, including long-range electron conduction between molecules and between microbes and molecules, local redox reactions between molecules, and hydrolysis (Pignatello et al. 2017). Reactions mediated by biochar can, thus, impact the biogeochemical fate of contaminants. To sum up, the impacts of biochar application to soil with respect to long-term persistence and bioavailability of organic contaminants are not clear (Dutta et al. 2017). Urgently needed are studies on long-term effects of biochar on contaminant stability in soil (Ahmad et al. 2014).

8.1.3 Biological Properties Relevant to Agricultural Application

Biochar may be habitat and food source for soil biota due to its porous structure and nutrient content, but little data are available for soil fauna (Lorenz and Lal 2014). For example, geophagous fauna such as the tropical earthworm *Pontoscolex corethrurus* ingest charcoal particles but less is known whether charcoal is digested during gut passage (Topoliantz and Ponge 2005). Otherwise, physicochemical change occurs as *P. corethrurus* ingests charcoal together with soil particles, mixes it with mucus secreted in the esophagus, finely grounds this mixture in the muscular gizzard, and excretes the mixture and deposits it on the soil surface as muddy paste (Ponge et al. 2006). However, charcoal and biochar may also contain compounds toxic to soil fauna. For example, it is hypothesized that genotoxic compounds in biomass burning residues introduce strand breaks in DNA of the earthworm *Eisenia fetida* (Cui et al. 2009). Whether soil fauna colonize biochar aside feeding on it and use it as a habitat is not known. However, the average body size of a soil protist is between 8 and 100 μm, while size ranges from 100 μm to 2 mm for soil microarthropods (Swift et al. 1979). Thus, only some soil protists potentially colonize biochar pores as pore diameters in a biochar particle are often smaller than 16 μm (Warnock et al. 2007).

Biochar may be a suitable habitat for soil microorganisms as presence and size distribution of pores protects them from predation and desiccation, and biochar provides C sources, energy, and mineral nutrients (Thies and Rillig 2009). In particular, bacteria, actinomycetes, and arbuscular mycorrhizal fungi may preferentially colonize biochar depending on its characteristics. The average sizes of soil bacteria and of fungal hyphae range from 1 to 4 μm and 2 to 64 μm, respectively (Swift et al. 1979). Many of the pores within a biochar particle may, thus, harbor microorganisms (Warnock et al. 2007). Furthermore, intrasurface accessibility to extracellular enzymes which are excreted by microorganisms may be a controlling factor of BC (biochar) degradation (Zimmerman 2010). Otherwise, enzymes may be size-excluded or sterically hindered from much of the BC surface.

Soil microorganisms incorporate and assimilate BC and biochar (Kuzyakov et al. 2009; Steinbeiss et al. 2009). For example, between 1.5 and 2.6% of BC input is incorporated in microorganisms after 624 days of incubating PyOC produced from *Lolium perenne* shoot litter mixed with soil (Kuzyakov et al. 2009). Steinbeiss et al. (2009) reported that soil microbial groups differ in their feeding preference for different biochar types. For example, fungi assimilate C from easily degradable biochar, whereas gram-negative bacteria assimilate C from relatively stable biochar during a 4 months incubation experiment. That microorganisms utilize BC and biochar is not surprising as the structure of coal is somehow related to that of charcoal, and some bacteria and fungi utilize coal with a high content of volatile (i.e., labile) OM as sole source of C and energy (Hofrichter and Fakoussa 2001). For example, Rumpel and Kögel-Knabner (2004) showed that lignite C is assimilated by soil microorganisms. Easily metabolized by microbes in biochar are, in particular, water-soluble compounds in pyrolysis condensates such as acids, alcohols, aldehydes, ketones, and sugars (Thies and Rillig 2009). Pyrolysis at lower temperatures produces biochar containing microbial oxidizable compounds such as dehydrated forms of cellulose, polysaccharides, and alcohols, whereas higher temperatures result in the formation of less microbial available condensed aryl and O-aryl furan structures (Baldock and Smernik 2002; Antal and Grønli 2003). Low-temperature biochar may also be more attractive for microbial decomposition because of narrower C:N ratios compared to biochar produced at higher temperatures. Also, biochar characterized by a low graphitic C content but with considerable N, O and S substitutions is more easily amenable to microbial oxidation (Knicker et al. 2008). Favorable microbial substrate is biochar rich in minerals and nutrients (Sohi et al. 2009). Biochar contains small amounts of nutrients that would be available to both soil biota and plant roots (Warnock et al. 2007). However, it may be unlikely that biochar is a source of either energy or cell C for microbiota in the long term after any initial bio-oils, condensates, or other labile compounds have been decomposed (Thies and Rillig 2009). Aside from the microbial food, however, pyrolysis condensates may also contain compounds toxic to bacteria and fungi (Thies and Rillig 2009). Further, biochar contains ethylene which is an important plant hormone as well as an inhibitor for soil microbial processes (Spokas et al. 2010). Thus, the observed ethylene production in biochar-amended soils may impact soil microbial and plant processes.

8.2 Biochar Effects on Carbon Sequestration in Agricultural Soils

Similar to charcoal, biochar can directly and indirectly affect SOC sequestration (Fig. 8.5; Laird 2008). Indirectly, biochar may alter C inputs to the soil by affecting NPP. Higher aboveground plant production after biochar application, and higher belowground production and increased root-derived C inputs to the soil may enhance SOC sequestration (Johnson et al. 2006; Sohi et al. 2009). Similarly, charcoal application has the long-term potential to increase C retention and storage in agricultural soils (Hernandez-Soriano et al. 2016). Charcoal provides a pool of C that is resistant to mineralization and that decreases the turnover rate of recent C due to better physical protection or C saturation of microbial processes. In combination with small increased annual inputs of fresh C from crop residues, charcoal has the capability to enhance SOC sequestration in an indirect way (Hernandez-Soriano et al. 2016). Otherwise, biochar may directly enhance SOC sequestration due to its effect on the average SOC residence time. Because it is considered to be relatively resistant against decomposition, biochar C residence times in soil may be in the range of 1,000 to 10,000 years with a common estimate of 5,000 years (Warnock et al. 2007). Less well understood are the priming effects of soil addition of biochar on SOC (Wang et al. 2017). Further, aging of biochar in soil occurs but the effects of aging on biochar properties, composition, and C sequestration are under discussion (de la Rosa et al. 2018). Biochar undergoes structural and chemical changes in soil leading to progressive formation of surface functional groups such as carboxyl, phenolic, and carbonyl groups (Mia et al. 2017). The O:C ratio, negative surface charge, and cation exchange capacity increase with increased level of aging. The surface-oxidized biochar and biochar-derived organic materials may interact with soil minerals, nutrients, and contaminants resulting in increased mineral-stabilized OM, cation retention, anion bioavailability, and reduced organic contaminants' sorption (Mia et al. 2017). In contrast, Paetsch et al. (2017) observed a shift from a soil system dominated by organo-mineral associations to particulate organic matter (POM)-dominated OC storage. In this study, the fate of high-temperature BC industrially produced by gasification of maize (*Zea mays* L.) pellets and applied to an agricultural soil was studied for one year. However, the results may depend strongly on the type of BC used (Paetsch et al. 2017).

8.2.1 Biochar Effects on Plant Growth and Yield

A positive effect of charcoal application on crop yield was already known to ancient cultures that unintentionally created *terra preta de Índio* soils (Soentgen et al. 2017; Glaser 2007). Similar, increases in crop yield are often observed after charcoal is applied (Oguntunde et al. 2004). The agricultural productivity often increases after biochar application but variability is high, and the impacts of soil properties, climatic conditions and plant species for the yield response are less well known (Blackwell et al. 2009). Meta-analyses have been hindered by missing and/or inconsistent reporting of soil properties, biochar properties, or other factors which may explain observed plant response (Crane-Droesch et al. 2013). Considering these limitations, Crane-Droesch et al. (2013) estimated an average crop yield increase of approximately 10% for 3 Mg ha^{-1} of biochar addition in the first year after application but variability was high. High soil CEC and SOC content were strong predictors of yield increases. The biochar pH, percentage C content, or temperature of pyrolysis did not predictors yield impacts. Globally, the largest potential yield increases to biochar were observed in areas with highly weathered soils, such as those characterizing much of the humid tropics. Richer soils characterizing much of the world's important agricultural areas appear to be less likely to benefit from biochar (Crane-Droesch et al. 2013). This observation was supported by a global-scale meta-analysis showing that biochar has, on average, no effect on crop yield in temperate latitudes (Jeffery et al. 2017). In contrast, yields increased by 25% in the tropics. Here, biochar increased yield through liming by biochar alkalis (Fidel et al. 2017), and fertilization, consistent with the low soil pH, low fertility, and low fertilizer inputs typical of arable tropical soils (Agegnehu et al. 2017). In contrast, arable soils in temperate regions are moderate in pH, higher in fertility, and generally receive higher fertilizer inputs, leaving little room for additional crop yield benefits from biochar (Jeffery et al. 2017). Compared to the control, biochar increased rice yield by 16% based on a meta-analysis on the effects of biochar on rice paddies studied in column, pot and field experiments (Awad et al. 2018). However, large applications of biochar, in particular, at a range of 41–50 Mg ha^{-1} were the best for rice productivity.

It has been hypothesized that reduction in soil acidity, improvements in CEC, pH, water-holding capacity and a favorable habitat for soil microorganisms are major reasons for productivity improvements (Table 8.2). Similarly, Scott et al. (2014) found evidence to suggest that biochar may improve nutrient availability indirectly through changes in pH, CEC, soil structure, improved fertilizer efficiency, decreased nutrient leaching, and may affect nutrient availability by changing gaseous N release and the soil microbial community. Animal-derived biochars may supply many more nutrients than plant-derived biochars. However, the longevity of beneficial biochar effects on soil fertility is less well known (Scott et al. 2014).

Temporal changes in crop productivity may occur, for example, through direct modification of soil chemistry (Sohi et al. 2009). These changes depend on the mineral nutrient content of fresh biochar and changes due to weathering processes as well as associated increases in CEC over time. As CEC is indicative of the capacity to retain essential nutrient cations in plant available form and of minimizing leaching loss, increases in CEC are often regarded as key factors for crop productivity improvements following biochar application. However, this may not always be the case as no changes in CEC in soil of meager fertility characteristics were observed after application of pecan (*Carya illinoinensis*) shell-based biochar, but soil fertility improved (Novak et al. 2009a). Otherwise, the soil fauna may also play a role in enhancing biochar effects on soil fertility. For example, *P. corethrurus*

Table 8.2 Examples for significant effects of black carbon on soil properties

Soil	Black carbon				Soil property changes	References
	Type	Source	Pyrolysis temperature (°C)	Application rate (Mg ha^{-1})		
Norfolk loamy sand (fine-loamy, kaolinitic, thermic typic Kandiudult)	Biochar	*Carya illinoinensis* shells	700	40–44	Increased pH, extractable Ca, K, P, Mn Reduced exchangeable acidity, extractable S, Zn	Novak et al. (2009a)
			700	40–44	Reduced penetration resistance, soil strength Improved soil water content during free drainage	Busscher et al. (2010)
		Arachis hypogaea hulls	400	40–44	Increased pH, exchangeable P, Na	Novak et al. (2009b)
			500	40–44	Increased pH, exchangeable P	
		Gallus domesticus litter	350 and 700	40–44	Increased pH, exchangeable P, Na	
		Panicum virgatum L	250	40–44	Increased pH	
			500	40–44	Increased pH, exchangeable P	
Xanthic Ferralsol	Charcoal	Secondary forest	Unknown	135.2	Increased total N, pH, exchangeable K, CEC, leaching K and NO_3^- Reduced extractable Al, leaching NH_4^+, N, Ca	Lehmann et al. (2003)

(continued)

Table 8.2 (continued)

Soil	Black carbon				Soil property changes	References
	Type	Source	Pyrolysis temperature (°C)	Application rate (Mg ha^{-1})		
Acid soil, reclaimed from grassland	Charcoal	*Acacia mangium* Willd. bark	260–360	37	Increased pH, total N, CEC, base saturation, colonization AM fungi, exchangeable Ca, Mg, K, Na Reduced available P, exchangeable acidity	Yamato et al. (2006)
Hartsells (fine-loamy, siliceous, subactive, thermic Typic Hapludult), Grady (fine, kaolinitic, thermic Typic Paleaquult)	Charcoal	*Triticum aestivum* L. straw	300–400	Unknown	Increased atrazine sorption capacity, non-desorbable atrazine fraction Reduced atrazine bioavailability	Loganathan et al. (2009)
Stuttgart silt loam (fine, smectitic, thermic Albaquultic Hapludalfs)	Charcoal	*Triticum aestivum* L. straw	300–400	Unknown	Increased benzonitrile sorption Reduced benzonitrile biodegradation	Zhang et al. (2004)
Alfisol (Chromosol)	Biochar	Grass clippings, cotton (*Gossypium* spp.) trash, plant prunings	450	100	Increased field capacity, pH, total N, extractable P, CEC, exchangeable Na, K, Ca Reduced tensile strength, exchangeable Al, microbial activity	Chan et al. (2007)
Ferrosol	Biochar	32.6% enhanced solids reductions sludge, 18.8% clarifier sludge, 48.6% waste wood chips	550	10	Increased pH, exchangeable Ca, CEC Increased microbial activity under *Glycine max* (L.) Merr.	Van Zwieten et al. (2010)

(continued)

Table 8.2 (continued)

Soil	Black carbon				Soil property changes	References
	Type	Source	Pyrolysis temperature (°C)	Application rate (Mg ha^{-1})		
					Increased *Eisenia fetida* numbers Reduced exchangeable Al	
		19.5% enhanced solids reductions sludge, 11.2% clarifier sludge, 69.3% waste wood chips	550	10	Increased pH, exchangeable Ca, CEC Increased *E. fetida* numbers Reduced microbial activity under *G. max*, *Raphanus sativus* var. Long Scarlet, and T. aestivum Reduced exchangeable Al	
Loamy Calcarosol		32.6% enhanced solids reductions sludge, 18.8% clarifier sludge, 48.6% waste wood chips	550	10	Decreased microbial activity under T. aestivum	
		19.5% enhanced solids reductions sludge, 11.2% clarifier sludge, 69.3% waste wood chips			Increased microbial activity under *G. max*	

potentially increases fertility in tropical soils under slash-and-burn practices by deposition of a reworked charcoal/soil mixture on the soil surface which favors the formation of stable humus (Ponge et al. 2006).

Depending on its persistence, biochar may affect crop productivity in the long term by providing chemically active surfaces that modify the dynamics of soil nutrients or catalyze useful reactions, and by modifying soil physical properties that benefit root growth and/or nutrient and water retention and acquisition. The effects of biochar (including char, BC, charcoal, pyrogenic organic matter, and agchar) application on root growth and traits were analyzed by meta-analysis (Xiang et al. 2017). Biochar application increased root biomass (+32%), root volume (+29%) and surface area (39%). Root morphological development to alleviate plant nutrient and water deficiency mainly benefited from biochar rather than biomass accumulation. Biochar application also affected root-associated microbes and significantly increased the number of root nodules (+25%). The responses of root traits to biochar application were generally greater in annual plants than in perennial plants, and were affected by soil texture and pH values. Biochar application did not change the root: shoot ratio but significantly increased root length and the number of root tips more strongly than root diameter. The increased number of thin roots may stimulate soil microbial activity due to the greater input of root exudates, and may have priming effects on SOM decomposition. Xiang et al. (2017) concluded that the potential SOC sequestration of biochar may be re-estimated depending on the tradeoff between biomass C inputs and SOC losses.

Long-term effects of biochar include also changes in water retention and soil bulk density (Sohi et al. 2009). In addition to CEC, soil moisture retention is another key factor in explaining biochar effects on crop yield. However, experimental evidence for changes in soil water retention capacity following charcoal application is scanty (Glaser et al. 2002). For example, soils under charcoal kilns in Ghana had higher saturated hydraulic conductivity, higher total porosity, and higher infiltration rates but lower bulk density than control (Oguntunde et al. 2008). These changes may increase water retention and decrease soil erosion and, thus, result in higher productivity of soils under charcoal kilns. Effects of biochar addition on soil physical properties depend on biochar properties (Table 8.2). For example, adding ground pecan shells pyrolyzed at 700 °C to a Norfolk sandy loam with poor physical characteristics reduced soil strength and improved soil water content during free drainage but neither improved aggregation nor the infiltration rate (Busscher et al. 2010). Supposedly, other biochar formulations would have been more effective in improving physical properties of the soil. Also, the water-holding capacity in this soil varied after applying biochars produced at temperatures from 250 to 700 °C from peanut (*Arachis hypogaea*) hulls, pecan shells, poultry litter, and switchgrass (Novak et al. 2009b). The magnitude and relative importance of the mechanisms affecting crop productivity depend on the slow process of biological, chemical, and physical modification of biochar in soil.

Published results about the effects of biochar on crop productivity are scanty (Blackwell et al. 2009). Published data about the biomass or crop response to biochar or charcoal additions compared to un-treated control in field experiments are few (Table 8.3). Biochar may improve productivity by mitigating losses of applied nutrients from the rooting zone. For example, improved corn grain yield after biochar application to an infertile, tropical soil could be explained by losses of Ca and Mg which were applied with dolomite (Major et al. 2010b). However, most experimental studies are often conducted on a small scale and over short periods of time and include methodological inconsistencies. Thus, firm conclusions about biochar effects on long-term crop yield are not possible (Sohi et al. 2009). Also, most studies have been conducted in tropical climates (tropical forest and savannah) in South America and Southeast Asia (Blackwell et al. 2009). Information is limited for other regions such as dry and temperate climates. Furthermore, the variety of plant species studied is small and often limited to field crops. Less information is available regarding the effects of biochar on productivity of pasture species, shrubs or trees. However, for some crop and soil combinations, moderate additions of biochar are beneficial to growth and only in very few cases have negative effects been observed (Chan and Xu 2009; Sohi et al. 2009). For example, charcoal additions increased crop yields at sites in Amazonia but declines in plant growth were also observed, especially at higher addition rates (Glaser 2007). However, it is unclear whether charcoal must be applied together with organic manures or compost to enhance soil fertility for sustainable management in the harsh tropical climates (Agegnehu et al. 2017).

Immediate beneficial effects of charcoal additions on crop productivity result from increase in availabilities of K, P, and Zn and, to some extent, from higher availabilities of Ca and Cu (Table 8.2; Lehmann et al. 2003). Application of poultry litter biochar results in strong increases in soil extractable P (Novak et al. 2009b). Otherwise, lower N and Mg uptakes after charcoal addition have also been observed which may cause decrease in crop growth. However, moderate charcoal additions are not a direct supplier of plant nutrients in the long term, but other effects of biochar on nutrient availability appear to be more important to crop yield responses (Glaser et al. 2002; Chan and Xu 2009). Specifically, biochar application may save nutrients which would have to be otherwise applied with fertilizer or may improve fertilizer-use efficiency and result in higher yield per unit of fertilizer applied (Chan and Xu 2009). The improved fertilizer-use efficiency may be caused by increases in soil CEC after biochar application which reduces nutrient losses by leaching. Furthermore, the soil fauna may also contribute to the improved efficiency. For example, adding charcoal together with P-rich cassava (*Manihot esculenta* Crantz) peels to soils inhabited by the earthworm *P. corethrurus* increased yield of yardlong beans (*Vigna unguiculata* subsp. *sesquipedalis* (L.) Verdc.) (Topoliantz et al. 2005).

Changes in soil microbial dynamics may contribute to higher nutrient availability after charcoal application (Glaser 2007). In addition, the promotion of beneficial soil microorganisms may contribute to the improved fertilizer-use efficiency (Warnock et al. 2007; Blackwell et al. 2009). For example, a higher colonization rate with arbuscular mycorrhizal fungi on corn roots was reported after bark charcoal application (Table 8.2; Yamato et al. 2006). However, the direct effects of biochar on soil microorganisms such as surface interactions with

Table 8.3 Effects of black carbon on biomass production or crop yield relative to un-treated control during field trials

Soil	Country	Black carbon		Crop/plant biomass			References
		Type	Source	Application rate (Mg ha^{-1})	Type	Biomass/yield response (%)	
Xanthic Ferralsol (FAO)	Brazil	Charcoal	Secondary forest	11	*Oryza sativa* L.	+75	Steiner et al. (2007)
Clay-loam oxisol (Typic Haplustox)	Colombia	Biochar	*Mangifera indica* L. wood	8	Native savanna	±0	Rondon et al. (2005)
				20		+31	
Acid soil, reclaimed from grassland	Indonesia	Charcoal	*Acacia mangium* Willd. bark	37	*Z. mays*	±0	Yamato et al. (2006)

microbial cell walls or capsular materials and indirect effects resulting from changes in adsorption of OM and effects on plant growth are less well known (Thies and Rillig 2009). Biochar may also moderate the environmental fate of pesticides by altering adsorption and desorption characteristics, and altering biodegradation. Loganathan et al. (2009) reported that the bioavailability of atrazine [6-chloro-*N*-ethyl-*N′*-(1-methylethyl)-1,3,5-triazine-2,4-diamine] was reduced in char-amended soil. Also, the biodegradation of benzonitrile was reduced in a soil containing char (Zhang et al. 2004). Further, biochar addition to a forest plantation soil with low SOC concentration may enhance the sorption of terbuthylazine (N^2*–tert*-butyl-6-chloro-N^4–ethyl-1,3,5-triazine-2,4-diamine; Wang et al. 2010). How interaction of soil biochar with pesticides alters C inputs from plants into the soil is not known but needs to be studied as biochar may persist for long periods of time in the soil and affect the efficiency of pesticides. Similarly, the enhanced adsorptivity of biochar containing soils for nonpolar and polar organic contaminants may affect their interaction with plant growth and SOC sequestration in agricultural soils (Chen et al. 2008). Long-term effects on SOC sequestration processes may be caused by the bioaccumulation of persistent organic pollutants adsorbed to biochar (Koelmans et al. 2006). However, only fresh biochar is a strong sorbent and may decrease the bioavailability, toxicity, and mobility of organic pollutants and potentially reduce the efficacy of pesticides and herbicides (Smernik 2009). Aged biochar, on the other hand, has limited sorption capacity and may be less effective.

The largest effects of charcoal application on productivity have been reported for tropical and irrigated systems on highly weathered and acid soils with low-activity clays (Blackwell et al. 2009). Reasons for the observed yield responses on these soils are increases in soil pH and alleviation of aluminum (Al) toxicity (Table 8.2; Yamato et al. 2006). Increases in soil pH are observed when the applied biochars are produced at higher pyrolysis temperature (Novak et al. 2009b). Toxic effects of available Al on crop root growth in acidic soils are reduced by pH increases related to the liming effect of biochar (Chan and Xu 2009). As a result of reduced Al toxicity, roots are able to better and more effectively explore even the acid soils to absorb nutrients and water and in increase crop yield. Other explanations discussed for increased productivity in biochar-amended soils are toxin neutralization, improved soil physical properties or reduced soil strength related to charge and surface area properties of biochar (Chan and Xu 2009). For example, reduction in tensile strength may have contributed to the increased fertilizer-use efficiency after biochar application to a hard-setting soil (Chan et al. 2007).

Negative yield responses to biochar applications happen when increase in pH exacerbates micronutrient deficiencies and calcifuge plant species are retarded by high Ca levels (Table 8.2; Chan and Xu 2009). Large additions of biochar and charcoal resulting in pH increases may also have detrimental effects on yield of pH-sensitive plants (Glaser et al. 2002). Also, addition of biochar to alkaline soils with high CEC may negatively affect crop biomass production (Van Zwieten et al. 2010). However, why dry matter production of radish (*Raphanus sativus* var. Long Scarlet) was reduced at the lowest biochar application rate but not affected at five- and ten-times higher rates remain unclear (Chan et al. 2007). Also, whether negative yield responses are possible from potentially harmful organic and inorganic amendments applied with biochar is not known.

In summary, at least for some crop and soil-type combinations, and in areas with highly weathered soils (i.e., in the humid tropics), moderate biochar additions have beneficial effects on yield. However, increases in crop production potential are highly variable and depend on crop type and probably also on addition of organic manure for sustainable fertile soils as was also observed for the Amazonian *terra preta de Índio* (Glaser 2007). Most importantly, a mechanistic understanding for interactions between crop, soil type and biochar type and its application rate is missing (Sohi et al. 2009). Also, whether biochar application promotes root biomass growth and root-derived C inputs to the soil and, thus, enhances SOC sequestration is not known. Major et al. (2010a) hypothesized that greater aboveground NPP after BC addition may translate into greater belowground NPP. Enhanced root growth and increases in amount of roots after char application are reported in both greenhouse and field experiments (Yamato et al. 2006; Magrini-Bair et al. 2009).

8.2.2 Biochar Effects on Soil Organic Carbon Sequestration

Soil application of biochar potentially affects SOC sequestration. For SOC sequestration to occur as a result of application of biochar to a land unit, all of the SOC sequestered must originate from the atmospheric CO_2 pool and be transferred into SOC through land unit plants, plant residues, and other organic solids (Olson et al. 2014). Further, the SOC stock present at end of a biochar application study must be greater than the pretreatment SOC stock levels in the same land unit.

Soils in fire-prone ecosystems and those receiving charcoal may contain some of the oldest SOC stock. This stabilization of the SOC stock (up to 50%) may be by inherent 'recalcitrance' of charcoal (Krull et al. 2006, 2008). However, little is known about the long-term effects of biochar on the SOC balance (Lorenz and Lal 2014). Recently, biochar effects on SOC have been assessed by meta-analysis of laboratory incubations, greenhouse pot and field studies (Table 8.4; Liu et al. 2016). Averaged across all studies, biochar enhanced SOC content by 40% (soil depth not specified), with the greatest enhancement in rice paddies compared to upland cropping systems and grasslands. Increases in biochar application rate, pyrolysis temperature, or C/N ratio of biochar, increased SOC content enhancement after biochar amendment. Biochars with pH of 8.1–9.0 had the greatest enhancement of SOC. Liu et al. (2016) recommended to take into account soil properties, land-use type, agricultural practice, and biochar characteristics to assess the practical potential of biochar for mitigating climate change. However, no studies ran more than 4 years, and only 21% of the observations showed results over a whole growing season with the presence of vegetation cover. Thus, field experiments with longer durations across a wider range of spatial and temporal scales are required (Liu et al. 2016).

Table 8.4 Relative change (%) in soil properties after biochar amendment compared to un-amended control treatments (modified from Liu et al. 2016)

Soil property	Factor		Relative change (%)	Comment
Soil carbon dioxide flux	Land use	All studies	+5	Not significant
		Rice paddies	+18	
	Experimental method	Laboratory incubations	+28	
		Pot experiments	−18	
		Field studies	+8	
	Soil texture	Coarse-textured soils	+24	No significant effect medium-textured soils
		Fine-textured soils	−7	
	Soil pH	pH 6.6–7.5	−24	No significant effect pH < 5.0 or > 7.5
		pH 5.0–6.5	+30	
	N-fertilizer	N-unfertilized	+21	
		N-fertilized	−5	No significant effect organic N-fertilizer (limited data waste compost and sludge)
		Synthetic N-fertilizer	+45	
	Biochar application rate	Application rate < 20 Mg ha^{-1}	+25	No significant effect application rate 20–40 Mg ha^{-1}
		Application rate 40–60 Mg ha^{-1}	−12	
	Feedstock	Crop residue	+5*	No significant effect wood
		Manure	+46	Limited data
		Biowaste	−12	
	Pyrolysis temperature	<400 °C	+24	No significant effect 400–500 °C or 500–600 °C
		>600 °C	−15	
	Biochar pH	pH < 7	+18	No significant effect pH 7–8, 8–9 or >10
		pH 9–10	−10*	
	Biochar C/N ratio	<50	+68	No significant effect C/N ratio 50–100 or 100–200
		200–300	−97	

(continued)

Table 8.4 (continued)

Soil property	Factor		Relative change (%)	Comment
Soil organic carbon content	Land use	All ecosystem types	+40	
		Rice paddies	+68	
	Experimental method	Laboratory incubations	+34*	
		Pot experiments	+71*	
		Field studies	+27*	
	Soil texture	Coarse-textured soils	+45*	
		Medium-textured soils	+46*	
		Fine-textured soils	+33*	
	Soil pH	pH < 5.0	+25	
		pH 5.0–6.5	+38*	
		pH 6.6–7.5	+49*	
		pH > 7.5	+53	
	N-fertilizer	N-unfertilized	+33*	
		N-fertilized	+44*	No pronounced N-fertilizer type effects
	Biochar application rate	Application rate < 20 Mg ha^{-1}	+23	
		Application rate 20–40 Mg ha^{-1}	+51*	
		Application rate 40–60 Mg ha^{-1}	+59	
	Feedstock	Wood	+48	
		Crop residue	+40*	
		Manure	+26*	Limited data
		Biowaste	+32*	
	Pyrolysis temperature	<400 °C	+32*	
		400–500 °C	+38*	

(continued)

Table 8.4 (continued)

Soil property	Factor		Relative change (%)	Comment
		500–600 °C	+47*	
		>600 °C	+56	Limited data
	Biochar pH	<7	+60*	Limited data
		7–8	+45*	
		8–9	+75	
		9–10	+21*	
		>10	+43*	
	Biochar C/N ratio	<50	+24*	Limited data C/N 200–300 and 300–400
		50–100	+32*	
		100–200	+40*	
Soil microbial biomass content	Experimental method	Laboratory incubations	−8*	
		Pot experiments	−20*	
		Field studies	+85*	
	Soil texture	Coarse-textured soils	+49*	No significant effect fine-textured soils (limited data medium-textured soils)
	Soil pH	pH < 5.0	+49	No significant effect pH > 7.5 (limited data pH 6.6–75)
	N-fertilizer	N-fertilized	+42	No significant effect N-unfertilized (limited data organic N-fertilizer and waste compost)
		Synthetic N-fertilizer	+27*	
		Waste compost	+77*	
	Biochar application rate	Application rate < 20 Mg ha^{-1}	+50	
		Application rate 20–40 Mg ha^{-1}	−8	
	Feedstock	Wood	+48	

(continued)

Table 8.4 (continued)

Soil property	Factor		Relative change (%)	Comment
		Crop residue	−8	
	Pyrolysis temperature	<400 °C	+64*	Limited data 400–600 °C
		>600 °C	−12*	
	Biochar pH	7–8	−19*	
		8–9	+68	
		>10	−12*	
	Biochar C/N ratio	50–100	+62	Limited data C/N 200–300
		100–200	−10*	

Biochar may enhance SOC sequestration due to intrinsic stability of some components but may also interact with the decomposition of SOC fractions. Specifically, SOC sequestration in agricultural soils occurs when biochar application increases the SOC stock in a specified area over time, and the stock of organic compounds with C residence time in the millennia range increases (Lorenz and Lal 2014). However, in natural soils, BC and biochar losses through decomposition, degradation, erosion and leaching occur in the long term as otherwise SOC would consist primarily of BC and biochar originating from biomass burning during wildfires (Kuhlbusch and Crutzen 1995; Lehmann et al. 2009). Degradation of BC (biochar) occurs abiotically (e.g., chemical oxidation, photooxidation, solubilization) and biotically (microbial incorporation or oxidative respiration) (Zimmerman 2010). A full mass balance is, thus, needed but often not available to account for the fate of soil-applied BC (Major et al. 2010a). Also, the contribution of the soil fauna to biotic degradation is unclear (Topoliantz and Ponge 2005). For example, *P. corethrurus* may stabilize charcoal-derived C in soil by favoring the formation of microaggregates within macroaggregates. These aggregates contain protected occluded C and their amount is enhanced by passage through the earthworm gut (Bossuyt et al. 2005; Ponge et al. 2006). Thus, earthworms may enhance SOC sequestration in agricultural soils after biochar application.

Biochar decomposition

BC and biochar are subject to decomposition in most surface soils as those are thermodynamically unstable under the oxidative prevailing conditions (Macías and Arbestain 2010). However, little is known about decomposition and stability under field conditions as most studies are conducted under laboratory conditions (Major et al. 2010a). Only ~6% of the initially added biochar was mineralized to CO_2 during 8.5 years' laboratory incubation, and condensed aromatic moieties were the most stable biochar fraction (Kuzyakov et al. 2014). However, direct estimations of biochar decomposition and transformations are difficult because (i) changes are too small for any relevant experimental period and (ii) due to methodological constraints. Even more challenging is studying BC decomposition in soils as all BC quantification methods are selective for the different BC phases and, thus, no single method for quantifying all char categories exists (Hammes et al. 2007). The most promising methods identified are ultraviolet (UV) or chemical oxidation with elemental and ^{13}C NMR analysis of residues, thermal analysis, and hydrogen pyrolysis (Hammes et al. 2007; De la Rosa et al. 2008; Ascough et al. 2009). Also, there are insufficient data to compare the short- and long-term decomposition of biochar under different climates and in different soils (Sohi et al. 2009). Nevertheless, slow and rapid decomposition rates of char have been reported after application to soil (Shneour 1966; Shindo 1991). For example, based on NMR using a molecular mixing model (Nelson and Baldock 2005), BC concentrations decreased rapidly to 30% of the initial contents during the first 30 years of cultivation of soils with corn that was grown on land cleared from previous forest by fire in western Kenya (Nguyen et al. 2008). Up to 100 years of cultivation, however, only small changes in BC contents and stocks occurred. Aside from the physical export, decomposition of BC also contributed to the observed losses. Major et al. (2010a) observed that less than 3% of BC produced from mango (*Mangifera indica* L.) prunings was lost as CO_2 after 2 years in a soil in Colombia. Three-quarters of the BC loss in this study had already occurred during the first year.

The loss of biochar C from soils by decomposition depends on inherent chemical stability of biochar and its physical structure, but also by its protection from microbially produced exoenzymes through soil physical structures (Keiluweit et al. 2010; Zimmerman 2010). Furthermore, the readily available phase in a biochar particle can be physically protected against decomposition by entrapment in a condensed and/or crystalline phase within the particle (Almendros et al. 2003). Co-composted biochar may contain nutrient-rich organic coating that covers the outer and inner (pore) surfaces of biochar particles (Hagemann et al. 2017). Fresh biochar, on the other hand, may also lose C abiotically by surface oxidation (Bruun et al. 2008).

The chemical stability of biochar depends on the aliphatic portion that is more readily decomposed and is in less abundance in biochar produced at higher temperatures (Lehmann et al. 2009). Chemical stability depends also on the aromatic portion that is decomposed more slowly, forming surficial, oxygen-containing functional groups including carboxylic acids. Historical BC samples are oxidized after 130 years in soil (Cheng et al. 2008a). Otherwise, adsorption of non-BC such as compost-derived OM may be important for surface chemistry of fresh biochar (Hagemann et al. 2017). In contrast, adsorption of non-BC such as humified macromolecules and/or microbes contributes to the much larger proportions of carboxylic and phenolic C forms surrounding the core and its surface even after thousands of years of decomposition in soil (Lehmann et al. 2005). The core of biomass-derived BC particles in this study is still highly aromatic and resembles fresh charcoal.

Some components of biochar are stable in the soil as is well known from the persistence of soil charcoal and its suitability from studies involving dating and paleo-environmental reconstruction (Titiz and Sanford 2007). Char in residues from forest fires may be up to 10,000 years old (Preston and Schmidt 2006). However, soil BC may also be relatively modern. For example, radiocarbon ages <50–400 years and a median age of 652 years have been reported for soil BC and charcoal in boreal forests, respectively (Kane et al. 2010; Ohlson et al. 2009). Otherwise, charcoal in *terra preta de Índio* soils had radiocarbon ages between 500 and 7,000 years BP (Neves et al. 2003). However, radiocarbon ages provide no quantitative information about the decomposition rate of biochar (Lehmann et al. 2009). Radiocarbon ages are only indicative of the average time elapsed since atmospheric CO_2 is fixed by photosynthesis in biomass which then formed feedstock for biochar. Additional information about the amount of biochar at deposition is needed to quantify the decay rate but this information is generally not available (Sohi et al. 2009). Nevertheless, BC, biochar, char, charcoal, and PyOC may be the only non-mineral-associated SOM component aside from fossil C that may be persistent in soil (Marschner et al. 2008).

Large differences have been observed in BC losses from soil, the mean residence time of soil BC, turnover times, and half-lives (Czimczik and Masiello 2007; Lehmann et al. 2009). For example, half-lives of less than 100 and up to 6,623 years have been reported for soil EC and black(pyrogenic) C (Bird et al. 1999; Preston and Schmidt 2006). Long-term mean residence times between 1,300 and 2,600 years are estimated for soil BC at sites where steady-state conditions of natural char production and disappearance occurred over long periods of time (Lehmann et al. 2008). Otherwise, based on analyzing modern and archived profile samples from a Russian steppe soil, Hammes et al. (2008) calculated a turnover time for BC of only 293 years. Murage et al. (2007) observed that the MRT of the free light fraction containing charcoal in soils is only about 20 years. Thus, soil BC can cycle on century timescales and may be less stable than previously thought.

Laboratory incubation studies have indicated that char C loss ranges between 0.3% (oak charred for 22 h at 800 °C) and 0.8% (maize charred at 350 °C for 2 h) during 60 days (Hamer et al. 2004). In mixture with soil, PyOC losses of 0.5% (rye grass charred briefly at 350 °C) and 3% (pine charred briefly at 350 °C) over 48 days, and BC losses of 4% (rye grass charred at 400 °C for 13 h) over 3 years have been reported (Hilscher et al. 2009; Kuzyakov et al. 2009). BC-C loss between 6% for oak wood charred at 350 °C under saturated water regime and of 21% for corn residues charred at the same temperature but incubated at unsaturated water regimes were observed in mixture with sand after 12 months (Nguyen and Lehmann 2009). Reasons for the rapid decomposition were high and consistent incubation temperatures, mixing with sand creating an oxygen-rich environment promoting rapid oxidation of BC, unavailability of BC-protecting mechanisms and significant amounts of an easily decomposable BC fraction. Oxidation processes, however, comprised a major mechanism for BC change and mineralization but what controls BC decomposition under submerged conditions requires additional research (Nguyen and Lehmann 2009). Incubation experiments only with BC from traditional charcoal production with wood as feedstock indicated that aging of BC occurs over a temperature range from −22 to 70 °C within a short period of 12 months (Cheng and Lehmann 2009). Higher temperatures and longer incubation time can enhance BC aging. The aged BC has higher oxygen concentrations, surface acidity, and negative surface charge but lower C concentrations, pH, surface basicity, point of zero net charge, and also a lower adsorption capacity of hydroquinone than fresh BC. Thus, decomposition studies with fresh BC may not be used as surrogate for studies about decomposition of naturally occurring BC (Cheng and Lehmann 2009).

Adding glucose to soil with BC stimulates its decomposition for a short period during incubation studies (Kuzyakov et al. 2009). This response indicates cometabolic decomposition (Hamer et al. 2004). Thus, microorganisms do not depend on BC utilization as a C or energy source but microbial enzymes produced for decomposition of other substrates such as rhizodeposits contribute to BC decomposition (Kuzyakov et al. 2009). The importance of this priming effect by labile substrates on BC decomposition in the field is largely unknown.

Biochar effects on the decomposition and stabilization of soil organic carbon

The effect of biochar on decomposition and stabilization of native SOC is poorly understood. Based on incubations studies, Wang et al. (2017) reported that biochar may enhance the physical protection of SOC in some soils by increasing the proportion of C stored within macroaggregates. This would be a novel mechanism by which biochar may contribute to soil aggregation and SOC sequestration. However, the C sequestration potential of biochar may be negated in certain soils by enhanced SOC losses due to priming. Thus, potential trade-offs of biochar serving as a promoter of both soil aggregation and C priming need to be better understood if the intent of adding biochar is to increase SOC sequestration (Wang et al. 2017).

Weng et al. (2017) monitored the fate of root-derived C for a 15-month period—commencing 8.2 years after biochar amended into a subtropical ferralsol. The biochar addition to this pasture soil built SOC by forming organo-mineral microstructures (<53 μm) that enhanced the retention of root-derived C by 20%. Negative priming was detected a decade after biochar incorporation as the biochar slowed SOC mineralization by 5.5%. Specifically, sorption of root exudates on biochar surfaces may have counteracted the dissolution of mineral-protected SOC catalyzed by root exudates (Weng et al. 2017). Otherwise, strong stimulation of humus decomposition has been observed in a boreal forest in the presence of charcoal (Wardle et al. 2008). This trend can be explained by preservation of some nutrients from leaching in charcoal and a subsequent increase of microbial activity induced humus decomposition. However, no significant effect of BC on decomposition of native SOC has been observed during other incubation studies (Kuzyakov et al. 2009). Similarly, no enhanced loss of existing SOC due to addition of BC has been observed in a field study (Major et al. 2010a). Otherwise, studies in *terra preta de Índio* suggest that biochar in soil may lead to a net stabilization of other SOC (Liang et al. 2010). Thus, effects of BC on decomposition of native SOC needs to be checked for a range of agroecosystems including calculating the C budget by measuring leached C and that released as CO_2, and identification of C sources (Lehmann and Sohi 2008).

Biochar and the priming effect

The decomposition rate of SOC can increase or decrease substantially after addition of easily decomposable C sources with biochar known as positive or negative priming effects (van der Wal and de Boer 2017). In addition, changes to soil properties due to biochar addition may affect both the direction and magnitude of priming effects (Luo et al. 2017). Based on meta-analysis, Maestrini et al. (2015) reported that on average PyOC induces a priming effect of similar magnitude but opposite direction on native SOC (positive priming effect) and fresh OM (negative priming effect). The positive priming effect occurred mostly in the short term, while negative priming effect appeared at a later stage. Negative priming effects were attributed to the physical protection of SOC, while positive priming effects were largely attributed to stimulation of microbial activity by the labile C contained within the biochar, or abiotic release of CO_2 from carbonates in the ash. However, many uncertainties regarding mechanisms involved exist, particularly on the parameters driving the amplitude and the direction of the priming effect. Maestrini et al. (2015) concluded that adding PyOC to the soil induces a cumulative positive priming effect on a yearly timescale on the native SOC, which may be counter-balanced by the negative priming effect observed on fresh OM.

Many studies have focused on the topsoil but the impact of biochar on subsoil SOC mineralization has hardly been assessed despite biochar's susceptibility to downward transport after soil amendment. For example, Naisse et al. (2015) reported that the priming effect on native SOC was dependent on the type of material added and also subsoil type specific. Biochar addition induced a small positive priming effect or a negative priming effect that was outbalanced by biochar degradation. Priming effects in subsoil may depend on subsoils' SOC content, sampling depth, and/or contribution of stabilized SOC compounds more than microbial biomass characteristics (Naisse et al. 2015).

Biochar loss by erosion and leaching

The effect of biochar on SOC sequestration in agricultural soils depends on the net changes in the pool of organic compounds in the soil profile of a specified agricultural area (Lorenz and Lal 2014). BC accumulates whenever it is removed from the soil surface by biological or physical mixing and stabilized by sorption to minerals (Czimczik and Masiello 2007). Thus, differences in erosion and leaching processes can partially explain biochar losses from a site (Lehmann et al. 2009; Sohi et al. 2009). Aside from the CO_2 losses after biochar application, the physical export of biochar particles by wind and water, mixing into the mineral soil and loss by leaching must, therefore, be determined.

The BC may migrate from the surface layer into the mineral soil (Leifeld et al. 2007). However, only few studies have quantified BC concentrations over the whole soil profile. For example, BC contents in soils under tropical slash-and-burn agriculture range between 5.5% of SOC in 0–20 cm and 4.1% of SOC in 35–60 cm depth (Rumpel et al. 2006a). Maximum profile BC concentrations in a Russian steppe soil and in Russian Chernozems have been observed between 30 and 50 cm depths (Rodionov et al. 2006; Hammes et al. 2008). Also, BC explained up to 50% of the subsoil SOC content for an arable land in Germany and downward migration is attributed to bioturbation and leaching (Brodowski et al. 2007). Translocation of BC to deeper soil depths may be promoted by oxidation processes which increase water solubility of BC (Guggenberger et al. 2008). The soluble BC transport may be favored by fragmentation and dissolution of charcoal BC through oxidation of the condensed aromatic structures. Hockaday et al. (2007) reported indirect evidence for a microbial BC dissolution mechanism in soil and, thus, BC-derived structures mostly condensed aromatic ring structures could be identified in soil pore water. High vertical BC transportation rates have been reported in peatland soil profiles, and related to large pore volumes and often saturated conditions (Leifeld et al. 2007). Thus, in deeper anaerobic peat soil horizons long-term BC accumulation may occur as microbial activity is reduced under water-saturated conditions. Furthermore, deeper horizons contained BC of higher thermal stability, most likely soot (Leifeld et al. 2007). Similarly, Rumpel et al. (2008) reported that long-term preservation of BC occurs in the deepest minerals soil horizons up to 80 cm depth under tropical slash-and-burn agriculture. Major et al. (2010a) reported that 1% of applied BC can be mobilized by percolating water over 2 years after application. Relatively more DOC than POC can be lost from BC. Thus, studies limited to surface horizons may miss the location of the most concentrated BC, where it contributes most to SOC, and the importance of downward migration and stabilization of BC in deeper soil horizons.

Physical export by erosion and surface runoff may be among major pathways for biochar loss from a site. For example, between 20 and 53% of applied BC may be lost from field plots by surface runoff during some intense rain events (Major et al. 2010a). Rumpel et al. (2006a) observed that a major part of BC under tropical slash-and-burn agriculture on steep slopes may not be associated with the mineral phase but as a part of the light fraction. Thus, BC is preferentially eroded compared to other SOC fractions (Rumpel et al. 2006b). Specifically, chemically the most 'recalcitrant' part of PyOC exhibits favorable floating behavior and splash erosion could easily remove it, mainly by sheetflow (Rumpel et al. 2009). Further, BC may be exported from a watershed entirely (Rumpel et al. 2006b). For example, Hockaday et al. (2006) reported evidence of charcoal-derived DOM in groundwater and a small river in a fire-impacted watershed. However, during transport in the dissolved phase, BC is reactive and chemical and physical changes may occur (Hockaday et al. 2007). Aside from BC's mobility in the dissolved phase, mobility in the colloidal phase is also an important pathway for its export from a watershed (Guggenberger et al. 2008). Once it becomes a component of riverine C, it is easily exported to the ocean and finally buried in deeper ocean sediments (Mitra et al. 2002; Masiello 2004). Prior to its ultimate deposition, delayed transmission of BC due to 'preaging' on millennial timescales in catchment soils occurs (Hanke et al. 2017). Thus, sequestration of BC can markedly lag its production.

Biochar and emissions of greenhouse gases from soils

Biochar application may alter the GHG balance of agricultural soils. Also, biochar application may interact with SOC sequestration through feedback mechanisms such as changes in temperature and precipitation when the radiative forcing is altered. Emission savings may arise indirectly from biochar application through (i) reduced need for fertilization due to enhanced fertilizer-use efficiency, (ii) avoided conversion of natural ecosystems for agriculture as crop yield may be higher on biochar-amended soil, (iii) reduced need for irrigation due to improved water-holding capacity, and (iv) reduced energy need for tillage by improved soil physical properties (Sohi et al. 2010). However, there is a little information on these effects. Similarly, only a limited number of studies have been conducted about the direct effects of biochar application on soil's GHG balance (Spokas and Reicosky 2009). Specifically, emissions of CO_2, CH_4, and N_2O from biochar-amended agricultural soil vary widely depending on the properties of the biochar, soil type, land use, and climate.

A meta-analysis of field studies indicated that biochar application to soil did not change soil CO_2 fluxes, but increased CH_4 fluxes by 25.4%, and decreased N_2O

fluxes by 33.0% and 28.8% in both N-fertilized and unfertilized soils, respectively (He et al. 2017). Thus, biochar application to soil did not affect global warming potential (GWP) of total soil GHG fluxes. The nonsignificant difference in CO_2 fluxes between control and biochar treatments was explained by low application rates and/or high biochar labile C leaching due to rainfall. The increased CH_4 fluxes were mainly reported from the treatments with N fertilization. This increase probably resulted from the stimulation of soil microbial activities, especially the methanogenic archaea and methanotrophic bacteria (He et al. 2017).

Jin et al. (2008) observed decreases in soil respiration with increasing application rates of maize-derived biochar to a mineral soil. In contrast, Steiner et al. (2008) reported no effects of wood char on the CO_2 emission of a highly weathered Amazonian upland soil. After adding BC produced from *Lolium perenne* residues, the cumulative CO_2 efflux from soil during an incubation study increased over 3.2 years but it decreased after applying BC to loess (Kuzyakov et al. 2009). Among fifteen biochars applied to an agricultural soil, two suppressed and five stimulated CO_2 respiration during an incubation study, while eight biochars did not affect respiration compared to the control (Spokas and Reicosky 2009). Increasing the amount of woodchip biochar added to an agricultural soil during an incubation study may suppress CO_2 production compared to the control (Spokas et al. 2009). However, the exact cause of such reduction is not understood. Biochar amendment had no significant effect on soil CO_2 fluxes when averaged across a range of studies assessed by meta-analysis (Table 8.4; Liu et al. 2016). However, a positive response of soil CO_2 fluxes to biochar amendment was found in rice paddies, soils without vegetation, and unfertilized soils. Further, responses of soil CO_2 fluxes to biochar amendment varied with soil texture and pH. The use of biochar in combination with synthetic N-fertilizer and waste compost fertilizer led to the greatest increases in soil CO_2 fluxes. Soil CO_2 fluxes responses to biochar amendment decreased with biochar application rate, pyrolysis temperature, or C/N ratio of biochar. Among different biochar feedstock sources, positive responses of soil CO_2 fluxes were the highest for manure and crop residue feedstock sources. Soil CO_2 flux responses to biochar amendment decreased with pH of biochar. However, to evaluate the effect of biochar on soil CO_2 fluxes field experiments with longer durations (>4 years) across a wider range of spatial and temporal scales are needed (Liu et al. 2016).

Similar to the effects of biochar addition on the soil CO_2 flux, the mechanisms through which biochar affects the CH_4 flux are poorly understood (Van Zwieten et al. 2009). Among fifteen biochars added to a soil and incubated in the laboratory, ten biochars reduced the net CH_4 oxidation rates (Spokas and Reicosky 2009). Thus, agricultural soil turned from a net CH_4 sink into a net CH_4 source. However, five biochars did not affect CH_4 oxidation rates of the agricultural soil. In another incubation study, adding woodchip biochar to an agricultural soil also reduced the net CH_4 oxidation (Spokas et al. 2009). Possible reasons are that the biochar contains a potential inhibitor for methanotrophs, but this microbial group probably also utilizes organic compounds sorbed on biochar vs. CH_4. Further, whether the suppression of CH_4 oxidation disappears once these organic compounds are metabolized requires longer incubation studies. Adding biochar to a forage grass stand in Colombia, on the other hand, completely suppressed CH_4 emissions (Renner 2007). Whether biochar application to soil inhibits CH_4 emission is, thus, not known.

Cayuela et al. (2014) performed a meta-analysis on the effects of 'biochar' (i.e., biochar, charcoal or BC) on soil N_2O emissions. Overall, 'biochar' reduced soil N_2O emissions by 54% with feedstock, pyrolysis conditions, and C/N ratio as key factors influencing emissions of N_2O, while a direct correlation occurred between the 'biochar' application rate and N_2O emission reductions (Cayuela et al. 2014). However, the average reduction in N_2O emissions was only 28% under field conditions (Cayuela et al. 2015). A substantially lower mitigation potential of biochar/charcoal of 9.2% reduction in N_2O emissions was reported in a subsequent meta-analysis of field studies by Verhoeven et al. (2017). However, N_2O emission reductions were not significantly affected by cropping system, biochar properties of feedstock, pyrolysis temperature, surface area, pH, ash content, application rate, or site characteristics of N rate, N form, or soil pH. The failure to detect effects was explained by uneven coverage in the range of these predictor variables in field studies. Thus, more field studies are needed to investigate effects of fertilizer N form, sustained and biologically relevant changes in soil moisture, multiple biochars per site, and time since biochar application (Verhoeven et al. 2017). Interactions between soil texture and 'biochar' and the chemical form of N-fertilizer applied with 'biochar' also had a major influence. Applying BCs derived from different feedstocks and pyrolysis temperatures reduced N_2O emissions in column, pot and field experiments, and various factors were responsible for this decrease (Awad et al. 2018). However, there is a significant lack in understanding of the key mechanisms which alter N_2O emissions (Cayuela et al. 2014).

A combination of biotic and abiotic factors may be involved in a decline in N_2O emissions from soil amended with biochar (Van Zwieten et al. 2009). Potential explanations include change in soil moisture potential, increase in potential for aerobic or anaerobic microsites within the biochar, increase in oxygen diffusion into soil due to reduced bulk density, abiotic reactions with N_2O and O_2 or metal oxides, or sorbed organic compounds toxic to the microbial community (Spokas et al. 2009). These factors may vary depending on soil type, land use, climate and biochar characteristics. Thus, biochar can be effective in mitigating N_2O emissions and reducing inorganic-N leaching from soils, depending on biochar type, biochar aging, soil type, and water-filled pore space (Singh et al. 2010). Further, any reduction in N_2O production after adding charcoal to a grassland soil during incubation experiments may be highly moisture dependent (Yanai et al. 2007). Adding wood biochar to a pasture soil may not influence the cumulative N_2O-N loss compared to the control (Clough et al. 2010). Also, the cumulative N_2O-N flux from bovine urine may not be reduced by biochar as soil inorganic-N pools available for N_2O production mechanisms are not always reduced by biochar application. Otherwise, the majority of biochars added to agricultural soil suppress N_2O emissions in laboratory incubations (Spokas and Reicosky 2009), probably because of a lower rate of N_2O production in the soil after biochar application and not a greater rate of N_2O reduction. In contrast, adding biochar containing nutrients and labile OM may increase N_2O production relative to the control. Spokas et al. (2009) reported that only high rates of woodchip biochar additions may lead to a net decrease in N_2O production rates compared to laboratory incubations with low amounts of biochar. Thus, reductions of N_2O emissions after biochar application to soil do not necessarily occur as different biochars vary in their effects on the activity of soil microorganisms responsible for N transformation (Van Zwieten et al. 2009).

During short-term laboratory incubation, adsorption of GHGs on biochar may affect the determination of fluxes. For example, a fraction of CO_2 respired after biochar addition may be sorbed on the biochar surface by chemisorption and not recovered by the assay to determine soil respiration (Thies and Rillig 2009). Thus, chemisorption can result in the reduction in the estimate of respiratory activity. Spokas and Reicosky (2009) reported that the rates of CO_2 consumption/production from fifteen different biochars incubated without soil were not correlated with biochar specific surface area. Thus, chemisorption may be of a minor importance as mechanism for reduced CO_2 release from biochar-amended soils (Thies and Rillig 2009). Furthermore, sorption/desorption interactions of CH_4 with different biochars may be minimal but how sorption processes potentially affect N_2O emissions from biochar needs additional research (Spokas and Reicosky 2009). Otherwise, sorption of N_2O on woodchip biochar is generally not observed as is neither consumption nor production of N_2O by the biochar itself (Spokas et al. 2009).

Black carbon (charcoal) stock in agricultural soils

The link between biochar properties and their efficacy for SOC sequestration is not well understood (Brewer et al. 2009). Estimates of the stock of BC in agricultural soils are scanty and highly uncertain due to analytical difficulties which affect the quantitative assessment of soil BC (Hammes et al. 2007). Nevertheless, substantially greater SOC stocks are reported for BC-containing soils from historical charcoal blast furnace sites, even 130 years after charcoal application (Cheng et al. 2008b). The abundance of charcoal C in most modern soils is largely unknown (Fang et al. 2010). Krull et al. (2008) predicted BC proportions for the first time in soil samples spanning all major climate zones and soil types across most parts of the world by MIR spectroscopy coupled with partial least squares (PLS) (Janik et al. 2007). While the BC can comprise between 0 and 50% of SOC stock, the variability is large. Soils richest in BC occur in latitudes of 20–30° in both hemispheres and are developed under tropical climates. Other BC-rich soils occur in higher latitudes in the Northern Hemisphere associated with an increase in SOC pool due to cool climates and low decomposition rates. However, soils at lower relief positions generally also accumulate BC through erosion and deposition. Furthermore, high amounts of expanding-lattice clays also favor BC accumulation regardless of the climate (Krull et al. 2008).

Stocks of terrestrial residual PyOC, i.e., PyOC initially remaining on the ground relatively close to the burning site, have been estimated at 86 Pg C (Landry and Matthews 2017). For soil PyOC, Reisser et al. (2016) produced the first worldwide inventory including more than 560 measurements from 55 studies. Global PyOC represented on average 13.7% of the SOC, ranging from 0 to 60%, with a stock estimation of 200 Pg C for the uppermost 2 m soil depth. Among different environmental explanatory variables, high soil pH and clay content were the most significant parameters explaining a high PyOC content in the SOC. Further, PyOC production parameters, such as fire or land use, did not well explain PyOC content patterns. Among agroecosystems, the highest content of PyOC in SOC was found in soils used for agriculture (16.0%), while contents in grasslands were lower (12.1%). However, the higher content of PyOC in the SOC in agricultural land might not be explained by the recent land use, but could be linked to stubble burning in the last century, or even much older agricultural practices from the Middle Ages or earlier, which may have included much more common use of fire. Another explanation might be the relative preservation of PyOC due to increased SOC turnover by tillage and other agricultural practices (Reisser et al. 2016).

Rodionov et al. (2010) estimated that between 5 and 30% of SOC stocks in grassland ecosystems of the world consist of BC. Thus, global steppe ecosystems may store between 4 and 17 Pg of BC. The BC contents can be estimated using benzene polycarboxylic acids (BPCAs) as molecular markers. BPCAs are as ubiquitous in soils as BC, and benzene hexacarboxylic acid (B6CA) is one of the most persistent and widely distributed natural organic compounds on Earth due to its unique resistance to oxidation and enormous water solubility (Haumaier 2010). However, other presently known precursors of soil BPCAs are fungal and plant pigments containing polycondensed aromatic moieties. For example, up to 25% of BC detectable in a soil by the BPCA method may have been produced in situ by soil microorganisms such as *Aspergillus niger*, without fire or charring (Glaser and Knorr 2008). Except for soils with low BC levels, natural products are of minor importance as source for soil BPCAs (Haumaier 2010). Similarly, humification processes in soils may form (pseudo-)melanoidins which contain chemical structures similar to BC (Brodowski et al. 2005). However, clear evidence for the existence of (pseudo-)melanoidins in soils and their distinction from BC is lacking (Rodionov et al. 2010). Llorente et al. (2018) proposed a simplification of the BPCA method for a great cost reduction (in time and expense) providing a rapid and accurate method for large soil sample sets. In the following section, some examples are given for charcoal C and BC contents in agricultural soils from different regions determined by a range of quantification methods (Table 8.5).

Historically, fire has been extensive in the North American grasslands and its legacy may contribute to charcoal C in US agricultural soils (Laird et al. 2008). Charcoal C of 10 to 35% of SOC in US agricultural soils (as determined by high energy oxidation and solid-state ^{13}C NMR with an empirical correction for underrepresented char) have been reported (Skjemstad et al. 2002). Based on quantitative ^{13}C NMR, Fang et al. (2010) estimated that between 27 and 36% of SOC is charcoal C in some agricultural soils in the USA.

Between 10 and 29% of SOC in soils under corn cultivation in western Kenya has been estimated to be BC as determined by NMR using a molecular mixing model (Nguyen et al. 2008). In tropical soils under slash-and-burn agriculture, BC contents estimated by a wet chemical method can range from 3.7 to 7.3% of SOC

Table 8.5 Black carbon/charcoal carbon contents in agricultural soils

Region	Method	Soil	Soil depth (cm)	Black carbon/charcoal carbon content		References
				g C kg^{-1}	% TOC	
USA	Ultraviolet (UV) oxidation followed by solid-state ^{13}C nuclear magnetic resonance (NMR) spectroscopy	Brennyville silt loam (coarse-loamy, mixed, superactive, frigid Aquic Glossudalfs)	0–20	1.8	9.7	Skjemstad et al. (2002)
		Elliott silt loam (fine, illitic, mesic Aquic Argiudoll)	0–20	6.6	23.0	
		Houston Black clay (fine, smectitic, thermic Udic Haplusterts)	0–20	7.6	20.6	
		Vallers silt clay loam (fine-loamy, mixed, superactive, frigid Typic Calciaquoll)	0–20	13.6	32.9	
		Walla Walla silt loam (coarse-silty, mixed, superactive, mesic Typic Haploxeroll)	0–20	3.6	35.0	
USA	Quantitative NMR	Dinsdale (fine-silty, mixed, superactive, mesic Typic Argiudoll; Phaeozem)	5–15		27	Fang et al. (2010)
		Maxfield (fine-silty, mixed, superactive, mesic Typic Endoaquoll; Stagnosol)	5–15		32	
		Clarion (fine-loamy, mixed, superactive, mesic Typic Hapludoll; Phaeozem)	5–15		34	
		Webster (fine-loamy, mixed, superactive, mesic Typic Endoaquoll; Stagnosol)	5–15		36	
Kenya	NMR coupled with molecular mixing model	Humic nitisols /Typic Palehumults	0–10	3.5–12.7 (calculated)		Nguyen et al. (2008)
Laos	Dichromate oxidation	Alfisols/Dystrochrepts	0–30, 20–35		7.3, 3.7	Rumpel et al. (2006a)
Australia	UV-NMR and mid-infrared (MIR) spectroscopy coupled	Major soil types	0–10 to 0–90 (A horizons)		32.9	Lehmann et al. (2008)

(continued)

Table 8.5 (continued)

Region	Method	Soil	Soil depth (cm)	Black carbon/charcoal carbon content		References
				g C kg^{-1}	% TOC	
	with partial least squares (PLS)					
Australia	UV-NMR and mid-infrared (MIR) spectroscopy coupled with partial least squares (PLS)	Two transects	0–30		13.6–31.5	
Argentina	Benzene polycarboxylic acids (BPCAs) and MIR-PLS	Chernozem, Kastanozem, Phaeozem	Profile		0.6–6.4	Bornemann et al. (2008)
China, Germany, Russia, USA		Calcisol, Cambisol, Chernozem, Greyzem, Kastanozem, Luvisol, Phaeozem	Variable		0.6–11.4	
Germany	BPCAs	Haplic Luvisol	0–30	0.58–0.62	3.8–4.6	Brodowski et al. (2006)
Germany	BPCAs	Haplic Chernozem	0–10, 90–120	2.90, 0.45	13.2, 31.5	Brodowski et al. (2007)
		Haplic Phaeozem	0–10, 100–128	1.71, 0.56	11.9, 50.0	
		Haplic Luvisol	0–10, 45–60	0.67, 0.39	2.7, 12.4	
Russia	BPCAs	Chernozem	0–5, 107–130	6.11, 0.01	8.2, 0.8	Hammes et al. (2008)

(Rumpel et al. 2006a). The BC content of soils quantified by UV oxidation followed by NMR spectroscopy may range from 0 to 82% of SOC in a continental-scale analysis of Australia (Lehmann et al. 2008).

Bornemann et al. (2008) reported that between 0.6 and 11.4% of SOC consisted of BC based on analyses of BPCAs combined with predictions by MIR-PLS in a range of arable lands from Argentina, China, Germany, Russia, and the USA. Based on analyses of BPCAs, Brodowski et al. (2006) reported that up to 7.2% of SOC located in the <53 μm fraction in arable lands in Germany can be BC. Physical inclusion, particularly within microaggregates is supposedly important to the long-term stabilization of BC. Between 11.9 and 13.2% of SOC in 0–10 cm depth consisted of BPCA-detectable BC at long-term arable sites in Germany which were influenced by fossil C inputs from brown coal dust, coal combustion, and soot (Brodowski et al. 2007). For an arable land use involving less fossil fuel combustion-derived C inputs, however, BC comprised of only 2.7% of SOC in the same soil depth. In subsoil, BC is increasingly connected to the coarse silt and sand minerals, and the heavy mineral fraction, and is thus preserved. While land-use change and mineral fertilization may have no effect on BC contents in the topsoil, farmyard manure application often increases BC contents as manure probably contains BC-like moieties (Brodowski et al. 2007). The BC analyzed by the BPCA method can be as much as 7% of SOC in the whole profile of Russian steppe soils with a maximum in 5–20 cm depth (Hammes et al. 2008). In summary, BC and charcoal may be found in agricultural soils globally, and, in particular, in soils with a significant C input from fire and biomass burning.

8.3 Research Needs

In agricultural soils and related land uses, soil properties, land-use types, agricultural practices, and biochar characteristics vary widely making generalizations about biochar effects on SOC sequestration highly uncertain. Much of the current understanding of the long-term dynamics of biochar is based on studies of charcoal from natural fire, new charcoal produced using traditional methods or analogous procedures undertaken in the laboratory (Sohi et al. 2009). However, natural charcoal and biochar may not be well suited as proxies for each other (Santín et al. 2017). A lack of long-term, well-designed field studies on the efficacy of biochar on different soil types and agro-climatic zones are limiting our current understanding of biochar's potential to enhance crop production and mitigate climate change (Agegnehu et al. 2017). For example, woody feedstock biochar, applied without organic or inorganic fertilizers were over-represented in European field experiments (Verheijen et al. 2017). Also, while fertile arable soils were over-represented, field experiments on shallow unfertile soils were underrepresented. Most biochar field studies in Europe focused on crop production and less on issues that can provide evidence of mechanisms and potential undesired side effects. Specifically, which contaminants or contaminant mixtures may potentially be remediated by soil application of different biochars needs additional research (Verheijen et al. 2017). Tammeorg et al. (2017) listed highest future research priorities regarding biochar effects in soils including (i) functional redundancy within soil microbial communities, (ii) bioavailability of biochar's contaminants to soil biota, (iii) SOM stability, (iv) GHG emissions, (v) soil formation, (vi) soil hydrology, (vii) nutrient cycling due to microbial priming and altered rhizosphere ecology, and (viii) soil pH buffering capacity. Life-cycle GHG assessments are particularly needed when using biochar as on-farm management tool for nutrient-rich biomass waste streams (Kammann et al. 2017).

Much less is known how to engineer the pyrolysis process conditions during energy production to produce the desired biochar properties for SOC sequestration in agricultural soils (Brewer et al. 2009). For example, higher pyrolysis temperatures result in greater surface area, elevated pH values and higher ash contents which may increase agricultural productivity and, thus, C inputs to the soil for SOC sequestration (Novak et al. 2009b). Thus, maximum temperature or 'Highest Treatment Temperature' (HTT) is the most frequently used parameter to describe PyOC formation conditions (Santín et al. 2017). Recently, charring intensity (CI) has been suggested as a more appropriate descriptor as it integrates both heating temperature and duration (Pyle et al. 2015). Future research should focus on (i) establishing methodology to precisely predict and design BC properties on the basis of pyrolysis and phase transformation of biomass; (ii) developing an assessment system to evaluate the long-term effect of BC on stabilization and bioavailability of contaminants, agrochemicals, and nutrient elements in soils; and (iii) elucidating the interaction of BC with plant roots, microorganisms, and soil components (Lian and Xing 2017). The harmonization of biochar legislation for production and soil application is needed. Currently, voluntary biochar quality standards are available in Europe with the European Biochar Certificate, in the UK with the Biochar Quality Mandate, and in the USA with the IBI Standard (Meyer et al. 2017). National legislation for fertilizers, soil improvers, and composts in EU countries does sometimes also consider biochar.

Probably most appropriate for SOC sequestration are biochars with a high aromatic C content. These can be produced by higher pyrolysis temperatures as volatile aliphatic compounds are then increasingly removed causing biochars to have higher percentages of C composed mostly of polycondensed aromatic moieties. Abiven et al. (2014) recommended to focus on a limited number of the most environmentally and economically promising biochar-based agricultural systems. Optimizing processing conditions to improve biochar properties for agricultural use remains a key issue (Tan et al. 2017).

As climate change is a global issue, outcomes of research about biochar effects on SOC sequestration must be applicable under diverse climates, agriculture and energy production systems (Sohi et al. 2009). The scientific knowledge about fundamental mechanisms by which biochar affects SOC sequestration in agricultural soils needs to be improved by studying (i) functional interactions of biochar with soil fauna and microbial communities, (ii) surface interactions, (iii) nutrient use efficiency, (iv) soil physical effects, (v) fate of biochar in the soil profile, watershed, and agricultural landscape, (vi) effects on CO_2, CH_4, and N_2O emissions, and (vii) plant physiological responses.

A classification system of biochars is necessary as the elemental composition, ash content and composition, density, water adsorbance, pore-size distribution, toxicity, ion adsorption and release, recalcitrance to biotic or abiotic decay, chemical or physical properties vary widely among different biochars (Joseph et al. 2009). Feedstock properties and production conditions may strongly affect biochar properties. Some of these properties may change soil C input by affecting plant growth and alter SOC sequestration directly by intrinsic stability in soil. These properties are (i) proportion of C forms stable against abiotic (chemical oxidation, photooxidation, solubilization) and biotic (microbial incorporation, oxidative respiration) degradation (Zimmerman 2010), (ii) presence and concentration of potentially harmful inorganic and organic compounds (Amonette and Joseph 2009; Sohi et al. 2009), (iii) porosity, pore-size distribution and total surface area (Downie et al. 2009), (iv) electrical conductivity and pH, (v) solubility of mineral elements, (vi) bulk density, (vii) particle size, (viii) compressive strength, (ix) type and amount of functional groups, (x) type and concentration of radicals, and (xi) oxidation and reduction potential (modified from Joseph et al. 2009).

Joseph et al. (2009) proposed four main biochar properties to be used for classification. These include (i) total C, H, and O contents, and the labile and stabile fraction of total C, (ii) contents of other elements and their solubility and availability to plants and mobility, (iii) surface area and pore-size distribution, and (iv) the ability to increase potential CEC and biochar pH. These properties should be measurable by wet and dry chemical methods and not require sophisticated equipment. Thus, biochar is classified as high in total C content when it contains >80% C, medium with 60–80% C, and low with 15–60% C. Joseph et al. (2009) defined the labile C content of biochar as the C content of the fraction which is mineralized abiotically and biotically to CO_2 within a short period of time. Classification is based on results of (i) heating biochar in an oxygen-free atmosphere at 350 °C and determination of the fractional C loss, (ii) heating in an inert atmosphere at 950 °C and determination the fractional C loss, and (iii) determination of the biochar C fraction dissolved in water to capture soluble C. The ratio of the volume of macropores to micropores was proposed to classify biochar with respect to major benefits for soil processes (Joseph et al. 2009). The development of biochar induced CEC in the soil may be suitable to classify its effects on plant growth after biochar application. However, a framework for classification of biochar can only be possible after more studies are conducted on characterization of the range of biochars and their effects on SOC sequestration in diverse soils under different climates and land uses. The most important biochar properties that determine SOC sequestration such as those affecting stability, plant growth, and soil health need to be precisely assessed. Thus, rigorous incubation and field experiments including detailed micro- and macroanalyses are required (Joseph et al. 2009).

Biochar must be studied in soil profiles because SOC sequestration occurs in deeper soil horizons (Lorenz and Lal 2014). Also, biochar losses through physical export in dissolved, gaseous, and particulate forms need to be quantified for a range of sites to fully address the contribution of biochar to SOC sequestration in an agricultural landscape. Thus, studies of agricultural watersheds are needed to assess the environmental fate of biochar in agricultural ecosystems as this soil component may persist for long periods of time (Sohi et al. 2009). Modeling the coupled C and N (and P?) cycles in soil with and without biochar is essential to understanding the fundamental mechanisms through which biochar affects SOC sequestration and the impact on soil-based GHG emissions. Most importantly, improved methods of quantification of biochar in soil are needed, along with the standardization of the pyrolysis process (Sohi et al. 2009). Finally, research gaps including the influence of biochar application on the short- and long-term fate of PAHs and VOCs, and the proper control tactics for biochar quality and associated risk must be addressed (Dutta et al. 2017). Long-term ecotoxicological studies to evaluate the environmental risk of biochar application are needed (Wang et al. 2017).

Physicochemical properties of highly weathered soils respond positively to biochar (Mehmood et al. 2017). Those soils are typical of humid tropics where agricultural productivity is low but may be raised by biochar application to improve food security and reduce poverty. Less developed countries are eager to improve soil fertility and agricultural productivity. However, the majority of biochar research is associated with highly developed countries. Thus, transfer and/or translation of biochar knowledge acquired in highly developed countries for application in less developed countries is needed (Mehmood et al. 2017).

8.4 Conclusions

Biochar has received increased attention recently as an amendment to agricultural soil for enhancing SOC sequestration. It is a by-product of heating biomass feedstock under oxygen-limited conditions during pyrolysis for energy production. Directly, it may provide a highly stable C form mainly due to contents of polycondensed aromatic structures with soil MRT in the millennial range. Indirectly, biochar may increase plant growth and, thus, above- and belowground litter input to the SOC stock. However, the biological, chemical, and physical properties among biochars vary widely depending on feedstock type and pyrolysis process conditions. Thus, interactions between soils and biochar are diverse and are challenging to predict. Effects of biochar on SOC sequestration in agricultural systems, in particular, differ among climate, soil types, and land uses. Thus, incubation and field experiments, and modeling studies are required to fully address the SOC sequestration potential by biochar application to agricultural soils.

8.5 Review Questions

1. What is biochar and how does it differ from activated carbon, BC, char, charcoal, elemental carbon, pyrogenic carbon, and soot?
2. What are major biological, chemical, and physical properties of biochar and how do they alter soil biological, chemical, and physical properties?
3. Contrast and compare feedstock properties and pyrolysis conditions and their effects on biochar properties.
4. Contrast and compare the use of char and charcoal by ancient cultures to increase soil fertility in depleted tropical soils with the effects of biochar application on SOC sequestration.
5. Describe the mechanisms how biochar application to soil potentially affects SOC sequestration.
6. Contrast and compare major quantification methods of soil biochar.
7. List major knowledge gaps before a widespread application of biochar to agricultural soils can be recommended.
8. Describe research projects needed for the global establishment of agricultural biochar systems.
9. List dominant controls of crop yield responses to biochar.
10. What soil types and climate regimes would be conducive to use of biochar as amendment?

References

Abiven S, Schmidt MWI, Lehmann J (2014) Biochar by design. Nat Geosci 7:326–327

Agblevor FA, Beiss S, Kim SS, Tarrant R, Mante NO (2010) Biocrude oils from the fast pyrolysis of poultry litter and hardwood. Waste Manage 30:298–307

Agegnehu G, Srivastava AK, Bird MI (2017) The role of biochar and biochar-compost in improving soil quality and crop performance: a review. Appl Soil Ecol 119:156–170

Ahmad M, Rajapaksha AU, Lim JE et al (2014) Biochar as a sorbent for contaminant management in soil and water: a review. Chemosphere 99:19–33

Almendros G, Knicker H, González-Vila FJ (2003) Rearrangement of carbon and nitrogen forms in peat after progressive thermal oxidation as determined by solid-state ^{13}C- and ^{15}N-NMR spectroscopy. Org Geochem 34:1559–1568

Amonette JE, Joseph S (2009) Characteristics of biochar: microchemical properties. In: Lehmann J, Joseph S (eds) Biochar for environmental management. Earthscan, London, U.K., pp 33–52

Antal MJ, Grønli M (2003) The art, science, and technology of charcoal production. Ind Eng Chem Res 42:1619–1640

Ascough PL, Bird MI, Brock F, Higham TFG, Meredith W, Snape CE, Vane CH (2009) Hydropyrolysis as a new tool for radiocarbon pre-treatment and the quantification of black carbon. Quat Geochronol 4:140–147

Awad YM, Wang J, Igalavithana AD et al (2018) Biochar effects on rice paddy: meta-analysis. Adv Agron. https://doi.org/10.1016/bs.agron.2017.11.005

Baldock JA, Smernik RJ (2002) Chemical composition and bioavailability of thermally altered *Pinus resinosa* (Red pine) wood. Org Geochem 33:1093–1109

Baveye PC, Berthelin J, Tessier D, Lemaire G (2018) The "4 per 1000" initiative: a credibility issue for the soil science community? Geoderma 309:118–123

Bird MI, Moyo C, Veenendaal EM, Lloyd J, Frost P (1999) Stability of elemental carbon in a savanna soil. Global Biogeochem Cycles 13:923–932

Blackwell P, Riethmuller G, Collins M (2009) Biochar application to soil. In: Lehmann J, Joseph S (eds) Biochar for environmental management. Earthscan, London, U.K., pp 207–226

Blanco-Canqui H (2017) Biochar and soil physical properties. Soil Sci Soc Am J 81:687–711. https://doi.org/10.2136/sssaj2017.01.0017

Bornemann L, Welp G, Brodowski S, Rodionov A, Amelung W (2008) Rapid assessment of black carbon in soil organic matter using mid-infrared spectroscopy. Org Geochem 39:1537–1544

Bossuyt H, Six J, Hendrix PF (2005) Protection of soil carbon by microaggregates within earthworm casts. Soil Biol Biochem 37:251–258

Brewer CE, Schmidt-Rohr K, Satrio JA, Brown RC (2009) Characterization of biochar from fast pyrolysis and gasification systems. Environ Prog Sus Energ 28:386–396

Bridle TR, Pritchard D (2004) Energy and nutrient recovery from sewage sludge via pyrolysis. Water Sci Technol 50:169–175

Brodowski S, Amelung W, Haumaier L, Zech W (2007) Black carbon contribution to stable humus in German arable soils. Geoderma 139:220–228

Brodowski S, John B, Flessa H, Amelung W (2006) Aggregate-occluded black carbon in soil. Eur J Soil Sci 57:539–546

Brodowski S, Rodionov A, Haumaier L, Glaser G, Amelung W (2005) Revised black carbon assessment using benzene polycarboxylic acids. Org Geochem 36:1299–1310

Brown RA, Kercher AK, Nguyen TH, Nagle DC, Ball WP (2006) Production and characterization of synthetic wood chars for use as surrogates for natural sorbents. Org Geochem 37:321–333

Bruun S, Jensen ES, Jensen LS (2008) Microbial mineralization and assimilation of black carbon: dependency on degree of thermal alteration. Org Geochem 39:839–845

Busscher WJ, Novak JM, Evans DE, Watts DW, Niandou MAS, Ahmedna M (2010) Influence of pecan biochar on physical properties of a Norfolk loamy sand. Soil Sci 175:10–14

Cayuela ML, Jeffery S, van Zwieten L (2015) The molar H: Corg ratio of biochar is a key factor in mitigating N2O emissions from soil. Agric Ecosyst Environ 202:135–138

Cayuela ML, van Zwieten L, Singh BP, Jeffery S, Roig A, Sánchez-Monedero MA (2014) Biochar's role in mitigating soil nitrous oxide emissions: a review and meta-analysis. Agric Ecosyst Environ 191:5–16

Cetin E, Moghtaderi B, Gupta R, Wall TF (2004) Influence of pyrolysis conditions on the structure and gasification reactivity of biomass chars. Fuel 83:2139–2150

Chan KY, Van Zwieten L, Meszaros I, Downie A, Joseph S (2007) Agronomic values of greenwaste biochar as a soil amendment. Austr J Soil Res 45:629–634

Chan KY, Xu Z (2009) Biochar: nutrient properties and their enhancement. In: Lehmann J, Joseph S (eds) Biochar for environmental management. Earthscan, London, U.K., pp 67–84

Chen B, Zhou D, Zhu L (2008) Transitional adsorption and partition of nonpolar and polar aromatic contaminants by biochars of pine needles with different pyrolytic temperatures. Environ Sci Technol 42:5137–5143

Cheng C-H, Lehmann J (2009) Ageing of black carbon along a temperature gradient. Chemosphere 75:1021–1027

Cheng C-H, Lehmann J, Engelhard MH (2008a) Natural oxidation of black carbon in soils: changes in molecular form and surface charge along a climosequence. Geochim Cosmochim Ac 72:1598–1610

Cheng C-H, Lehmann J, Thies JE, Burton SD (2008b) Stability of black carbon in soils across a climatic gradient. J Geophys Res 113:G02027. https://doi.org/10.1029/2007JG000642

Cheng C-H, Lehmann J, Thies JE, Burton SD, Engelhard MH (2006) Oxidation of black carbon by biotic and abiotic processes. Org Geochem 37:1477–1488

Clough TJ, Bertram JE, Ray JL, Condron LM, O'Callaghan M, Sherlock RR, Wells NS (2010) Unweathered wood biochar impact on nitrous oxide emissions from a bovine-urine-amended pasture soil. Soil Soc Am J 74:852–860

Crane-Droesch A, Abiven S, Jeffery S, Torn MS (2013) Heterogeneous global crop yield response to biochar: a meta-regression analysis. Environ Res Lett 8:044049. https://doi.org/10.1088/1748-9326/8/4/044049

Cui X, Wang H, Lou L, Chen X, Yu Y, Shi J, Xu L, Khan MI (2009) Sorption and genotoxicity of sediment-associated pentachlorophenol and pyrene influenced by crop residue ash. J Soils Sediments 9:604–612

Currie LA, Benner BA, Kessler JD, Klinedinst DB, Klouda GA, Marolf JV, Slater JF, Wise SA, Cachier H, Cary R, Chow JC, Watson J, Druffel ERM, Masiello CA, Eglinton TI, Pearson A, Reddy CM, Gustafsson O, Hartmann PC, Quinn JG, Hedges JI, Prentice KM, Kirchstetter TW, Novakov T, Puxbaum H, Schmid H (2002) A critical evaluation of interlaboratory data on total, elemental, and isotopic carbon in the carbonaceous particle reference material, NIST SRM 1649a. J Res Natl Inst Stand Technol 107:279–298

Czimczik CI, Masiello CA (2007) Controls on black carbon storage in soils. Global Biogeochem Cycles 21, GB3005. https://doi.org/10.1029/2006gb002798

De la Rosa JM, Knicker H, López-Capél E, Manning DAC, González-Pérez JA, González-Vila FJ (2008) Direct detection of black carbon in soils by Py-GC/MS, carbon-13 NMR spectroscopy and thermogravimetric techniques. Soil Sci Soc Am J 72:258–267

De la Rosa JM, Rosado M, Paneque M, Miller AZ, Knicker H (2018) Effects of aging under field conditions on biochar structure and composition: implications for biochar stability in soils. Sci Tot Environ 613–614:969–976

Downie A, Crosky A, Munroe P (2009) Physical properties of biochar. In: Lehmann J, Joseph S (eds) Biochar for environmental management. Earthscan, London, U.K., pp 13–32

Dutta T, Kwon E, Bhattacharya SS et al (2017) Polycyclic aromatic hydrocarbons and volatile organic compounds in biochar and biochar-amended soil: a review. GCB Bioenerg 9:990–1004. https://doi.org/10.1111/gcbb.12363

Fang X, Chua T, Schmidt-Rohr K, Thompson ML (2010) Quantitative ^{13}C NMR of whole and fractionated Iowa Mollisols for assessment of organic matter composition. Geochim Cosmochim Ac 74:584–598

Fidel RB, Laird DA, Thompson ML, Lawrinenko M (2017) Characterization and quantification of biochar alkalinity. Chemosphere 167:367–373

Garcia-Perez M (2008) The formation of polyaromatic hydrocarbons and dioxins during pyrolysis. Washington State University Biological Systems Engineering Department, Washington State University Extension Energy Program

Glaser B (2007) Prehistorically modified soils of central Amazonia: a model for sustainable agriculture in the twenty-first century. Phil Trans R Soc B 362:187–196

Glaser B, Knorr K-H (2008) Isotopic evidence for condensed aromatics from non-pyrogenic sources in soils—implications for current methods for quantifying soil black carbon. Rapid Commun Mass Sp 22:935–942

Glaser B, Lehmann J, Zech W (2002) Ameliorating physical and chemical properties of highly weathered soils in the tropics with charcoal. Biol Fertil Soils 35:219–230

Griscom BW, Adams J, Ellis PW, et al (2017) Natural climate solutions. Proc Natl Acad Sci USA Early Edition

Guggenberger G, Rodionov A, Shibistova O, Grabe M, Kasansky OA, Fuchs H, Mikheyeva N, Zrazhevskaya G, Flessa H (2008) Storage and mobility of black carbon in permafrost soils of the forest tundra ecotone in Northern Siberia. Glob Change Biol 14:1367–1381

Guo M, He Z, Uchimiya SM (2016) Introduction to biochar as an agricultural and environmental amendment. In: Guo M, He Z, Uchimiya SM (eds) Agricultural and environmental applications of biochar: advances and barriers. SSSA Special Publication 63, SSSA, Madison, WI, USA, pp 1–14. https://doi.org/10.2136/sssaspecpub63.2014.0034

Hagemann N, Joseph S, Schmidt HP et al (2017) Organic coating on biochar explains its nutrient retention and stimulation of soil fertility. Nature Commun 8:1089. https://doi.org/10.1038/s41467-017-01123-0

Hamer U, Marschner B, Brodowski S, Amelung W (2004) Interactive priming of black carbon and glucose mineralization. Org Geochem 35:823–830

Hammes K, Abiven S (2013) Identification of black carbon in the earth system. In: Belcher CM (ed) Fire phenomena and the earth system: an interdisciplinary guide to fire science. Wiley, Hoboken, NJ, USA, pp 157–176

Hammes K, Schmidt MWI, Smernik RJ, Currie LA, Ball WP, Nguyen TH, Louchouran P, Houel S, Gustafsson Ö, Elmquist M, Cornelissen G, Skjemstad JO, Masiello CA, Song J, Peng P'A, Mitra S, Dunn JC, Hatcher PG, Hockaday WC, Smith DM, Hartkopf Fröder C, Böhmer A, Lüer B, Huebert BJ, Amelung W, Brodowski S, Huang L, Zhang W, Gschwend PM, Xana Flores-Cervantes D, Largeau C, Rouzaud J-N, Rumpel C, Guggenberger G, Kaiser K, Rodionov A, Gonzalez-Vila FJ, Gonzalez-Perez JA, de la Rosa JM, Manning DAC, López-Capél E, Ding L (2007) Comparison of quantification methods to measure fire-derived (black/elemental) carbon in soils and sediments using reference materials from soil, water, sediment and the atmosphere Global Biogeochem Cycles 21, GB3016. https://doi.org/10.1029/2006gb002914

Hammes K, Torn MS, Lapenas AG, Schmidt MWI (2008) Centennial black carbon turnover in a Russian steppe soil. Biogeosciences 5:1339–1350

Hanke UM, Reddy CM, Braun ALL et al (2017) What on earth have we been burning? Deciphering sedimentary records of pyrogenic carbon. Environ Sci Technol 51:12972–12980. https://doi.org/10.1021/acs.est.7b03243

Haumaier L (2010) Benzene polycarboxylic acids—a ubiquitous class of compounds in soils. J Plant Nutr Soil Sci in press

He Y, Zhou X, Jiang L et al (2017) Effects of biochar application on soil greenhouse gas fluxes: a meta-analysis. Glob Change Biol 9:743–755

Hedges JI, Eglinton G, Hatcher PG, Kirchman DL, Arnosti C, Derenne S, Evershed RP, Kögel-Knabner I, de Leeuw JW, Littke R, Michaelis W, Rullkötter J (2000) The molecularly-uncharacterized component of nonliving organic matter in natural environments. Org Geochem 31:945–958

Hernandez-Soriano MC, Kerré B, Goos P, Hardy B, Duffey J, Smolders E (2016) Long-term effect of biochar on the stabilization of recent carbon: soils with historical inputs of charcoal. GCB Bioenerg 8:371–381

Hilber I, Mayer P, Gouliarmou V et al (2017) Bioavailability and bioaccessibility of polycyclic aromatic hydrocarbons from (post-pyrolytically treated) biochars. Chemosphere 174:700–707

Hileman B (2007) Arsenic in chicken production. Chem Eng News 85:34–35

Hilscher A, Heister K, Siewert C, Knicker H (2009) Mineralisation and structural changes during the initial phase of microbial degradation of pyrogenic plant residues in soil. Org Geochem 40:332–342

Hockaday WC, Grannas AM, Kim S, Hatcher PG (2006) Direct molecular evidence for the degradation and mobility of black carbon in soils from ultrahigh-resolution mass spectral analysis of dissolved organic matter from a fire-impacted forest soil. Org Geochem 37:501–510

Hockaday WC, Grannas AM, Kim S, Hatcher PG (2007) The transformation and mobility of charcoal in a fire-impacted watershed. Geochim Cosmochim Ac 71:3432–3445

Hofrichter M, Fakoussa RM (2001) Microbial degradation and modification of coal. In: Hofrichter M, Steinbüchel A (eds) Biopolymers-volume 1: lignin, humic substances and coal. Wiley-VCH, Weinheim, Germany, pp 393–429

Janik LJ, Skjemstad JO, Sheperd KD, Spouncer LR (2007) The prediction of soil carbon fractions using mid-infrared-partial least square analysis. Aust J Soil Res 45:73–81

Jeffery S, Abalos D, Prodana M et al (2017) Biochar boosts tropical but not temperate crop yields. Environ Res Lett 12:053001

Jegajeevagan K, Mabilde L, Gebremikael MT, Ameloot N, De Neve S, Leinweber P, Sleutel S (2016) Artisanal and controlled pyrolysis-based biochars differ in biochemical composition, thermal recalcitrance, and biodegradability in soil. Biomass Bioenerg 84:1–11

Jin H, Lehmann J, Thies JE (2008) Soil microbial community response to amending maize soils with maize stover charcoal. In: Proceedings of the 2008 conference of international biochar initiative, 8–10 September 2008, Newcastle, UK

Johnson JMF, Allmaras RR, Reicosky DC (2006) Estimating source carbon from crop residues, roots and rhizodeposits using the national grain-yield database. Agron J 98:622–636

Joseph S, Peacocke C, Lehmann J, Munroe P (2009) Developing a biochar classification and test methods. In: Lehmann J, Joseph S (eds) Biochar for environmental management. Earthscan, London, U.K., pp 107–126

Kammann C, Ippolito J, Hagemann N et al (2017) Biochar as tool to reduce the agricultural greenhouse-gas burden—knowns, unknowns and future research needs. J Environ Eng Landsc 25:114–139

Kane ES, Hockaday WC, Turetsky MR, Masiello CA, Valentine DW, Finney BP, Baldock JA (2010) Topographic controls on black carbon accumulation in Alaskan black spruce forest soils: implications for organic matter dynamics. Biogeochemistry 100:39–56

Keiluweit M, Nico PS, Johnson MG, Kleber M (2010) Dynamic molecular structure of plant biomass-derived black barbon (biochar). Environ Sci Technol 44:1247–1253

Kleber M, Rößner J, Chenu C, Glaser B, Knicker H, Jahn R (2003) Prehistoric alteration of soil properties in a Central German Chernozemic soil: in search of pedologic indicators for prehistoric activity. Soil Sci 168:292–306

Knicker H, Hilscher A, González-Vila FJ, Almendros G (2008) A new conceptual model for the structural properties of char produced during vegetation fires. Org Geochem 39:935–939

Koelmans AA, Jonker MTO, Cornelissen G, Bucheli TD, Van Noort PCM, Gustafsson Ö (2006) Black carbon: the reverse of its dark side. Chemosphere 63:265–277

Koutcheiko S, Monreal CM, Kodama H, McCracken T, Kotlyar L (2007) Preparation and characterization of activated carbon derived from the thermo-chemical conversion of chicken manure. Bioresource Technol 98:2459–2464

Krull ES, Baldock JA, Skjemstad JO, Smernik RJ (2009) Characteristics of biochar: organo-chemical properties. In: Lehmann J, Joseph S (eds) Biochar for environmental management. Earthscan, London, U.K., pp 53–65

Krull E, Lehmann J, Skjemstad J, Baldock J, Spouncer L (2008) The global extent of black C in soils: is it everywhere? In: Schröder HG (ed) Grasslands: ecology, management and restoration. Nova Science Publishers, New York, pp 13–17

Krull ES, Swanston CW, Skjemstad JO, McGowan JA (2006) Importance of charcoal in determining the age and chemistry of organic carbon in surface soils. J Geophys Res 111: G04001. https://doi.org/10.1029/2006JG000194

Kuhlbusch TAJ (1998) Black carbon and the carbon cycle. Science 280:1903–1904

Kuhlbusch TA, Crutzen PJ (1995) Toward a global estimate of black carbon in residues of vegetation fires representing a sink of atmospheric CO_2 and a source of O_2. Global Biogeochem Cycles 9:491–501

Kuzyakov Y, Bogomolova I, Glaser B (2014) Biochar stability in soil: Decomposition during eight years and transformation as assessed by compound-specific ^{14}C analysis. Soil Biol Biochem 70:229–236

Kuzyakov Y, Subbotina I, Chen H, Bogomolova I, Xu X (2009) Black carbon decomposition and incorporation into soil microbial biomass estimated by ^{14}C labeling. Soil Biol Biochem 41:210–219

Laird DA (2008) The charcoal vision: a win-win-win scenario for simultaneously producing bioenergy, permanently sequestering carbon, while improving soil and water quality. Agron J 100:178–181

Laird DA, Chappell MA, Martens DA, Wershaw RL, Thompson M (2008) Distinguishing black carbon from biogenic humic substances in soil clay fractions. Geoderma 143:115–122

Landry JS, Matthews HD (2017) The global pyrogenic carbon cycle and its impact on the level of atmospheric CO_2 over past and future centuries. Glob Change Biol 23:3205–3218. https://doi.org/10.1111/gcb.13603
Lehmann J (2007) A handful of carbon. Nature 447:143–144
Lehmann J, Czimczik CI, Laird DA, Sohi S (2009) Stability of biochar in the soil. In: Lehmann J, Joseph S (eds) Biochar for environmental management. Earthscan, London, U.K., pp 183–205
Lehmann J, da Silva JP, Steiner C, Nehls T, Zech W, Glaser B (2003) Nutrient availability and leaching in an archaeological Anthrosol and a Ferralsol of the Central Amazon basin: fertilizer, manure and charcoal amendments. Plant Soil 249:343–357
Lehmann J, Joseph S (2009) Biochar for environmental management: an introduction. In: Lehmann J, Joseph S (eds) Biochar for environmental management. Earthscan, London, U.K., pp 1–12
Lehmann J, Liang B, Solomon D, Lerotic M, Luizão F, Kinyangi J, Schäfer T, Wirick S, Jacobsen C (2005) Near-edge X-ray absorption fine structure (NEXAFS) spectroscopy for mapping nano-scale distribution of organic carbon forms in soil: application to black carbon particles. Global Biogeochem Cyc 19, GB1013. https://doi.org/10.1029/2004gb002435
Lehmann J, Skjemstad J, Sohi S, Carter J, Barson M, Falloon P, Coleman K, Woodbury P, Krull E (2008) Australian climate-carbon cycle feedback reduced by soil black carbon. Nature Geosci 1:832–835
Lehmann J, Sohi S (2008) Comment on "Fire-derived charcoal causes loss of forest humus". Science 321:1295c
Leifeld J, Fenner S, Müller M (2007) Mobility of black carbon in drained peatland soils. Biogeosciences 4:425–432
Li Y, Hu S, Chen J et al (2017) Effects of biochar application in forest ecosystems on soil properties and greenhouse gas emissions: a review. J Soils Sediments. https://doi.org/10.1007/s11368-017-1906-y
Lian F, Xing B (2017) Black carbon (biochar) in water/soil environments: Molecular structure, sorption, stability, and potential risk. Environ Sci Technol 51:13517–13532. https://doi.org/10.1021/acs.est.7b02528
Liang B, Lehmann J, Sohi SP, Thies JE, O'Neill B, Trujillo L, Gaunt J, Solomon D, Grossman J, Neves EG, Luizão FJ (2010) Black carbon affects the cycling of non-black carbon in soil. Org Geochem 41:206–213
Liang B, Lehmann J, Solomon D, Kinyangi J, Grossman J, O'Neill B, Skjemstad JO, Thies J, Luizao FJ, Petersen J, Neves EG (2006) Black carbon increases cation exchange capacity in soils. Soil Sci Soc Am J 70:1719–1730
Liu S, Zhang Y, Zong Y, Hu Z, Wu S, Zhou J, Jin Y, Zou (2016) Response of soil carbon dioxide fluxes, soil organic carbon and microbial biomass carbon to biochar amendment: a meta-analysis. GCB Bioenerg 8:392–406
Llorente M, Turrión MB, Glaser B (2018) Rapid and economical quantification of black carbon in soils using a modified benzene polycarboxylic acids (BPCA) method. Org Geochem 115:197–204
Loganathan VA, Feng Y, Sheng GD, Clement TP (2009) Crop-residue-derived char influences sorption, desorption and bioavailability of atrazine in soils. Soil Sci Soc Am J 73:967–974
Lorenz K, Lal R (2014) Biochar application to soil for climate change mitigation by soil organic carbon sequestration. J Plant Nutr Soil Sci 177:651–670
Luo Y, Zang H, Yu Z et al (2017) Priming effects in biochar enriched soils using a three-source partitioning approach: ^{14}C labelling and ^{13}C natural abundance. Soil Biol Biochem 106:28–35
Lutfalla S, Abiven S, Barré P et al (2017) Pyrogenic carbon lacks long-term persistence in temperate arable soils. Front Earth Sci 5:96. https://doi.org/10.3389/feart.2017.00096
Macías F, Arbestain MC (2010) Soil carbon sequestration in a changing global environment. Mitig Adapt Strateg Glob Change 15:511–529
Maestrini B, Nannipieri P, Abiven S (2015) A meta-analysis on pyrogenic organic matter induced priming effect. GCB Bioenergy 7:577–590. https://doi.org/10.1111/gcbb.12194

Magrini-Bair KA, Czernik S, Pilath HM, Evans RJ, Maness PC, Leventhal J (2009) Biomass derived, carbon sequestering, designed fertilizers. Ann Environ Sci 3:217–225
Major J, Lehmann J, Rondon M, Goodale C (2010a) Fate of soil-applied back carbon: downward migration, leaching and soil respiration. Glob Change Biol 16:1366–1379
Major J, Rondon M, Molina D, Riha SJ, Lehmann J (2010b) Maize yield and nutrition during 4 years after biochar application to a Colombian savanna oxisol. Plant Soil in press
Marschner B, Brodowski S, Dreves A, Gleixner G, Gude A, Grootes PM, Hamer U, Heim A, Jandl G, Ji R, Kaiser K, Kalbitz K, Kramer C, Leinweber P, Rethemeyer J, Schäffer A, Schmidt MWI, Schwark L, Wiesenberg GLB (2008) How relevant is recalcitrance for the stabilization of organic matter in soils? J Plant Nutr Soil Sci 171:91–110
Masiello CA (2004) New directions in black carbon organic geochemistry. Mar Chem 92:201–213
Mehmood K, Garcia EC, Schirrmann M et al (2017) Biochar research activities and their relation to development and environmental quality. A meta-analysis. Agron Sustain Dev 37:22. https://doi.org/10.1007/s13593-017-0430-1
Meyer S, Genesio L, Vogel I et al (2017) Biochar standardization and legislation harmonization. J Environ Eng Landsc 25:175–191
Mia S, Dijkstra FA, Singh B (2017) Long-term aging of biochar: a molecular understanding with agricultural and environmental implications. Adv Agron 141:1–50
Mitra S, Bianchi TS, McKee BA, Sutula M (2002) Black carbon from the Mississippi River: quantities, sources, and potential implications for the global carbon cycle. Environ Sci Technol 36:2296–2302
Mukherjee A, Lal R (2014) The biochar dilemma. Soil Res 52:217–230
Murage EW, Voroney P, Beyaert RP (2007) Turnover of carbon in the free light fraction with and without charcoal as determined using the ^{13}C natural abundance method. Geoderma 138:133–143
Muralidhara HS (1982) Conversion of tannery waste to useful products. Resour Conserv 8:43–59
Naisse C, Girardin C, Davasse B, Chabbi A, Rumpel C (2015) Effect of biochar addition on C mineralisation and soil organic matter priming in two subsoil horizons. J Soils Sediments 15:825–832. https://doi.org/10.1007/s11368-014-1002-5
Nelson PN, Baldock JA (2005) Estimating the molecular composition of a diverse range of natural organic materials from solid-state ^{13}C NMR and elemental analysis. Biogeochemistry 72:1–34
Neves EG, Petersen JB, Bartone RN, Silva CAD (2003) Historical and socio-cultural origins of Amazonian Dark Earths. In: Lehmann J, Kern DC, Glaser B, Woods WI (eds) Amazonian Dark Earths: origin, properties, management. Kluwer Academic Press, Dordrecht, The Netherlands, pp 29–50
Nguyen BT, Lehmann J (2009) Black carbon decomposition under varying water regimes. Org Geochem 40:846–853
Nguyen BT, Lehmann J, Kinyangi J, Smernik R, Riha SJ, Engelhard MH (2008) Long-term black carbon dynamics in cultivated soil. Biogeochemistry 89:295–308
Novak JM, Busscher WJ, Laird DA, Ahmedna M, Watts DW, Niandou MAS (2009a) Impact of biochar amendment on fertility of a Southeastern coastal plain soil. Soil Sci 174:105–112
Novak JM, Lima I, Xing B, Gaskin JW, Steiner C, Das KC, Ahmedna M, Rehrah D, Watts DW, Busscher WJ, Schomberg H (2009b) Characterization of designer biochar produced at different temperatures and their effects on a loamy sand. Ann Environ Sci 3:195–206
Oguntunde PG, Abiodun BJ, Ajayi AE, van de Giesen N (2008) Effects of charcoal production on soil physical properties in Ghana. J Plant Nutr Soil Sci 171:591–596
Oguntunde PG, Fosu M, Ajayi AE, van de Giesen N (2004) Effects of charcoal production on maize yield, chemical properties and texture of soil. Biol Fertil Soils 39:295–299
Ohlson M, Dahlberg B, Økland T, Brown KJ, Halvorsen R (2009) The charcoal carbon pool in boreal forest soils. Nature Geosci 2:692–695
Olson KR, Al-Kaisi MM, Lal R, Lowery B (2014) Experimental consideration, treatments, and methods in determining soil organic carbon sequestration rates. Soil Sci Soc Am J 78:348–360. https://doi.org/10.2136/sssaj2013.09.0412

Paetsch L, Mueller CW, Rumpel C et al (2017) A multi-technique approach to assess the fate of biochar in soil and to quantify its effect on soil organic matter composition. Org Geochem 112:177–186

Pignatello JJ, Mitch WA, Xu W (2017) Activity and reactivity of pyrogenic carbonaceous matter toward organic compounds. Environ Sci Technol 51:8893–8908. https://doi.org/10.1021/acs.est.7b01088

Ponge JF, Topoliantz S, Ballof S, Rossi JP, Lavelle P, Betsch JM, Gaucher P (2006) Ingestion of charcoal by the Amazonian earthworm *Pontoscolex corethrurus*: a potential for tropical soil fertility. Soil Biol Biochem 38:2008–2009

Ponomarenko EV, Anderson DW (2001) Importance of charred organic matter in Black Chernozem soils of Saskatchewan. Can J Soil Sci 81:285–297

Preston CM, Schmidt MWI (2006) Black(pyrogenic) carbon: a synthesis of current knowledge and uncertainties with special consideration of boreal regions. Biogeosciences 3:397–420

Purevsuren B, Avida B, Gerelmaa T, Davaajava Y, Morgan TJ, Herod AA, Kandiyoti R (2004) The characterization of tar from the pyrolysis of animal bones. Fuel 83:799–805

Pyle LA, Hockaday WC, Boutton T et al (2015) Chemical and isotopic thresholds in charring: implications for the interpretation of charcoal mass and isotopic data. Environ Sci Technol 49:14057–14064. https://doi.org/10.1021/acs.est.5b03087

Qi F, Kuppusamy S, Naidu R et al (2017) Pyrogenic carbon and its role in contaminant immobilization in soils. Crit Rev Environ Sci Technol. https://doi.org/10.1080/10643389.2017.1328918

Qiu G, Guo M (2010) Quality of poultry litter-derived granular activated carbon. Bioresour Technol 101:379–386

Raveendran K, Ganesh A, Khilart KC (1995) Influence of mineral matter on biomass pyrolysis characteristics. Fuel 74:1812–1822

Reisser M, Purves RS, Schmidt MWI, Abiven S (2016) Pyrogenic carbon in soils: a literature-based inventory and a global estimation of its content in soil organic carbon and stocks. Front Earth Sci 4:80. https://doi.org/10.3389/feart.2016.00080

Renner R (2007) Rethinking biochar. Environ Sci Technol 41:5932–5933

Rodionov A, Amelung W, Haumaier L, Urusevskaja I, Zech W (2006) Black carbon in the zonal steppe soils of Russia. J Pant Nutr Soil Sci 169:363–369

Rodionov A, Amelung W, Peinemann N, Haumaier L, Zhang X, Kleber M, Glaser B, Urusevskaya I, Zech W (2010) Black carbon in grassland ecosystems of the world. Global Biogeochem Cyc 24, GB3013. https://doi.org/10.1029/2009gb003669

Rondon MA, Molina D, Hurtado M, Ramirez J, Amezquita E, Major J, Lehmann J (2005) Enhancing the productivity of crop and grasses while reducing greenhouse gas emissions through bio-char amendments to unfertile tropical soils. CIAT annual report Cali, Colombia

Rumpel C, Alexis M, Chabbi A, Chaplot V, Rasse DP, Valentin C, Mariotti A (2006a) Black carbon contribution to soil organic matter composition in tropical sloping land under slash and burn agriculture. Geoderma 130:35–46

Rumpel C, Chaplot V, Planchon O, Bernadou J, Valentin C, Mariotti A (2006b) Preferential erosion of black carbon on steep slopes with slash and burn agriculture. CATENA 65:30–40

Rumpel C, Ba A, Darboux F, Chaplot V, Planchon O (2009) Erosion budget and process selectivity of black carbon at meter scale. Geoderma 154:131–137

Rumpel C, Chaplot V, Chabbi A, Largeau C, Valentin C (2008) Stabilisation of HF soluble and HCl resistant organic matter in sloping tropical soils under slash and burn agriculture. Geoderma 145:347–354

Rumpel C, Kögel-Knabner I (2004) Microbial use of lignite compared to recent plant litter as substrates in reclaimed coal mine soils. Soil Biol Biochem 36:67–75

Santín C, Doerr SH, Merino A et al (2017) Carbon sequestration potential and physicochemical properties differ between wildfire charcoals and slow-pyrolysis biochars. Sci Rep 7:11233. https://doi.org/10.1038/s41598-017-10455-2

Schmid E-M, Skjemstad JO, Glaser B, Knicker H, Kögel-Knabner I (2002) Detection of charred organic matter in soils from a Neolithic settlement in Southern Bavaria, Germany. Geoderma 107:71–91

Schmidt MWI, Noack AG (2000) Black carbon in soils and sediments: analysis, distribution, implications, and current challenges. Glob Biogeochem Cyc 14:777–793

Scott HL, Ponsonby D, Atkinson CJ (2014) Biochar: an improver of nutrient and soil water availability—what is the evidence? CAB Reviews 9:019. https://doi.org/10.1079/PAVSNNR20149019

Shindo H (1991) Elementary composition, humus composition, and decomposition in soil of charred grassland plants. Soil Sci Plant Nutr 37:651–657

Shneour EA (1966) Oxidation of graphitic carbon in certain soils. Science 151:991–992

Sigmund G, Huber D, Bucheli TD et al (2017) Cytotoxicity of biochar: a workplace safety concern? Environ Sci Technol Lett 4:362–366. https://doi.org/10.1021/acs.estlett.7b00267

Singh BP, Hatton BJ, Singh B, Cowie AL, Kathuria A (2010) Influence of biochars on nitrous oxide emission and nitrogen leaching from two contrasting soils. J Environ Qual in press

Skjemstad JO, Reicosky DC, Wilts AR, McGowan JA (2002) Charcoal carbon in U.S. agricultural soils. Soil Sci Soc Am J 66:1249–1255

Smernik RJ (2009) Biochar and sorption of organic compounds. In: Lehmann J, Kern DC, Glaser B, Woods WI (eds) Amazonian dark earths: origin, properties, management. Kluwer Academic Press, Dordrecht, The Netherlands, pp 289–300

Soentgen J, Hilbert K, von Groote-Bidlingmaier C et al (2017) Terra preta de índio: commodification and mythification of the Amazonian Dark Earths. GAIA 26:136–143

Sohi S, Krull E, Lopez-Capel E, Bol R (2010) A review of biochar and its use and function in soil. Adv Agron 105:47–82

Sohi S, Lopez-Capel E, Krull E, Bol R (2009) Biochar's roles in soil and climate change: a review of research needs. CSIRO Land and Water Science Report 05/09

Sombroek WG, Nachtergaele FO, Hebel A (1993) Amounts, dynamics and sequestering of carbon in tropical and subtropical soils. Ambio 22:417–426

Song J, Peng P'A (2010) Characterisation of black carbon materials by pyrolysis–gas chromatography–mass spectrometry. J Anal Appl Pyrol 87:129–137

Spokas KA, Baker JM, Reicosky DC (2010) Ethylene: potential key for biochar amendment impacts. Plant Soil in press

Spokas KA, Koskinen WC, Baker JM, Reicosky DC (2009) Impacts of woodchip biochar additions on greenhouse gas production and sorption/degradation of two herbicides in a Minnesota soil. Chemosphere 77:574–581

Spokas KA, Reicosky DC (2009) Impacts of sixteen different biochars on soil greenhouse gas production. Ann Environ Sci 3:179–193

Steinbeiss S, Gleixner G, Antonietti M (2009) Effect of biochar amendment on soil carbon balance and soil microbial activity. Soil Biol Biochem 41:1301–1310

Steiner C, Das KC, Garcia M, Förster B, Zech W (2008) Charcoal and smoke extract stimulate the soil microbial community in a highly weathered xanthic Ferralsol. Pedobiologia 51:359–366

Steiner C, Teixeira WG, Lehmann J, Nehls T, de Macêdo JLV, Blum WEH, Zech W (2007) Long term effects of manure, charcoal and mineral fertilization on crop production and fertility on a highly weathered Central Amazonian upland soil. Plant Soil 291:275–290

Swift MJ, Heal OW, Anderson JM (1979) Decomposition in terrestrial ecosystems. University of California Press, Berkeley

Tammeorg P, Bastos AC, Jeffery S et al (2017) Biochars in soils: towards the required level of scientific understanding. J Environ Eng Landsc 25:192–207

Tan Z, Lin CSK, Ji X, Rainey TJ (2017) Returning biochar to fields: a review. Appl Soil Ecol 116:1–11

Thies JE, Rillig MC (2009) Characteristics of biochar: biological properties. In: Lehmann J, Joseph S (eds) Biochar for environmental management. Earthscan, London, U.K., pp 85–105

Titiz B, Sanford RL (2007) Soil charcoal in old-growth rain forests from sea level to the continental divide. Biotropica 39:673–682

Topoliantz S, Ponge JF (2005) Charcoal consumption and casting activity by *Pontoscolex corethrurus* (Glossoscolecidae). Appl Soil Ecol 28:217–224

Topoliantz S, Ponge JF, Ballof S (2005) Manioc peel and charcoal: a potential organic amendment for sustainable soil fertility in the tropics. Biol Fertil Soils 41:15–21

van der Wal A, de Boer W (2017) Dinner in the dark: illuminating drivers of soil organic mat-ter decomposition. Soil Biol Biochem 105:45–48

Van Zwieten L, Kimber S, Morris S, Chan KY, Downie A, Rust J, Joseph S, Cowie A (2010) Effects of biochar from slow pyrolysis of papermill waste on agronomic performance and soil fertility. Plant Soil 327:235–246

Van Zwieten L, Sing B, Joseph S, Kimber S, Cowie A, Chan KY (2009) Biochar and emissions of non CO_2 greenhouse gases from soil. In: Lehmann J, Joseph S (eds) Biochar for environmental management. Earthscan, London, U.K., pp 227–249

Verheijen FGA, Jeffery S, Bastos AC, van der Velde M, Diafas I (2009) Biochar application to soils—a critical scientific review of effects on soil properties, processes and functions. EUR 24099 EN, Office for the Official Publications of the European Communities, Luxembourg, 149pp

Verheijen FGA, Mankasingh U, Penizek V et al (2017) Representativeness of European biochar research: Part I—field experiments. J Environ Eng Landsc 25:140–151

Verhoeven E, Pereira E, Decock C et al (2017) Toward a better assessment of biochar–nitrous oxide mitigation potential at the field scale. J Environ Qual 46:237–246

Wang C, Wanga Y, Herath HMSK (2017a) Polycyclic aromatic hydrocarbons (PAHs) in biochar—their formation, occurrence and analysis: A review. Org Geochem 114:1–11

Wang D, Fonte SJ, Parikh SJ, Six J, Scow KM (2017b) Biochar additions can enhance soil structure and the physical stabilization of C in aggregates. Geoderma 303:110–117

Wang H, Lin K, Hou Z, Richardson B, Gan J (2010) Sorption of the herbicide terbuthylazine in two New Zealand forest soils amended with biosolids and biochars. J Soils Sediments 10:283–289

Wardle DA, Nilsson M-C, Zackrisson O (2008) Fire-derived charcoal causes loss of forest humus. Science 320:629

Warnock DD, Lehmann J, Kuyper TW, Rillig MC (2007) Mycorrhizal responses to biochar in soil—concepts and mechanisms. Plant Soil 300:9–20

Weng Z, Van Zwieten L, Singh BP et al (2017) Biochar built soil carbon over a decade by stabilizing rhizodeposits. Nat Clim Change 7:371–379

Wornat MJ, Hurt RH, Yang NYC (1995) Structural and compositional transformations of biomass chars during combustion. Combustion Flame 100:131–143

Xiang Y, Deng Q, Duan H, Guo Y (2017) Effects of biochar application on root traits: a meta-analysis. GCBBioenerg 9:1563–1572. https://doi.org/10.1111/gcbb.12449

Yamato M, Okimori Y, Wibowo IF, Anshori F, Ogawa M (2006) Effects of the application of charred bark of *Acacia mangium* on the yield of maize, cowpea and peanut, and soil chemical properties in South Sumatra, Indonesia. Soil Sci Plant Nutr 52:489–495

Yanai Y, Toyota K, Okazaki M (2007) Effects of charcoal addition on N_2O emissions from soil resulting from rewetting air-dried soil in short-term laboratory experiments. Soil Sci Plant Nutr 53:181–188

Zhang P, Sheng G, Wolf DC, Feng Y (2004) Reduced biodegradation of benzonitrile in soil containing wheat-residue-derived ash. J Environ Qual 33:868–872

Zimmerman AR (2010) Abiotic and microbial oxidation of laboratory-produced black carbon (biochar). Environ Sci Technol 44:1295–1301

Chapter 9
Importance of Soils of Agroecosystems for Climate Change Policy

Contents

Abstract Until now, international climate change policy has neglected the potential of agroecosystems to contribute to climate change mitigation by reduction in net greenhouse gas (GHG) emissions, and by enhancing soil inorganic carbon (SIC) and soil organic carbon (SOC) sequestration. Proven agricultural practices for creating a positive soil C balance such as those under the umbrella of climate-smart agriculture (CSA) are readily available. In the future, the importance of soil C stocks for climate change mitigation will be addressed by the target of Land Degradation Neutrality (LDN) of United Nations Convention to Combat Desertification (UNCCD), and land- and soil-related targets for achieving United Nations Sustainable Development Goals (SDGs). Further, many nationally determined contributions (NDCs) submitted to United Nations Framework Convention on Climate Change (UNFCCC) after the 2015 Paris Agreement include references to soil enhancement and to the necessity of better soil management practices. Similarly, an ambitious goal to increase global soil C stocks of managed agricultural land by 0.4% per year in the topsoil was set by the '4 Per 1000 Initiative: Soils

K. Lorenz and R. Lal, *Carbon Sequestration in Agricultural Ecosystems*,
https://doi.org/10.1007/978-3-319-92318-5_9

for Food Security and Climate' (4p1000). Based on this, spatially diversified strategies need to be developed to sequester SOC in agroecosystems by addressing limitations imposed by nutrient (i.e., nitrogen and phosphorus) availability, and due consideration of economic and social aspects. Thus, meaningful actions that help to restore and protect soils of agroecosystems including their SOC stocks have been initiated recently. It is hoped that soils of agroecosystems will increasingly be managed to address climate climate change while at the same time contributing to food security under increasing pressure from population growth and dietary changes. This chapter begins with an overview over some practices to enhance soil C stocks in agroecosystems for climate and food security. This is followed by a description of previously proposed climate change mitigation policies and their reference to agroecosystems soil C stocks. Then, a brief overview over CSA, carbon farming, and the soil security framework is given. This chapter concludes with a section on payments for ecosystem services (ESs).

Keywords Rio conventions · Sustainable development goals · 4 per 1000 initiative · Climate-smart agriculture · Carbon farming · Soil security Payments for ecosystem services

9.1 Management of Soil Carbon of Agroecosystems and Climate Change Policy

The previous chapters have shown that agriculture and, especially, soils of agroecosystems both contribute to climate change by net greenhouse gas (GHG) emissions and are affected by climate change. The Intergovernmental Panel on Climate Change (IPCC) considers agricultural GHG emissions from croplands and pasture (excluding livestock systems) mostly in the form of nitrous oxide (N_2O) emissions from agricultural soils (from fertilizer and manure application, and crop residue management), methane (CH_4) emissions from paddy cultivation, and carbon dioxide (CO_2) emissions through energy use (i.e., for fertilizer production and machinery use) (Porter et al. 2014). However, soils of agroecosystems can mitigate some of the GHG emissions as discussed in the previous chapters, but this potential has been previously overlooked (Paustian et al. 2016). This neglect applies, in particular, to international climate change policy (Streck and Gay 2017). Land-based sequestration efforts generally receive only about 2.5% of climate mitigation dollars (Buchner et al. 2015). However, conservation, restoration, and improved land management actions that increase C storage and/or avoid GHG emissions across global forests, wetlands, grasslands, and agricultural lands are natural climate solutions (Griscom et al. 2017). Among those solutions is 'climate-smart agriculture (CSA)' initially proposed by FAO in 2010 and projected to yield adaptation and mitigation benefits while increasing productivity (FAO 2013b). CSA will be discussed in a section below. Otherwise, 'climate resilient agriculture (CRA)' encompasses the incorporation of adaptation and resilient practices in agriculture which increases the capacity of the agroecosystem to respond to various climate-related disturbances by resisting damage and ensures quick recovery (Rao et al. 2016). Among disturbances are drought, flood, heat/cold wave, erratic rainfall pattern, pest outbreaks, and other threats caused by changing climate. Resilience is the ability of the system to bounce back and essentially involves judicious and improved management of natural resources, land, water, soil, and genetic resources through adoption of best management practices (BMPs). For example, crop diversification is a CRA strategy, but its implementation in the US Corn Belt faces many challenges (Roesch-McNally et al. 2018). CRA seeks to support countries and other actors in securing the necessary policy, technical, and financial conditions to enable them to: (i) sustainably increase agricultural productivity and incomes in order to meet national food security and development goals, (iii) build resilience and the capacity of agricultural and food systems to adapt to climate change, and (iii) seek opportunities to mitigate GHG emissions and increase carbon (C) sequestration. These three conditions (food security, adaptation, and mitigation) are referred to as the 'triple win' of CRA (Rao et al. 2016).

Some examples of proven practices to create a positive soil C balance, i.e., when the rate of soil C inputs is higher than that of soil C losses, are summarized in the following section while in-depth discussions can be found in the previous chapters.

9.1.1 Managing Soil Carbon Stocks in Agroecosystems for Climate Change Mitigation

Agroecosystems contain both soil inorganic carbon (SIC) and soil organic carbon (SOC) stocks, and management of both is needed for climate change mitigation by C sequestration. With regard to increasing SOC stocks of agricultural soils, increasing organic matter (OM) inputs (plant residue, manure or other organic waste) and reducing decomposition rates (e.g., by reducing soil disturbance) are proven strategies (Paustian et al. 2016). Practices that increase C inputs include (i) improved varieties or species with greater root mass C input to deposit C in deeper layers where turnover is slower, (ii) adopting crop rotations that provide greater soil C inputs, (iii) more residue retention, and (iv) cover crops during fallow periods to provide year-round soil C inputs. Other practices to increase net primary productivity (NPP) and, thus, soil C inputs include irrigation in water-limited systems, and additional fertilizer input in low-yielding, nutrient-deficient agroecosystems. Further, SOC losses in many croplands can be reduced through less intensive tillage, particularly no tillage, owing to less disruption of soil aggregate structure which reduces decomposition rates. A change from annual to perennial crops typically increases belowground C inputs and reduces soil disturbance, leading to SOC sequestration. In grasslands, SOC sequestration can be increased through optimal stocking/grazing density. Avoiding conversion and degradation of native grassland ecosystems, in particular, is a strong mitigation alternative as SOC stocks are much higher than those of their agricultural counterparts. Conversely, restoration of marginal or degraded lands to grassland increases SOC stocks. The SOC sequestration rates on land maintained in agricultural use are, however, lower than for land restoration to grassland (Paustian et al. 2016). Toensmeier (2016) summarized lifetime SOC stocks for various carbon farming systems. Lifetime SOC gains per hectare ranged from 30 to 50 Mg for improved annual cropping systems to up to 300 Mg for multi-strata agroforestry systems. The general trend was that systems that incorporate trees sequester more C. Less well studied are, however, practices to increase subsoil SOC stocks which may be more important than surface SOC stocks for C sequestration and climate change mitigation (Lorenz and Lal 2005).

Addition of plant-derived C from external, offsite sources such as composts or biochar can also increase SOC stocks, and result in net CO_2 removals from the atmosphere (Paustian et al. 2016). However, because the OM originates from outside the agroecosystem 'boundary,' net life-cycle emissions will largely depend on whether the biomass used as a soil amendment would have otherwise been burnt (either for fuel or as waste disposal), added to a landfill, or left in place as living biomass or detritus. In contrast, often observed is mineralization of existing SOC in response to amendments ('priming') immediately following biochar addition. Further, soil N_2O and CH_4 emissions may also be affected by soil amendments. Important for more persistent amendments such as biochar are effects on plant growth in amended soils and resultant feedbacks to SOC (Paustian et al. 2016).

Generally overlooked are the effects of agricultural land use and management on the SIC stock although practices enhancing SOC sequestration may also alter the SIC balance. For example, soil carbonate formation may be enhanced by increased Ca and Mg input to soil, and by increased soil CO_2 concentrations through increased OM inputs. Otherwise, OM inputs may result in soil acidification and, thus, potential SIC losses. SIC sequestration in soils of agroecosystems may occur, in particular, when Ca^{2+} is released directly from silicates (Monger et al. 2015). However, whether SIC formation can contribute to the overall mitigation potential of soils of agroecosystems is unclear (Sanderman 2012).

The implementation of soil-based GHG mitigation activities is at an early stage, and accurately quantifying emissions and reductions is a substantial challenge (Paustian et al. 2016). International standards are needed to account for changes in SOC stocks and GHG emission fluxes from soils to ensure quality and comparability of data (Bispo et al. 2017). Such standards can contribute strongly to improve the reliability of climate models and mitigation measures and, thus, facilitate effective policy developments. How soils receive currently attention in international climate negotiations will be discussed in the following section.

9.1.2 Addressing Soil Carbon Stocks of Agroecosystems in Climate Change Policy

Management and protection of soils of agroecosystems are among important land-based practices to address climate change. The set of rules and procedures adopted under the UN Framework Convention on Climate Change (UNFCCC) provides the overarching framework for national and international collaborative efforts to address climate change (Streck and Gay 2017). However, there are other important international instruments offering opportunities for soil conservation and for addressing the importance of soil C stocks for climate change mitigation. Among important policy instruments are the United Nations Convention to Combat Desertification (UNCCD; UNGA 1994), and the United Nations Sustainable Development Goals (SDGs) as part of the '2030 Agenda for Sustainable Development' (UNGA 2015). All initiatives appear to be mandated to take C sequestration in soil through agricultural, pastoral, and forestry practices into account (Boer et al. 2017). Increased SOC sequestration and healthy soil C cycling underpin several SDGs (Lorenz and Lal 2016; Smith et al. 2018). However, other than the enhancement of C reservoirs of terrestrial vegetation (e.g., by UNFCCC), soil C sequestration is not on the political agenda, and was not formally discussed at the Bonn Conference of the Parties (COP) 23 meeting in Germany in November 2017 (Rumpel et al. 2018).

The UNCCD has adopted recently the clear overarching goal and quantitative target of Land Degradation Neutrality (LDN; Lal et al. 2012; Minelli et al. 2017). Land degradation was defined by UNCCD as 'reduction or loss, in arid, semiarid, and dry subhumid areas, of the biological or economic productivity and complexity of rain-fed cropland, irrigated cropland, or range, pasture, forest, and woodlands' (UNGA 1994). Further, LDN was defined by UNCCD as 'a state whereby the amount and quality of land resources necessary to support ecosystem functions and services and enhance food security remain stable or increase within specified temporal and spatial scales and ecosystems' (UNCCD 2016). The goal of LDN is maintaining or enhancing the land resource base, i.e., the stocks of natural capital associated with land resources, and the ecosystem services that flow from them (Orr et al. 2017). In addition, LDN may also serve as a target and indicator of Sustainable Land Management (SLM; Kust et al. 2017). Land degradation may cause a decline in C sequestration potential and reductions in the multiple goods and services provided by ecosystems. Degradation of agricultural land may, thus, reduce soil C stocks which would diminish the role of soils of agroecosystems for climate change mitigation, and also food security.

The three Rio Conventions UNCCD, UNFCCC, and the Convention on Biological Diversity (CBD) all play key roles in shaping the international LDN governance and implementation context (Akhtar-Schuster et al. 2017). Synergies in UNCCD, CBD, and UNFCCC actions can be used to advance toward LDN. For example, the UNFCCC advocates for cooperation in research, training, information compilation, and sharing on Nationally Appropriate Mitigation Actions (NAMAs) to be implemented by developing country Parties, thereby considering both restoration and rehabilitation. This indicates that the UNFCCC consciously differentiates between restoration and rehabilitation, offering opportunities to pursue joint activities with the UNCCD (rehabilitation) or the CBD (restoration). Developing synergies between the three Rio Conventions in moving toward LDN may pragmatically focus more on rehabilitation as restoration measures cannot durably enhance ecosystem resilience due to climate change (Akhtar-Schuster et al. 2017).

To address all lands on earth rather than just those in the drylands, LDN was accepted as target of a SDG aimed at neutralizing the rate of lands coming under degrading use of their productivity (Safriel 2017). This is highlighted in SDG 15 ('Protect, restore, and promote sustainable use of terrestrial ecosystems, sustainably manage forest, combat desertification, and halt and reverse land degradation, and halt biodiversity loss') (UNGA 2015). Target 15.3 'By 2030, combat desertification, restore degraded land and soil… strive to achieve a land degradation-neutral world' refers to LDN as crosscutting issue that must be addressed to achieve sustainable development. However, a target of completely halting land degradation processes by 2030 may not be realistic, but offsetting ongoing degradation by restoring productivity of lands in degraded state may be achievable (Safriel 2017).

Monitoring achievement of neutrality will quantify the balance between area of gains (positive changes in LDN indicators = improvement) and area of losses (negative changes in LDN indicators = degradation) (Orr et al. 2017). The LDN indicators and associated metrics are land cover (land-cover change), land productivity (net primary production) and C stocks (SOC), and others as needed. However, quantitative evidence for the link between changes in SOC stocks and the numerous land degradation drivers and processes is scanty (Lorenz and Lal 2016). Nevertheless, areas with increasing land productivity and C stocks (SOC) can generally be interpreted as improving (Minelli et al. 2017). LDN would significantly contribute to mitigating climate change (Safriel 2017). Thus, implementing practices in agroecosystems toward achieving LDN may also contribute to climate change mitigation by increasing SOC stocks. However, LDN needs further scientific research and development of effective methods to measure the balance between different terrestrial ecosystems' qualities, functions, and services (Kust et al. 2017). Further, achieving multiple SDG targets at the national scale for the land sector faces a myriad of challenges (Gao and Bryan 2017).

Similar to the Kyoto Protocol (Kyoto Protocol 1997), neither UNFCCC nor the Paris Agreement adopted by 196 Parties of UNFCCC (Paris Agreement 2015), deal with soil explicitly (Streck and Gay 2017). For example, the Clean Development Mechanism (CDM) of the Kyoto Protocol does exclude accounting for soil C (FAO 2013a). However, countries ratifying climate agreements and protocols are mandated to adopt policies and measures that reduce GHG emissions. Thus, soils are relevant as their restoration and protection ensure the continuous delivery of ecosystem services in a changing climate including C sequestration, while use of land for crops and livestock results in high levels of emissions, in particular, CH_4 and N_2O. Mitigation and adaptation activities will, therefore, pertain to the agricultural sector. Incentives for better cropland and pasture management were previously limited to developed countries, and even there SOC was often not appropriately considered in national inventories. Only since 2010 is agriculture discussed by UNFCCC climate negotiators as a sector that requires special attention. It is increasingly recognized that improved protection and enhancement of SOC stocks, restoring degraded land, improving efficiency and productivity of agricultural systems in a sustainable manner can at the same time build resilience to climate change impacts (Streck and Gay 2017). At Cop23, from November 6–17, 2017, in Bonn, Germany, agriculture was finally included in formal UNFCCC processes (UNFCCC SBSTA 2017).

The treatment of agriculture including that of soil C stocks of agroecosystems has evolved over the lifetime of the IPCC, as tracked by the assessment reports with the first one published in 1990 (Porter et al. 2017). Increasing C sinks in agricultural soils has already been considered as a mitigation option in the first report. In the second report, estimates of global technical mitigation potentials were made, with an emphasis on soil C sequestration. However, Porter et al. (2017) suggested that impacts of climate change and adaptation on net GHG budgets, particularly on soil C levels, are either missing or underrepresented among key research points on food security after the latest assessment report published 2013–2014.

A new treaty under the Paris Agreement and with the framework of the UNFCCC will enter into force in 2020 (Streck and Gay 2017). It relies on countries making non-binding dynamic, bottom-up, and ever-more ambitious mitigation pledges to hold the increase in the global average temperature to well below 2 °C and further limiting increases to 1.5 °C. These temperature references can only be understood as changes in climatological averages attributed to human activity excluding natural variability (Rogelj et al. 2017). Long-term temperature targets provide guidance for short-, middle-, and long-term global mitigation actions and can be translated into specific C budgets over different time periods that can inform mitigation requirements. For example, the Paris Agreement requires complete decarbonization of both the energy and land use, land-use change and forestry (LULUCF) sectors before the end of the twenty-first century (Walsh et al. 2017), which has implications for the agricultural sector. However, the role of the land sink is often left implicit in C budget analyses, and the Paris Agreement itself leaves ambiguous the definition of 'net-zero emissions'; i.e., it does not specify whether sinks need be directly anthropogenic or natural. Nevertheless, the C flux from the atmosphere into biomass and soil will be essential to the magnitude and timing of economic and technological interventions necessary to realize the climate targets aside the C flux from the atmosphere into the oceans. Most importantly, the emissions' mitigation potential of global agricultural systems needs to be fully exploited (Walsh et al. 2017). Aggressive GHG emissions reduction must be accompanied by sustainable extensive terrestrial CO_2 removal including agricultural practices and established immediately to avoid temperature 'overshoot' (Boysen et al. 2017). However, even if all pledges are fulfilled, they will deliver only about one-third of the emissions reductions needed to be on a least-cost pathway for the goal of staying well below 2 °C (UNEP 2017). Thus, without tougher targets, it is extremely unlikely that the goal of holding warming below 2 °C can be reached.

Envisioned by the Paris Agreement is a global process of engagement, follow-up, regular stock-take exercises, and cooperative action including on the provision of funding for developing countries from a floor of 100 billion USD per year (Streck and Gay 2017). Among the achievements of the Paris Agreement is that it provides a common framework within which individual countries (or alliances of countries) are invited to define action plans or 'Intended Nationally Determined Contributions' (INDCs) taking into account the overall goal of the UNFCCC and the Paris Agreement as well as their own capacities (FAO 2016). INDC of a Party will become its first Nationally Determined Contribution (NDC) upon submitting its instrument of ratification for the Paris Agreement. INDCs/NDCs outline countries' climate change priorities for the post-2020 period, and include targets and concrete strategies for addressing the causes of climate change and responding to its effects. The agriculture sectors (crops, livestock, forestry, fisheries, and aquaculture) feature prominently in meeting national mitigation and adaptation goals (FAO 2016). However, while most NDCs indicate inclusion of mitigation by the land sector, only 38 of 160 NDCs assessed by Forsell et al. (2016) specify land sector mitigation contributions. Despite these limitations, analyses indicate that if NDCs were fully implemented, natural climate solutions would contribute 20% of climate change mitigation by 2030 (Griscom et al. 2017). Solutions include 20 conservation, restoration, and improved land management actions that increase C storage and/or avoid GHG emissions across global forests, wetlands, grasslands, and agricultural lands.

Developing countries—particularly the least-developed countries (LDCs)—put a strong emphasis on the agriculture sectors (FAO 2016). The land sector is one of the largest contributors to national GHG emissions of developing countries (Streck and Gay 2017). Many of them have already pledged to reduce emissions from agriculture. Thus, about 80% of the intended NDCs include references to agriculture and C sequestration. Some also include references to soil enhancement and to the necessity of better soil management practices. For example, Nigeria's NDCs include the adoption of better soil management practices in the crops and livestock sector, Botswana's NDCs refer to CSA as part of Adaptation Plan to reduce soil erosion, and Uruguay's NDCs are aimed at the protection of croplands by development of soil use and management plans to reduce erosion and preservation of OM in croplands. However, developing countries made it clear that adaptation takes priority over mitigation. Otherwise, there is limited understanding about potential complementarity between management practices that promote adaptation and mitigation in developing countries, and limited basis to account for GHG emission reductions in this sector (Ogle et al. 2014). Many activities, such as pasture management, agroforestry, and sustainable intensification (SI), combine adaptation and mitigation benefits while increasing yields. Such 'triple-win' activities are, thus, considered as CSA practices (Streck and Gay 2017).

Other than countries, international climate initiatives can play an important role for reducing global emissions (Graichen et al. 2017). For example, when 19 assessed initiatives meet their targets, they could reduce global emissions by 5–11 $GtCO_{2e}$ in 2030 beyond the national targets by countries. In addition, these initiatives help governments in meeting their NDCs. Many initiatives have targets that are more ambitious than those of national governments. If the national governments would take all of the actions into account, they could be more ambitious in their NDCs. Initiatives with active involvement of non-governmental organizations (NGOs) tend to lead to higher reductions and more co-benefits. Another common element of many successful initiatives is a permanent secretariat (Graichen et al. 2017).

9.1.3 'Les Sols Pour La Sécurité Alimentaire et Le Climat'—4 Per 1000—Soils for Food Security and Climate

In relation to the Paris Agreement, France as part of the Lima-Paris Action Agenda (LPAA) launched the '4 Per 1000 Initiative: Soils for Food Security and Climate' (4p1000) declaration, which emphasizes the important role of soils and SOC as a mitigation strategy (http://4p1000.org). The Executive Secretariat of this international voluntary multi-stakeholder coalition is hosted by the Consultative Group on International Agricultural Research (CGIAR). Overall aim is to contribute to three complementary goals: (i) the improvement of food security, (ii) the adaptation of agriculture to climate change, and (iii) the mitigation of climate change (Aubert et al. 2017). The initiative provides a guideline for locally increasing SOC stocks in the topsoil (0–40 cm depth) by 0.4% per year, implying an annual SOC sequestration rate proportional to the initial SOC stock (Soussana et al. 2017). With this aspirational target, degraded soils with low SOC stocks need to sequester also less C. The goal of 4p1000 is to engage stakeholders in a transition toward a productive, resilient agriculture, based on a sustainable soil management and generating jobs and incomes, hence ensuring sustainable development. The Government of France has committed to ensure that at least 50% of French agricultural holdings adopt an agroecological approach by 2020, while Australia has pledged to build the resilience of 3 million km^2 of Australia's inland and northern rangelands in the front line and at risk of degradation (Streck and Gay 2017).

Chambers et al. (2016) concluded that the majority of US cropland and grassland soils have the capacity to increase C stocks consistent with the 4p1000 initiative. However, it would be important to focus on topsoil (0–20 cm) C stocks, and prioritize degraded lands, cropland, and grassland. Globally, Minasny et al. (2017) estimated for twenty regions that there is some scope to increase SOC stocks, and this potential is mostly on managed agricultural lands. High C sequestration rates (up to 1.0%) can be achieved, in particular, for soils with low initial SOC stock (topsoil less than 30 Mg C ha^{-1}), and at the first twenty years after implementation of best management practices (Minasny et al. 2017). Long-term field experiments indicate that annual increases in SOC of 0.4% are possible for certain soil types under specific land-use and management practices, but only for a limited period (Johnston et al. 2017). Sanderman et al. (2017) concluded that there are identifiable regions which can be targeted for SOC restoration efforts. Globally, assuming a 4p1000 increase in the top 1 m of agricultural soils, SOC sequestration rates may range between 2 and 3 Pg C yr^{-1}, which would effectively offset 20–35% of global anthropogenic GHG emissions (Minasny et al. 2017). Croplands worldwide may sequester between 0.90 and 1.85 Pg C yr^{-1} to 30 cm depth, i.e., 26–53% of the 4p1000 target (Zomer et al. 2017). Of importance are intensively cultivated regions such as North America, Europe, India and intensively cultivated areas in Africa, such as Ethiopia. However, an achievable potential is more likely to be limited to $\sim$1 Pg C yr^{-1} (Smith 2016).

There are many scientific, political, and socioeconomical issues with the large-scale implementation of 4p1000 (Chabbi et al. 2017). For example, additional benefits to farmers from increased SOC storage in terms of improved income through enhanced soil productivity, lower fertilizer requirements, and sustained yields remain largely unquantified. Further, there is a paucity of information demonstrating the benefits of increasing SOC stocks across a range of different soils, agroecosystems, and climatic zones. One practical way to help reach the 4p1000 goal is may be to demonstrate the quantitative co-benefits of SOC and their economic value to landowners. Soils and agroecosystems most suitable to increase SOC sequestration must be identified. Further, the economic benefits of additional C storage for different soil types, climate zones, production systems, and farming capabilities must be assessed. The implementation success will depend, in particular, on the ability to improve global governance and regional collaboration between actors and across sectors, and to communicate the benefits of SOC sequestration to farmers, landowners and through the consumer supply chain (Chabbi et al. 2017).

Among the scientific challenges for implementing 4p1000 is how to maximize the residence time of additional SOC (Dignac et al. 2017). Another challenge is how to balance higher SOC stabilization with respect to C mineralization due to antagonist biotic mechanisms. On one hand, soil biota consumes and transforms SOC, but on the other hand, soil biota ultimately contributes to SOC stabilization (Dignac et al. 2017). Further limitations for increasing SOC stocks are imposed by nutrient availability as soil organic matter (SOM) contains also nutrients such as nitrogen (N) and phosphorus (P) (van Groenigen et al. 2017a). The capability of terrestrial ecosystems to fix C is constrained by nutrient availability (Augusto et al. 2017). N limitation is not linked to soil parent materials, but best explained by climate. Specifically, ecosystems under harsh (i.e., cold and or dry) climates are more N limited than ecosystems under more favorable climates. In contrast, P limitation is not driven by climate, but by soil parent materials (Augusto et al. 2017). Since global N:P stoichiometry is increasing, P limitation could become more critical for global SOC sequestration in the long term than N limitation (Soussana et al. 2017). Degraded soils immobilize less N and P than soils rich in OM. However, at the global scale, many soils with a large SOC stock are (extensively) grazed grasslands, which typically have small inputs of N and P inputs, and small surpluses. In contrast, many intensively managed arable soils, which typically have lower SOC stocks, have large inputs of N and P leading to large surpluses (van Groenigen et al. 2017b). Clearly, the 4 per 1000 target has to be spatially differentiated in terms of N and P requirements (Soussana et al. 2017).

Chabbi et al. (2017) proposed that intensification of waste organic residue recycling and taking into account the associated nutrient budgets may be a promising strategy for increasing SOC sequestration. Efforts should concentrate on soils currently having a low SOC stock, and where nutrients and residues are available. Examples are soils that have become degraded due to long periods of intensive arable cropping or overgrazed grasslands in cool, temperate, or Mediterranean climatic regions especially in Asia, Europe, and North America (van Groenigen et al. 2017a). Old arable soils, degraded lands, and mine wastes have a greater potential for increasing SOC than soils with high SOC stocks (Van Groenigen et al. 2017b). Thus, management practices adapted to local conditions are needed to increase soil C input fluxes and stabilization for enhancing SOC stocks to meet the 4p1000 goal (Dignac et al. 2017).

In response to thought-provoking comments on Minasny et al. (2017) by de Vries (2018), VandenBygaart (2018), White et al. (2018), and Baveye et al. (2018), Minasny et al. (2018) reiterated that 4p1000 is: (i) more of a concept than the actual number itself; (ii) not achievable everywhere and finite in time/duration; (iii) a timely and useful initiative; and (iv) an aspirational target. Also, the economic and social aspects of 4p1000 need further investigation (Minasny et al. 2018). Similarly, Lal (2016) emphasized that the 4p1000 proposal should be more about the concept than any specific numbers. The concept that soils and agriculture are solutions to the global issues of climate change, food insecurity, and environmental pollution may, however, be a major paradigm shift of historic significance (Lal 2016). Baveye et al. (2018) cautioned that 4p1000 should not repeat the mishap of the biochar community which presented biochar as a panacea, 'win-win,' 'win-win-win,' or even 'winfinity' solution. Specifically, claims that climate change could be mitigated through the incorporation of massive amounts of biochar into soils were not backed by research a decade ago when those claims were made. Actually, the idea that biochar could help resolve practically the climate change challenge is mentioned by an ever shrinking portion of the relevant literature (Baveye et al. 2018).

While 4p1000 aims at topsoil, SOC below 30–40 cm depth in agroecosystems may also respond to changes in land use and management as deep SOC may be more dynamic than previously thought (Dignac et al. 2017; Trumbore et al. 1995). For example, subsoil SOC has been shown to be reactive to agricultural land-use changes on decadal timeframes despite ^{14}C ages of several millennia, which may lead to large losses of young C from the subsoil (Hobley et al. 2017). Subsoil SOC may be lost, in particular, by the priming effect when labile substrates enter deeper soil depths causing an increase in decomposition rates (Heitkötter et al. 2017). This possible risk that could be incurred in the implementation of 4p1000 projects needs to be taken into account (Aubert et al. 2017).

Aside topsoil and subsoil SOC, SIC stocks may also be considered in the implementation of 4p1000 practices. Specifically, SIC stocks may be more reactive as previously assumed, and change by as much as 1 Mg C ha^{-1} yr^{-1} in agricultural soils (Sanderman 2012). Management practices can influence SIC stocks over a time span of months to years (Wang et al. 2016). Further, there are complex interactions between SIC and SOC. For example, Dong et al. (2017) reported antagonistic interactions between SOC and SIC levels while increasing SOC may also lead to an enhancement in SIC formation, particularly in the subsoil and at soil pH > 8 (Shi et al. 2017). A prerequisite for C sequestration as SIC is that calcium is directly derived from silicate minerals (termed silicatic pedogenic carbonate; Monger et al. 2015). However, a mechanistic understanding of SIC dynamic is rather incomplete (Zamanian et al. 2016). Nevertheless, the risks of increasing SOC stocks for changes in SIC stocks by implementation of 4p1000 practices, particularly in the subsoil, need to be assessed.

If the 4p1000 initiative is to deliver simultaneously on all its objectives, Aubert et al. (2017) proposed a set of indicators to contribute to the development of an implementation framework for 4p1000 projects. Three main recommendations were made. First, two types of indicators need to be distinguished, i.e., those that can be considered as safeguards to help to assess the possible negative impacts of a 4p1000 project on crucial social and environmental aspects, and that will serve as exclusionary principles. Other indicators that need to be distinguished are those for which projects have to clarify the hypothesis they rely upon regarding the intended changes, and the expected impacts they will have on the objectives of 4p1000. Secondly, safeguards' indicators should cover land tenure, human rights, and ecosystems integrity. Thirdly, indicators should help to uncover how 4p1000 projects intend to impact upon: (i) soil C stocks and global GHG emissions at the project level, and (ii) agrarian structures and the four pillars of food and nutrition security (Aubert et al. 2017).

9.1.4 The Initiative for the Adaptation of African Agriculture to Climate Change

Launched upstream of COP22 organized in Morocco in 2016, the initiative for the Adaptation of African Agriculture (AAA) aims to reduce the vulnerability of Africa and its agriculture to climate change (http://www.aaainitiative.org/initiative). AAA promotes and fosters the implementation of specific projects to improve soil management, agricultural water control, climate risk management, and capacity building and funding solutions. Similarly to 4p1000, the AAA initiative is an important response not only to climate change, but also to food insecurity.

AAA has identified core axes for clarifying the scientific and technical challenges of AAA (Working group report of the scientific committee of the initiative for the Adaptation of African Agriculture to Climate Change 2016). Those include soil management, i.e., (i) soil fertility and crop fertilization, (ii) fruit farming, rangeland management, and agroforestry, and (iii) agroecological innovations and carbon sequestration. Further challenges are associated with agricultural-water control, and climate-risk management.

Soil management is a major challenge while African soils offer an important potential of C sequestration, which may contribute to significantly mitigating GHG emissions (Working group report of the scientific committee of the initiative for the Adaptation of African Agriculture to Climate Change 2016). Importantly, Africa is home to 60% of the globally unexploited arable lands offering a huge potential for C storage. Soil C stocks in Africa are low because of anthropic factors (inadequate farming practices, burn-beating, overgrazing, deforestation, and non-incorporation of crop residues on soil), and natural factors (bad weather conditions, low precipitation, high temperatures, loss of vegetation cover, extreme rainfalls, and high mineralization). Thus, African soils lose SOC, largely in arid and semiarid zones. However, most of African agroecological zones are naturally characterized by soils with decreased levels of SOM and nutrients. This restricts their productivity and endangers food security. The AAA working group recommended to develop an agroecology adaptable to the agropedo-climatic contexts of African countries, and driven by the optimization of the balance between productivity and preservation.

AAA proposed several key concrete and realistic agroecological and C sequestration solutions (Working group report of the scientific committee of the initiative for the Adaptation of African Agriculture to Climate Change 2016). Among those is to implement a conservation-oriented agricultural model based on the minimal use of mechanized tillage (no plowing or direct seeding), the permanent coverage of soil by organic, C-rich mulch (straw and/or other crop residues), crop rotations and associations (including N-fixing legumes). Other key solutions are: (i) to introduce intercropping (between tree rows), develop composting, crop-residue incorporation, and convert organic waste into compost; (ii) develop bio-intensive micro-agriculture; (iii) undertake fallowing; (iv) develop oases sustainably; and (v) develop agropastoral zones sustainably. AAA suggested that the first step to implement these solutions is to share and disseminate good agricultural practices and knowledge by researchers and professionals.

In conclusion, while there is some publicity on accelerated voluntary efforts to contribute to climate change mitigation, it remains to be seen whether the Paris Agreement triggers meaningful action by countries and initiatives, including action that helps to restore and protect soils of agroecosystems including their SOC stocks.

9.1.5 South–South Cooperation (SSC)

The serious problem of soil degradation in the Global South will be exacerbated by the increasing population, changes in consumption patterns, and the projected climate change. Adoption of BMPs to increase SOC sequestration in agroecosystems for improving soil health and enhancing ecosystem services (ESs) in southern countries may be promoted by the concept of South–South Cooperation (SSC; Lal 2018). Some newly emerging economies (i.e., Brazil, India, South Africa, China) have been successful in enhancing agricultural production through adoption of improved and proven technology. Involving those developing countries in SSC may be promising for advancing agriculture in other developing countries in the Global South through adoption of BMPs. For example, Brazil has been successful in a widespread adoption of conservation agriculture (CA). Brazil's experience can be a great asset in promoting the adoption of CA in Asia and Africa through SSC, triangular cooperation (TAC), on South–South-–North Cooperation (SSNC). A similar strategy may be useful in promoting the concept of SI and other BMPs in developing countries. To be successful and effective, the cooperations must be science-based approaches to promoting adoption of proven technologies developed and widely used in some southern countries. The traditional system of North–South Cooperation must be revisited. Among barriers to implement those cooperations are weak institutional and human resources, gender inequality, lack of access to credit and market, non-availability of scientific information, and poor enabling environment such as payments for ESs. It is essential to promote stewardship of soil and other natural resources as integral component of education and communication in addition to building human resources (Lal 2018).

9.2 Climate-Smart Agriculture

The previous chapters have highlighted both the importance of agriculture as a major contributor to climate change by enhancing emissions of GHGs including those from SOC losses but also as a potential climate change mitigation strategy by, e.g., SOC sequestration (Smith et al. 2014). The often negative effects of climate change on crop production and the food system exert major pressure on the need to feed a growing and more affluent global population whose dietary choices are also changing. Thus, adaptation to climate change and lower GHG emission intensities per output will be necessary instead of a business-as-usual scenario to achieve high productivity at low emissions (FAO 2013b). Agricultural stakeholder views and decisions on climate change relate to decisions about adaptation and mitigation practices (Chatrchyan et al. 2017). For example, although the majority of US farmers believe the climate is changing, many remain skeptical of the issue and uncertain about the anthropogenic causes of climate change. US farmers' climate change mitigation and adaptation decisions vary widely, and are often correlated with belief or other factors such as personal experience with extreme weather, costs of change, or fear of regulation (Chatrchyan et al. 2017).

The three objectives of CSA are: (i) sustainably increasing agricultural productivity to support equitable increases in incomes, food security, and development; (ii) adapting and building resilience to climate change from the farm to national levels; and (iii) developing opportunities to reduce GHG emissions from agriculture compared with past trends (Lipper et al. 2014). CSA can potentially contribute to the mitigation of climate change by reducing GHG emissions per unit of land and/or agricultural product, and by increasing C sinks. By jointly addressing food security and climate challenges, CSA integrates the three dimensions of sustainable development (economic, social, and environmental). Further, CSA is an opportunity to incorporate mitigation practices across the agricultural sector (Dickie et al. 2014). Agricultural ministries, agribusinesses, and financial institutions and donors, all need to create and adopt best practices for an integrated climate mitigation and productivity agenda. For example, the World Bank has committed to support CSA (Dickie et al. 2014). Among the most promising climate financing sources are: (i) the Adaptation Fund, an innovative financing mechanism that focuses on the needs of the most vulnerable communities and the possibility of direct access; (ii) the Global Environment Fund (GEF); and (iii) the Green Climate Fund (GCF) (Lipper et al. 2014). However, CSA is regularly perceived as addressing only adaptation, and not mitigation and food security (Saj et al. 2017). An eco- and sociological approach to CSA may be required to promote inclusive development and participate to collective efforts to manage agriculture and food systems under climate change.

CSA aims to (i) sustainably increase agricultural productivity and incomes, (ii) adapt and build resilience to climate change, and (iii) reduce and/or remove GHG emissions where possible. This includes also plant selection and modification at multiple scales and may involve engineering novel trait combinations or improved stress tolerance into agricultural species of interest (Way and Long 2015). At regional and larger scales, CSA impacts on volatile organic carbon (VOC) emissions and the biophysical properties of agroecosystems may be important (Devaraju et al. 2015; Rosenkranz et al. 2015). Implementation of CSA has also been linked to soil health in the USA (USDA 2016). The goal is to promote soil conservation practices that improve SOM, reduce GHG emissions from soils, and promote healthier soils. The recommended practices include residue and tillage management, no till, reduced tillage, contour buffer strips, grassed waterways, filter strips, field borders, conservation crop rotation, vegetative barriers, herbaceous wind barriers, and cover crops (USDA 2016).

Three CSA pillars can be distinguished, i.e., (i) agronomic and economic productivity, (ii) resilience and adaptive capacity, and (iii) climate change mitigation in the developing world (Rosenstock et al. 2016). CSA is an integrated approach to the implementation of agricultural development policies and programmes that strive to improve food security, livelihoods, and resilience under climate change, while at the same time capturing mitigation co-benefits where possible. While the CSA concept has been rapidly adopted, the empirical evidence base for CSA has not been assembled. Little empirical evidence has been put forth so far, in particular, to systematically evaluate the outcomes of CSA practices. Thus, Rosenstock et al. (2016) evaluate current knowledge on the effectiveness of CSA to achieve its intended benefits and to inform discourse on food, agriculture, and climate change. This will represent the largest meta-analysis of CSA practices to date. The derived improved knowledge will benefit action by public, private, and civil society stakeholders at the international to local levels in the areas: (i) building evidence and assessment tools; (ii) strengthening national and local institutions; (iii) developing coordinated and evidence-based policies; and (iv) increasing financing and its effectiveness (Lipper et al. 2014)

Examples of CSA practices on the supply side include integrated crop, livestock, aquaculture, and agroforestry systems; improved pest, water, and nutrient management; landscape approaches; improved grassland and forestry management; practices such as reduced tillage and use of diverse varieties and breeds; integrating trees into agricultural systems; restoring degraded lands; improving the efficiency of water and N fertilizer use; and manure management (Lipper et al. 2014). These changes need to be complemented by efforts to change consumption patterns, reduce waste, and create positive incentives along the production chain. CSA measures on the supply side are likely to be insufficient or ineffective if not accompanied by changes in consumer behavior, as CSA will affect the supply of agricultural commodities and require changes on the demand side in response (Scherer and Verburg 2017). For example, higher food prices related to CSA require willingness of the consumers to pay more or reduce their consumption to avoid burden shifting to another location. Otherwise, a higher demand for local food or meat analogs requires changes in the supply chain. Optimizing agricultural production must be interlinked with demand management such as reduced food waste and dietary changes (Scherer and Verburg 2017).

Some examples of success stories on CSA implementation are available from Tanzania (agroforestry), Peru (genetic diversity), and China (sustainable grazing) among other national initiatives (FAO 2013b). As part of the Mitigation of Climate Change in Agriculture (MICCA) Programme, smallholder CSA pilot projects were implemented in an integrated crop–livestock–tree farming system in Kenya and in a cereal-based upland farming system in the United Republic of Tanzania (Rioux et al. 2016). CA practices were promoted to improve agricultural productivity and to increase the C content in the soil of the agroforestry system. Improved manure management and soil nutrients have been shown to improve SOM in Kenya benefitting the CSA objective food productivity. Further, improved livestock management increased soil C sequestration and benefited the CSA objective mitigation. Similarly, increased SOC and improved soil fertility by CA practices in Tanzania can result in increased productivity contributing to the CSA objectives food productivity and mitigation. Connecting research, practice, and policy has been shown to be critical for the effective scaling up of CSA. Further, National Adaptation Plans (NAPs) and NAMAs can be used to deliver a broad range of co-benefits to farmers that extend beyond climate change adaptation and mitigation, and to support sustainable agricultural development and the implementation of INDCs introduced by the Paris Agreement (Rioux et al. 2016).

9.3 Carbon Farming

Aside CSA, there are other approaches and initiatives to mitigate climate change by management of agroecosystems. Among them is 'carbon farming' which refers to farming practices that sequester C. However, there is no single definition. Lal defined carbon farming as a system of increasing C in terrestrial ecosystems for adaptation and mitigation of climate change to enhance ecosystem goods and services, and trade carbon credits for economic gains (Toensmeier 2016). Carbon farming practices and species have been identified for annual cropping systems, livestock systems, perennial cropping systems, and agroforestry (Toensmeier 2016).

Large-scale carbon farming has been advocated by some to be the best option for stabilizing or reducing atmospheric CO_2 levels (Becker and Lawrence 2014). Carbon farming is one way for farmers to reduce GHG emissions and has been proposed as an important component in Australia's efforts to mitigate GHG emissions (Dumbrell et al. 2016). The Carbon Farming Initiative (CFI) was introduced in 2011 (Parliament of the Commonwealth of Australia 2011). The CFI allows the creation of tradable Australian Carbon Credit Units (ACCUs) derived from across the ecosystem sector via project-level baseline and credit activities (van Oosterzee 2012). It is the first national offset scheme in the world to broadly include farming and forestry projects. CFI has the capacity to significantly reduce the cost of meeting Australia's mitigation targets and promote more sustainable land-management practices (Macintosh and Waugh 2012).

In 2014, the CFI was merged into the new policy framework: the Emissions Reduction Fund (ERF; Parliament of the Commonwealth of Australia 2014). The ERF operates as a reverse auction scheme; i.e., farmers are invited to submit project bids that specify the carbon farming practices they are willing to undertake, and the required price per Mg of emissions reductions or sequestration to undertake the practice(s). The government purchases the lowest cost projects. The carbon farming practices include: protecting native vegetation from being cleared; sequestering soil C in grazing systems; reducing emissions from livestock through changes in feed; and reducing emissions through improved fire management in northern savanna regions (Dumbrell et al. 2016). Since 2017, farmers in Victoria and South Australia are taking part in a world-first carbon-capture scheme to generate a new source of income (http://www.abc.net.au/news/rural/2017-08-11/carbon-farmers-criticise-government/8796256). For soil carbon, farmers are paid roughly AU$10 for each carbon credit based on how much carbon they can sequester in their soil over a 10-year period.

Among obstacles to the success of CFI are carbon price uncertainty, transaction costs, path dependencies, and integrity- and perverse impact-related restrictions (Macintosh 2013). To address these barriers, several opportunities for technical improvements to be made to the CFI have been proposed. In particular, the 100-year rule (i.e., all sequestration projects are required to maintain the relevant C stores for 100 years) that is designed to address the risk of non-permanence could be replaced by discount-based approach for Kyoto offset projects. Proponents would then be able to select different permanence periods and a permanence deduction would then be calculated on the basis of the project characteristics and length of the period. Other suggestions for improvements are modifying the risk of reversal buffer and leakage deduction processes to improve returns to project proponents (Macintosh 2013). However, biodiversity considerations are not integrated into the CFI (van Oosterzee 2012).

Participation in emissions abatement policies is voluntary, and, thus, it is unclear whether farmers are willing to participate in carbon farming (Dumbrell et al. 2016). Among the factors that are important in farmers' decisions to change agricultural management practices are the (monetary and non-monetary) investment costs of the new practice, the impacts of the new practice on farm profitability, whether the practice 'fits' in the current farming system, the farmer's financial situation and personal values, the social context in which the farmer operates, as well as the public co-benefits generated by adopting the practice (Dumbrell et al. 2016). Tang et al. (2016) reviewed economic analyses published between 1995 and 2014 on the economic impacts of policies that incentivize agricultural GHG mitigation. The cost estimates varied between \$3 and \$130 per Mg CO_{2eq} in 2012 US dollars, depending on the mitigation strategies, spatial locations, and policy scenarios considered. However, most studies assessed only the consequences of a single, rather than multiple, mitigation strategies, and few considered the co-benefits of carbon farming. Tang et al. (2016) recommended to improve biophysical knowledge about carbon farming co-benefits, predict the economic impacts of employing multiple strategies and policy incentives, and develop the associated integrated models, to estimate the full costs and benefits of agricultural GHG mitigation to farmers and the rest of society.

Dryland cropping and mixed crop-livestock farmers in Western Australia ranked improved soil quality and reduced soil erosion as the most important potential co-benefits of carbon farming (Dumbrell et al. 2016). Factors discouraging farmers from participating in carbon farming contracts were policy and carbon price uncertainty, and the uncertain impact of carbon farming practices on productivity and profitability. The farmers in Western Australia had strong preferences for stubble retention and no-till cropping practices as carbon farming strategies. The practices that farmers preferred least were applying biochar and planting trees. Dumbrell et al. (2016) concluded that it is important to communicate potential co-benefits (rather than opportunities to earn compensation or carbon credits) to increase farmers' engagement in carbon sequestration activities and carbon farming.

9.4 Soil Security

Operationalizing soil security will affect the SOC stock of agroecosystems and climate change policy. Soil security refers to the maintenance and improvement of the world's soil resources so that they can continue to provide food, fiber, and freshwater, make major contributions to energy and climate sustainability, and help maintain biodiversity and the overall protection of ecosystem goods and services (Koch et al. 2012). The concept of soil security is multi-dimensional and acknowledges the five dimensions of capability, condition, capital, connectivity, and codification, of soil entities which encompass the social, economic, and biophysical sciences and recognize policy and legal frameworks (McBratney et al. 2014).

The capability of a soil refers to its potential functionality (Morgan et al. 2017). Soils differ in their ability to provide soil functions such as the seven distinguished by the Soil Protection Strategy of the European Union (i.e., biomass production, filtering nutrients, source of biodiversity, cultural environment, raw materials, carbon pool, heritage). The dimension of soil condition refers to the current state of a soil, reflects human management of soil, and how state and management alter or enhance soil functions. The dimension of soil capital refers to the economic and natural capital value of the soil resource. Placing a monetary value on the asset soil resource enables a society to value or secure the soil resource. Connectivity refers to the connection of individual land managers/farmers with the soil they manage, and the broader connection of soil to society and with society to soil. Connectivity also encompasses issues of knowledge, education, training, and awareness. Codification refers to the policies, regulations and governance arrangements, in both the public and private sectors that enable soil security (Morgan et al. 2017).

Soil security can only be achieved when the soil resource is maintained and/or improved, requiring a reversal of current degradation processes (Bouma 2015). Based on the five dimensions of soil security, Morgan et al. (2017) summarized goals to secure soil in the next two decades so that it can contribute to solving other global issues. Goals included: (i) Fifty percent of soil is used according to its capability by 2030; (ii) Soil condition is optimally managed according to the inherent capability in 50% of managed soil systems by 2030; (iii) Increase annual capital value of soil ecosystem services by 5% per annum by 2030 and commercial land values based on full economic value of soil capability and condition, by 2020; (iv) Ninety percent awareness and understanding of soil security among the general public by 2030; and (v) Fifty percent of national governments recognise soil security in their laws and regulations by 2025 (Morgan et al. 2017). Similarly, the SDGs need to be achieved by 2030 (UNGA 2015), but making soil security a SDG in its own right has not been successful. Nevertheless, soil security is homologous to the constructs of food, water, energy, and climate security (McBratney and Field 2015). Thus, soil security provides a conceptual framework that puts soil at the center of addressing some of the major problems that face humanity (Koch et al. 2013).

SOC (soil carbon) has been identified as one of the significant universal indicators to monitor soil condition globally and, thus, as an exemplar indicator of soil security (Koch et al. 2013). Carbon is also easily understood by policymakers and the wider community. Thus, soil carbon may be adopted as an indicator for soil change (McBratney and Field 2015). Agricultural management practices that are known to increase SOC are key mechanisms for improving and maintaining the functionality of soil and its ability to support ecosystem service delivery. However, many nations suffer from a policy disconnect between soil and agriculture at the national and state government levels (Koch 2017). Governments are in a unique position to address this disconnect by investing in research and innovation in soil management. Based on digital soil mapping, telemetry, and sensing technology, farmers will be able to optimize the performance of their soils. This will lead to a coherent soil security strategy that places agriculture to achieve the win-win of increased productivity of food, fiber and fuel, and ecosystem service delivery from the management of the precious soil resource (Koch 2017).

9.5 Payments for Ecosystem Services

Ecosystem services are the ecological characteristics, functions, or processes that directly or indirectly contribute to human well-being, i.e., the benefits that people derive from functioning ecosystems (Costanza et al. 1997). Among the regulating ES is climate regulation and, more specifically, C sequestration with changed agricultural practices contributing to the growth of the role of agroecosystems for this ES (MA 2005). However, focusing only on the C sequestration service of ecosystems may sometimes reduce the overall value of the full range of ESs (Costanza et al. 2017).

Assessing the monetary value of ESs is a major challenge (Costanza et al. 1997). For example, the societal value of soil carbon refers to the monetary equivalent of ESs provisioned by a unit amount of SOC (Lal 2014). Some ESs of SOC include increasing NPP and agronomic yield, improving plant available water capacity in the root zone, reducing water runoff and soil erosion, minimizing sedimentation and non-point source pollution, offsetting anthropogenic emissions and mitigating/ adapting to climate change, denaturing pollutants and purifying water, and enhancing biodiversity. However, it is difficult to assess the real societal value of SOC. While soil may be the most 'priceless' gift of nature, it is not possible to assess soils' true worth in monetary terms (Lal 2014). Otherwise, the inherent value of SOC can be estimated as the so-called 'hidden cost' of all inputs, including crop residues/hay, fertilizers, and labor. The inherent cost of SOC must, however, not be confused with its societal value. Nevertheless, inherent or the societal value of SOC can be computed on the basis of the hidden costs of the inputs involved (Lal 2014).

Farmers often do not receive a price signal for enhancing C sequestration (Ribaudo et al. 2010). One possible way to increase private investment in C sequestration is to create a market for them. The two key market instruments are: (i) the markets for ES (based on the polluter pays principle) that address negative environmental externalities, and (ii) the payments for ecosystem services (PES) or 'steward earns principle,' based on positive environmental externalities (Grima et al. 2016).

Among the important terms regarding carbon pricing is cap and trade where a government sets a ceiling, or cap, on the total amount of carbon emissions it will allow (Green 2017). Companies trade carbon allowances with each other to meet their targets. Carbon allowances are sold by the government or distributed for free. Allowances give firms a right to emit a share of the total carbon emissions, as set by the cap. Carbon offset credits may also be bought by companies. These count as carbon emissions' reductions by enabling emitters to pay for decarbonizing activities. Carbon trading is buying and selling allowances or offsets. Companies that exceed their carbon emissions quota can purchase allowances from others. Companies can sell or bank spare allowances for future use. Carbon markets are where allowances and credits are traded. Carbon taxes are paid by companies to the government for each Mg of CO_2 emitted. They are simpler to implement than markets and favored by many economists, but they are politically unappealing (Green 2017).

Markets for ES may include markets for water quality trading, wetland mitigation, carbon cap-and-trade, over-the-counter carbon, eco-labeling, and fee hunting (Ribaudo et al. 2010). Cap-and-trade programs are examples for emission trading, i.e., an artificial market organized around the creation of a private good (e.g., agroecosystem SOC stock) related to the provision of an ES (e.g., C sequestration). The regulatory agency creates a good closely linked with the ES, i.e., a discharge allowance or an offset credit. This good has private good characteristics: it is rival and non-excludable (Ribaudo et al. 2010). The government then stimulates demand for the good (e.g., agroecosystem SOC stock) by requiring regulated firms to have enough allowances to meet a regulatory requirement (e.g.,, the discharge of CO_2) and enforces property rights (Tietenberg 2006). Regulated firms are allowed to trade discharge allowances among themselves, thus reducing the overall cost of achieving pollution (i.e., CO_2 emission) control goals (Ribaudo et al. 2010).

The Chicago Climate Exchange (CCX) was a legally binding, voluntary cap, and trade program for reducing net GHG emissions but ceased trading carbon credits at the end of 2010 due to inactivity in the US carbon markets (Reuters Staff 2010). Members voluntarily joined and agreed to reduce their GHG emissions. Members could meet their obligations by purchasing offsets from qualifying emissions reductions projects, including C sequestration in agricultural soils, CH_4 destruction, and conversion of cropland to permanent grass or trees (Ribaudo et al. 2010). Soil sequestration projects contributed about half of the 35 Gg CO_2 eq. traded in 2003–2010 (Hamilton et al. 2008).

Over-the-counter markets are purely voluntary (Ribaudo et al. 2010). Markets are based on bilateral deals between producers and consumers, and operate outside of an exchange. The retail carbon market is the most visible over-the-counter market where ES from agriculture can be sold. Consumers can offset their carbon footprint from a variety of activities. The market is highly fragmented and operates without commonly accepted standards, either for how GHG 'footprints' are calculated or for selecting eligible offsets (Trexler and Kosloff 2006). In 2007, about 7% of the offsets were produced on farms (soil C sequestration and CH_4 destruction) (Hamilton et al. 2008).

The market instrument PES has emerged as a concept to compensate and encourage landowners to improve land management practices for the maintenance and provision of ES (Grima et al. 2016). The central idea is that the stewards of the ES should be compensated by those who benefit. During the 1990s, the widespread expansion of PES as an integrated conservation and development mechanism began to emerge (Wunder et al. 2008). Hundreds of PES schemes (mostly local level arrangements) are now being implemented around the world covering four main ES: biodiversity conservation, watershed services, C sequestration, and landscape beauty (Arriagada and Perrings 2013).

Some authors have made efforts to evaluate the performance of PES on the ground. For example, Sattler et al. (2013) introduced a multi-classification system for PES based on different PES characteristics, such as PES type, ES addressed, payment modus, involved actors, their roles in the schemes, among others. This approach was applied to a sample of 22 PES cases from the US and Germany that were assessed as successful by expert judgment. Important for PES success were: (i) intermediary involvement, (ii) involvement of governmental actors, (iii) long-term contracting, (iv) co-benefits, (v) voluntariness in entering the PES agreement, and (vi) output-based PES. However, additional characteristics might be necessary to sufficiently describe PES for developing countries (Sattler et al. 2013). For example, Grima et al. (2016) building on 40 PES cases in Latin America summarized what contributes to a successful PES. In particular, PES schemes that secured the continued provisioning and quality of a critical resource while positively contributing to local livelihoods were successful. Local and regional scales were most widely used, both with high degrees of success. PES projects operating within a period between 10 and 30 years were regarded as most successful. The use of in-kind contributions reduced the probability of PES failure. Those transactions were preferable rather than using only cash payments. Finally, there was a dominance of successful PES schemes where mostly private actors were involved. Also, PES schemes with no intermediaries between the buyers and the sellers tended to be more successful (Grima et al. 2016). Common features of PES with low degrees of success in Latin America were: (i) that the implementations of the PES schemes did not manage to reduce the pressures on the ecosystem, (ii) that the investors did not see the added value of their investments (i.e., additionality was not achieved), (iii) that the opportunity costs were not met, (iv) that the local livelihoods did not improve, (v) existing land rights arrangements and power structures were threatened, and (vi) where unfair practices in the distribution of benefits were observed (Grima et al. 2016).

PES is market-based instruments which also include taxes and subsidies, mitigation or species banking, and certification (Pirard 2012). These could either be used to increase SOC sequestration in agricultural soils or to penalize farmers and land managers whose land-use strategies do not actively stop SOC losses (Djanibekov and Villamor 2017). The latter could be realized through the implementation of a punitive carbon tax as it would oblige CO_2 emitters to pay for the alleged damage to the climate (Pirard 2012). However, it might not be the best possible tool to enhance farmers' SOC sequestration efforts. Agri-environmental measures within the Common Agricultural Policy (CAP) are subsidies as the EU provides payments for farmers based on their ES (Pirard 2012). Comparable mechanisms have also been developed in the USA (Claassen et al. 2008). Subsidies might be promising in terms of voluntary adoption of SOC sequestration practices as farmers are rewarded for additional effort without being obliged to participate (Pirard and Lapeyre 2014).

While agricultural GHGs are not considered in the Emission Trading Scheme (ETS) in Europe, Australia's Carbon Farming Initiative allows farmers to earn carbon credits by adopting techniques that reduce SOC losses and promote SOC sequestration (Hermann et al. 2017). These credits can then be sold to parties wishing to balance their emissions (Australian Government 2012). Pirard (2012) indicated that the markets of tradeable permits are very dependent on the current political will, and, thus, often only provide voluntary agreements which are not credible. Only a few studies focus on market-based instruments from the farmer's perspective (Hermann et al. 2017). Thus, it is unknown which policy measures are likely to effectively encourage increases in SOC levels on farms.

Based on a study of German farmers, Hermann et al. (2017) suggested that farmers tended to reduce their efforts to build SOC over time with no political intervention to promote SOC maintenance of enhancement. In contrast, increased SOC-promoting efforts were observed when policy measures with payments to promote SOC were applied. However, when policy measures were designated as 'certificates,' heterogeneity of SOC promotion increased compared to a designation of such payments as 'subsidies.' Labeling a policy measure as a 'certificate' and including an uncertain and unpredictable payment structure increased farmers' efforts to apply SOC techniques. Most farmers could be motivated to promote SOC in the future if subsidies or certificates with a rather uncertain but market- and incentive-based payment structure would be implemented. However, it was unclear whether decision-making behavior is country-specific in the EU and influenced by other climate policy goals. Hermann et al. (2017) suggested that future research should investigate which SOC promoting measures farmers would prefer to implement, and whether the willingness to participate in a SOC-increasing program is influenced by the farmer's attitude toward climate protection.

Currently, only 2% of climate finance flows to natural solutions (Buchner et al. 2017), including soil C sequestration in agroecosystems. Among the issues is that climate change has been described as a well-known energy problem combined with a lesser-known land problem (Turner 2018). However, improved land management can reduce emissions and remove atmospheric CO_2 while safeguarding food security and biodiversity (Griscom et al. 2017), and land managers must be rewarded financially for implementing improved practices. The estimated cost of increasing ecosystem (biomass and soil) carbon varies vastly (Larjavaara et al. 2018). Thus, effectiveness of an ecosystem carbon subsidy would depend a lot on the policy's ability to target globally the sites with the largest potential to make a difference. Rangelands, wastelands, and other land uses spared from intensive crop production may be prime land uses to enhance C sequestration by ecosystem carbon subsidies (Larjavaara et al. 2018).

9.6 Conclusions

Recently, the soil C stock of agroecosystems and its contribution to climate change adaptation and mitigation, and its importance for food security has received more attention in the international policy arena. Practices to enhance soil C stocks in agroecosystems are well known and their implementation will be supported in the future by UNCCD's LDN, land- and soil-related SDGs, INDCs submitted to UNFCCC, and 4p1000. Concepts such as CSA and carbon farming are promising for increasing soil C stocks and associated ESs in agroecosystems. However, it is critically important that farmers and land managers are rewarded for soil C enhancing practices as these are associated with additional costs. Incentives may include payments for ESs such as taxes, subsidies, and certifications. However, which policy measures effectively encourage increases in SOC stocks on farms, especially, by incentivizing the large number of smallholders needs additional research. Implementation and scaling up of proven technologies can be enhanced globally through SSC, TAC, and SSNC.

9.7 Review Questions

1. Describe well-proven practices to enhance and increase SOC stocks in agroecosystems.
2. How do the Rio Conventions consider SOC sequestration?
3. Why are there no SDGs directly related to soil and SOC?
4. Discuss the potential of 4p1000 to contribute to climate change adaptation and mitigation, and enhancing food security.
5. Contrast and compare CSA versus climate-resilient agriculture.
6. What is carbon farming?
7. How does the soil security framework consider soils of agroecosystems and their SOC stock?
8. How can farmers be rewarded for SOC enhancing farming practices?
9. What are major obstacles that land managers including smallholders are not implementing SOC-smart practices more widely?
10. How can the recognition of the value of agroecosystem SOC for practitioners, stakeholders, policymakers and the general public be improved?

References

Akhtar-Schuster M, Stringer LC, Erlewein A et al (2017) Unpacking the concept of land degradation neutrality and addressing its operation through the Rio Conventions. J Environ Manage 195:4–15

Arriagada R, Perrings C (2013) Making payments for ecosystem services work. In: Kumar P, Thiaw I, Barker T (eds) Values, payments and institutions for ecosystem management: a developing country perspective. Edward Elgar Publishing Cheltenham, UK

Aubert PM, Treyer S, Sablé AL, et al (2017) Implementing the "4 per 1000" initiative: contribution for the establishment of a reference/normative framework. IDDRI Policy Brief N° 02/17 JANUARY 2017

Augusto L, Achat DL, Jonard M, Vidal D, Ringeval B (2017) Soil parent material—a major driver of plant nutrient limitations in terrestrial ecosystems. Glob Change Biol 23:3808–3824. https://doi.org/10.1111/gcb.13691

Australian Government (2012) Carbon Credits (Carbon farming initiative) Act 2011. Australian Government, Canberra

Baveye PC, Berthelin J, Tessier D, Lemaire G (2018) The "4 per 1000" initiative: A credibility issue for the soil science community? Geoderma 309:118–123

Becker K, Lawrence P (2014) Carbon farming: the best and safest way forward? Carbon Manage 5:31–33

Bispo A, Andersen L, Angers DA et al (2017) Accounting for carbon stocks in soils and measuring GHGs emission fluxes from soils: do we have the necessary standards? Front Environ Sci 5:41. https://doi.org/10.3389/fenvs.2017.00041

Boer BW, Ginzky H, Heuser IL (2017). International soil protection law: history, concepts and latest developments. In: Ginzky H, et al (eds) International Yearbook of Soil Law and Policy 2016. Springer International Publishing AG, pp 49–72

Bouma J (2015) Reaching out from the soil-box in pursuit of soil security. Soil Sci Plant Nutr 61:556–565

Boysen LR, Lucht W, Gerten D et al (2017) The limits to global-warming mitigation by terrestrial carbon removal. Earth's Future 5:463–474. https://doi.org/10.1002/2016EF000469

Buchner BK, Oliver P, Wang X, et al (2017) Global landscape of climate finance 2017. Climate policy initiative. https://climatepolicyinitiative.org/wp-content/uploads/2017/10/2017-Global-Landscape-of-Climate-Finance.pdf

Buchner BK, Trabacchi C, Mazza F, Abramskiehn D, Wang D (2015) Global landscape of climate finance 2015. https://climatepolicyinitiative.org/publication/global-landscape-of-climate-finance-2015/

Chabbi A, Lehmann J, Ciais P et al (2017) Aligning agriculture and climate policy. Nat Clim Change 7:307–309

Chambers A, Lal R, Paustian K (2016) Soil carbon sequestration potential of US croplands and grasslands: implementing the 4 per Thousand Initiative. J Soil Water Conserv 71:68A–74A. https://doi.org/10.2489/jswc.71.3.68A

Chatrchyan AM, Erlebacher RC, Chaopricha NT et al (2017) United States agricultural stakeholder views and decisions on climate change. WIREs Clim Change 8:e469. https://doi.org/10.1002/wcc.469

Claassen R, Cattaneo A, Johansson R (2008) Cost-effective design of agri-environmental payment programs: U.S. experience in theory and practice. Ecol Econ 65:737–752

Costanza R, Arge R, de Groot R et al (1997) The value of the world's ecosystem services and natural capital. Nature 387:253–260

Costanza R, de Groot R, Braat L et al (2017) Twenty years of ecosystem services: How far have we come and how far do we still need to go? Ecosyst Serv 28:1–16

Devaraju N, Bala G, Nemani R (2015) Modeling the influence of land-use changes on biophysical and biochemical interactions at regional and global scales. Plant Cell Environ. https://doi.org/10.1111/pce.12488

De Vries W (2018) Soil carbon 4 per mille: a good initiative but let's manage not only the soil but also the expectations. Geoderma 309:111–112

Dickie A, Streck C, Roe S, et al (2014) Strategies for mitigating climate change in agriculture: abridged report. Climate Focus and California Environmental Associates, prepared with the support of the Climate and Land Use Allinace. www.agriculturalmitigation.org

Dignac MF, Derrien D, Barré P et al (2017) Increasing soil carbon storage: mechanisms, effects of agricultural practices and proxies. A review. Agron Sustain Dev 37:14

Djanibekov U, Villamor GB (2017) Market-based instruments for risk-averse farmers: rubber agroforest conservation in Jambi Province, Indonesia. Environ Dev Econ 22:133–155

Dong X, Hao Q, Li G, Lin Q, Zhao X (2017) Contrast effect of long-term fertilization on SOC and SIC stocks and distribution in different soil particle-size fractions. J Soils Sediments 17:1054–1063. https://doi.org/10.1007/s11368-016-1615-y

Dumbrell NP, Kragt ME, Gibson FL (2016) What carbon farming activities are farmers likely to adopt? Abest–worst scaling survey. Land Use Pol 54:29–37

FAO (2013a) Agriculture, forestry and other land use mitigation project database. Second assessment of the current status of land-based sectors in the carbon market. http://www.fao.org/docrep/017/i3176e/i3176e.pdf

FAO (2013b) Climate-smart agriculture sourcebook. FAO, Rome

FAO (2016) The agriculture sectors in the intended nationally determined contributions: analysis. In: Strohmaier R, Rioux J, Seggel A, Meybeck A, Bernoux M, Salvatore M, Miranda J, Agostini A (eds) Environment and natural resources management Working Paper No. 62. Rome

Forsell N, Turkovska O, Gusti M et al (2016) Assessing the INDCs' land use, land use change, and forest emission projections. Carbon Balance Manage 11:26. https://doi.org/10.1186/s13021-016-0068-3

Gao L, Bryan BA (2017) Finding pathways to national-scale land-sector sustainability. Nature 544:217–235

Graichen J, Healy S, Siemons A, et al (2017) International climate initiatives—a way forward to close the emissions gap? Umweltbundesamt [German Environment Agency] CLIMATE CHANGE 22/2017, https://www.umweltbundesamt.de/en/publikationen/international-climate-initiatives-a-way-forward-to-0

Green JF (2017) Don't link carbon markets. Nature 543:484–486

Grima N, Singh SJ, Smetschka B, Ringhofer L (2016) Payment for ecosystem services (PES) in Latin America: analysing the performance of 40 case studies. Ecosyst Serv 17:24–32

Griscom BW, Adams J, Ellis PW et al (2017) Natural climate solutions. Proc Natl Acad Sci USA 114:11645–11650

Hamilton K, Sjardin M, Marcello T, Xu G (2008) Forging a frontier: state of the voluntary carbon markets 2008. Ecosystem Marketplace & New Carbon Finance, New York and Washington, DC

Heitkötter J, Heinze S, Marschner B (2017) Relevance of substrate quality and nutrients for microbial C-turnover in top and subsoil of a Dystric Cambisol. Geoderma 302:89–99

Hermann D, Sauthoff S, Mußhoff O (2017) Ex-ante evaluation of policy measures to enhance carbon sequestration in agricultural soils. Ecol Econ 140:241–250

Hobley E, Baldock J, Hua Q, Wilson B (2017) Land-use contrasts reveal instability of subsoil organic carbon. Glob Change Biol 23:955–965

Johnston AE, Poulton PR, Coleman K, Macdonald AJ, White RP (2017) Changes in soil organic matter over 70 years in continuous arable and ley–arable rotations on a sandy loam soil in England. Eur J Soil Sci 68:305–316

Koch A (2017) Soil security for agricultural productivity: the policy disconnect and a promising future. In: Field DJ, Morgan CLS, McBratney AB (eds) Global Soil Security, Progress in Soil Science. Springer International Publishing Switzerland, pp 425–435. https://doi.org/10.1007/978-3-319-43394-3_39

Koch A, McBratney A, Adams M et al (2013) Soil security: solving the global soil crisis. Glob Pol 4:434–441

Koch A, McBratney A, Lal R (2012) Global soil week: put soil security on the global agenda. Nature 492:186

Kust G, Andreeva O, Cowie AL (2017) Land degradation neutrality: concept development, practical applications and assessment. J Environ Manage 195:16–24

Kyoto Protocol (1997) Kyoto protocol to the United Nations framework convention on climate change adopted at COP3 in Kyoto, Japan, 11 Dec 1997

Lal R (2014) Societal value of soil carbon. J Soil Water Conserv 69:186A–192A. https://doi.org/10.2489/jswc.69.6.186A

Lal R (2016) Beyond COP21: potential and challenges of the "4 per Thousand" initiative. J Soil Water Conserv 71:20A–25A. https://doi.org/10.2489/jswc.71.1.20A

Lal R (2018) Promoting "4 per thousand" and "adapting African agriculture" by south-south cooperation: conservation agriculture and sustainable intensification. Soil Till Res. https://doi.org/10.1016/j.still.2017.12.015

Lal R, Safriel U, Boer B (2012) Zero net land degradation. UNCCD Position Paper for Rioþ20. Bonn, Germany. http://www.unccd.int/Lists/SiteDocumentLibrary/Rioþ20/UNCCD_PolicyBrief_ZeroNetLandDegradation.pdf

Larjavaara M, Kanninen M, Gordillo H et al (2018) Global variation in the cost of increasing ecosystem carbon. Nat Clim Change 8:38–42

Lipper L, Thornton P, Campbell BM et al (2014) Climate-smart agriculture for food security. Nat Clim Change 4:1068–1072

Lorenz K, Lal R (2005) The depth distribution of soil organic carbon in relation to land use and management and the potential of carbon sequestration in subsoil horizons. Adv Agron 88:35–66

Lorenz K, Lal R (2016) Soil organic carbon—an appropriate indicator to monitor trends of land and soil degradation within the SDG Framework? Umweltbundesamt [German Environment Agency] Text 77/2016. https://www.umweltbundesamt.de/publikationen/soil-organic-carbon-an-appropriate-indicator-to

Macintosh A (2013) The carbon farming initiative: removing the obstacles to its success. Carbon Manage 4:185–202

Macintosh A, Waugh L (2012) An introduction to the carbon farming initiative: key concepts and principles. Environ Plan Law J 29:439–461

McBratney A, Field DJ (2015) Securing our soil. Soil Sci Plant Nutr 61:587–591

McBratney A, Field DJ, Koch A (2014) The dimensions of soil security. Geoderma 213:203–213

Minasny B, Arrouays D, McBratney AB et al (2018) Rejoinder to comments on Minasny et al., 2017 Soil carbon 4 per mille Geoderma 292, 59–86. Geoderma 309:124–129

Minasny B, Malone BP, McBratney AB et al (2017) Soil carbon 4 per mille. Geoderma 292:59–86

Minelli S, Erlewein A, Castillo V (2017) Land degradation neutrality and the UNCCD: From political vision to measurable targets. In: Ginzky H, et al. (eds) International yearbook of soil law and policy 2016. Springer International Publishing AG, pp 85–104

Monger HC, Kraimer RA, Khresat S, Cole DR, Wang X, Wang J (2015) Sequestration of inorganic carbon in soil and groundwater. Geology 43:375–378

Morgan CLS, McBratney AB, Field DJ, et al (2017) Synthesis: goals to achieve soil security. In: Field DJ, Morgan CLS, McBratney AB (eds) Global soil security, progress in soil science. Springer International Publishing Switzerland, pp 457–462. https://doi.org/10.1007/978-3-319-43394-3_43

Ogle SM, Olander L, Wollenberg L et al (2014) Reducing greenhouse gas emissions and adapting agricultural management for climate change in developing countries: providing the basis for action. Glob Change Biol 20:1–6. https://doi.org/10.1111/gcb.12361

Orr BJ, Cowie AL, Castillo Sanchez VM et al (2017) Scientific conceptual framework for land degradation neutrality. A report of the science-policy interface. United Nations Convention to Combat Desertification (UNCCD), Bonn, Germany

Paris Agreement (2015) Adoption of the Paris Agreement. FCCC/CP/2015/L.9/Rev.1

Parliament of the Commonwealth of Australia (2011) Carbon credits (Carbon Farming Initiative) bill. Explanatory Memorandum, Parliament of the Commonwealth of Australia, Canberra, ACT

Parliament of the Commonwealth of Australia (2014) Carbon farming initiative amendment bill. Explanatory Memorandum. Parliament of the Commonwealth of Australia, Canberra, ACT

Paustian K, Lehmann J, Ogle S et al (2016) Climate-smart soils. Nature 532:49–57

Pirard R (2012) Market-based instruments for biodiversity and ecosystem services: a lexicon. Environ Sci Pol 19–20:59–68

Pirard R, Lapeyre R (2014) Classifying market-based instruments for ecosystem service: a guide to the literature jungle. Ecosyst Serv 9:106–114

Porter JR, Howden M, Smith P (2017) Considering agriculture in IPCC assessments. Nat Clim Change 7:680–683

Porter JR, Xie L, Challinor AJ, et al (2014) Food security and food production systems. In: IPCC. 2014. Climate change 2014: impacts, adaptation, and vulnerability. Part A: global and sectoral aspects. Contribution of working group II to the fifth assessment report of the intergovernmental panel on climate change. Cambridge University Press, Cambridge, UK and New York, USA, pp 485–533

Rao CS, Gopinath KA, Prasad JVNS, Prasannakumar Singh AK (2016) Climate resilient villages for sustainable food security in tropical India: concept, process, technologies, institutions, and impacts. Adv Agron 140:101–214

Reuters Staff (2010) ICE cuts staff at Chicago climate exchange-sources. https://www.reuters.com/article/carbon-ccx-layoffs/ice-cuts-staff-at-chicago-climate-exchange-sources-idUSLDE6791WI20100812

Ribaudo M, Greene C, Hansen L, Hellerstein D (2010) Ecosystem services from agriculture: steps for expanding markets. Ecol Econ 69:2085–2092

Rioux J, Juan MGS, Neely C, et al (2016) Planning, implementing and evaluating climate-smart agriculture in smallholder farming systems. The experience of the MICCA pilot projects in Kenya and the United Republic of Tanzania. Mitigation of Climate Change in Agriculture Series 11, FAO, Rome

Roesch-McNally GE, Arbuckle JG, Tyndall JC (2018) Barriers to implementing climate resilient agricultural strategies: The case of crop diversification in the U.S. Corn Belt. Glob Environ Change 48:206–215

Rogelj J, Schleussner CF, Hare W (2017) Getting it right matters: Temperature goal interpretations in geoscience research. Geophys Res Lett 44:10,662–10,665. https://doi.org/10.1002/2017GL075612

Rosenkranz M, Pugh TAM, Schnitzler JP, Arneth A (2015) Effect of land-use change and management on biogenic volatile organic compound emissions—selecting climate-smart cultivars. Plant Cell Environ. https://doi.org/10.1111/pce.12453

Rosenstock TS, Lamanna C, Chesterman S, et al. (2016) The scientific basis of climate-smart agriculture: a systematic review protocol. CCAFS Working Paper no. 138. Copenhagen, Denmark: CGIAR Research Program on Climate Change, Agriculture and Food Security (CCAFS). https://www.ccafs.cgiar.org

Rumpel C, Lehmann J, Chabbi A (2018) Boost soil carbon for food and climate. Nature 553:27

Safriel U (2017) Land Degradation Neutrality (LDN) in drylands and beyond—where has it come from and where does it go. Silva Fennica vol 51 no. 1B article id 1650. 19pp. https://doi.org/10.14214/sf.1650

Saj S, Torquebiau E, Hainzelin E, Pages J, Maraux F (2017) The way forward: An agroecological perspective for climate-smart agriculture. Agric Ecosyst Environ 250:20–24

Sanderman J (2012) Can management induced changes in the carbonate system drive soil carbon sequestration? A review with particular focus on Australia. Agric Ecosyst Environ 155:70–77

Sanderman J, Hengl T, Fiske GJ (2017) Soil carbon debt of 12,000 years of human land use. Proc Natl Acad Sci USA 114:9575–9580

Sattler C, Trampnau S, Schomers S, Meyer C, Matzdorf B (2013) Multi-classification of payments for ecosystem services: how do classification characteristics relate to overall PES success? Ecosyst Serv 6:31–45

Scherer L, Verburg PH (2017) Mapping and linking supply- and demand-side measures in climate-smart agriculture. A review. Agron Sustain Dev 37:66. https://doi.org/10.1007/s13593-017-0475-1

Shi HJ, Wang XJ, Zhao YJ et al (2017) Relationship between soil inorganic carbon and organic carbon in the wheat-maize cropland of the North China Plain. Plant Soil 418:423–436. https://doi.org/10.1007/s11104-017-3310-1

Smith P (2016) Soil carbon sequestration and biochar as negative emission technologies. Glob Change Biol 22:1315–1324

Smith P, Bustamante M, Ahammad H, et al (2014) Agriculture, forestry and other land use (AFOLU). In: Edenhofer O, Pichs-Madruga R, Sokona Y et al (eds) Climate change 2014: mitigation of climate change. Contribution of working group III to the fifth assessment report of the intergovernmental panel on climate change. Cambridge University Press, Cambridge, pp 811–922

Smith P, Lutfalla S, Riley WJ et al (2018) The changing faces of soil organic matter research. Eur J Soil Sci 69:23–30. https://doi.org/10.1111/ejss.12500

Soussana JF, Lutfalla S, Smith P, et al (2017) Letter to the editor: Answer to the viewpoint "Sequestering soil organic carbon: A nitrogen dilemma". Environ Sci Technol 51:11502–11502. https://doi.org/10.1021/acs.est.7b03932

Streck C, Gay A (2017) The role of soils in international climate change policy. In: Ginzky H et al (eds) International yearbook of soil law and policy 2016. Springer International Publishing AG, pp 105–128

Tang K, Kragt ME, Hailu A, Ma C (2016) Carbon farming economics: what have we learned? J Environ Manage 172:49–57

Tietenberg TH (2006) Emissions trading: principles and practice. Resources for the Future, Washington, DC

Toensmeier E (2016) The carbon farming solution: a global toolkit of perennial crops and regenerative agriculture practices for climate change mitigation and food security. Chelsea Green Publishing, White River Junction, Vermont, USA

Trexler MC, Kosloff LH (2006) Selling carbon neutrality. The Environmental Forum, March/April, pp 34–39

Trumbore SE, Davidson EA, de Camargo PB et al (1995) Belowground cycling of carbon in forests and pastures of Eastern Amazonia. Glob Biogeochem Cycles 9:515–528

Turner WR (2018) Looking to nature for solutions. Nat Clim Change 8:14–21

UNEP (2017) The emissions gap report 2017. United Nations Environment Programme (UNEP), Nairobi

United Nations Convention to Combat Desertification (UNCCD) (2016) Report of the Conference of the Parties on its twelfth session, held in Ankara from 12 to 23 October 2015. Part two: Actions. ICCD/COP(12)/20/Add.1. United Nations Convention to Combat Desertification (UNCCD), Bonn, Germany

United Nations Framework Convention on Climate Change, Subsidiary Body for Scientific and Technological Advice UNFCCC SBSTA (2017) Issues relating to agriculture. FCCC/SBSTA/2017/L.24/Add.1

United Nations General Assembly (UNGA) (1994) Elaboration of an international convention to combat desertification in countries experiencing serious drought and/or desertification, particularly in Africa. Final text of the Convention. A/AC.241/27. http://www.unccd.int/Lists/SiteDocumentLibrary/conventionText/conv-eng.pdf

United Nations General Assembly (UNGA) (2015) Draft outcome document of the United Nations summit for the adoption of the post-2015 development agenda. Resolution A/69/L.85, United Nations General Assembly, New York

U.S. Department of Agriculture USDA (2016) Implementation Plan and Progress Report USDA Building Blocks for Climate Smart Agriculture and Forestry May 2016. Washington, D.C.

VandenBygaart AJ (2018) Comments on soil carbon 4 per mille by Minasny et al. 2017. Geoderma 309:113–114

van Groenigen JW, van Kessel C, Hungate BA et al (2017a) Sequestering soil organic carbon: a nitrogen dilemma. Environ Sci Technol. https://doi.org/10.1021/acs.est.7b01427

Van Groenigen JW, van Kessel C, Hungate BA et al (2017b) Response to the letter to the edi-tor regarding our viewpoint "Sequestering soil organic carbon: a nitrogen dilemma". Envi-ron Sci Technol 51:11503–11504. https://doi.org/10.1021/acs.est.7b04554

van Oosterzee (2012) The integration of biodiversity and climate change: a contextual assessment of the carbon farming initiative. Ecol Manage Restore 13:238–244. https://doi.org/10.1111/emr.12001

Walsh B, Ciais P, Janssens IA et al (2017) Pathways for balancing CO_2 emissions and sinks. Nat Commun 8:14856

Wang J, Monger C, Wang X, Serena M, Leinauer B (2016) Carbon Sequestration in response to grassland–shrubland–turfgrass conversions and a test for carbonate biomineralization in desert soils, New Mexico, USA. Soil Sci Soc Am J 80:1591–1603

Way DA, Long SP (2015) Climate-smart agriculture and forestry: maintaining plant productivity in a changing world while minimizing production system effects on climate. Plant Cell Environ 38:1683–1685

White RE, Davidson B, Lam SK, Chen D (2018) A critique of the paper 'Soil carbon 4 per mille' by Minasny et al. (2017). Geoderma 309:115–117

Working group report of the scientific committee of the initiative for the Adaptation of African Agriculture to Climate Change (2016) The initiative for the adaptation of African agriculture to climate change (AAA)—White Paper http://www.aaainitiative.org/sites/aaainitiative.org/files/AAA_livre_blanc_ENG.pdf

WunderS Engel S, Pagiola S (2008) Taking stock: a comparative analysis of payments for environmental services programs in developed and developing countries. Ecol Econ 65:834–852

Zamanian K, Pustovoytov K, Kuzyakov Y (2016) Pedogenic carbonates: forms and formation processes. Earth Sci Rev 157:1–17

Zomer RJ, Bossio DA, Sommer R, Verchot LV (2017) Global sequestration potential of increased organic carbon in cropland soils. Sci Rep 7:15554. https://doi.org/10.1038/s41598-017-15794-8

Conclusions: The Way Forward

Agricultural activities are directly affecting the carbon (C) dynamics on 40% of Earth's ice-free land area. By conversion of natural terrestrial to agroecosystems—and by managing croplands, grasslands, grazing lands, and wetlands—human activities have caused the release of large amounts of biomass and soil C in the form of carbon dioxide (CO_2) into the atmosphere. The associated increase in atmospheric CO_2 concentrations contributes to accelerated climate change—already observed and even more worryingly jeopardizing the future well-being of our descendants for millennia. Further, the human-caused decrease in soil organic carbon (SOC) stocks of agroecosystems threatens the provision of food, feed, fuel, and fiber of an ever-increasing, more affluent, and more animal products consuming global population. The precious natural resource SOC needs our stewardship to ensure well-being of future generations to come. Therefore, the objective of writing this book was to provide an overview of what we know about the effects of management practices in agroecosystems on SOC stocks and how C sequestration processes may be enhanced through conversion to a judicious land use and adoption of the best management practices in agroecosystems. The book also discusses the often overlooked soil inorganic carbon (SIC) stock which plays an unknown and probably underappreciated role in climate change mitigation.

Chapter 1 provided an overview on the history of agriculture and its effects on natural terrestrial ecosystems, the global C cycle, and climate change caused by emission of greenhouse gases (GHGs). Also, discussed were the Green Revolution (GR) and recent concepts such as sustainable intensification (SI). Agroecosystems can be managed for enhancing SOC sequestration for improving soil health through the adoption of soil-/site-specific recommended management practices (RMPs).

Chapter 2 presented an extensive and detailed overview on the processes, factors, and causes influencing SIC and SOC dynamics with reference to soils of agroecosystems. This included the process of capture of atmospheric CO_2 by photosynthesis and the fate of the fixed C in terrestrial environments and, particularly, in soil. While scientific understanding of SOC protection and stabilization mechanisms has greatly improved, the discussion of the importance of biochemical

K. Lorenz and R. Lal, *Carbon Sequestration in Agricultural Ecosystems*,
https://doi.org/10.1007/978-3-319-92318-5

recalcitrance (e.g., lipids and waxes in anaerobic microsites) versus physicochemical protection (e.g., organo-mineral associations in soil aggregates) is ongoing. Another section was devoted to the importance of SOC for soil-derived ecosystem services (ESs), soil quality, and food security. It is obvious that we know enough how to increase SOC sequestration in agroecosystems for improving soil health and enhancing ESs, and it is the right time to translate science into action.

Chapter 3 took a closer look at the processes of SIC and SOC sequestration in croplands. Croplands have a large potential for SOC storage as their soils lost appreciable amounts of C when those were converted from natural ecosystems, and the widespread use of tillage contributes to depletion of SOC through accelerated erosion and mineralization. Croplands can be recarbonized through adoption of RMPs such as conservation agriculture (CA), agroforestry. The potential of sequestering C in the subsoil layers of deep mineral soil profiles under croplands must be further explored to strengthen climate change adaptation and mitigation.

Chapter 4 discussed the importance of SIC and SOC sequestration in grasslands and how management-induced soil C inputs can be increased. Root-derived C inputs are critical for recarbonization of grasslands as these soils are relatively undisturbed, and perennial grass species develop deep and extended root systems. Grazing management must be targeted toward SOC sequestration due to the large global grazing land area and potential for considerable rates of increase in SOC stock. The dairy (i.e., Danone) and food industry has recently shown increased interest to improve SOC stocks and soil health to reduce their GHG footprint and also to enhance the resilience of their supply chain against climate perturbations.

Chapter 5 highlighted the disproportional large C loss to the atmosphere when wetlands are managed for agricultural production. Soils of wetlands and, in particular, peatlands accumulated large amounts of organic C over millennia. Thus, sustainable management of wetlands for agriculture has been and remains to be a major challenge. Restoring drained agricultural peatlands to flooded conditions improved fertilizer and water management of paddy fields, breeding of new crop cultivars better adapted to anoxic wetland soil conditions, and paludiculture of wetlands are promising and suitable agricultural practices with the co-benefit of C sequestration. The strategy is to reduce the rate of oxidation of drained soils while also reducing the risks of emission of methane (CH_4).

Chapter 6 addressed the importance of agroforestry systems for SOC sequestration. The agroforestry options involve integration of trees with crops and livestock as silvopastoral, agrosilvicultural, and agrosilvopastoral systems. Their greater ability to capture and utilize natural resources (i.e., light, nutrients, water) compared to single-species systems may also enhance soil C inputs. Compared to soils of croplands, soils under agroforestry systems are less disturbed and the perennial plants (i.e., trees) transfer C deep into the subsoil where the SOC sequestered has a long mean residence time for climate change mitigation. There is significant potential for more widespread adoption of agroforestry practices as the low-cost environmental benefits of agroforestry are at present underappreciated and underexploited.

Chapter 7 discussed the effects of biomass production in agroecosystems to replace petroleum as feedstock for the chemical industry and for traditional and advanced biofuels. While dedicated bioenergy plantations show some promise as sustainable production systems, removing the large volumes of organic residues from soils and their use as feedstock may deplete SOC stocks and degrade soil health. Sustainable residue removal rates must be assessed for site-specific conditions. However, a major challenge that needs to be addressed is how to divert resources for biofuel production under the conditions of ever-increasing human demands for food, feed, fiber, and other essential ESs. Inanimately harvested solar energy has the much higher solar radiation conversion efficiency and, thus, may be the major renewable energy source leaving bioenergy for specific niches.

Chapter 8 intensively discussed biochar defined as charcoal for application to soils and its potential to contribute to SOC sequestration in agroecosystems. Applying biochar to the soil potentially improves soil productivity, SOC storage, and infiltration of percolating soil water in the long term through its porous but stable structure. However, interactions between soils and biochar are diverse and difficult to predict. Faced with the dilemma, generalizations about benefits of biochar on agroecosystems must be avoided. Long-term field experiments are particularly needed to elucidate the full potential of soil application of biochar for improving soils and adapting and mitigating climate change by SOC sequestration.

Chapter 9 concluded the book with a critical appraisal of the importance of SOC stocks of agroecosystems in the context of climate change policy. This included the target of Land Degradation Neutrality (LDN) of United Nations Convention to Combat Desertification (UNCCD) and land- and soil-related targets for achieving United Nations Sustainable Development Goals (SDGs). Many nationally determined contributions (NDCs) submitted to United Nations Framework Convention on Climate Change (UNFCCC) include references to soil enhancement and to the necessity of better soil management practices. Similarly, the ambitious goal to increase global soil C stocks of managed agricultural land by 0.4% per year to 40 cm depth was set by the "4 Per Thousand" Initiative: Soils for Food Security and Climate (4p1000). Concepts such as climate-smart agriculture (CSA) and carbon farming are promising for increasing SOC stocks and associated ESs in agroecosystems. It is critically important that farmers and land managers are rewarded for SOC enhancing practices as these are associated with additional costs.

To sum up, the above chapters highlighted an in-depth synthesis of the available knowledge on effects of soil and land-use management practices on SOC stocks of agroecosystems. RMPs for agroecosystems are readily available and must be implemented to enhance food security and adapt to and mitigate climate change. Land managers must be rewarded for SOC stewardship through just and fair payments for provisioning ESs as this will improve the well-being of humans and the health of soils, the environment, and of the planet Earth.

Index

K. Lorenz and R. Lal, *Carbon Sequestration in Agricultural Ecosystems*,
https://doi.org/10.1007/978-3-319-92318-5

The manufacturer's authorised representative in the EU is Springer Nature Customer Service Centre GmbH, Europaplatz 3, 69115 Heidelberg, Germany. If you have any concerns regarding our products, please contact ProductSafety@springernature.com

Printed and bound by CPI Group (UK) Ltd, Croydon, CR0 4YY
25/09/2026
02222853-0007